# Design of Reinforced Concrete Buildings for Seismic Performance

# Design of Reinforced Concrete Buildings for Seismic Performance

## Practical Deterministic and Probabilistic Approaches

Mark Aschheim, Enrique Hernández-Montes, and Dimitrios Vamvatsikos

CRC Press
Taylor & Francis Group
Boca Raton  London  New York

CRC Press is an imprint of the
Taylor & Francis Group, an **informa** business

A SPON PRESS BOOK

MATLAB® is a trademark of The MathWorks, Inc. and is used with permission. The MathWorks does not warrant the accuracy of the text or exercises in this book. This book's use or discussion of MATLAB® software or related products does not constitute endorsement or sponsorship by The MathWorks of a particular pedagogical approach or particular use of the MATLAB® software.

CRC Press
Taylor & Francis Group
6000 Broken Sound Parkway NW, Suite 300
Boca Raton, FL 33487-2742

First issued in paperback 2020

© 2019 by Taylor & Francis Group, LLC
CRC Press is an imprint of Taylor & Francis Group, an Informa business

No claim to original U.S. Government works

ISBN 13: 978-0-367-73055-0 (pbk)
ISBN 13: 978-0-415-77881-7 (hbk)

---

**Library of Congress Cataloging-in-Publication Data**

---

Names: Aschheim, Mark, author. | Hernandez-Montes, Enrique, author. |
Vamvatsikos, Dimitrios, author.
Title: Design of reinforced concrete buildings for seismic performance :
practical deterministic and probabilistic approaches / Mark Aschheim,
Enrique Hernandez-Montes and Dimitrios Vamvatsikos.
Description: Boca Raton : Taylor & Francis, a CRC title, part of the Taylor & Francis imprint, a
member of the Taylor & Francis Group, the academic division of T& F Informa, plc, [2019] |
Includes bibliographical references and index. |
Identifiers: LCCN 2018051257 (print) | LCCN 2018055289 (ebook) | ISBN 9781315354811 (ePub) |
ISBN 9781315335759 (Mobipocket) | ISBN 9781482266924 (Adobe PDF) | ISBN 9780415778817
(hardback) | ISBN 9781315375250 (ebook)
Subjects: LCSH: Earthquake resistant design. | Buildings, Reinforced concrete—Earthquake effects.
Classification: LCC TA658.44 (ebook) | LCC TA658.44 .A767 2019 (print) | DDC 693.8/52—dc23
LC record available at https://lccn.loc.gov/2018051257

---

Visit the Taylor & Francis Web site at
http://www.taylorandfrancis.com

and the CRC Press Web site at
http://www.crcpress.com

To the faculty and practicing engineers who have helped us to improve our understanding of earthquake-resistant structural engineering, and to those who perished in past earthquakes.

# Contents

## SECTION IV
## Reinforced Concrete Systems                                           319

## 15  Component proportioning and design based on ACI 318              321

# Acknowledgments

The authors are grateful to their families for their support and patience throughout the many years of meetings, writing, and editing needed to bring our vision to reality.

In addition, the authors thank Mr. Charis Lyritsakis and Mr. Fotis Andris, who helped to develop the structural models for the moment-resistant frame case studies, and especially Ms. Akrivi Chatzidaki, who finalized the modeling and undertook the analysis and reporting of results of the six case studies in Chapter 19. The authors are also grateful to Dr. Tjen Tjhin for his contributions on shear-wall systems.

MATLAB® is a registered trademark of The MathWorks, Inc. For product information, please contact:

The MathWorks, Inc.
3 Apple Hill Drive
Natick, MA 01760-2098 USA
Tel: 508-647-7000
Fax: 508-647-7001
E-mail: info@mathworks.com
Web: www.mathworks.com

# Authors

**Mark Aschheim** is the Peter Canisius S.J. Professor in the Department of Civil Engineering at Santa Clara University, California. After earning his B.S., M.Eng., and Ph.D. at U.C. Berkeley, he taught and conducted research at the University of Illinois, Urbana. He participated in post-earthquake reconnaissance teams after the Northridge, Great Hanshin, Izmit, Gujarat, and Fukushima earthquakes. He served on two code-writing bodies: the 318-H subcommittee of the American Concrete Institute (2002–2008) and the Provisions Update Committee of the Building Seismic Safety Council (2005–2009). His research interests include development of earthquake-resistant structural systems and design methods, and the use of sustainable structural materials.

**Enrique Hernández Montes** is a Professor in the School of Civil Engineering at the University of Granada, Spain. He earned his engineer's and doctoral degrees at the University of Granada. He has served as a visiting faculty member at the Politecnico di Milano and Santa Clara University. He has been a regular consultant for Prointec and ERSIGroup.

**Dimitrios Vamvatsikos** is an Assistant Professor in the School of Civil Engineering at the National Technical University of Athens. He earned his Diploma in Civil Engineering from the National Technical University of Athens and his M.Sc. and Ph.D. from Stanford University, where he focused on probabilistic methods, seismic performance, and risk assessment. He has co-operated with leading structural engineering firms (ARUP, Halcrow/CH2M), the oil & gas industry (Shell, ExxonMobil), and catastrophe risk modelers (AIR Worldwide, REDRisk), while his research has been funded by the Applied Technology Council, the Federal Emergency Management Agency, the US National Institute of Standards and Technology, the International Organization for Standardization, the Global Earthquake Model Foundation, and the European Commission.

# Section 1

# Introduction

# Chapter 1

# Introduction

## 1.1 HISTORICAL CONTEXT

Earthquake-resistant structural engineering has evolved to encompass a sophisticated body of knowledge in the last 100 years. From the earliest notions that lateral forces should be proportional to building mass (adopted in Italy in 1909) and recognition of the need to provide ductile details (1960s) for structures designed on the basis of linear elastic analysis, we now have sophisticated models of seismic hazard, robust tools for nonlinear structural analysis, and probabilistic bases that provide an explicit framework to judge not only the mean annual frequency of exceeding a seismic performance objective but also expected losses in terms of potential casualties, repair costs, and repair time.

This progress is partially reflected in building codes, which have gradually begun allowing the use of nonlinear structural analysis. Yet, because the linear elastic methods of these codes are used routinely, courses in earthquake-resistant structural engineering emphasize these methods, sometimes at the expense of a deeper, more comprehensive understanding of inelastic response that is intended with current design approaches.

More recently, displacement-based approaches have been developed, which allow for the design and evaluation of seismic performance in terms of relative distortions of structural and nonstructural components. Such displacement-based approaches naturally allow for a change in index of seismic demand from the conventional focus on vibration period to our emphasis on yield displacement. The yield displacement is a kinematic quantity, derived from geometry and material properties (yield strain), and is a more stable parameter in design than period. When coupled with plastic mechanism analysis, design iterations based on an estimated yield displacement can progress rapidly, leading quickly to a design that meets the seismic performance objectives. The design approaches described herein can be applied simply as an alternative to approximate the equivalent lateral force method of many building codes, or in a more sophisticated manner to achieve designs with acceptably low mean annual frequencies of exceeding specified performance limits.

## 1.2 PURPOSE AND OBJECTIVES

The purpose of this book is to offer simple and effective approaches for performance-based seismic design that are anchored in theory (and observation) and which promote intuition about structural behavior. First, we provide a deterministic, displacement-based design approach that uses plastic mechanism analysis in conjunction with the stability of the yield displacement, and then we extend this approach to consider uncertainties associated with the full seismic hazard environment.

We have written this book for both students and practitioners of earthquake-resistant structural engineering, aiming to provide both a solid foundation in the field and the guidance needed to implement performance-based design and assessment. While we have tried to provide the breadth needed to understand the essentials of the field, we have not attempted to provide encyclopedic coverage of all relevant work. References are provided for those who seek a deeper understanding.

## 1.3 KEY ELEMENTS

The design methods described herein are built upon the following key observations and constructs:

- Displacement response is predominantly in the first mode. Therefore, limits on displacement and ductility demand can be achieved using a first-mode representation of the structure.
- The yield displacement of a structure in a first-mode pushover analysis is nearly invariant during design iterations in which strength and stiffness are adjusted to achieve the desired seismic performance.
- Seismic demands can be portrayed as a function of yield displacement, in the form of yield point spectra (YPS) for deterministic analysis and in the form of yield frequency spectra (YFS) for representation of the seismic hazard. These graphic aids embody estimates of inelastic response, relative to a given elastic demand (or hazard). These graphic tools allow quick estimation of the system strength required to achieve the desired seismic performance.
- Plastic mechanism analysis is useful for achieving an intended mechanism, proportioning components of the lateral force-resisting system, and promoting intuition about system behavior. We use mechanism analyses to design the yielding components to have the strengths required for the system to achieve the desired seismic performance.
- The strengths required of non-yielding components (that is, force-protected members) generally are determined by evaluating strength demands using nonlinear dynamic analyses.

We see design and analysis as fundamentally different but complementary tasks. The many assumptions that must be made to model a structure using modern analysis software almost ensure that the model will deviate from the actual structure in some way. Hence, in our view, the purpose of a design-oriented analysis is to establish a structure that we can have confidence in, rather than providing a precise characterization of what is to be expected in any one strong shaking event. The shift toward nonlinear analysis has removed much of the doubt that was present when only linear analysis was available; as modeling abilities continue to improve, uncertainties with nonlinear analysis will reduce and analytical predictions will become increasingly accurate.

There have been many developments in design approaches in recent years, and no contribution lives in isolation. To provide some context for the present work, we recognize some elements that are shared in the following approaches:

- Goel and Chao (2008) also use mechanism analysis to proportion the lateral force-resisting system in design. We provide a slightly different approach to the determination of moments in beams and columns, and use YPS or YFS for determining the required base shear strength rather than applying an energy-balance concept.

- Priestley et al. (2009) also use the yield displacement as a basis for design. We employ a plastic mechanism analysis, use modal parameters to establish the "equivalent" single-degree-of-freedom (SDOF) system, represent inelastic response demands as a function of yield displacement, and employ direct relationships (e.g., so-called $R$–$\mu$–$T$ relationships) to establish the inelastic spectra rather than using equivalent damping.

We note that other risk-based approaches have been proposed for design. For limiting the risk of collapse, Sinković et al. (2016) propose guidelines for adjusting a design based on pushover analyses, followed by the use of nonlinear dynamic analysis for collapse assessment. Franchin et al. (2018) limit the mean annual frequency of exceeding one or more limit states by use of a gradient-based constrained optimization technique applied to an equivalent linear multi-degree-of-freedom (MDOF) model.

## 1.4 ILLUSTRATION OF DESIGN APPROACH

A simple example is described in this section to illustrate the design approach and allow for comparisons to a conventional, period-based, design approach. The example consists of the seismic design of a single-column pier of a highway bridge for transverse response (Figure 1.1). We seek to limit the peak displacement to 2% of the height of the structure (0.02(19.8 ft) = 0.40 ft = 4.75 in. = 0.121 m). In this case, the drift limit is specified with respect to a specific ground motion, the NS component of the 1940 El Centro record,[1] along the lines reported by Aschheim and Hernández-Montes (2003).

The pier has a circular cross section and is subjected to a service dead load of 1,380 kips (6,140 kN). Considering similar bridge piers, the diameter of the column is initially selected to be 48 in. (1.22 m). Grade 60 steel reinforcement is used, having a minimum yield strength $f_y$ = 60 ksi (414 MPa). The concrete compressive strength is 4,000 psi (27.6 MPa). Expected material strengths according to Caltrans (Section 17.4) are $1.2f_y$ = 72 ksi (496 MPa) and $1.3f_c'$ = 5,200 psi(35.9 MPa).

The effective yield curvature, $\phi_y$, for this circular reinforced concrete column cross section can be estimated (Hernández-Montes and Aschheim, 2003) as

$$\phi_y = \frac{2.3\varepsilon_y}{d}$$

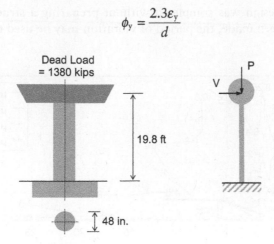

Figure 1.1  Bridge column and SDOF model.

[1] Note that a smoothed inelastic design spectrum (in the YPS format) could have been used instead.

where $\varepsilon_y$ is the yield strain of the reinforcement and $d$ is the depth to the centroid of the extreme tension reinforcing bar. For this column, $\varepsilon_y = f_y/E_s = 72/29{,}000 = 0.00248$ and $d \approx 48 - 2.5 = 45.5$ in. (1.156 m). Thus, we estimate $\phi_y = 0.000126$ rad/in.

The corresponding yield displacement, $u_y$, for a cantilever column subjected to lateral load applied at the end, is estimated (assuming shear deformations are negligible) as

$$u_y = \phi_y \frac{L^2}{3}$$

or 2.37 in. (0.0602 m) for this column, having length $L = 19.8$ ft (6.04 m).

Thus, the ductility limit associated with a drift limit of 2% of the height is 4.75 in./ 2.37 in. = 2.0.

The yield point spectra for the 1940 NS El Centro record is plotted in Figure 1.2— curves are plotted that show the required yield strength coefficient, $C_y$, as a function of yield displacement, for a given constant ductility response. For a yield displacement of 6.02 cm, the base shear coefficient at yield that limits the ductility demand to 2.0 is given by $C_y = 0.14$. Superimposed on the YPS is a schematic capacity curve, illustrating the yield and peak displacements.

For this SDOF example, the required base shear strength at yield is given by $V = C_y \cdot W = 0.14 \cdot W = 193$ kips = 859 kN.

A plastic mechanism will form when a plastic hinge forms at the base of the column. Thus, the required strength of the plastic hinge is given by $M = V \cdot h = 3{,}820$ kip$\cdot$ft = 5,180 kN$\cdot$m. A 48-in. (1.22-m) diameter cross section with 18 No. 10 bars (32-mm diameter), for which $\rho_g = 1.26\%$, has adequate nominal strength to resist these actions under the given axial dead load.

One may observe that the required base shear strength was determined as a function of the yield displacement, without having made any reference to the period of vibration. The period is the outcome of the design process, and may be determined as

$$T = 2\pi \sqrt{\frac{u_y}{C_y g}} = 2\pi \sqrt{\frac{2.37 \text{ in.}}{0.14 \cdot 386.1 \text{ in./s}^2}} = 1.32 \text{ s}$$

Although this initial design was completed without preparing a structural model, once a structural model has been made, the period of vibration may be used to confirm the design

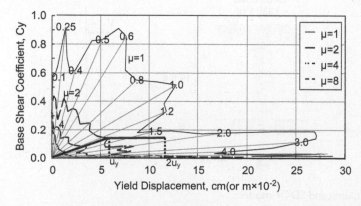

Figure 1.2 YPS for the 1940 NS El Centro motion. A bilinear response is superimposed on the spectra to schematically illustrate the estimation of peak displacement response for the case that $\mu = 2$.

assumptions. The flexural stiffness of the column in the structural model should be consistent with the assumed yield displacement (and may be estimated as $EI = M_n/\phi_y$), rather than using a fixed percentage of the gross stiffness, because the steel content affects both strength and stiffness.

This design process recognizes the stability of the yield displacement (for a given column diameter) and allowed the required strength (and reinforcement) to be determined directly, that is, without the need for iteration. For comparison, a similar example was considered by Chopra and Goel (2001). The influence of changes in stiffness on the period of vibration and yield displacement were considered in this example, and required as many as five iterations to reach convergence.

Finally, had we wished to limit the mean annual frequency of exceeding a specified ductility demand, we would have used the YFS representation of the hazard. This is illustrated in Section 13.5.3.

## 1.5  ORGANIZATION OF BOOK

The 19 chapters of this book are organized into five major sections:

- Section I consists of this *Introduction* (Chapter 1).
- Section II addresses *Seismic Demands*, beginning with the ground motion excitation (Seismology, Chapter 2), continuing with the dynamic response of SDOF and MDOF systems (Chapters 3–5), and concluding with the characterization of dynamic response using Principal Components Analysis (Chapter 6) and the development of so-called "equivalent" SDOF systems (Chapter 7).
- Section III addresses *Essential Concepts Of Earthquake-Resistant Design*, in relation to the design approaches described herein. These include general principles (Chapter 8), an elaboration on the stability of the yield displacement (Chapter 9), seismic performance objectives (Chapter 10), plastic mechanism analysis (Chapter 11), system proportioning (Chapter 12), probabilistic considerations (Chapter 13), and system modeling issues (Chapter 14).
- Section IV addresses *Reinforced Concrete Systems* in particular, identifying the most prominent design requirements in American codes (Chapter 15) and European codes (Chapter 16), along with guidance for component modeling and acceptance criteria (Chapter 17).
- Finally, Section V addresses *Design Methods And Examples*, with several distinct design methods identified in Chapter 18, which are illustrated by a set of examples in Chapter 19.

## REFERENCES

Aschheim, M., and Hernández-Montes, E. (2003). The representation of P-Δ effects using yield point spectra, *Engineering Structures*, 25:1387–1396.

Chopra, A., and Goel, R. (2001). Direct displacement-based design: Use of inelastic vs. elastic design spectra, *Earthquake Spectra*, Earthquake Engineering Research Institute, 17(1):47–64.

Franchin, P., Petrini, F., and Mollaioli, F. (2018). Improved risk-targeted performance-based seismic design of reinforced concrete frame structures, *Earthquake Engineering & Structural Dynamics*, 47(1):49–67.

Goel, S. C., and Chao, S. H. (2008). *Performance-Based Plastic Design: Earthquake-Resistant Steel Structures*, International Code Council, Washington, DC.

Hernández-Montes, E., and Aschheim, M. (2003). Estimates of the yield curvature for design of reinforced concrete columns, *Magazine of Concrete Research*, 55(4):373–383.

Priestley, M. J. N., Calvi, G. M., and Kowalsky, M. J. (2007). *Direct Displacement-Based Seismic Design of Structures*, IUSS Press, Pavia.

Sinković, N. L., Brozovič, M., and Dolšek, M. (2016). Risk-based seismic design for collapse safety, *Earthquake Engineering & Structural Dynamics*, 45(9):1451–1471.

# Section II

# Seismic Demands

# Chapter 2

# Seismology and site effects

## 2.1 PURPOSE AND OBJECTIVES

This chapter reviews seismology and site effects to provide background for understanding ground motion characteristics and their influence on the behavior of structures. Topics include earthquake sources, magnitude, and magnitude-recurrence relationships, the propagation of ground motion to a site, near-source effects, site amplification, and shaking intensity and its representation using response spectra. A more comprehensive coverage of seismological issues from an engineering perspective is available from Kramer (1996).

## 2.2 EARTHQUAKE SOURCES AND WAVE PROPAGATION

Seismology is the field of science that studies earthquakes and the propagation of waves through the earth. These waves arise from tectonic, volcanic, or even man-made sources (e.g., mining, explosives, fracking, and seismic vibrators).

According to plate tectonic theory, the lithosphere (the rigid outermost shell of the earth) is broken up into plates (Figure 2.1a), which float on top of the mantle, a mostly plastic, yet non-molten, rock that may be considered to behave like a very viscous fluid when viewed over the scale of thousands of years (geologic time). The excess heat that exists in the core and the mantle (e.g., due to nuclear decay/fission of heavy isotopes) causes convection currents to develop within the mantle, i.e., loops of rising hot rock and sinking cold rock. This convective movement within the mantle drags the lithospheric plates that rest upon it, causing them to move relative to each other. Where the hot mantle rises, the overlying plates diverge and new lithosphere is formed (Figure 2.1b), typically in a mid-oceanic ridge, such as the one running the length of the Atlantic ocean from north to south. Where the cold mantle sinks, the old lithosphere is being destroyed, typically as a denser plate sinks below a lighter one (Figure 2.1c). This, for example, occurs underneath the Mediterranean, where the African plate dips below the Eurasian plate as the two converge.

Faults are planar fractures or discontinuities where relative displacement or movement has occurred. Faults are classified based on the different types of movement that can appear (Figure 2.2). Strike-slip faults are characterized by relative horizontal displacements as the two segments appear to be sheared relative to each other. Normal and reverse faults display vertical movement whereby one segment is pulled or pushed, respectively, against the other along a sloping fault plane. Thrust faults are essentially reverse faults with a nearly horizontal fault plane.

Earthquakes are caused by sudden slip occurring between the opposed fault surfaces, and result when the state of stress exceeds that which can be withstood. The sudden slip associated with fault rupture releases stored strain energy and generates an impulsive movement

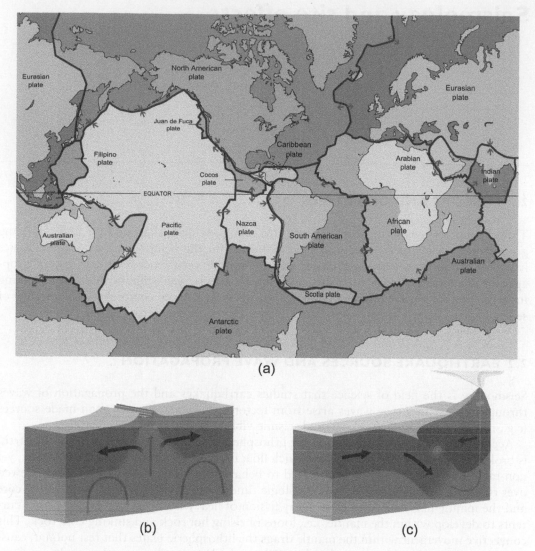

*Figure 2.1* (a) Earth's lithospheric plates and their movement due to convection, (b) diverging plates and generation of new crust by rising magma, and (c) converging plates and subduction of the lighter oceanic crust beneath the continental crust. (http://geomaps.wr.usgs.gov and By domdomegg - Own work, CC BY 4.0, https://commons.wikimedia.org/w/index.php?curid=45871555 and 50772217.)

that is transmitted in the form of radial waves (seismic waves of different types). The waves can be classified as body waves and surface waves (Figure 2.3).

P-waves and S-waves (P from primary and S from secondary) are body waves that travel through subsurface rock. P-waves are essentially compression-tension waves that, much like sound, can travel through both solids and fluids. S-waves are shear waves, which can only travel through solids. Love and Rayleigh waves are surface waves, generated by the interaction of the P- and S-waves with the surface (and a layer of soil in the case of Love waves) and they only travel along the earth's surface. These waves appear most prominently in alluvial soil basins and generally decay more slowly than body waves with distance. The sum of all waves produces ground motion, which is sampled and recorded at any point by a seismograph (see Figure 2.4).

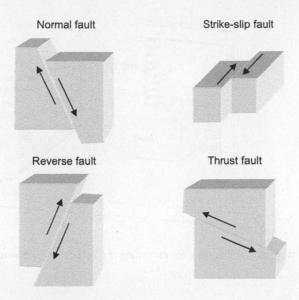

Figure 2.2 Types of faults: normal, strike-slip, reverse, and thrust.

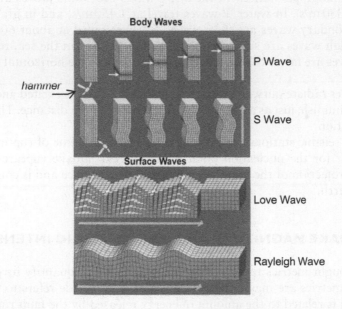

Figure 2.3 Types of waves. A hammer is depicted to help to understand P- and S-wave motion. (Adapted from http://earthquake.usgs.gov.)

Because seismographs are attached to a moving body (the ground surface of the earth), a fixed reference is not available. Velocity or displacement cannot be measured directly by seismographs, as the recording is triggered by a change in acceleration. Instead, velocity and displacement histories are obtained indirectly by integration of the ground acceleration record.

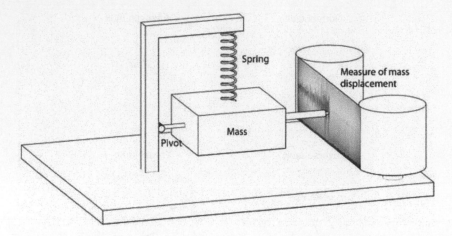

*Figure 2.4* Basis of a seismograph (in this case, for recording the vertical component of a ground motion).

The velocity of each wave depends on the type of wave and the density and stiffness of the medium. On the crust, the velocity ranges between 2 and 8 km/s, and reaches 13 km/s in the deep mantle. Primary waves are pressure waves and travel the fastest, being the first to arrive at a seismograph station. In the air, P-waves are sound waves and travel at the speed of sound (330 m/s). In water, P-waves travel at 1,450 m/s, and in granite they travel at 5,000 m/s. Secondary waves are shear waves. S-waves travel at about 60% of the speed of P-wave. Rayleigh waves are similar to the superficial waves in the sea, rolling along the surface. Love waves are faster than Rayleigh waves, and produce horizontal motion moving from side to side.

As seismic waves radiate outward, more and more rock mass is excited and thus the wave amplitude must diminish, just as a light bulb appears dimmer with distance. This phenomenon is termed attenuation.

Three or more seismic stations are needed to determine the time of rupture and location of the hypocenter (or the nucleation point where the earthquake rupture initiated). The epicenter is the projection of the hypocenter to the earth's surface and is commonly plotted on maps of the earth.

## 2.3 EARTHQUAKE MAGNITUDE AND MACROSEISMIC INTENSITY

Scientists have sought metrics to measure a seismic event and quantify its properties. The most prominent metrics are magnitude and intensity. Magnitude refers to the earthquake rupture itself and is related to the amount of energy released by the fault rupture. Intensity describes the effect of a ground motion at a particular location, and thus depends not only on the magnitude of the event but also on the site of interest. An event of a given magnitude can be felt with different intensities in different places.

There are different measures of intensity, some quantitative (Section 2.6) and some qualitative. Macroseismic intensity scales provide a qualitative description of the effects of an earthquake. Different macroseismic intensity scales are used around the world, such as the Shido scale in Japan and the MSK scale (by Medvedev, Sponheur, and Karnik) in Russia. The most famous is the Modified Mercalli Intensity (MMI) scale, which ranges from level I (not felt) to level XII (total destruction)—see Figure 2.5. MMI VIII is a useful point of

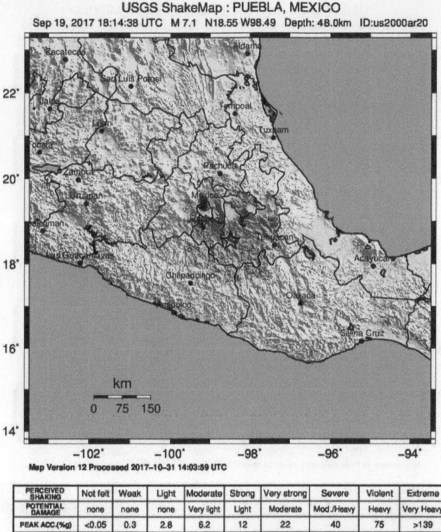

USGS ShakeMap : PUEBLA, MEXICO
Sep 19, 2017 18:14:38 UTC   M 7.1   N18.55 W98.49   Depth: 48.0km   ID:us2000ar20

Map Version 12 Processed 2017–10–31 14:03:59 UTC

| PERCEIVED SHAKING | Not felt | Weak | Light | Moderate | Strong | Very strong | Severe | Violent | Extreme |
|---|---|---|---|---|---|---|---|---|---|
| POTENTIAL DAMAGE | none | none | none | Very light | Light | Moderate | Mod./Heavy | Heavy | Very Heavy |
| PEAK ACC.(%g) | <0.05 | 0.3 | 2.8 | 6.2 | 12 | 22 | 40 | 75 | >139 |
| PEAK VEL.(cm/s) | <0.02 | 0.1 | 1.4 | 4.7 | 9.6 | 20 | 41 | 86 | >178 |
| INSTRUMENTAL INTENSITY | I | II–III | IV | V | VI | VII | VIII | IX | X+ |

Scale based upon Worden et al. (2012)

*Figure 2.5* Macroseismic intensity shakemap for the Puebla M7.1 event of September 19, 2017. The star indicates the epicenter, yet heavy damage occurred at Mexico City, nearly 100 km northwest of the epicenter. (From USGS, https://earthquake.usgs.gov/earthquakes/eventpage/us2000ar20#shakemap.)

reference, described in Wikipedia (2019) as "Damage slight in specially designed structures; considerable damage in ordinary substantial buildings with partial collapse. Damage great in poorly built structures. Fall of chimneys, factory stacks, columns, monuments, walls. Heavy furniture overturned." Peak ground accelerations for the same event are plotted in Figure 2.6.

Magnitude is a more quantitative way to measure an earthquake without reference to a specific location. It can be thought of as a measure of energy released in the event, and can be compared to the energy released by a nuclear bomb, in terms of kilotons of TNT. Several magnitude scales have been defined. The Richter magnitude, also known as local

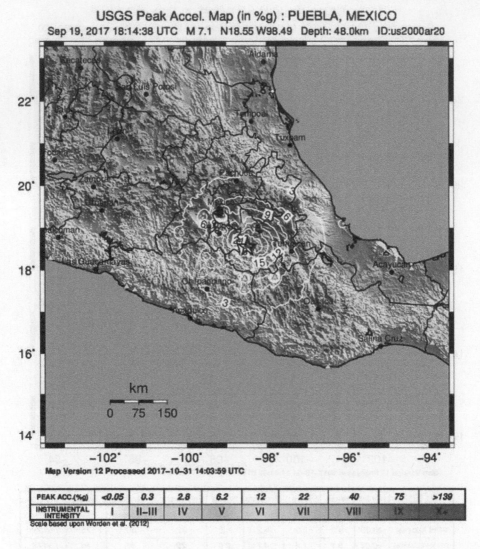

Figure 2.6 PGA shakemap for the Puebla M7.1 event of September 19, 2017. (From USGS, https://earthquake. usgs.gov/earthquakes/eventpage/us2000ar20#shakemap.)

magnitude ($M_L$), is based on the maximum amplitude of motion recorded by a seismograph. It is defined by the following expression:

$$M_L = \log \frac{A}{A_0} \tag{2.1}$$

where $A$ is the peak amplitude measured in a Wood-Anderson seismograph (in micrometers, μm) at 100 km from the epicenter of the earthquake, and $A_0$ is the peak amplitude of the zero-magnitude earthquake at the same distance, defined as 1 μm.

Other measures of magnitude are the surface-wave magnitude ($M_s$) and the body-wave magnitude ($M_b$). However, these magnitudes ($M_L$, $M_s$, and $M_b$) are not able to satisfactorily measure the size of the largest earthquakes—for such earthquakes their values tend to be

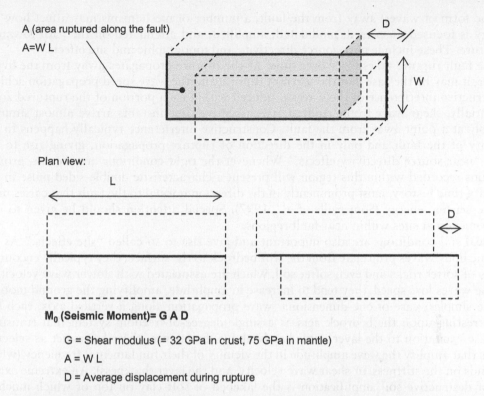

A (area ruptured along the fault)
A=W L

Plan view:

$M_0$ (Seismic Moment)= G A D

G = Shear modulus (= 32 GPa in crust, 75 GPa in mantle)

A = W L

D = Average displacement during rupture

Figure 2.7 Definition of the seismic moment, $M_0$.

between 8.0 and 8.9 and do not account for the size of the ruptured fault area and how much energy is released. In technical parlance, they tend to "saturate."

Another measure of magnitude, the moment magnitude ($M_w$), overcomes the mentioned limitations. $M_w$ is based on the concept of seismic moment, and it is uniformly applicable to all sizes of earthquakes. The seismic moment ($M_0$) is a measure of the size of an earthquake based on the area of fault rupture, the average displacement across the fault rupture surface, and the force that was required to overcome the friction across the fault surface (Figure 2.7). Seismic moment can also be calculated from the amplitude spectra of seismic waves.

Based on the seismic moment (expressed in N·m), the moment magnitude ($M_w$) is defined as:

$$M_W = \frac{2}{3} \log M_0 - 6.0 \tag{2.2}$$

It can be observed that $M_w$ is similar to $M_L$ in the range of 3–5, and always greater than $M_L$ for $M_w > 7.5$.

## 2.4 NEAR-SOURCE, TOPOGRAPHIC, AND SITE EFFECTS ON GROUND MOTION

A fault rupture should be thought of as a source of energy (e.g., a lamp or a loudspeaker) that radiates energy with greater intensity in the direction of rupture. As energy propagates (in the form of waves) away from the fault, a number of mechanisms may affect how the energy is focused, dispersed, modulated, amplified, and attenuated on the path toward a structure. These include near-source directivity and topographic and site effects.

The fault rupture takes place over time. As the rupture propagates away from the hypocenter, it may be the case that the speed of rupture and the wave speed propagation achieve constructive interference of the waves produced within each portion of the ruptured zone. Essentially, shear wave fronts emitted at consecutive time instants arrive almost simultaneously at a point away from the fault. Constructive interference typically happens in the vicinity of the fault and only in the direction of rupture propagation, giving rise to the term "near-source directivity effects." Whenever the right conditions are present, ground motions recorded within this region will present a characteristic double-sided pulse in the velocity time history, most prominently in the direction normal to the fault that caries most of the seismic energy (Somerville et al., 1997). Special attention should be given to this phenomenon at sites within near-fault regions.

Local soil conditions are also important and give rise to so-called "site effects." As the seismic body waves propagate from the stiff bedrock to the surface, they typically encounter layers of softer rocks and even softer soil, which are associated with slower wave velocities. As the waves lose speed, they tend to increase in amplitude, amplifying the ground motion. In the simplest case of one-dimensional wave propagation along a vertical axis, each soil layer resting upon the bedrock acts as a single-degree-of-freedom system that transmits its base excitation to the layer immediately above. In essence, these layers act as selective filters that amplify the wave amplitude in the vicinity of their fundamental frequency (which depends on the stiffness, or shear wave velocity, and the layer thickness). An extreme example of destructive soil amplification is the lakebed of soft clay on top of which much of Mexico City is founded. This caused very strong shaking at the surface at the fundamental period of the soil layer and resulted in catastrophic collapses of buildings that resonated at this period, during the 1985 and 2017 events.

Beyond this simple 1D view of soil effects, the three-dimensional topography of the site may also play a major role. For example, when seismic waves enter into a sediment-filled basin they can be trapped, producing reverberations that increase the duration of the earthquake and further amplify the shaking intensity. Similarly, canyons can generate strong constructive interference of waves, while rock outcroppings and hilltops cause the waves to focus toward the peak, in all cases creating locally higher intensities.

## 2.5 GEOLOGICAL AND GEOTECHNICAL HAZARDS

While structural engineers tend to focus on damage to buildings from ground shaking, various geological and geotechnical hazards are worth noting.

- Ground failures such as landslides and debris flows induced by earthquakes. Landslides include a wide range of ground movement; slopes that are barely stable statically or which may be slowly creeping under static conditions may be triggered to fail suddenly by earthquake shaking.
- Liquefaction is a specific phenomenon where the flux of water causes increased pore pressure that provokes the loss of contact among the grains of the soil, causing the grain structure to fail suddenly. Liquefaction can also be implicated in bearing capacity failures (e.g., the city of Adapazari, in the 1999 Izmit Earthquake).
- Surficial features due to fault slip. Surface soils may undergo compressional or extensional movement (e.g., Northridge 1994), displacing supported structures. Soils beneath

buildings may slump (e.g., Ford factory in Gölcük, in the 1999 Izmit Earthquake). Slip of faults beneath buildings, bridges, pipelines, and other structures will induce damage or failure of the overlying structures (e.g., in the city of Gölcük during the same earthquake).

## 2.6  QUANTITATIVE MEASURES OF INTENSITY BASED ON GROUND MOTION RECORDS

Although magnitude is a useful metric of the energy of an earthquake, it does not provide location-dependent information on the severity of the shaking. For that, we employ the three translational components (two horizontal and one vertical, usually oriented north-south, east-west and up-down) of the acceleration time history, as recorded by seismographs. Some seismographs also measure rotations, although the measurement and use of the three rotational components of ground motions is still a developing area.

A ground motion measured at a seismic station is just one point of a spatial wavefield. A set of stations—sometimes configured in what is called a seismic array—needs to be deployed in an area to properly characterize the spatial distribution of earthquake ground motion. Examples include the arrays in Parkfield, California, and Lotung, Taiwan.

The first waves that travel from the hypocenter to any place—for example, the recording station—are body waves, because they take the shortest path. P-waves are longitudinal waves and are fastest, after which the first S waves arrive, followed by the surface waves (which are slowest and take a longer path along the surface), as illustrated in Figure 2.8.

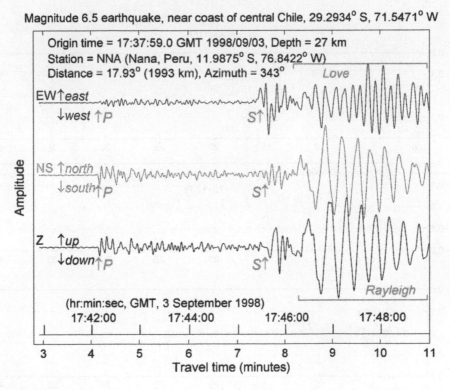

*Figure 2.8* Seismograms recorded by a three-component seismograph at Nana, Peru. P-waves appear first, followed by S-waves and finally surface waves: Love-waves appear in the horizontal components while Rayleigh waves appear in the vertical component. (Adapted from *A Guide for Teachers*, by Prof. Lawrence, University of Purdue.)

The actual raw data recorded by a seismograph are not used directly. Instead, a number of signal processing techniques (filtering, baseline correction, etc.) are applied to remove noise and sensor artifacts in order to generate the processed acceleration time histories. An example of the processed acceleration time histories is given in Figure 2.9. Notice that the vertical component (lowest panel of Figure 2.9) appears to have higher frequencies and lower amplitude when compared to the two horizontal ones. The former is a general characteristic of the vertical component, while the latter has a lot to do with the distance from the rupture. Because high-frequency waves attenuate faster with distance, most such far-field recordings will have a lower amplitude vertical component. The combination of these two features as well as the inherent capacity of most buildings for resisting vertical loads makes the vertical ground motion component less important for the seismic response of typical structures. Still, in the near field and especially when investigating systems that are sensitive to the vertical ground motion (e.g., structures with long spans and rocking systems), one should not disregard vertical motion without careful consideration.

By integrating the ground acceleration $\ddot{u}_g(t)$ over time, one may recover the corresponding ground velocity and ground displacement time histories (Figures 2.10 and 2.11, respectively):

$$\dot{u}_g(t) = \int_0^t \ddot{u}_g(t)\ dt$$

$$u_g(t) = \int_0^t \dot{u}_g(t)\ dt \tag{2.3}$$

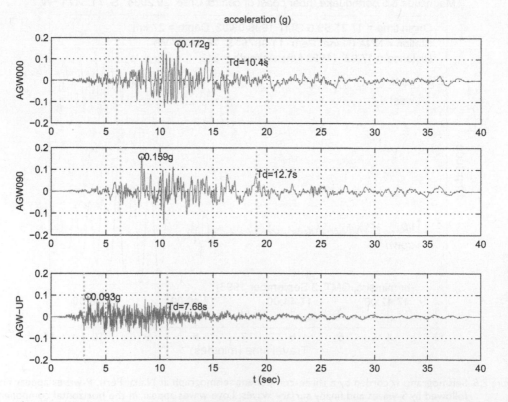

*Figure 2.9* Ground acceleration time histories for the three components (NS, EW, UP) of the Loma Prieta Earthquake (October 17, 1989), Agnews State Hospital Station.

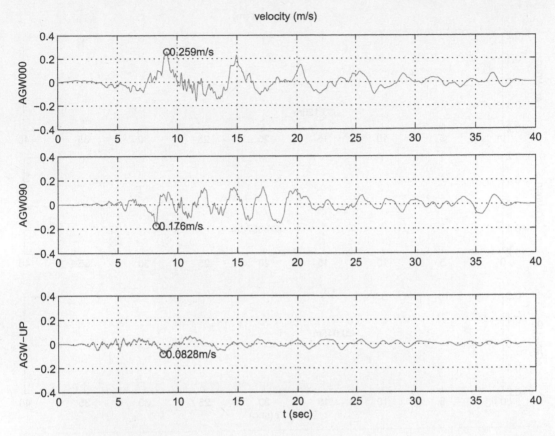

*Figure 2.10* Ground velocity time histories for the three components (NS, EW, UP) of the Loma Prieta Earthquake (October 17, 1989), Agnews State Hospital Station.

Using such time histories, one may derive useful scalar measures of the ground motion intensity and quantify its effects on structures. The most typical measures are the peak absolute values of each trace, correspondingly termed peak ground acceleration (PGA), peak ground velocity (PGV), and peak ground displacement (PGD), indicated with a small circle in each of Figures 2.9–2.11:

$$PGA = \max_{t} \left| \ddot{u}_g(t) \right|$$

$$PGV = \max_{t} \left| \dot{u}_g(t) \right| \qquad\qquad (2.4)$$

$$PGD = \max_{t} \left| u_g(t) \right|$$

As discussed in Chapter 3, PGA better correlates with structural response of short-period systems, PGV with medium-period systems, and PGD with long-period systems.

Duration is another scalar measure of interest, especially when discussing the behavior of cyclically degrading structures (Chapter 4), where the many cycles of long duration motions may be detrimental if they cause degradation of lateral strength. Still, not all 40 s of ground motion shown in Figure 2.9 are of interest, as most cycles are of low amplitude. Perhaps the simplest measure of strong motion duration is the bracketed duration, $T_d$, defined as the

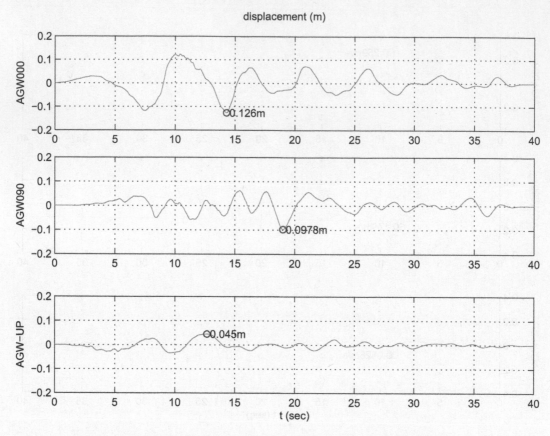

displacement (m)

*Figure 2.11* Ground displacement time histories for the three components (NS, EW, UP) of the Loma Prieta Earthquake (October 17, 1989), Agnews State Hospital Station.

time between the first and last point to exceed $0.05\,g$ (in terms of absolute values) where $g$ is the acceleration of gravity. The dotted boundary lines indicate these points in Figure 2.9, showing the portion of strong motion of greatest interest.

Finally, the most useful measures of intensity, the so-called spectral quantities of acceleration, velocity, and displacement, are presented in Chapter 3, as they are related to the response of linear oscillators of a given period. Of these, $S_a$ (named spectral pseudo-acceleration) is by far the most used intensity measure in modern earthquake engineering.

Orbital plots are also used in the study of ground motions. These plots show the movement of a point in plan view, whether in terms of displacements or accelerations. Figure 2.12 shows an orbital plot for the 1994 Northridge earthquake recorded in the Rinaldi Receiving Station.

Given that the ground motion records for any given event and site also depend on the orientation of the recording axes, so do the intensity measures extracted from them. In general, if one rotates the components $a_x$ and $a_y$ recorded along the axes $x$ and $y$ by an angle $\theta$ (in radians) in the horizontal plane, then the new ground motion components $x_a$ and $y_a$ are given by:

$$x_a = a_x \cos\theta + a_y \sin\theta$$
$$y_a = -a_x \sin\theta + a_y \cos\theta$$

(2.5)

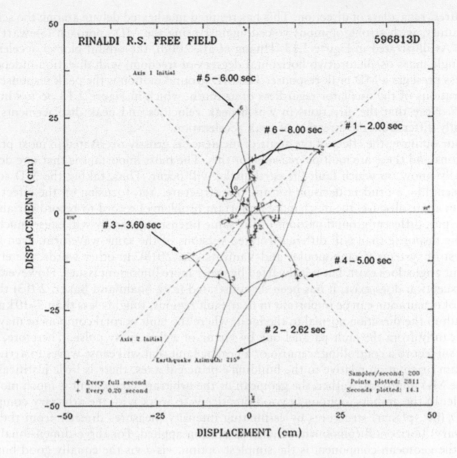

*Figure 2.12* Orbital displacement plot for the Rinaldi Receiving Station (RRS), showing the times of the first six peak amplitudes. (Adapted from Porter and Leeds, 2004.)

Given that one generally does not know in advance the direction (or angle of incidence) from which seismic waves may arrive, selecting a horizontal direction to extract intensity values from can present some issues. Three main approaches to take into account the incident angle have appeared in the literature:

- Select one of the two horizontal components (either as recorded or randomly rotated) and extract the intensity. This is the so-called arbitrary component approach.
- Use both components (either as recorded or randomly rotated), estimate the corresponding intensities of interest and take their geometric mean, i.e., the square root of their product. This is the geomean approach.
- Use both components and estimate their corresponding geomean intensity after rotating at multiple angles between 0° and 90° (e.g., at constant steps of 1°). Take the median of all estimates. This is the GMrot50 approach.

The last approach was proposed by Boore et al. (2006) to offer an orientation-independent measure of intensity. Still, for all practical purposes such values are generally considered to be numerically equivalent to the plain geomean ones (Beyer and Bommer, 2006).

In addition to the above, the 2009 NEHRP Provisions and Commentary (BSSC, 2009) introduced the definition of maximum direction (MD) for use in the seismic design of structures, regardless of direction. This has resulted in a heated debate among the scientific community, with strong opinions voiced against using the MD approach (Stewart et al., 2011). As illustrated in Figure 2.13 (Huang et al., 2008), the orbital plot of accelerations of a single mass oscillator (two horizontal degrees of freedom) with direction-independent stiffness presents a MD peak response. The MD ground motion is the peak response of the accelerations of the oscillator regardless of azimuth, which in Figure 2.13 occurs in direction Y'. Note that the directions in which peak velocities and peak displacements occur generally differ from the direction of peak acceleration.

In our opinion, the effect of the angle of incidence is grossly overstated in most practical situations and there are multiple reasons for this. The most important is that one does not generally know on which fault the earthquake will occur. Thus, taking the MD solution may seem like a prudent decision for any one structure. Yet, focusing on the effect of the incident angle obscures the much more important problem of record-to-record variability—simply put, different ground motions of the same intensity generally will cause much more variable response than will different rotated versions of the same waveform, even for the simplest of systems (Giannopoulos and Vamvatsikos, 2018). In other words, the effect of incident angle does exist, but it is dwarfed by other more important issues. However, a specific exception does exist. It has been demonstrated (e.g., Shahi and Baker, 2013) that the effect of orientation can be important in near-fault regions, roughly less than 5–10 km from the fault in the direction normal to the fault, where the fault normal component may differ significantly from the fault parallel one by virtue of a directivity pulse. Therefore, unless one is subject to a controlling scenario of a nearby fault that will cause waves to arrive with a certain orientation relative to the building's principal axes, there is little justification to use the MD approach. Rather, the geomean or the arbitrary component is much more reasonable. In the authors' opinion, a good practice is to work with the arbitrary component for 2D (i.e., planar) structures by estimating intensity measures directly from the single (arbitrary) horizontal component of ground motion applied. For three-dimensional structures the geomean component is the simplest option, vis-à-vis the equally good but more complex GMrot50.

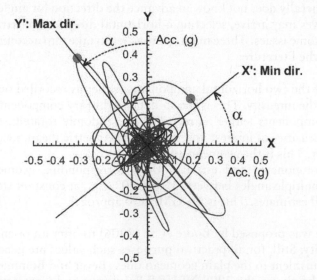

*Figure 2.13* Example of acceleration orbit of a two-degree-of-freedom oscillator used to compute minimum and maximum spectral demand (Huang et al., 2008).

# REFERENCES

Beyer, K., and Bommer, J. J. (2006). Relationships between median values and between aleatory variabilities for different definitions of the horizontal component of motion, *Bulletin of the Seismological Society of America*, 96:1512–1522.

Boore, D. M., Watson-Lamprey, J., and Abrahamson, N. A. (2006). Orientation-independent measures of ground motion, *Bulletin of the Seismological Society of America*, 96:1502–1511.

BSSC (2009). NEHRP Recommended Provisions for Seismic Regulations for New Buildings and Other Structures. FEMA P-749, Building Seismic Safety Council, Federal Emergency Management Agency, Washington, DC.

Giannopoulos, D., and Vamvatsikos, D. (2018). Ground motion records for seismic performance assessment: To rotate or not to rotate? *Earthquake Engineering and Structural Dynamics*. DOI: 10.1002/eqe.3090.

Huang, Y. N., Whittaker, A. S., and Luco, N. (2008). Maximum spectral demands in the near-fault region, *Earthquake Spectra*, 24(1):319–341.

Kramer, S. L. (1996). *Geotechnical Earthquake Engineering*, Prentice Hall, Upper Saddle River, NJ.

Porter, L. D., and Leeds, D. J. (2004). Correlation of seismic data types across strong motion arrays, Paper No. 2363, *13th World Conference on Earthquake Engineering*, Vancouver, BC, Canada, August 1–6, 2004.

Shahi, S. K., and Baker, J. W. (2013). NGA-West 2 Models for Ground-Motion Directionality, Technical report number 2013/10, Pacific Earthquake Engineering Research Center, Berkeley, CA.

Somerville, P. G., Smith, N. F., Graves, R. W., and Abrahamson, N. A. (1997). Modification of empirical strong ground motion attenuation relations to include the amplitude and duration effects of rupture directivity, *Seismological Research Letters*, 68:199–222.

Stewart, J. P., Abrahamson, N. A., Atkinson, G. M., Baker, J. W., Boore, D. M., Bozorgnia, Y., Campbell, K. W., Comartin, C. D., Idriss, I. M., Lew, M., Mehrain, M., Moehle, J. P., Naeim, F., and Sabol, T. A. (2011). Representation of bidirectional ground motions for design spectra in building codes, *Earthquake Spectra*, 27(3):927–938.

Wikipedia (2019). Modified Mercalli intensity scale. https://en.wikipedia.org/wiki/Modified_Mercalli_intensity_scale. Last modified March 8, 2019.

# Chapter 3

# Dynamics of linear elastic SDOF oscillators

## 3.1 PURPOSE AND OBJECTIVES

This chapter addresses the dynamic response of single-degree-of-freedom (SDOF) oscillators in the linear elastic domain, with emphasis on ground-induced excitation. The equation of motion is developed, solution approaches are discussed, and the representation of response using pseudo-acceleration, velocity, and displacement spectra is presented. While fundamentally important in its own right, this material also supports extensions that address the nonlinear response of SDOF systems (Chapter 4) and so-called "equivalent" SDOF systems (Chapter 7) and the linear and nonlinear response of multi-degree-of-freedom (MDOF) systems (Chapter 5).

## 3.2 EQUATION OF MOTION

### 3.2.1 Newton's first and second laws of motion

The equations governing structural dynamics are easily established based on Newton's first two laws of motion. Newton's first law of motion describes the behavior of bodies that are not subjected to unbalanced forces. This law can be stated as a body at rest tends to stay at rest and a body in motion tends to stay in motion at the same speed, unless acted upon by a force. The force referred to in the preceding sentence is the vector sum of forces acting on the body; the first law implies that any unbalanced system of forces (nonzero vector sum) causes a change in momentum.

The momentum of a body is a vector quantity given by the product of mass and velocity. Mass is a scalar, denoted by m, and velocity is a vector, denoted by $\mathbf{v}$, which may be expressed with respect to the Cartesian coordinates $x$, $y$, and $z$. Thus, the momentum, $\mathbf{q}$, may be expressed as[1]:

$$\mathbf{q} = m\mathbf{v} = m\frac{d\mathbf{x}}{dt} = m\dot{\mathbf{x}}$$ (3.1)

where $\mathbf{x}$ is a vector that indicates the position of the body.[2]

Newton's second law of motion states: *The force acting on a body and causing its movement is equal to the rate of change of momentum in the body.* In earthquake engineering, mass is typically invariant over time. Thus,

---

[1] Bold type is used to indicate vectors and matrices.
[2] A single dot represents the first derivative with respect to time; two dots represent the second derivative with respect to time.

$$\mathbf{F} = \frac{d\mathbf{q}}{dt} = \frac{d}{dt}(m\mathbf{v}) = m\frac{d\mathbf{v}}{dt} = m\frac{d\dot{\mathbf{x}}}{dt} = m\ddot{\mathbf{x}} = m\mathbf{a} \quad \text{or simply} \quad \mathbf{F} = m\mathbf{a} \tag{3.2}$$

where **a** is a vector representing the acceleration of the body.

According to D'Alembert, the second law of motion can be reformulated to state:

$$\mathbf{F} - m\mathbf{a} = 0 \tag{3.3}$$

which may be interpreted as a statement of static equilibrium if the body is considered to exhibit an inertial resistance to acceleration, described by a force equal to $-m\mathbf{a}$. One can imagine the schoolyard experience of throwing a baseball to recognize the inertial resistance exhibited by a body that is being accelerated by an applied force. This inertial resistance is a force acting in a direction opposite to the acceleration; by Newton's second law, this force is equal to $-m\mathbf{a}$. D'Alembert's Principle allows the equilibrium of the system to be expressed using the usual equations of statics provided that inertial resistance to acceleration is expressed as a force.

### 3.2.2 Free-body diagram for SDOF systems

In mechanics, degrees of freedom (DOFs) are the independent displacements and/or rotations that specify completely the displaced or deformed position and orientation of the body or system. A particle (or point) that moves in three-dimensional space has three DOFs that consist of translational displacement components in Cartesian space. A rigid body has six DOFs consisting of the three independent translations and three independent rotations (about each of the three Cartesian coordinates). Translation is displacement along an axis without any rotation, while rotation is angular motion about an axis without displacement.

The classical example of an SDOF system in structural dynamics is shown under various conditions in Figure 3.1, where the mass, $m$, at point P is shown with positive displacement $u > 0$ and positive acceleration ($\ddot{u} > 0$) in the $x$-direction. In this case, $u(t)$ describes the position of $P$ at time $t$.

The system may be damped or undamped, and may be subjected to an excitation consisting of (i) an externally applied force, $p(t)$, applied to point P in Figure 3.1, or (ii) shaking represented by motion of the frame of reference (represented by ground displacement $u_g(t)$, ground velocity $\dot{u}_g(t)$, or ground acceleration $\ddot{u}_g(t)$). In the absence of an external excitation, an oscillating system is said to be in free vibration.

In the classic case, resistance to the motion of the mass is developed by three forces:

- a spring force $F_S(t)$ provided by a massless spring. Linear elastic springs develop resistance proportional to displacement ($F_S(t) = ku(t)$, where $k$ is the elastic spring stiffness). However, $F_S(t)$ can be a nonlinear function of $u(t)$, due to softening, yielding, and possibly outright failure of the spring; this is discussed in Chapter 4 on nonlinear response. The resisting force is a function of the displacement of point P relative to the base, denoted by $u$ and sometimes described as a "relative displacement" because it represents deformation of the structure, irrespective of movement of the base.
- a damping force $F_D(t)$. Damping is used to account for the empirical observation that systems in free vibration do not oscillate in perpetuity. Damping reduces the amplitude of oscillations over time. Traditionally, damping is modeled as linearly proportional to velocity, ($F_D(t) = c\dot{u}(t)$, where $c$ is the damping coefficient); this is known as linear viscous damping. The velocity term is a "relative velocity" because it represents

the motion of point P relative to the base. Simple solutions are available when linear viscous damping is used to establish the equation of motion.

- an inertial force $F_I(t)$, which acts opposite to the direction that the mass is accelerating (based on D'Alembert's Principle), and is proportional to the absolute acceleration of the mass: $F_I(t) = m\ddot{u}(t)$, where $\ddot{u}(t)$ is the absolute acceleration of the mass (which includes the acceleration of the base).

The equation of motion can be established by considering equilibrium. Figure 3.1 illustrates masses that are accelerating in the $+x$ direction. Initially, when the mass accelerates to the right ($\ddot{u} > 0$), velocity $\dot{u}$, and displacement $u$, are also positive. This condition will be assumed when developing the equation of motion, although the relationship of the signs of these quantities will differ at other instants in the response. Different cases result depending on the nature of the excitation and resisting forces: undamped free vibration (Section 3.3) is used to establish system characteristics (such as the natural period of vibration of the system); linear elastic response (Section 3.3) occurs for $F_S(t) = ku(t)$, and nonlinear response (Chapter 4) occurs for other definitions of $F_S(t)$.

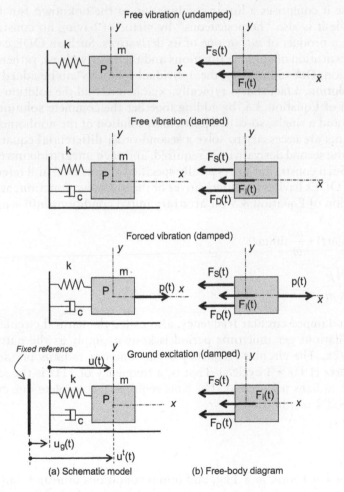

*Figure 3.1* Applied force $p(t)$ and resisting forces ($F_S(t)$, $F_I(t)$, and $F_D(t)$) used to develop the equation of motion. Forces are shown in the direction that they act. (a) Schematic model and (b) free-body diagram.

## 3.3 UNDAMPED FREE VIBRATION OF LINEAR ELASTIC SYSTEMS

As may be seen in Figure 3.1 the only forces acting on the undamped system in free vibration are the spring force, $F_S(t)$, and the inertial force, $F_I(t)$. The inertial force resists the acceleration to the right, and thus acts to the left, at the instant shown. Thus, horizontal equilibrium requires

$$F_I(t) + F_S(t) = 0 \tag{3.4}$$

For linear elastic behavior, $F_S(t) = ku(t)$. Substitution allows the equation of motion to be expressed as:

$$m\ddot{u}(t) + ku(t) = 0 \tag{3.5}$$

Equation 3.5 is a homogeneous second-order ordinary differential equation (ODE). "Ordinary" signifies the terms are functions of only one independent variable (time). "Differential" indicates that one or more derivatives are present. "Second order" signifies that derivatives up to and including the second one are present. Moreover, Equation 3.5 is linear because it comprises a linear combination of the unknown function, $u(t)$, and its derivatives, while it is also "homogeneous" by virtue of having no constant term, i.e., no term that is not a product of $u(t)$ or one of its derivatives. Such an ODE can be solved analytically (by substitution of suitable functions and their derivatives), rather than requiring a numerical solution. Due to linearity, the individual solutions can be added together to form the complete solution, a fact that is typically exploited to find the solution to the nonhomogeneous version of Equation 3.5, by adding together the complete solution of the homogeneous equation and a single, so-called particular, solution of the nonhomogeneous ODE.

Two constraints are necessary to solve a second-order differential equation, because two integrations of the second derivative are required, and each integration introduces a constant of integration. Such constraints are typically specified at time $t = 0$ and referred to as "initial conditions" for ODEs having time-derivatives of the unknown function, as in Equation 3.5. The exact solution of Equation 3.5 for arbitrary initial conditions $u(0) = u_0$ and $\dot{u}(0) = \dot{u}_0$ is

$$u(t) = u_0 \cos(\omega t) + \frac{\dot{u}_0}{\omega} \sin(\omega t) \tag{3.6}$$

where $\omega = \sqrt{\dfrac{k}{m}} \tag{3.7}$

is the natural undamped circular frequency, also called the natural circular frequency. The number of oscillations per unit time period is known simply as the natural frequency, $f$, given by $f = \omega/2\pi$. The circular frequency $\omega$ has units of radians per second, while $f$ is measured in Hertz (1 Hz = 1 cycle/s). That is, a frequency of 1 Hz is the same as a circular frequency of $2\pi$ radians per second. The time required to complete one cycle of motion is the natural period $T$ of vibration:

$$T = \frac{1}{f} = \frac{2\pi}{\omega} \tag{3.8}$$

For example, for $k = 1\,\text{N/m}$, $m = 1\,\text{kg}$, and initial conditions of $u(0) = 1\,\text{m}$ and $\dot{u}(0) = 0$, the displacement response over time (obtained using Equation 3.6) is represented by the thin line plotted in Figure 3.2, which is a sinusoidal wave having period $T = 2\pi/\omega = 2\pi$ sec.

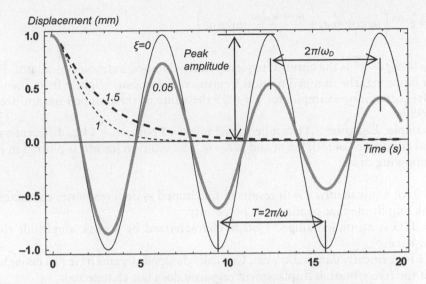

Figure 3.2 Effect of damping on free vibration response of a linear elastic oscillator.

## 3.4 DAMPED FREE VIBRATION OF LINEAR ELASTIC SYSTEMS

Damping introduces a force that opposes the velocity. With reference to Figure 3.1, this force acts to the left, opposing the increase in displacement. Horizontal equilibrium of the free body at the instant shown (Figure 3.1, damped free vibration) requires:

$$F_I(t) + F_D(t) + F_S(t) = 0 \tag{3.9}$$

Substituting appropriate terms for these forces results in the equation of motion for free vibration response of a damped linear elastic system:

$$m\ddot{u}(t) + c\dot{u}(t) + ku(t) = 0 \tag{3.10}$$

Equation 3.10 is known as the equation of damped harmonic motion. Dividing each term of Equation 3.10 by $m$ allows this equation to be restated in the following form:

$$\ddot{u}(t) + 2\xi\omega\dot{u}(t) + \omega^2 u(t) = 0 \tag{3.11}$$

where $\omega$ is the natural circular frequency (Equation 3.7) of the oscillator and $\xi$ is a constant called the damping ratio.

$$\xi = \frac{c}{2m\omega} \tag{3.12}$$

Most systems of interest in earthquake-resistant structural engineering are "underdamped," meaning that $\xi < 1$. A system is considered to be critically damped if $\xi = 1$, and overdamped if $\xi > 1$. Equation 3.12 establishes that the damping coefficient for critical damping is given by $c_c = 2m\omega$.

Equations 3.10 and 3.11 are homogeneous second-order linear ODEs that can be solved analytically. The general solution to Equation 3.11 for arbitrary initial conditions (given by $u(0) = u_0$ and $\dot{u}(0) = \dot{u}_0$) is:

$$u(t) = e^{-\xi\omega t}\left( u_0 \cos(\omega_{\mathrm{D}}t) + \frac{\dot{u}_0 + u_0\xi\omega}{\omega_{\mathrm{D}}} \sin(\omega_{\mathrm{D}}t) \right) \tag{3.13}$$

where $\omega_{\mathrm{D}} = \omega\sqrt{1-\xi^2}$ is the natural frequency of the damped harmonic oscillator. For typical values of damping, the damped natural frequency differs only slightly from the undamped natural frequency; for example, for $\xi = 0.05$ the value of the damped natural frequency is $\omega_{\mathrm{D}} = 0.9987\omega$.

For example, Equation 3.11 can be solved for $k = 1\,\mathrm{N/m}$, $m = 1\,\mathrm{kg}$, different values of $\xi$, and initial conditions of $u(0) = 1\,\mathrm{m}$ and $\dot{u}(0) = 0$. The solution for $u(t)$ is plotted in Figure 3.2 for the following cases:

- $\xi = 0$ (or equivalently, $c = 0$) results in undamped system response, characterized by a peak amplitude that continues in perpetuity.
- $\xi = 0.05$ is an underdamped system, characterized by a peak amplitude that decays over time.
- $\xi = 1$ is a critically damped system. Critically damped systems have just enough damping that the free vibration displacement response does not change sign.
- $\xi = 1.5$ is an overdamped system, having response shown by a thick dashed line.

## 3.5 FORCED VIBRATION OF LINEAR ELASTIC SYSTEMS AND RESONANCE

Equilibrium of the free body in Figure 3.1 (forced vibration-damped) for a system beginning at rest and accelerating to the right ($\ddot{u} > 0$) under the action of an applied force $p(t)$ requires:

$$F_{\mathrm{I}}(t) + F_{\mathrm{D}}(t) + F_{\mathrm{S}}(t) = p(t) \tag{3.14}$$

Substituting appropriate terms for these forces results in the equation of motion for forced vibration response of a damped linear elastic system, a nonhomogeneous second-order linear ODE:

$$m\ddot{u}(t) + c\dot{u}(t) + ku(t) = p(t) \tag{3.15}$$

One special case of interest is when the applied force consists of harmonic loading, i.e., $p(t) = p_0 \sin(\Omega t)$. This may occur where a motor spinning at angular frequency $\Omega$ is unbalanced, producing a harmonic excitation on the system.[3] In this case Equation 3.15 can be expressed as:

$$\ddot{u}(t) + 2\xi\omega\dot{u}(t) + \omega^2 u(t) = \frac{p_0}{m}\sin\Omega t \tag{3.16}$$

Equation 3.16 can be solved analytically. The general solution depends on two complex conjugated constants ($c_1$ and $c_2$) that can be determined based on the initial conditions[4]:

---

[3] A harmonic load acting only in the $x$-direction can be obtained by using two eccentric mass vibrators set to spin equal masses in opposite directions, thereby causing the $y$-components of the centripetal forces imposed on the SDOF system to cancel one another, while the $x$-components are in phase and add together.

[4] For the sake of simplicity, we have not introduced the initial conditions into Equation 3.17, as we did before with Equation 3.13. This may complicate the expression, but the motivated reader can easily solve such cases with Mathematica® or similar software.

$$u(t) = \underbrace{e^{-\xi\omega t}(c_1\cos(\omega_D t) + c_2\sin(\omega_D t))}_{\text{Transient}} + \underbrace{\frac{p_0\left(-2\xi\omega\,\Omega\cos(\Omega t) + (\omega^2 - \Omega^2)\sin(\Omega t)\right)}{m\left(\omega^4 + 2(2\xi^2 - 1)\omega^2\Omega^2 + \Omega^4\right)}}_{\text{Steady-state}} \qquad (3.17)$$

The first term of the general solution involves something multiplied by $\exp[-\xi\omega t]$, and thus it vanishes over time—this term is known as the transient part of the solution. The second term does not vanish with time, and is known as the steady-state part of the solution. The transient solution is also the solution of the homogeneous Equation 3.11 for given initial conditions (compare with Equation 3.13).

To illustrate the solution given by Equation 3.17, consider the case of $k = 1, m = 1, \Omega = 0.5\omega$, $p_0 = 2, \xi = 0.05$ with initial conditions $u(0) = 0$ and $\dot{u}(0) = 0$. Figure 3.3 shows the transient response (black line), the steady response (dashed line), and their superposition, i.e., the complete response (thick gray line):

The peak amplitude of the steady-state response can be determined by applying the generic trigonometric relationship $a\sin\theta + b\cos\theta = \sqrt{a^2 + b^2}\sin(\theta + \phi)$ to the steady-state term of Equation 3.17, resulting in

$$\max u_{\text{steady-state}} = \frac{p_0/k}{\sqrt{1 - 2\left(\dfrac{\Omega}{\omega}\right)^2(1 - 2\xi^2) + \left(\dfrac{\Omega}{\omega}\right)^4}} \qquad (3.18)$$

The peak amplitude or maximum response of the steady-state term, given by Equation 3.18, can be plotted in terms of $p_0/k$ and as function of the forcing frequency, $\Omega$, normalized by the natural circular frequency, $\omega$. Because $p_0/k$ is the displacement resulting from the static application of $p_0$, this result is termed the steady-state dynamic magnification factor. The steady-state dynamic magnification factor depends strongly on the damping ratio and normalized forcing frequency, as shown in Figure 3.4.

Resonance is the tendency of a system to oscillate with large peak amplitude under forced vibration at particular frequencies, known as the system's resonant frequencies. Resonance occurs for an undamped SDOF system if the system's natural frequency ($\omega$) coincides with the frequency of harmonic loading (i.e., $\Omega = \omega$). Response amplitudes approach infinity

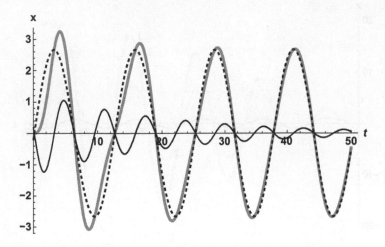

*Figure 3.3* Illustration of transient, steady, and complete response.

for undamped systems, but practically speaking, as the amplitude increases a point will be reached where the physical properties of the system change, thus creating a nonlinear problem. For damped SDOF systems, resonance occurs at frequencies $\Omega$ slightly less than $\omega$, as can be observed in Figure 3.4. While damping causes a reduction in peak response amplitudes relative to the undamped case (e.g., Figure 3.4), resonant amplitudes can still be quite large, compared with other cases of forced vibration.

Consider three cases defined by the dashed line in Figure 3.4, and represented in Figure 3.5: $\xi = 1.0$, $\xi = 0.1$, and the undamped case $\xi = 0$, for $\Omega = 1.1\omega$, $k = 1$, $m = 1$, and $p_0 = 2$.

The peak steady-state displacement response was represented in Figure 3.4, while the complete response (steady state + transient) is shown in Figure 3.5. Figure 3.6 shows the ratio of the peak response amplitude obtained for the complete solution to that obtained for the

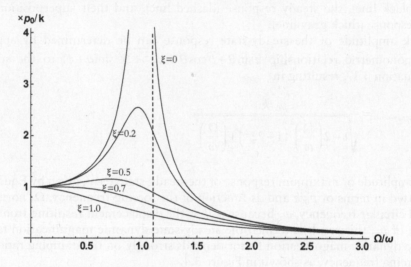

Figure 3.4 Variation of the steady-state dynamic magnification factor (max $u_{\text{steady-state}}/(p_0/k)$) with damping and normalized forcing frequency ($\Omega/\omega$).

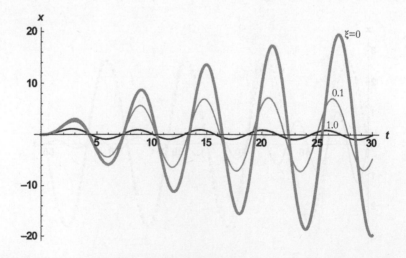

Figure 3.5 Example of dynamic amplification (complete solution).

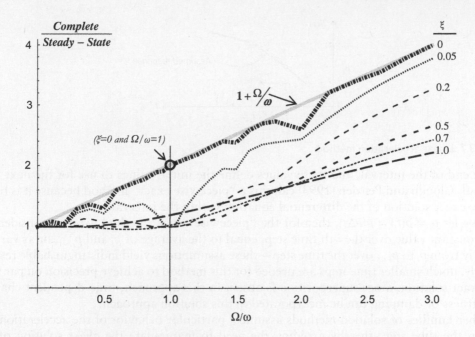

Figure 3.6 Ratio of peak displacement response for complete solution to peak displacement response for steady-state solution.

steady-state solution. Since this ratio is always greater than one, the magnification factor for the complete response is always greater than the magnification factor of the steady-state response. The ratio tends to increase with an increase in $\Omega/\omega$ and a reduction in damping. Gil-Martín et al. (2012) have shown that this ratio can be expressed simply for the undamped case as:

$$\frac{\max u_{complete}(\xi = 0, \Omega/\omega)}{\max u_{steady\text{-}state}(\xi = 0, \Omega/\omega)} = \frac{\Omega}{\omega} + 1 \tag{3.19}$$

while for the damped case the ratio is smaller.

As discussed by Gil-Martín et al. (2012), an interesting indeterminacy exists for $\xi = 0$ and $\Omega/\omega = 1$. If this condition is approached by changing the normalized forcing frequency for the case of zero damping, the ratio of peak displacements obtained for the complete and steady-state responses is 2, while if this condition is approached by reducing damping to zero for the case of $\Omega/\omega = 1$, this ratio is 1.

## 3.6 NUMERICAL SOLUTIONS OF DAMPED FORCED VIBRATION

Analytical solutions such as those developed in Section 3.5 (Equations 3.16 and 3.17) are not available for forced vibration under general (or arbitrary) loading. There are many options to obtain the solution when the loading is not defined by an analytical expression, as is the case of seismic actions. We choose a numerical solution approach, because this approach will also be applicable where nonlinearities are present (in later portions of this book).

The differential equation (Equation 3.15) can be evaluated at discrete time-steps (intervals) of $\Delta t$. If values of displacement and velocity are known at the beginning of each interval, the differential equation can be solved to determine the values of displacement and velocity

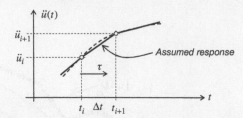

*Figure 3.7* Linear acceleration method.

at the end of the interval, and these values define the initial values to use for the next time interval. Clough and Penzien (1993) call this a "piecewise exact" method because it is based on the exact solution of the differential equation within the interval.

If we let $p_i = p(t_i) = p(i \cdot \Delta t)$, then for the "piecewise exact" method we may consider $p(t)$ as a constant value over the $i$-th time step, equal to the average of $p_i$ and $p_{i+1}$, or as varying linearly from $p_i$ to $p_{i+1}$ over the time step—these assumptions yield indistinguishable results. Clearly, much smaller time steps are needed for this method to achieve precision on par with the exact solution. Since superposition is absent in this approach, time-dependent changes in stiffness or damping can be incorporated in this solution approach.

Other families of solution methods assume a particular behavior of the acceleration $\ddot{u}(t)$ within the time step, thereby avoiding the need to manipulate the exact solution of the differential equation. For example, one can assume constant average acceleration or linear acceleration, a concept originally developed by Newmark (Bathe, 1996). If linear acceleration is assumed during the time interval, the acceleration varies linearly with time $\tau$, where $\tau = 0$ at the start of the interval at time $t = t_i$ (see Figure 3.7):

$$\ddot{u}(\tau) = \ddot{u}_i + \left( \frac{\ddot{u}_{i+1} - \ddot{u}_i}{\Delta t} \right) \tau \tag{3.20}$$

Equation 3.20 is represented in Figure 3.7.

Expressions for the variation of velocity and displacement can be obtained by integration of the former equation:

$$\dot{u}(\tau) = \dot{u}_i + \ddot{u}_i \tau + \frac{\ddot{u}_{i+1} - \ddot{u}_i}{\Delta t} \frac{\tau^2}{2} \tag{3.21}$$

$$u(\tau) = u_i + \dot{u}_i \tau + \ddot{u}_i \frac{\tau^2}{2} + \frac{\ddot{u}_{i+1} - \ddot{u}_i}{\Delta t} \frac{\tau^3}{6}$$

Evaluating both equations at $\tau = \Delta t$:

$$\dot{u}_{i+1} = \dot{u}_i + \frac{\ddot{u}_{i+1} + \ddot{u}_i}{2} \Delta t \tag{3.22}$$

$$u_{i+1} = u_i + \dot{u}_i \Delta t + \ddot{u}_i \frac{\Delta t^2}{3} + \ddot{u}_{i+1} \frac{\Delta t^2}{6}$$

The equilibrium equation (3.15), evaluated at $t_{i+1}$ is:

$$\ddot{u}_{i+1} + 2\xi\omega\dot{u}_{i+1} + \omega^2 u_{i+1} = \frac{p(t_{i+1})}{m} \tag{3.23}$$

Introducing both expressions from Equation 3.22 into Equation 3.23 and solving for $\ddot{u}_{i+1}$:

$$\ddot{u}_{i+1} = \frac{6\dfrac{p(t_{i+1})}{m} - \omega u_i - (2\xi\omega + \omega^2\Delta t)\dot{u}_i - \omega\Delta t\left(\xi + \dfrac{\omega\Delta t}{3}\right)\ddot{u}_i}{1 + \xi\omega\Delta t + \dfrac{\omega^2\Delta t^2}{6}} \tag{3.24}$$

For given initial conditions, the solution can be obtained with Equations 3.22 and 3.24.

*Example:*

Response to an impulse can be partitioned into a forced vibration response during the impulse and free vibration response commencing at the end of the impulse. Mathematically, the solution consists of two ODEs, coupled through the boundary conditions. For damped response to a load of duration $\Delta t = t_1$, the governing equations of motion and boundary conditions are:

Equation of motion
$$\begin{cases} \ddot{u}_1(t) + 2\xi\omega\dot{u}_1(t) + \omega^2 u_1(t) = p(t) & \text{for} \quad t \le t_1 \\ \ddot{u}_2(t) + 2\xi\omega\dot{u}_2(t) + \omega^2 u_2(t) = 0 & \text{for} \quad t > t_1 \end{cases}$$

Boundary conditions
$$\begin{cases} u_1(0) = 0 \\ \dot{u}_1(0) = 0 \\ u_2(t_1) = u_1(t_1) \\ \dot{u}_2(t_1) = \dot{u}_1(t_1) \end{cases} \tag{3.25}$$

Consider an impulse consisting of a half sine wave (Figure 3.8a), where $p(t) = p_0 \sin[\Omega t]$. The displacement response is plotted in Figure 3.8b for the following parameter values: $k = 1\,\text{N/m}$, $m = 1\,\text{kg}$, $\Omega = 1.1\omega$, $p_0 = 2\,\text{N}$, and $\xi = 0.05$. The thick gray line is the damped forced motion and the thin black line is the damped free vibration response.

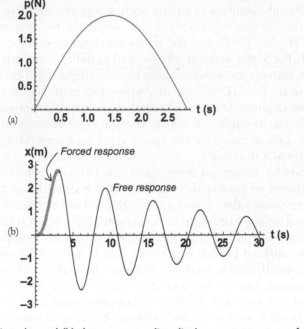

Figure 3.8 (a) Half-sine impulse and (b) the corresponding displacement response of a damped system.

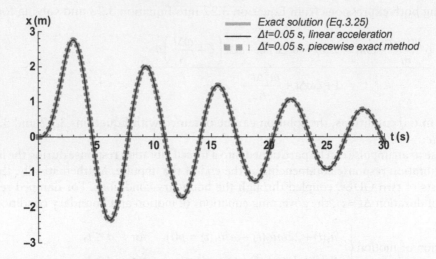

*Figure 3.9* Solution by piecewise exact method and linear acceleration method for the half-sine impulse forcing function of Figure 3.8a.

Figure 3.9 compares the solution given in Figure 3.8 with the solution solved using the piecewise exact and the linear acceleration methods. In both cases, $p(t)$ has been assumed constant within the time interval, equal to the mean, $(p_i + p_{i+1})/2$.

The linear acceleration method is equivalent to the use of splines of class $C^2$ (i.e., displacement, velocity, and acceleration are continuous). The piecewise exact method only equates displacements and velocities (as shown in Equation 3.25), and is equivalent to the use of splines of class $C^1$; for this reason the so-called "piecewise exact" method can be less accurate for a given time discretization.

Stability refers to the solution not becoming unbounded due to accumulation of errors; stability is not a sufficient condition to ensure accuracy of the solution. The constant average acceleration method is unconditionally stable, while the linear acceleration method is conditionally stable (Bathe, 1996); i.e., the linear acceleration method is stable as long as $\Delta t/T < \sqrt{3}/\pi = 0.551$. For SDOF systems, the issue of stability is of little importance as one can easily satisfy the aforementioned constraint by selecting a small enough time step. This becomes more important for MDOF systems, where the existence of very short vibration periods may make the required $\Delta t$ extremely small and thus computationally expensive (see Chapter 5). Nevertheless, as explained above, continuity of the second derivative (relative acceleration) improves the accuracy of the linear acceleration method, thus making it the preferred approach for SDOF systems.

In general, methods for numerical integration can be classified as explicit or implicit. Explicit methods attempt to predict the response of the structure at the end of the time step on the basis of response values determined at the end of the preceding time step. Small time steps are required for most ordinary structures, which contain relatively stiff elements. Because numerical error accumulates as the solution progresses, explicit methods can be unstable (error grows without bound) if the time step is not small enough. Thus, they are referred to as conditionally stable methods because the stability depends on the size of the time step. By contrast, implicit methods attempt to satisfy the differential equation of motion at the end of the time step. Obviously, programming explicit methods is simpler than programming implicit methods, which require iteration in most cases (and especially for nonlinear problems). Implicit methods may be conditionally or unconditionally stable.

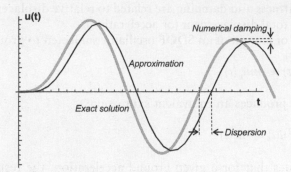

*Figure 3.10* Numerical damping and dispersion as measures of the discrepancy between the exact and the numerical solutions for free vibration response.

The Newmark constant average or linear acceleration schemes presented earlier are examples of implicit schemes that result in non-iterative (and thus easy to use) formulas for the special case of linear systems.

In the case of linear systems, the accuracy of a numerical method is evaluated by comparison with the exact solution, typically of an undamped SDOF system in free vibration. The deviation of both amplitude and period are considered: amplitude decay is also referred to as numerical damping, while period elongation is referred to as dispersion in the literature (see Figure 3.10). The latter should not be confused with the dispersion of random variables due to uncertainty (discussed in Chapter 13).

## 3.7 EARTHQUAKE-INDUCED GROUND EXCITATION

### 3.7.1 Equation of motion for linear elastic response

One of the most important excitations in seismic regions is induced by ground motion. The last panel of Figure 3.1 illustrates an SDOF system subjected to such motion. The ground displacement is $u_g(t)$. The displacement of the structure is $u^t(t)$ and is termed the "total displacement." The total displacement consists of the relative displacement between the mass and the ground, $u(t)$, associated with deformation of the structure, plus the ground displacement $u_g(t)$:

$$u^t(t) = u(t) + u_g(t) \qquad (3.26)$$

The first and second derivatives of Equation 3.26 with respect to time provide parallel relationships for total and relative velocities, as well as total and relative accelerations. We refer to total and absolute acceleration interchangeably.

The free-body diagram for the SDOF system is illustrated in the last panel of Figure 3.1, for which equilibrium requires that:

$$F_I(t) + F_s(t) + F_D(t) = 0$$

where:

$$F_I(t) = m\ddot{u}^t(t) = m\left(\ddot{u}(t) + \ddot{u}_g(t)\right) \qquad (3.27)$$

$$F_s(t) = ku(t)$$

$$F_D(t) = c\dot{u}(t)$$

We note again that stiffness and damping are related to relative displacement while the inertial force is related to total displacement (or acceleration).

Thus, the equation of motion of an SDOF oscillator subjected to ground motion $\ddot{u}_g(t)$ is:

$$m\ddot{u}(t) + c\dot{u}(t) + ku(t) = -m\ddot{u}_g(t) \tag{3.28}$$

Normalizing by mass produces an equivalent expression:

$$\ddot{u}(t) + 2\xi\omega\dot{u}(t) + \omega^2 u(t) = -\ddot{u}_g(t) \tag{3.29}$$

Equation 3.29 indicates that for a given ground acceleration, the response of the system depends only on the damping ratio $\xi$ and on the natural frequency $\omega$ of the system (or period, $T$).

### 3.7.2 Response history

Earthquake ground motion is conveniently defined by acceleration records obtained in three orthogonal directions (Cartesian coordinates) from a seismograph. For example, Figure 3.11 shows an accelerogram that was obtained from the Corralitos station (horizontal plane in a direction with azimuth of 83.5 with respect to the epicenter) during the 1989 Loma Prieta Earthquake. The peak ground acceleration (PGA) of this record is 0.799$g$, where $g = 9.81$ m/s is the acceleration of gravity; the peak occurs at 2.02 s.

Accelerogram data usually consist of series data provided at a constant time increment such as 0.01 or 0.02 s. Thus, for a given accelerogram, the relative displacement response history can be calculated for a given natural frequency, $\omega$ (or period, $T = 2\pi/\omega$), and damping ratio, $\xi$. The relative displacement response history indicates the deformation of the structure that takes place, and hence is usually of greater interest than the total displacement. The term "displacement history" thus usually refers to the relative displacement history. For the Loma Prieta accelerogram of Figure 3.11 and an SDOF oscillator having period, $T$, of 0.25 s, and damping ratio of 5% ($\xi = 0.05$), the displacement response history of Figure 3.12 is obtained. By inspection it becomes obvious how the SDOF system "filters" the excitation and mostly vibrates in its first-mode period, each second of motion roughly consisting of four cycles of oscillation.

The time derivative of the relative displacement time history is the relative velocity time history; its time derivative is the relative acceleration time history. Usually, when considering acceleration, the total or absolute acceleration is of interest, as this indicates the accelerations that sensitive components and other objects within the structure are subjected to. Total accelerations may be obtained by adding to the relative acceleration the ground acceleration at the relevant instant in time.

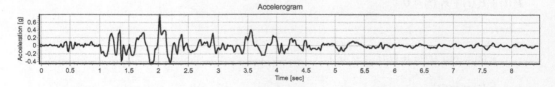

*Figure 3.11* Accelerogram recorded at the Corralitos station in the Loma Prieta Earthquake of October 17, 1989 (normalized by the acceleration of gravity).

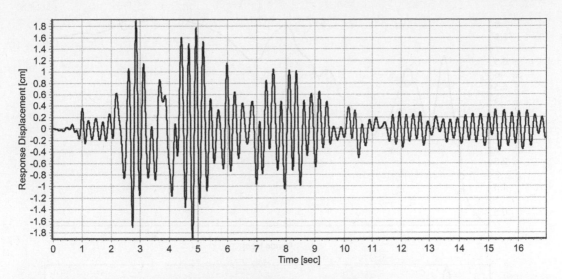

*Figure 3.12* Displacement response history for an SDOF system (5% damping and period of 0.25 s) subjected to the accelerogram shown in Figure 3.11.

### 3.7.3 Elastic response spectrum

The response spectrum provides a way to characterize peak response quantities that occur during the response history. Results are plotted for a range of oscillator periods and for one or more damping levels. The displacement response spectrum arises naturally from Equation 3.29. The (relative) displacement response spectrum for the Corralitos record is plotted in the top panel of Figure 3.13. To illustrate the genesis of this figure, note that the peak displacement shown in Figure 3.12 is −1.906 cm (at 4.81 s). The absolute value of the peak displacement is plotted in the top panel of Figure 3.13 at a period of $T = 0.25$ s. Other points plotted in this panel correspond to the absolute values of the peak relative displacements that occur for oscillators having different periods. Absolute values are used because we rarely care about the sign of the response.

Velocity and acceleration response spectra are also of interest. Successive differentiation of the relative displacement response histories produces the relative velocity and relative acceleration histories. Adding the ground acceleration history to the latter produces the total (or absolute) acceleration history. The absolute values of the peak values of the relative velocity and total acceleration histories are plotted in the middle and lower panels of Figure 3.13, respectively, for the Corralitos record.

To develop some intuition, it is useful to consider limiting values of the spectra at $T = 0$ (i.e., an infinitely stiff system, akin to a rigid monolith perfectly fixed to the ground), and $T = \infty$ (an infinitely flexible system, akin to a frictionless slider or a helium balloon attached to the ground by a string). The monolith experiences zero relative displacement, velocity, and acceleration. Consequently, its absolute acceleration is equal to that of the ground, the PGA. Thus, the displacement and velocity spectra of Figure 3.13 start from zero, while the acceleration spectrum starts at the PGA. The balloon stays motionless in absolute coordinates (having zero absolute displacement, velocity, and acceleration); thus, its relative displacement and velocity are equal to the peak ground displacement (PGD) and peak ground velocity (PGV), respectively. Hence, as $T$ grows the displacement and velocity spectra converge to PGD and PGV, respectively, while the (total) acceleration spectrum would

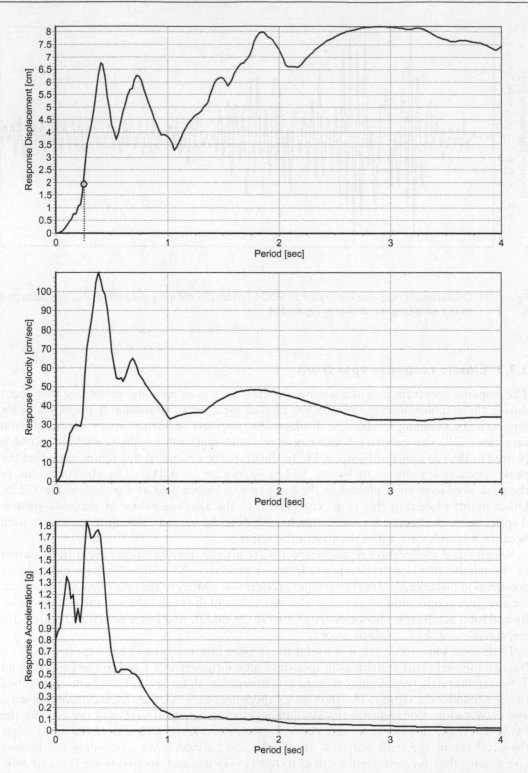

*Figure 3.13* Displacement response spectrum, velocity response spectrum, and acceleration response spectrum, for one component of the Corralitos record of the Loma Prieta Earthquake of October 17, 1989, for 5% damping.

reduce to zero. These trends are already visible in the panels of Figure 3.13, despite showing responses up to a period of only $T = 4\,\mathrm{s}$.

Calculation of the velocity and acceleration spectra (e.g., Figure 3.13) is rarely done. Rather, two other spectral representations are much more common: the pseudo-velocity and pseudo-acceleration response spectra. The last of these is used most often, and is often called an acceleration response spectrum for brevity. The term "pseudo" signifies that these are not the actual relative velocity and relative acceleration response spectra, but approximations thereof.

The fundamental quantity of interest usually is the peak displacement relative to the base, observed in the response of the SDOF oscillator (e.g., using Equation 3.29). When plotted on a displacement response spectrum, this peak relative displacement is termed the spectral displacement $S_d(T)$, of an oscillator having period, $T$, and specified damping. The product of $S_d(T)$ and the circular frequency, $\omega$ $(= 2\pi/T)$, is the spectral pseudo-velocity, $S_v(T)$, and has units of velocity. The product of $S_v(T)$ and $\omega$ is the spectral pseudo-acceleration, $S_a(T)$, and has units of acceleration. To summarize:

$$S_d(T) = u_{max}(t) = \max|u(t)|$$

$$S_v(T) = \omega S_d(T) \tag{3.30}$$

$$S_a(T) = \omega S_v(T) = \omega^2 S_d(T)$$

To obtain the internal force in the oscillator spring, only peak displacement, represented by $S_d(T)$, is needed.

Figure 3.14 presents the pseudo-velocity and pseudo-acceleration response spectra for one component of the Corralitos record.

Historically, it has been simplest to present seismic design in terms of forces, on par with the forces generated by dead, live, and wind loads. Thus, the force to be applied statically at the mass that causes a peak displacement equal to that achieved dynamically is known as the equivalent static force, $F_{S1}$:

$$F_{S1} = k u_{max}(t) = \omega^2 m u_{max}(t) = \omega^2 m S_d(T) = m S_a(T) = \frac{w}{g} S_a(T) \tag{3.31}$$

In Equation 3.31, $w$ is the weight of the mass and $g$ is the acceleration of gravity.

The SDOF system may be represented by the oscillator of Figure 3.15b, having mass $m$ supported on a massless column having stiffness, $k$, and damping, $c$. This representation more closely resembles a building although it is mathematically equivalent to that shown in Figure 3.15a. Horizontal equilibrium requires that the peak base shear, $V_b$, equals the equivalent static force. The peak base shear is a basic design quantity and is readily determined using pseudo-acceleration spectra (Equation 3.31).

Other peak quantities may be of interest, such as the peak strain energy. The peak value of the strain energy stored by an SDOF system is related to the spectral pseudo-velocity, $S_v$:

$$E_S = \frac{1}{2} k u_{max}^2(t) = \frac{1}{2} \omega^2 m u_{max}^2(t) = \frac{1}{2} m S_v(T)^2 \tag{3.32}$$

Another graphic representation of response spectra is the *tripartite* spectrum, which presents the displacement response spectrum together with the pseudo-velocity and pseudo-acceleration response spectra on a single plot. This is possible because of the fixed

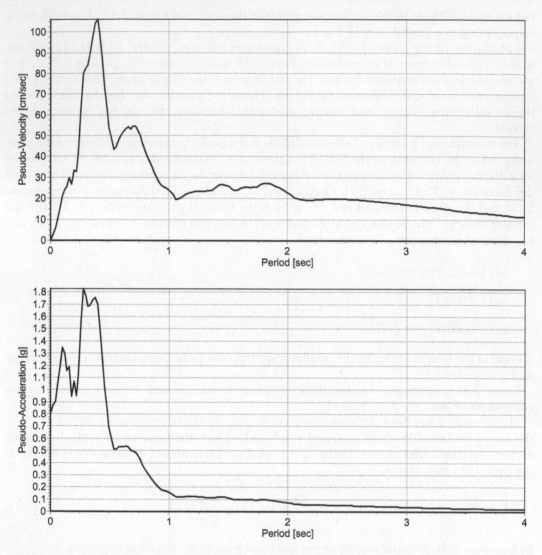

*Figure 3.14* Pseudo-velocity and pseudo-acceleration response spectra for the Corralitos record of the Loma Prieta Earthquake of October 17, 1989, for 5% damping.

relationships among $S_a(T)$, $S_v(T)$, and $S_d(T)$. Taking logarithms of the second and third equations of Equation 3.30:

$$\log S_v(T) = \log 2\pi - \log T + \log S_d(T)$$

$$\log S_a(T) = \log 2\pi - \log T + \log S_v(T) \rightarrow \log S_v(T) = -\log 2\pi + \log T + \log S_a(T)$$

(3.33)

The first equation indicates that on a spectral plot of pseudo-velocity (i.e., $S_v(T)$ versus $T$ using logarithmic scales), curves of constant spectral displacement ($S_d$) are lines with constant negative slope. The second equation indicates that on the same spectral plot of pseudo-velocity, curves of constant spectral pseudo-acceleration ($S_a$) are lines with constant positive slope. The tripartite spectrum of seven known records for 5% damping are plotted

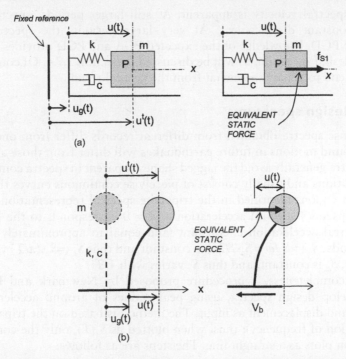

Figure 3.15 (a) Schematic model and equivalent static forces, (b) SDOF "lollipop" idealization and equivalent static forces.

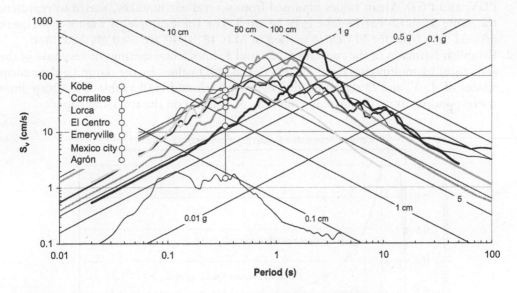

Figure 3.16 Tripartite response spectra of seven records, for 5% damping.

in Figure 3.16. Note that the Corralitos and Emeryville records are from the same event (1989 Loma Prieta Earthquake) at two different stations.

The tripartite spectrum displays a "hat"-shaped behavior. At low periods, the spectrum converges to constant acceleration equal to the PGA. At slightly larger periods, the spectral amplitudes display nearly constant pseudo-acceleration. Over a range of still larger periods, nearly constant spectral velocity is apparent. At still larger periods, spectral amplitudes display nearly constant displacement. At very large periods, the spectral amplitudes correspond to the PGD. Knowledge of the expected PGA and PGD provides a basis to construct an elastic design spectrum, as will be discussed in Section 3.7.4. Of course, individual ground motions tend to depart somewhat from the general trend.

### 3.7.4 Elastic design spectrum

In general, response spectra obtained from different records differ from one another, and the spectra of ground motions in future earthquakes will differ from those already known. Thus, design spectra generally avoid the jagged shapes apparent in spectra computed for individual ground motions and usually consist of piecewise continuous curves that correspond to the hat-shaped regions identified in the tripartite spectral representation (Figure 3.17). Specifically, the (pseudo) spectral acceleration at $T = 0$ corresponds to the PGA. At short periods, the spectral acceleration is constant (and equal to approximately 2.5·PGA). At intermediate periods, $S_v$ ($= S_a/\omega = S_a \cdot T/2\pi$) is constant and thus $S_a$ ($= S_v \cdot 2\pi/T$) varies with $1/T$. At longer periods, $S_d$ is constant, and thus $S_a$ varies with $1/T^2$.

For historical completeness, a procedure proposed by Newmark and Hall (1982) is described to develop design spectra, using peak values of ground acceleration, ground velocity, and ground displacement as input. The method is based on the tripartite representation (as a function of frequency); thus, when plotted as $S_a(T)$, only the constant pseudo-acceleration region plots as a straight line. The steps are as follows:

1. For the site and soil conditions under consideration, estimate the PGA, PGV, and PGD. Using tripartite paper, draw the lines corresponding to constant values of PGA, PGV, and PGD. Mean values obtained from several earthquakes, scaled to represent the same scenario earthquake may be used. For the example of Figure 3.18, peak ground data from the El Centro motion of May 18, 1940 are used (dashed lines).
2. Establish estimates of the constant $S_a$, $S_v$, and $S_d$ lines (representing the response of the structure) by multiplying the PGA, PGV, and PGD values by the damping-dependent values of $A$, $V$, and $D$ recommended by Newmark and Hall (Table 3.1). Draw lines corresponding to these constant values ($S_a$, $S_v$, and $S_d$) on the tripartite paper.

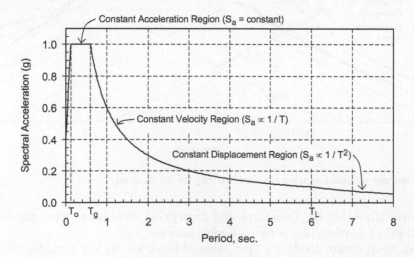

*Figure 3.17* Example of a smoothed design response spectrum, showing periods at the borders of the constant acceleration, velocity, and displacement regions.

*Table 3.1* Newmark–Hall amplification factors for the construction of elastic design spectra

| Damping (%) | Mean value (50th percentile) | | | Mean + one standard deviation (84.1th percentile) | | |
|---|---|---|---|---|---|---|
| | A | V | D | A | V | D |
| 0.5 | 3.68 | 2.59 | 2.01 | 5.10 | 3.84 | 3.04 |
| 1 | 3.21 | 2.31 | 1.82 | 4.38 | 3.38 | 2.73 |
| 2 | 2.74 | 2.03 | 1.63 | 3.66 | 2.92 | 2.42 |
| 3 | 2.46 | 1.86 | 1.52 | 3.24 | 2.64 | 2.24 |
| 5 | 2.12 | 1.65 | 1.39 | 2.71 | 2.30 | 2.01 |
| 7 | 1.89 | 1.51 | 1.29 | 2.36 | 2.08 | 1.85 |
| 10 | 1.64 | 1.37 | 1.20 | 1.99 | 1.84 | 1.69 |
| 20 | 1.17 | 1.08 | 1.01 | 1.26 | 1.37 | 1.38 |

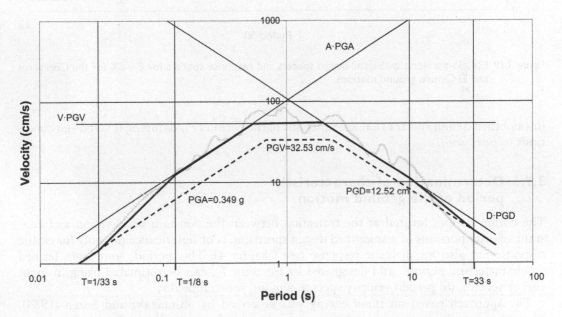

*Figure 3.18* Construction of the elastic design spectrum for $\xi = 5\%$ of El Centro ground motion, according to mean values proposed by Newmark and Hall (1982).

3. Draw vertical lines corresponding to periods of 1/33 s, 1/8 s, 10 s, and 33 s.
4. With reference to Figure 3.18, construct the hat-shaped tripartite design spectrum by drawing transition lines to connect the constant $S_a$ and PGA lines and the constant $S_d$ and PGD lines, using the periods of (Step 3) to define the end points of these transition lines. Connect the PGA, PGV, and PGD lines in the interval [1/8, 10] s.

This method was applied to the Corralitos and El Centro motions (Figure 3.19) to construct design spectra for 5% damping. It can be observed that the design spectrum does not fully envelope the response spectrum for the Corralitos record. Of course, design spectra should consider more than just the values of peak ground motion observed in past earthquakes. Modern design spectra are generated from a probabilistic seismic hazard analysis that is formulated to consider all potential rupture scenarios at a given site (see Chapter 13), resulting

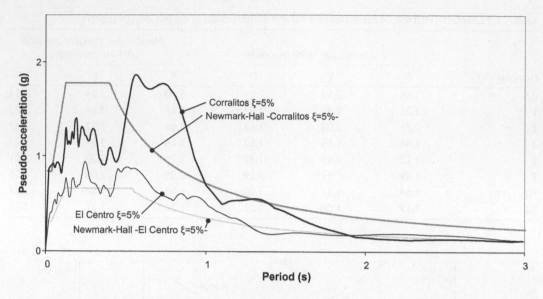

*Figure 3.19* Pseudo-acceleration elastic design spectra and response spectra for $\xi = 5\%$ for the Corralitos and El Centro ground motions.

in smoothed design spectra that are functions of the period ($T$), damping ($\xi$), and site conditions (type of soil).

### 3.7.5 Determination of characteristic period of the ground motion

The corner period, located at the transition between the constant acceleration and constant velocity portions of a smoothed design spectrum, is of significance not only for elastic response but also for inelastic response (see Chapter 4). This period, sometimes termed a "characteristic period" and designated by the term $T_g$, can be estimated using (i) input energy spectra, (ii) pseudo velocity spectra, and (iii) spectral peaks.

The approach based on input energy was described by Shimazaki and Sozen (1984). Equivalent-velocity spectra were determined for linear elastic oscillators having 5% damping. The equivalent velocity, $V_m$, is related to input energy, $E_m$, and ground acceleration and response parameters by the following expression:

$$\frac{1}{2} m V_m^{\,2} = E_m = m \int \ddot{u}_g \dot{u}\, dt \tag{3.34}$$

where $m$ = mass of the SDOF oscillator, $\ddot{u}_g$ = the ground acceleration, and $\dot{u}$ = the relative velocity of the oscillator mass. The characteristic period is estimated as the first (lowest-period) peak of the equivalent-velocity spectrum, and, at the same time, the period at which the transition occurs between the constant-acceleration and constant-velocity portions of a smooth design spectrum fitted to the 5% damped spectrum (Shimazaki and Sozen, 1984; Qi and Moehle, 1991; Lepage, 1997).

The approach based on pseudo-velocity spectra was described by Miranda (1993). In essence, the corner period is estimated as the period at which the 5% damped pseudo-velocity spectrum is a maximum.

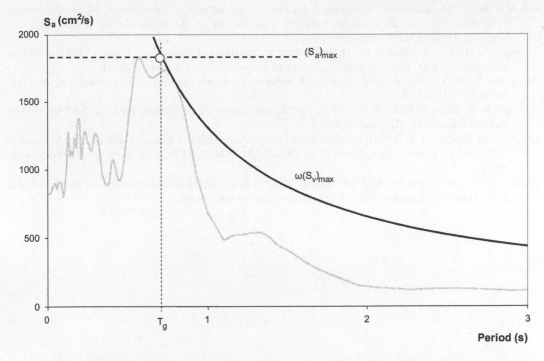

*Figure 3.20* Identification of $T_g$ for the Corralitos motion, $\xi = 5\%$, according to the method of Cuesta and Aschheim (2001).

Finally, the approach based on spectral peaks was described by Cuesta and Aschheim (2001). Very simply, the corner period, $T_g$, is the period at the transition between the "constant acceleration" and the "constant velocity" portions of the damped elastic spectrum, applicable to both smoothed design spectra and the response spectra of real ground motions. In both cases, $T_g$ may be estimated by:

$$T_g = 2\pi \frac{(S_v)_{\max}}{(S_a)_{\max}} \tag{3.35}$$

where $S_v$ and $S_a$ are the elastic pseudo-velocity and pseudo-acceleration, respectively. A graphical interpretation of this is shown in Figure 3.20 for the Corralitos ground motion. Curves of $S_a(T)$ corresponding to different values of $S_v$ are shown. The maximum pseudo-velocity occurs at the point indicated $(S_v)_{\max}$. The largest $(S_a)_{\max}$ occurs at the point indicated. The intersection of these two curves $[S_a = (S_a)_{\max}$ and $S_a = \omega(S_v)_{\max} = 2\pi(S_v)_{\max}/T]$ occurs at the period determined by Equation 3.35. For this ground motion, $(S_v)_{\max} = 208.27$ cm/s and $(S_a)_{\max} = 1.86g$, yielding an estimate of $T_g = 0.71\,\text{s}$.

## REFERENCES

Bathe, K. J. (1996). *Finite Element Procedures*, Prentice-Hall, Englewood Cliffs, NJ.

Clough, R. W., and Penzien, J. (1993). *Dynamics of Structures*. 2nd Edition, McGraw-Hill, New York.

Cuesta, I., and Aschheim, M.A. (2001). *Using Pulse R-Factors to Estimate Structural Response to Earthquake Ground Motions*, CD release 01-03, Mid-America Earthquake Center, Urbana, IL.

Gil-Martín, L. M., Carbonell-Márquez, J. F., Hernández-Montes, E., Aschheim, M., and Pasadas-Fernández, M. (2012). Dynamic magnification factors of SDOF oscillators under harmonic loading, *Applied Mathematics Letters*, 25(1):38–42.

Lepage, A. (1997). A Method for Drift Control in Earthquake Resistance Design of RC Building Structures. *Ph.D. Dissertation*, University of Illinois, Urbana-Champaign, IL.

Miranda, E. (1993). Site-dependent strength reduction factors, *Journal of Structural Engineering*, 119(12):3503–3519.

Newmark, N. M., and Hall, W. J. (1982). *Earthquake Spectra and Design*, Earthquake Engineering Research Institute, Oakland, CA.

Qi, X., and Moehle, J. P. (1991). Displacement Design Approach for Reinforced Concrete Structures Subjected to Earthquakes, Report No. UCB/EERC-91/02, EERC, UC Berkeley, Berkeley, CA, 186 pp.

Shimazaki, K., and Sozen, M. A. (1984). *Seismic Drift of Reinforced Concrete Structures*, Hazama-Gumi Ltd., Tokyo (in Japanese), and draft research report (in English).

# Chapter 4

# Dynamics of nonlinear SDOF oscillators

## 4.1 PURPOSE AND OBJECTIVES

This chapter addresses the equation of motion for the response of nonlinear single-degree-of-freedom (SDOF) oscillators, its solution, common hysteretic models, inelastic response trends and estimates, inelastic spectra, and the influence of $P$-$\Delta$ effects on response. Nonlinear SDOF oscillators provide an excellent basis for estimating the displacement response of multi-degree-of-freedom (MDOF) systems, as emphasized in Chapter 7 and utilized in the design approaches of this book.

## 4.2 INTRODUCTION

Elastic systems are conservative or non-dissipating systems and do not dissipate (or consume) energy during any quasi-static (i.e., very slow) complete cycle of loading (an excursion from and return to any point on the load–displacement curve). The load–displacement response of an elastic system may be linear or nonlinear; the same load–displacement path is followed for both loading and unloading. Energy dissipation occurs when loading occurs dynamically (i.e., at speed), often modeled as viscous damping. In contrast, typical inelastic systems dissipate energy by damage to materials (e.g., yielding of reinforcement and damage to concrete).

For elastic systems, deformations during response fully recover after cessation of the loading, leaving no residual deformation. For linear elastic systems, a load applied slowly at a point is linearly proportional to the displacement that results at that point. Since this is true for every point, the deformation throughout the member (or system) is a linear function of the applied load. Figure 4.1a shows a linear elastic system that is deformed from ① to ② and then returns to ③, recovering all deformation in a linear way. Figure 4.1b shows a nonlinear elastic system, which also returns to the origin. In contrast to these elastic systems, Figure 4.1c shows an inelastic system. Inelastic systems experience damage during loading, which ordinarily results in nonzero (or residual) deformation upon removal of the load. Nevertheless, so-called "self-centering" systems exist, which may exhibit near-zero residual deformation while sustaining damage (see the discussion on "flag-shaped" hysteretic systems in Section 4.10.6.

During nonlinear response, the displacement of a point or the deformation within the system is not proportional to the load (e.g., Figure 4.1b). For inelastic systems, a given displacement may not correspond to a unique value of load, because the unloading curve generally deviates from the loading curve. Consequently, the force-deformation history must be tracked to be able to establish the load at a given displacement. Of interest for inelastic systems (see Figure 4.1c) are the yield point ($u_y$) or proportional limit (the point where the

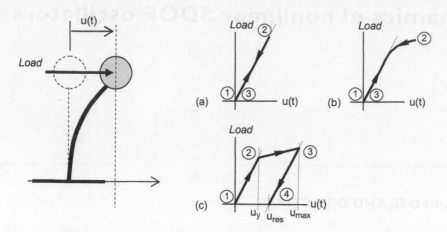

Figure 4.1 Load–displacement response of a (a) linear elastic, (b) nonlinear elastic, and (c) an inelastic system.

system ceases to be linear), the peak displacement ($u_{max}$), and the residual (or permanent) displacement ($u_{res}$).

Linear elastic response implies that there is no damage to the structural system. In highly seismic regions, linear elastic response would require the structural system to resist a lateral load approaching or exceeding the weight of the building—this can be thought of equivalently as turning a building on its side so that it cantilevers out from an imaginary fixed support, without sustaining structural damage. Few buildings can be designed to be this strong at reasonable cost. Fortunately, because the lateral excitations of earthquake ground motion are of short duration and tend to reverse quite often (relative to the fundamental period of vibration of the building), the structural system typically can be designed for much lower forces, as long as it can sustain the resulting inelastic deformations. Doing so will result in structural damage in the rare events associated with strong shaking. The frequent reversals of the ground acceleration limit the deformations and damage sustained by the structural members. While this approach is feasible for most buildings, note that as the fundamental period of the building reduces (e.g., for shorter, stiffer, and lighter buildings), the ground accelerations have longer duration relative to the period of the building, and the loading begins to approach that of a statically applied load. In such cases, the loading is sustained long enough that a reduction of lateral strength (from that required for linear elastic response) will tend to result in large inelastic deformations. Thus, the rapidly reversing nature of seismic excitation allows typical structural systems to be designed economically, with yield strength less than that required for linear elastic response, but the structural members must be able to sustain inelastic deformation without failing.

## 4.3 HYSTERETIC BEHAVIOR

Hysteretic diagrams are load–displacement curves that represent the quasi-static response of a structural component (or structure). Hysteretic response is usually modeled without dependence on velocity.

A typical structural member subjected to quasi-static cyclic loads of gradually increasing amplitude initially responds linearly and then gradually softens as damage occurs (such as yielding of steel and/or cracking or crushing of concrete). The hysteretic response of representative steel and concrete beams is illustrated in Figure 4.2. Structural damage is evident in a series of loops that develop under reversed cyclic loading. These loops are known as hysteretic loops. A hysteretic loop, representing a single cycle of response, is shaded for both test specimens in Figure 4.2.

The steel beam (Figure 4.2a) shows very gradual softening and a relatively full "hysteretic loop." The stiffness upon unloading from a peak is similar to the initial stiffness.

In the case of the reinforced concrete beam, the hysteretic loops are not as full, and the stiffnesses during loading and unloading show greater departure from the initial stiffness. For use in subsequent modeling, different stiffnesses can be identified and quantified. An initial stiffness, before cracking, is given by $k_0$. Although cracking occurs as early as Point B, the cracked stiffness ($k_{cr}$) may be defined at reinforcement yield (Point C). The strain hardening stiffness, $k_s$, is identified in Figure 4.2b as the line from D to D''. The unloading stiffness, $k_u$, varies from D to E, D' to E', D'' to E'', H to F, H' to F', and H'' to F'', decreasing as the maximum displacement increases. A degraded stiffness, $k_r$, may be identified as the line from E'' to F''.

Loss of stiffness occurs due to softening and cracking of concrete, yielding of reinforcement, slip along cracked surfaces, and bond slip. The low stiffness from E'' to F'' is because cracks that developed in the loading from H' to D'' have not closed at this point. The phenomenon is known as pinching, due to the characteristic pinch of the hysteresis loops evident near the origin in Figure 4.2b, relative to the fuller loops of Figure 4.2a.

The area enclosed by each loop in Figure 4.2 represents strain energy expended in one loading cycle, that is, work that is dissipated through inelastic deformation of the structural material (e.g., movement of dislocations within the steel grain structure, breaking of bonds within concrete or masonry materials, or dissipation of energy through friction) and, of course, is not recoverable.

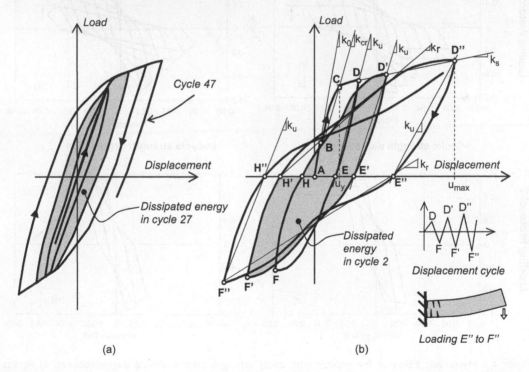

*Figure 4.2* Load–displacement response of (a) a steel girder (Adapted from Popov and Bertero, 1973) and (b) a reinforced concrete beam (Adapted from Riddell and Newmark, 1979).

Comparison of the hysteretic behavior evident in these tests shows the steel girder to have relatively "full" hysteretic loops while reinforced concrete loops typically display pinching behavior. For similar levels of yield strength and overall deformation, one might infer that the fuller hysteresis loops of the steel girder of Figure 4.2a would result in more energy dissipation and therefore better seismic performance compared with the pinched reinforced concrete loops. Having full or pinched loops simply characterizes the behavior of different systems under quasi-static cyclic loading, without necessarily indicating anything about seismic performance, as discussed later. Note also, in some instances, steel girders can display pinching behavior.

Far more important than the pinching (or not) of hysteresis loops is the strength degradation phenomena that may be observed along the loading phase of each cycle. Cyclic degradation appears as loss of strength and stiffness from one cycle to the next, as in Figure 4.3a. In contrast, the loss of strength and the appearance of negative stiffness within a single cycle are characteristics of in-cycle degradation (Figure 4.3b). Although the two systems of Figure 4.3 appear to have the same force-deformation envelope, that is, an outer boundary that is initially rising elastically and then dropping with a negative stiffness, their dynamic response is very different. In-cycle degradation will always be triggered beyond a certain level of deformation, regardless of the number of cycles, and the resulting negative stiffness can lead to unstable dynamic response resulting in much larger displacements and possibly collapse. On the contrary, the performance of the cyclically degrading system highly

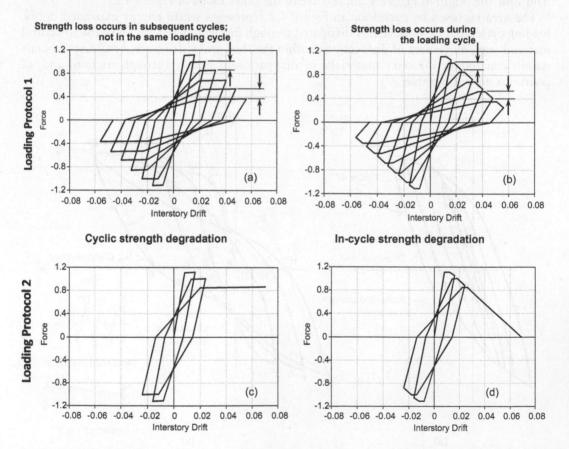

Figure 4.3 Hysteretic behavior for models with cyclic strength (and stiffness) degradation (a, c) versus in-cycle degradation (b, d), both subjected to the two different loading protocols of Figure 4.4. (From FEMA P440A.)

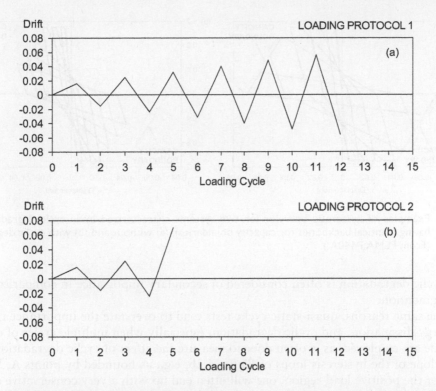

*Figure 4.4* Loading protocols I (a) and 2 (b) used in Figure 4.3 to illustrate cyclic versus in-cycle strength degradation. (From FEMA P440A.)

depends on the number of cycles that it is subjected to. This becomes apparent if we employ a different loading protocol with fewer cycles and a high amplitude final loading cycle. Generally speaking, the cyclically degrading system in Figure 4.3c behaves in a much more stable manner compared to the in-cycle degrading system of Figure 4.3d. In the former case, no cycle will impart actual negative stiffness (although in a realistic system this would eventually happen), thus the large deformation of the final cycle is absorbed along a ductile plateau. Instead, in Figure 4.3d, the negative stiffness of the system quickly leads to zero strength and eventual sidesway collapse as lateral forces can no longer be resisted.

Realistic systems typically exhibit both types of degradation, albeit to different degrees. One way to account for these types of degradation is to establish a force-deformation capacity boundary (or backbone) via monotonic loading, effectively capturing the in-cycle degradation, and then add the effect of cycles. When cyclic degradation is not accounted for, as in the example of Figure 4.5a, the system always reaches the capacity boundary regardless of the preceding number of cycles. When cyclic degradation is present, as in Figure 4.5b, the system can no longer reach the capacity boundary after a small number of cycles have occurred.

Contrary to artificially imposed quasi-static loading cycles, the vast majority of realistic ground motions impart only a limited number of cycles that are strong enough to drive a system to inelasticity. Therefore, unless one is considering long duration motions (typical of rare large-magnitude events characteristic of subduction zones) or earthquake swarms (such as the 2016 multiple foreshock/aftershock events of Central Italy) combined with short-period systems with high levels of cyclic degradation, there are typically not enough loading-unloading reversals to cyclically impart a significant strength reduction.

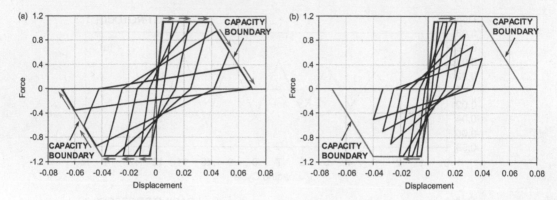

*Figure 4.5* Examples of hysteretic behavior for two systems characterized by in-cycle degradation and having identical backbones (or capacity boundaries), (a) without and (b) with cyclic degradation. (From FEMA P440A.)

Thus, cyclic degradation is often considered of secondary importance in comparison to in-cycle degradation.

For the same reasons, quasi-static cyclic tests tend to overstate the importance of pinching, energy dissipation, and cyclic degradation, especially when multiple cycles of constant amplitude are employed, as is often done to measure the effect of cyclic degradation. Using the envelope of the hysteresis loops of Figure 4.2b, e.g., as bounded by points A, B, C, D, D′, D″ for the positive load region, one will often end up with a very conservative estimate for the force-deformation capacity boundary of the system. On the other hand, a force-deformation backbone derived from monotonic loading will generally show considerably higher strength at each deformation as Figure 4.5b shows. The actual force-deformation envelope experienced by the system in a realistic seismic event will lie between the two extremes, typically closer to the monotonic one unless long durations and/or short periods are present. A good representation of system behavior under all possible loading cases can only be achieved if both monotonic and cyclic tests are available and the monotonic backbone is coupled with proper cyclic degradation rules. An example of such a model is the Ibarra–Medina–Krawinkler (IMK, discussed in Section 4.6.2).

## 4.4 INFLUENCE OF HYSTERETIC FEATURES ON DYNAMIC RESPONSE

We can consider some general characteristics of hysteretic response using the idealized model of Figure 4.6.

During the initial linear elastic portion of the response, even with no external motion the oscillator may have enough velocity to continue beyond the yield point (point ② in Figure 4.6a). Considering the equation of motion at that instant (e.g., Equation 3.28, repeated for convenience as Equation 4.1 below) and imagining that both damping and the ground acceleration $\ddot{u}_g(t)$ are zero at this instant, the internal spring force is positive while the relative acceleration of the mass $\ddot{u}(t)$ must be negative. At the instant that the stiffness reduces from $k$ to $\alpha k$, the relative acceleration must also reduce, reducing the rate at which the mass decelerates.

$$m\ddot{u}(t) + c\dot{u}(t) + ku(t) = -m\ddot{u}_g(t) \qquad (4.1)$$

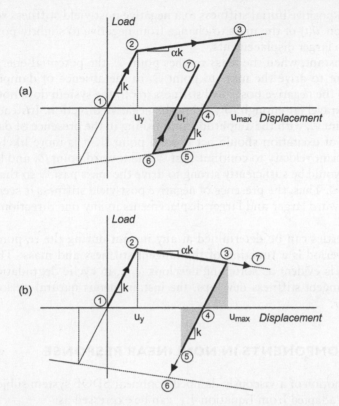

Figure 4.6 Unloading response for a bilinear model with (a) positive post-yield stiffness and (b) negative post-yield stiffness.

At some later instant, the mass may be displaced to a position such as point ③ in Figure 4.6a. Imagine for a moment that the increase in relative displacement ceases ($\dot{u} = 0$) and the ground acceleration abruptly stops at this point. The potential energy at this point, also known as the recoverable (or stored) strain energy, is given by the area of the triangle below the label ③. This energy is released and drives the mass toward the origin. At the abscissa (point ④), the stored elastic strain energy is zero and the mass is moving with velocity $\dot{u}$. If damping forces are neglected, the kinetic energy at this point, $m(\dot{u})^2/2$, is equal to the potential energy at ③. In the absence of further excitation and damping, this potential energy would be enough to drive the oscillator beyond point ⑤, until the kinetic energy has been absorbed (in a combination of stored elastic strain energy and energy expended on inelastic work) as shown by the shaded area below the abscissa. At point ⑥, the stored strain energy is sufficient to drive the oscillator to point ⑦. The mass would then oscillate around point $u_r$; in the presence of damping the oscillations diminish and the mass comes to rest with residual displacement $u_r$. If the excitation were to continue for a few random pulses beyond point ③, the mass would come to rest to one side or the other of $u_r$. Thus, for the case with positive post-yield stiffness, there is a tendency for the mass to return (in increments) toward the origin ($|u_r| < |u_{max}|$).

We can repeat the same thought experiment for the case that the post-yield stiffness is negative (as may happen if there is significant in-cycle degradation and/or $P$-$\Delta$ effects present). Considering the instant that the mass reaches and continues beyond point ② in Figure 4.6b, and again, imagining that the excitation and damping are zero at this instant, the change from a positive initial stiffness to a negative post-yield stiffness requires that the relative acceleration $\ddot{u}(t)$ of the mass to change from negative to slightly positive. This will propel the mass to larger displacements.

At some later instant, when the mass reaches point ③, the potential energy at this point would be sufficient to drive the mass to point ⑤, in the absence of damping and further excitation. Due to the negative post-yield stiffness ($\alpha < 0$) the system does not have sufficient kinetic energy to travel far enough to yield in the negative direction. Instead, the mass will oscillate about point ④, with the amplitude diminishing in the presence of damping. If some random impulses of excitation should act beyond point ③, it is more likely that the mass will develop sufficient velocity to continue past ③ and toward point ⑦, and less likely that a random impulse would be sufficiently strong to drive the mass past ⑤ so that it reaches and travels past point ⑥. Thus, the presence of negative post-yield stiffness is seen to introduce a systematic bias toward larger and larger displacements in any one direction (either positive or negative).

Dynamic properties can be determined at any instant during the response. The instantaneous natural period is a function of the tangent stiffness and mass. Thus, substantial period elongation is evident as softening develops (e.g., in cyclic degradation). In the case that a negative tangent stiffness develops, the instantaneous natural period would be an imaginary number.

## 4.5 ENERGY COMPONENTS IN NONLINEAR RESPONSE

The equation of motion of a viscously damped nonlinear SDOF system subjected to ground acceleration $\ddot{u}_g(t)$, adapted from Equation 4.1, can be expressed as:

$$m\ddot{u}(t) + c\dot{u}(t) + f_S = -m\ddot{u}_g(t) \tag{4.2}$$

In order to evaluate the energy balance during a ground excitation, we multiply the terms of Equation 4.2 by a differential of relative displacement $du$ and integrate to obtain the equation of motion in terms of relative energy (Uang and Bertero, 1988):

$$\int m\ddot{u}(t)\,du + \int c\dot{u}(t)\,du + \int f_S\,du = -\int m\ddot{u}_g(t)\,du$$

$$E_K + E_D + E_S = E_i \tag{4.3}$$

The system is excited by the ground excitation $u_g(t)$; hence, the integral on the right-hand side of Equation 4.3 defines the input energy, $E_i$.

The relative kinetic energy is:

$$E_K = \int m\ddot{u}(t)\,du = \int m\frac{d\dot{u}}{dt}\,du = \int m\dot{u}\,d\dot{u} = m\frac{\dot{u}^2}{2} \tag{4.4}$$

The damping energy, associated with the linear viscous damping term in Equation 4.2, is:

$$E_D = \int c\frac{du}{dt}\,du = \int c\left(\frac{du}{dt}\right)^2 dt \tag{4.5}$$

In order to calculate $E_D$ it is necessary to know $u(t)$ (and its derivative), which can be determined by solving Equation 4.1 for the damped system, as we will see later in this chapter.

Energy is absorbed during the deformation of a structure and released during unloading. For linear elastic systems displaced slowly from 0 to $u$, the external work is given by $W_E = \frac{1}{2}Fu$ (the area under the load–displacement curve). Elastic systems are conservative, meaning that the work done in moving a particle between two points is independent of the path taken. Thus, the external work (done by the external force moving through a displacement) equals the internal work associated with strain in the material as the structure deforms.

Inelastic systems have strain energy and dissipated energy (expended strain energy plus damping energy) as shown in Figure 4.7. For inelastic systems, a portion of the strain energy is recoverable strain energy. The remaining strain energy (strain energy less recoverable strain energy) is termed expended strain energy and is associated with the work done on the material; this work causes permanent damage associated with inelastic deformation and typically results in residual displacement upon unloading.

In general, the dissipation of energy during elastic response comes solely from viscous damping. In the post-yield range, unless one specifically employs actual dampers, $E_{SE}$ nearly always dwarfs $E_D$. Thus, dissipation in the inelastic range is mainly provided through hysteretically absorbed energy. This observation has typically led to the notion that dissipation of energy is the main mechanism through which a ductile structure takes advantage of ductility to resist seismic loads while having a lower strength than what an elastic system of the same period would require. However, it has been shown that strain energy dissipated either in a quasi-static cyclic test or during earthquake loading is poorly correlated to the maximum or residual displacement response (see, for example, Priestley, 1993). For the four systems appearing in Figure 4.8, the quasi-statically dissipated energy cannot even predict the dynamically dissipated one. As shown in Table 4.1, the seemingly full loops of the elastic-plastic kinematic hardening system (Figure 4.8a) dissipated less energy in realistic

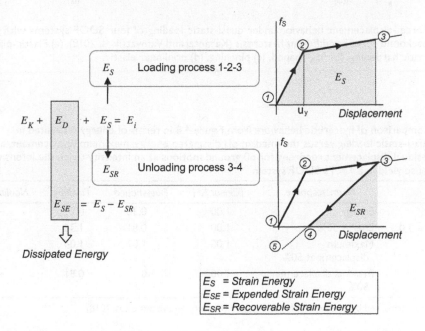

*Figure 4.7* Decomposition of strain energy in a half cycle of response of a nonlinear system.

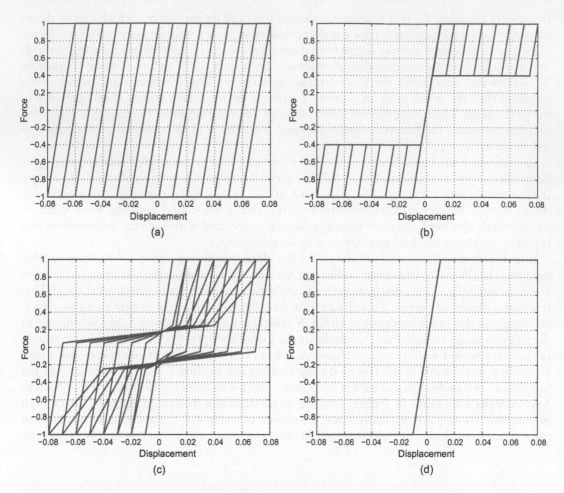

*Figure 4.8* Force–displacement behavior under quasi-static loading of four SDOF systems with the same backbone curve but different hysteresis (Kazantzi and Vamvatsikos, 2018). (a) Elastic-plastic kinematic hardening, (b) flag-shaped, (c) pinching, (d) nonlinear elastic.

*Table 4.1* Comparison of hysteretic behaviors from Figure 4.8 in terms of energy dissipated in quasi-static loading versus the (median of) dissipated energy, maximum displacement, and residual displacement recorded for 60 ground motions at an intensity twice the intensity to cause yield in a $T = 1$ s SDOF system

| Loading | Norm. response | Bilinear KH | Flag-shaped | Pinching | Nonlinear elastic |
|---|---|---|---|---|---|
| Quasi-static | Energy | 1.00 | 0.30 | 0.29 | ~0 |
| Dynamic $R = 2.0$ | Energy 50% | 1.00 | 0.81 | 1.36 | ~0 |
| | Maximum displacement 50% | 1.00 | 1.07 | 1.04 | 1.33 |
| | Residual displacement 50% | 1.00 | ~0 | 0.81 | ~0 |

The results are normalized by those of the bilinear system (Kazantzi and Vamvatsikos, 2018).

conditions compared to the skinny loops of the pinching system (Figure 4.8c). Still, both systems have about the same maximum displacement, while the pinching system, due to the shape of the unloading branch, has markedly lower residual displacement. The flag-shaped and the nonlinear elastic system also show practically zero residual displacements. Still, the very low levels of energy dissipated by both seem to result in a larger peak displacement: while the flag-shaped system has only a 10% increase in peak displacement relative to the bilinear kinematic hardening system, the nonlinear elastic system has a 33% increase in peak displacement. Thus, energy dissipation can be helpful, but is not sufficient for explaining the behavior of hysteretic systems.

## 4.6 HYSTERETIC MODELS

Various hysteretic models are available to represent in an idealized, approximate manner the response observed in experimental tests. These are referred to as phenomenological models in that they represent the observed phenomenon at a macro (empirical) level, rather than relying on very detailed modeling of the component materials and their interactions to produce the observed macro behavior.

A variety of models have been developed to represent the nonlinear load–displacement[1] response of different structural materials; trade-offs exist between simplicity and accuracy, and the more complex models require the specification of a larger number of modeling parameters for their calibration.

Figure 4.9 illustrates some hysteretic models that are commonly used in earthquake engineering. The basic models are described here, while more specialized models are described in Sections 4.6.1–4.6.3.

Figure 4.9a illustrates the elastic-plastic model. This is the simplest of the available nonlinear models, and is completely specified by any two of the following three parameters: yield strength, $F_y$, yield displacement, $u_y$, and stiffness, $k = F_y/u_y$. Response is always bounded by $F_y$ and $-F_y$, and movement between these two boundaries occurs on a line with slope $k$.

Figure 4.9b illustrates the bilinear model. This model differs from the elasto-plastic model by inclusion of a post-yield stiffness ($\alpha k$), termed kinematic hardening. Response is bounded by lines passing through the yield points having slope $\alpha k$.

In steel and concrete structures the post-yield stiffness of the member generally is associated with strain-hardening of the steel, whether a hot-rolled shape or a reinforcing bar.

The elasto-plastic and bilinear models have "full" hysteretic loops, for which energy dissipation in a displacement cycle is maximized. These models are often used to represent the idealized response of steel members (e.g., axially loaded bars that do not buckle) and may be used to represent steel components in fiber models. However, even axially loaded steel bars typically do not display sharp yield points; rather, as yielding progresses through the member, the presence of residual stresses causes gradual softening (e.g., Figure 4.2a). The Ramberg–Osgood (1943) hysteretic model provides a gradual transition between linear and plastic zones (Figure 4.9c).

---

[1] These are general terms: "load" may refer to a force or moment; and "displacement" may refer to translation or rotation.

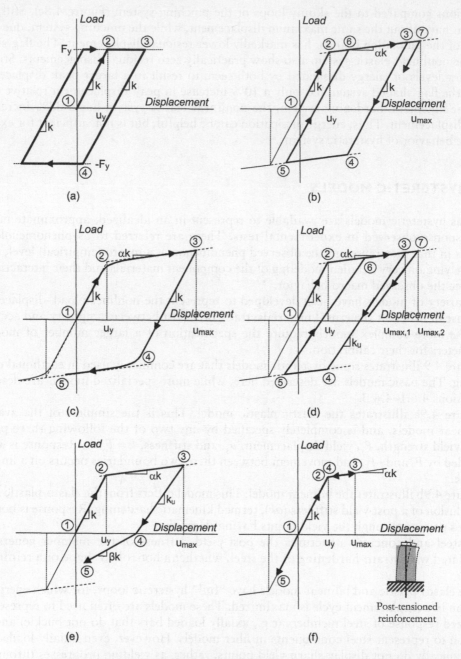

Figure 4.9 Some basic hysteretic models: (a) elasto-plastic, (b) bilinear, (c) Ramberg–Osgood, (d) bilinear with unloading stiffness degradation, (e) trilinear with unloading stiffness degradation, and (f) bilinear elastic.

Reinforced concrete members have additional sources of inelastic behavior beyond steel yielding, typically associated with cracking of concrete, slip of reinforcing bars, softening and crushing of concrete, sliding across cracked surfaces, and buckling and fracture of reinforcing steel. Thus, stiffness degradation often is modeled for reinforced concrete structures. Both bilinear and trilinear degrading models are available (see Figure 4.9d,e). In these models, the stiffnesses depend on the current maximum displacement reached ($u_{max}$) and on the unloading stiffness degradation index, $a$, (generally $a$ is in the range 0.0–0.5 for reinforced concrete members). The unloading stiffness, $k_u$, is given by

$$k_u = k \left| \frac{u_y}{u_{max}} \right|^a \tag{4.6}$$

where $u_{max}$ is the maximum displacement amplitude reached up to this point in the response. The same value of $k_u$ is used for unloading and reloading, for the same absolute value of $u_{max}$.

This approach to representing stiffness degradation is incorporated into the Takeda model, which is described in the next section.

Other models such as the peak-oriented models (Clough and Johnston, 1966) introduce stiffness degradation by means of the reloading path, which may target the previous maximum displacement in the relevant quadrant.

As mentioned in Section 4.3, reinforced concrete members typically display so-called "pinching" behavior, a term that refers to the hysteretic loops appearing to be pinched in the vicinity of the origin (as if someone squeezed together the full hysteretic loops to obtain the pinched loops). The pinching phenomenon can be incorporated in the hysteretic model (e.g., the trilinear degrading model of Figure 4.9e). The pinched hysteretic loops display softening at low loads, and stiffening as cracks close or reinforcement reengages the concrete after undergoing bond-slip. The load–displacement response softens as the section capacity is reached.

Another hysteretic model of interest is the bilinear elastic model, which is characteristic of gap-opening or rocking behavior (for which the nonlinearity is associated with changes in the contact area between two surfaces). A simple representation of this is given by the bilinear elastic model of Figure 4.9f. Bilinear elastic models are often used to represent the lateral response of a wall under self-weight or in the presence of additional vertical load induced by prestressing. With sufficient lateral force, the preexisting compression is overcome at one edge of the wall, causing a change in the contact area at the base. Associated with the opening of a gap at the base of the wall (rocking), there is a change in stiffness, as shown in Figure 4.9f. Usually only damping losses are represented (e.g., energy losses due to impact upon gap closure may be ignored). While the complete lack of hysteretic yielding within the model may produce larger displacements than systems with hysteretic losses, displacements of wall buildings tend to be low due to their inherent stiffness. The presence of a restoring force returns the system to its original position after the cessation of lateral loading. The lack of permanent deformation results in a building that can be repaired more quickly and at lower cost, allowing for quicker resumption of use.

Detailed descriptions of three specific hysteretic models (Takeda, Ibarra, and flag) are provided in the following subsections.

## 4.6.1 Takeda model

Takeda et al. (1970) developed a hysteretic model for representing the pinched response of reinforced concrete members. The model defines a trilinear primary curve for initial loading (Figure 4.6a) and a set of rules for reversal and reloading based on data obtained from specimens tested in an earthquake simulator.

The primary curve is defined by two points, the cracking point ($F_{cr}$, $u_{cr}$) and the yield point ($F_Y$, $u_Y$). The slope of the third segment is determined by the strain-hardening of the reinforcement. The set of rules for the construction of the hysteretic loops is provided in Box 4.1, extracted almost literally from the original paper of Takeda et al. (1970). These rules are implemented in the program SAKE (Otani, 1974).

## BOX 4.1 SPECIFIC RULES FOR THE TAKEDA MODEL

**Rule I. Condition.** The cracking load, $F_{cr}$, has not been exceeded in one direction. The load in reversed from a load $F$ in the other direction. The load is smaller than $F_Y$.

**Rule.** Unloading follows a straight light from the position at load $F$ to the point of cracking in the other direction.

**Example.** Segment 3 in Figure 4.10b (if unloading occurs before deformations represented by segment 2, the rules provide no hysteretic loop).

**Rule 2. Condition.** A load $F_1$ is reached in one direction on the primary curve such that $F_1$ is larger than $F_{cr}$ but smaller than $F_Y$. The load is then reversed to $-F_2$ such that $F_2 < F_1$.

**Rule.** Unload parallel to the loading curve for that half cycle.

**Example.** Segment 5 parallel to segment 3 in Figure 4.10b.

**Rule 3. Condition.** A load $F_1$ is reached in one direction such that $F_1$ is larger than $F_{cr}$ but not larger than $F_Y$. The load is then reversed to $-F_3$ such that $F_3 > F_1$.

**Rule.** Unloading follows a straight line joining the point of return and the cracking point in the other direction.

**Example.** Segment 10b in Figure 4.10b.

**Rule 4. Condition.** One or more loading cycles have occurred. The load is zero.

**Rule.** To construct the loading curve, connect the point at zero load to the point reached in the previous cycle, if that point lies on the primary curve or on a line aimed at a point on the primary curve. If the previous loading cycle contains no such point, go to the preceding cycle and continue to process until such a point is found. Then connect that point to the point at zero load.

**Exception.** If the yield point has not been exceeded and if the point at zero load is not located within the horizontal projection on the primary curve for that direction of loading, connect the point at zero load to the yield point to obtain the loading slope.

**Examples.** Segment 12 in Figure 4.10b represents the exception. It is aimed at the yield point rather than at the highest point on segment 2. Segment 8 in Figure 4.10b represents a routine application, while segment 20 represents a case where the loading curve is aimed at the maximum point of segment 12.

**Rule 5. Condition.** The yield load, $F_Y$, is exceeded in one direction.

**Rule.** Unloading curve follows the slope given by Equation 4.6, where in Takeda's model: $k_u$ is the slope of the unloading curve; $k$ is the slope of a line joining the yield point in one direction to the cracking point in the other direction; $u_{max,i}$ is the maximum deflection attained in the direction of loading; $u_Y$ is the deflection at yield; and exponent $a$ is set equal to 0.4.

**Example.** Segment 4 in Figure 4.10c.

**Rule 6. Condition.** The yield load is exceeded in one direction but the cracking load is not exceeded in the opposite direction.

**Rule.** Unloading follows Rule 5. Loading in the other direction continues as an extension of the unloading line up to the cracking load. Then the loading curve is aimed at the yield point.

**Example.** Segments 4 and 5 in Figure 4.10c.

**Rule 7. Condition.** One or more loading cycles have occurred.

**Rule.** If the immediately preceding quarter-cycle remained on one side of the zero-load axis, unload at the rate based on rules 2, 3, or 5, whichever governed in the previous loading history. If the immediately preceding quarter-cycle crossed the zero-load axis, unload at 70% of the rate based on rules 2, 3, or 5, whichever governed in the previous loading history, but not at a slope flatter than the immediately preceding loading slope.

**Example.** Segment 7 in Figure 4.10c.

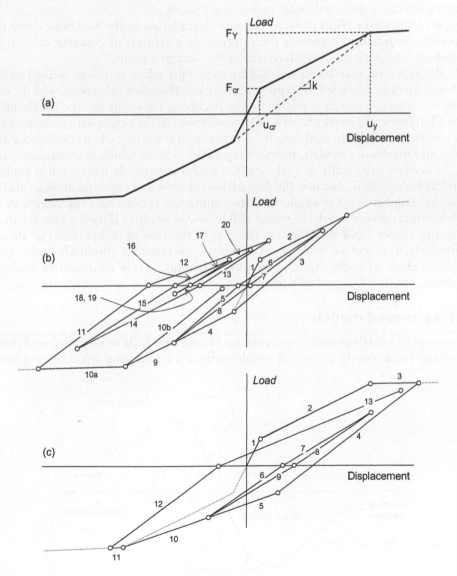

*Figure 4.10* Takeda model—(a) primary curve, and (b, c) rules governing changes in stiffness. (Adapted from Takeda et al., 1970.)

### 4.6.2 Ibarra–Medina–Krawinkler model

Ibarra et al. (2005) introduced one of the more popular hysteretic models that include both cyclic and in-cycle degradation of strength and stiffness. This model is suggested for use for components made of materials such as reinforced concrete, steel, or plywood. The IMK model is defined by (i) a backbone curve, (ii) a set of rules that define the basic characteristics of the hysteretic behavior between the reference bounds defined by the backbone curve, and (iii) a set of rules that define various modes of cyclic deterioration of the backbone curve.

The backbone curve, portrayed in Figure 4.11, is a reference force-deformation relationship that defines the bounds within which the hysteretic response of the component is confined. The parameters $\alpha_s$, $u_c/u_y$, $\alpha_c$, and $\lambda$ are obtained either from tabulated values (Ibarra et al., 2005) or by calibration of the hysteretic response to that obtained in experiments. If no cyclic deterioration occurs, the backbone curve is close to (but not necessarily identical to) the curve obtained under monotonic loading.

Figure 4.12 shows the effect of cyclic strength degradation to the backbone curve for the IMK model, calibrated to represent the response of a reinforced concrete column under cyclic loading. The peak point is referred to as the "capping point."

Cyclic deterioration may be represented by two approaches: (i) use of a fixed backbone curve that is already degraded, to represent cyclic deterioration inherently, and (ii) explicit modeling of cyclic deterioration relative to the backbone curve. In (ii), the backbone curve serves as a reference; the envelope curve then moves toward the origin with successive cycles, as shown in Figure 4.12. Physical data from experimental testing, when combined with first principles and mechanics models, provides information from which deterioration parameters can be derived empirically or analytically. Caution is generally warranted in calibrating such models via option (i), because the degradation observed in experiments is a function of the cyclic loading history. For example, if the calibration is done on experiments in which gradual deterioration occurred, the model will be unconservative if used where the imposed cyclic history causes rapid deterioration (such as in the case of compression or shear failures). Similarly, as discussed in Section 4.3, multi-cycle constant amplitude cyclic tests can overstate the effect of cyclic degradation and may result in very conservative models. For such reasons, option (ii) is generally preferred.

### 4.6.3 Flag-shaped models

Systems composed of self-centering components should have little or no permanent deformation resulting from seismic actions. A simple method for achieving self-centering behavior

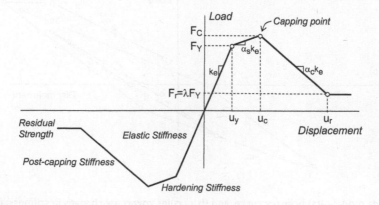

*Figure 4.11* **The backbone curve of the IMK model.**

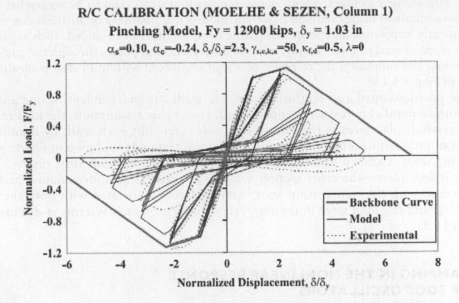

**Figure 4.12** IMK model for a reinforced concrete column. (Adapted from Ibarra and Krawinkler, 2005.)

in reinforced concrete systems is to reinforce a wall with post-tensioned reinforcement that is designed to remain elastic during the seismic response. Elastic deformations take place within the wall; when sufficient lateral force develops, the wall rocks about its base. The development of rocking is accompanied by a large reduction in stiffness, neglecting any damage that may occur, this behavior is often idealized as bilinear elastic (Figure 4.13a). Impact can be expected upon gap closure, and hence details at the contact surfaces are important both for minimizing damage, as well as preventing lateral slip at the contact surface.

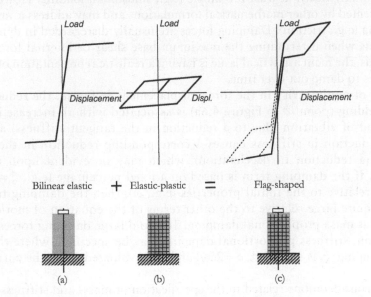

**Figure 4.13** Idealized flag-shaped hysteresis response. (a) Post-tensioned wall, (b) reinforced concrete wall, (c) post-tensioned reinforced wall.

Peak displacements obtained with nonlinear elastic systems tend to be somewhat larger than those obtained using common hysteretic models (e.g., bilinear or stiffness degrading). Consequently, supplemental energy dissipating components may be added, such as yielding devices made of steel or more simply, mild reinforcement that passes through the gap opening portion. For simplicity, the response of a wall reinforced with mild steel is idealized as shown in Figure 4.13b.

If the post-tensioned and reinforced concrete walls are independent of one another, they work in parallel to resist the applied load. Under this assumption, the load resisted at any given displacement is the sum of the loads carried by each wall. By limiting the amount of mild reinforcement, the combined system responds as shown in Figure 4.13c, producing what is known as a "flag-shaped" hysteretic model. In reality, the two walls are not independent—the cross section has a single neutral axis, determined based on equilibrium, with the compression zone carrying forces associated with both the post-tensioning and the mild steel in tension. Proportioning of such systems is discussed in Section 15.9.

## 4.7 DAMPING IN THE NONLINEAR RESPONSE OF SDOF OSCILLATORS

Oscillators in free vibration ultimately come to rest. The energy of vibration is diminished by resistance to movement through the air, by radiating energy back to the supporting medium, and by the conversion of vibrational energy to heat and sound. In elastic response, this loss of energy is only represented by damping, which may be considered as a fictitious force that opposes the motion of the mass, causing the motion amplitude to diminish over time.

Chapter 3 introduced the conventional damping model in which the damping force is linearly proportional to and in opposition to the relative velocity of oscillator mass. This model of damping is pragmatic as the resulting second-order differential equation of motion can be solved analytically, at least for simple (e.g., sinusoidal) loading. However, damping may be represented by other mathematical formulations and may address a variety of physical phenomena (e.g., friction). Damping forces are usually disregarded in dynamic analysis results, such as when determining the maximum base shear or internal forces within the structure. Thus, the main analytical issue is having a realistic representation of the tendency for oscillations to damp out over time.

In the case of nonlinearity in the force–displacement response, the reduction in stiffness upon yielding (point ② in Figure 4.6a) is associated with an increase in the instantaneous period of vibration (due to a reduction in the tangent stiffness) and reduction in $\omega$. This reduction in stiffness causes a corresponding reduction in the acceleration of the mass (a reduction in deceleration), which may be evident upon inspection of Equation 4.1. If the damping term is based on a fixed percentage (e.g., $\xi = 5\%$) of critical damping relative to the initial properties (e.g., $\omega$), then the damping term, $c = 2\xi\omega$, may become quite large relative to the other terms of the equation of motion. This case is referred to as mass-proportional damping. To avoid large damping forces in the incremental solution, stiffness proportional damping may be specified, where the percentage of critical damping, $\xi$, is specified, $c = 2\xi\omega$, and $\omega$ is evaluated using the current (tangent) stiffness.

Additional considerations related to the specification of mass- and stiffness-proportional damping arise in the case of MDOF models, and are discussed in Section 5.2.8.

## 4.8 RESPONSE OF INDIVIDUAL OSCILLATORS

### 4.8.1 Equation of motion

In Chapter 3, we considered the response of linear elastic systems to dynamic loads $p(t)$ and base excitation. This section focuses on the response of nonlinear systems to base excitation. For this case, the equation of dynamic equilibrium is formulated for small increments in time, thereby avoiding difficulties associated with expressing the nonlinear relation $f_S(u)$ in global coordinates. Considering the free body in Figure 4.14 between two positions, at times $t_1$ and $t_2$, respectively:

$$\Delta f_I(t) + \Delta f_D(t) + \Delta f_S(t) = 0$$

where

$$\Delta f_I(t) = m\big(\ddot{u}_2(t) - \ddot{u}_1(t)\big) = m(\ddot{u}_2(t) + \ddot{u}_{g2}(t) - \ddot{u}_1(t) - \ddot{u}_{g1}(t)) = m\Delta\ddot{u} + m\Delta\ddot{u}_g$$

$$\Delta f_S(t) = k\,\Delta u$$

$$\Delta f_D(t) = c\,\Delta\dot{u}$$

The increments of the spring and damping forces, $\Delta f_S$ and $\Delta f_D$, are expressed as functions of the increments of the relative displacements, $\Delta u(t)$, and relative velocity, $\Delta\dot{u}(t)$, respectively, concluding:

$$m\Delta\ddot{u} + c\Delta\dot{u} + k\Delta u = -m\Delta\ddot{u}_g, \text{ or}$$

$$\Delta\ddot{u} + 2\xi\omega\Delta\dot{u} + \omega^2\Delta u = -\Delta\ddot{u}_g$$

(4.7)

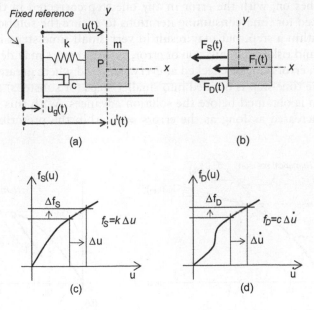

(a)

(b)

(c)

(d)

*Figure 4.14* Nonlinear dynamic equilibrium. (a) Schematic model, (b) free-body diagram, (c) spring force, and (d) damping force.

Equation 4.7 is presented in terms of dynamic equilibrium over a time step. During this time step, properties (such as stiffness or damping) may change as a function of relative displacement and relative velocity, as illustrated in Figure 4.14.

## 4.8.2 Solution approaches

Superposition relies upon linearity, and is implicit in the use of other mathematical approaches such as frequency domain solutions. Nonlinearity in the spring and possibly damping forces makes superposition inapplicable, leaving numerical methods as the only option for solving Equation 4.7. The stability and accuracy of the numerical solution of linear dynamic systems was discussed in Section 3.6. For nonlinear systems such as the one represented in Equation 4.7, modifications to the numerical solution procedure are required to account for the nonlinear and time-dependent behavior of the stiffness and possibly damping terms. Additional sources of error are present in solutions of the nonlinear equation of motion, primarily due to two related sources. The first is that the stiffness or damping may not be constant throughout the time step, that is, stiffness $k_t(u_i)$ at the beginning of the interval $u_i = u(t_i)$ is known, but because $u(t_{i+1})$ is unknown at this point, the stiffness at the end of the interval is not known (see Figure 4.15a) until a solution for the time step has been determined. The second source is associated with the difficulty or computational cost of precisely identifying the point within a time step that a sudden change in loading, stiffness, or damping occurs (see Figure 4.15b), thus leading to approximations that induce error. Where these parameters change gradually (expressed as curvilinear functions rather than piecewise continuous functions), deviation within a time step is expected and an inherent trade-off exists between precision and computational cost.

Different strategies exist for the solution of nonlinear systems (see also Section 3.6 for linear systems). In all practical cases, time steps are always less than or equal to the time step used to express the excitation, and interpolation of the excitation may be required. In some strategies, a small fixed time step is used (on the order of $T/100$ or smaller) and the solution marches on, with the error in any one step corrected in the subsequent step. This avoids the need for time-consuming iterations to ensure the solution is within a given error tolerance within a step, but may result in very small time steps being used over the entire excitation, and risks accumulation of error, as large errors may develop within a time step. Alternatively, errors can be checked at every step, and where greater than a prescribed error tolerance, the time step is divided into smaller steps (or substeps) as needed to ensure adequate precision is obtained before the solution advances. With this approach, step size is subsequently increased as long as the errors are within the prescribed error tolerance.

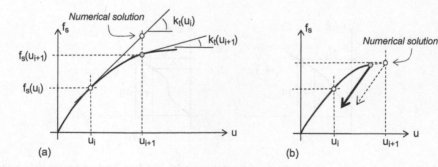

*Figure 4.15* Errors in the solution of nonlinear response: (a) stiffness variation (b) time step does not detect changes in loading or yielding.

This approach avoids the computational cost of using small time steps where larger ones would suffice, and naturally avoids large errors within a step, but has some computational overhead. Users' manuals should be consulted for guidance relevant to the particular algorithms implemented in nonlinear dynamic analysis programs of interest.

## 4.8.3 Solution by linear acceleration method

In this section, the linear acceleration method is applied to determine the response of the nonlinear oscillator over an increment of time; repeated application of this approach determines the response over the desired time period. Expressing Equation 3.22 in terms of increments and evaluating at the end of the interval (see Figure 3.7):

$$\Delta \dot{u} = \ddot{u}_i \Delta t + \Delta \ddot{u} \frac{\Delta t}{2}$$

$$\Delta u = \dot{u}_i \Delta t + \ddot{u}_i \frac{\Delta t^2}{2} + \Delta \ddot{u} \frac{\Delta t^2}{6}$$

(4.8)

Solution method:

- At the beginning of the interval, $u_i$ and $\dot{u}_i$ are known, and consequently $f_{Si}$ and $f_{Di}$ are known. Then $\ddot{u}_i$ can be deduced using the equation of motion (Equation 4.2):

$$\ddot{u}_i = \frac{1}{m}\left(-f_{Si} - f_{Di} - m\ddot{u}_{gi}\right)$$

(4.9)

- Equations 4.8 are introduced in Equation 4.7; solving for $\Delta \ddot{u}$ results in:

$$\Delta \ddot{u} = -\frac{3\left(2\Delta \ddot{u}_g + 4\ddot{u}_i \Delta t \xi \omega + 2\dot{u}_i \Delta t \omega^2 + \ddot{u}_i \Delta t^2 \omega^2\right)}{6 + 6\Delta t \xi \omega + \Delta t^2 \omega^2}$$

- The value of $\Delta \ddot{u}$ is introduced in Equation 4.8 to obtain the information at the end of the interval: $u_{i+1}$ and $\dot{u}_{i+1}$.

Obviously, $f_{Si}$, $f_{Di}$, $\xi$ and $\omega$ depend on the hysteretic model used; their values need to be tracked for use in the computations above. $\xi$ and $\omega$ are updated (based on tangent properties) when the stiffness changes.

Figure 4.16 shows the time history response for a bilinear degrading hysteretic model (see Figure 4.9d) determined using the linear acceleration method. The unloading stiffness degradation index is 0.4 (see Equation 4.6), kinematic hardening is given by $\alpha = 0.1$, viscous damping $\xi = 0.05$, mass $m = 1$ kg, period $T = 0.25$ s (for the elastic portion of the response), and yield displacement $u_y = 0.01$. The response of this oscillator is identical to that of the linear elastic oscillator over most of the ground motion record, but deviates significantly over a small portion of the record where the excitation is strong enough to cause inelastic response. The nonlinear response was computed using a time step of 0.02 s; the peak displacement $u_{max} = 0.022$ m and occurs at 4.4 s. Permanent (or residual) displacement is apparent at the end.

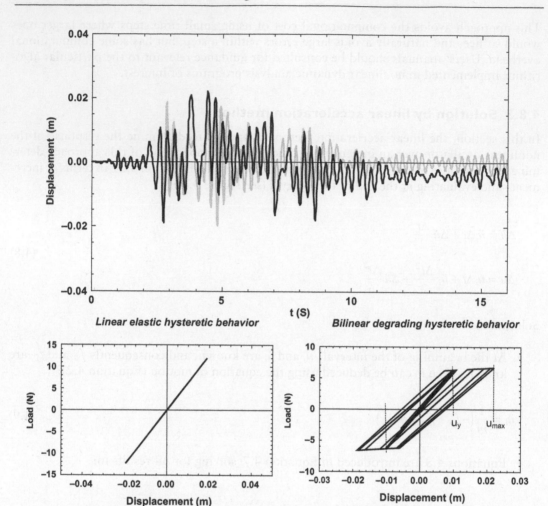

*Figure 4.16* Displacement history for an SDOF system subject to the Corralitos record of the Loma Prieta Earthquake of October 17, 1989, for 5% damping and period of 0.25 s. Elastic response (gray line) and bilinear degrading model (black line).

### 4.8.4 Nondimensional response parameters

The response of an oscillator undergoing inelastic response is often characterized using nondimensional ratios. Figure 4.17 illustrates schematically the yield displacement and the peak displacement occurring under dynamic response; the many intermediate cycles (e.g., Figure 4.16) were omitted for clarity.

Since we are usually interested in the peak displacement regardless of sign, $u_{max}$ is typically taken as the maximum of the absolute values of the peak displacements in the positive and negative directions. The peak displacement is sometimes called the inelastic displacement in the literature, and is comprised of elastic and inelastic components.

The ductility demand is a nondimensional parameter defined as the peak displacement normalized by the yield displacement (see Figure 4.17a):

$$\mu = \frac{u_{max}}{u_y} \tag{4.10}$$

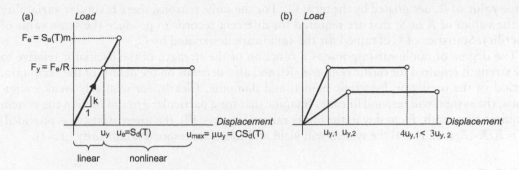

Figure 4.17 (a) Parameters defining peak displacement response of a nonlinear SDOF system and (b) illustration that larger ductility demands do not necessarily correspond to larger displacement demands.

In the previous example of Figure 4.16 the ductility demand is $\mu = 2.2$. The ductility demand can then be compared to the ductility capacity of the particular member or system under consideration.

Ductility demand is normalized (and thus, is unitless). Equation 4.10 defines ductility demand in terms of the peak and yield displacements of the SDOF system. When comparing the response of two systems, the one with larger peak displacement may have larger or smaller ductility demand (Figure 4.17b).

Other definitions of ductility are sometimes used: strain ductility is defined at the material level ($\mu_\varepsilon = \varepsilon_{max}/\varepsilon_y$), curvature ductility is defined at the section level ($\mu_\phi = \phi_{max}/\phi_y$), and chord rotation ductility ($\mu_\theta = \theta_{max}/\theta_y$) is defined at the member level.

The strength required for the system (Figure 4.17a) to remain in the linear range is $F_e$. Nevertheless, the system has a yield strength $F_y$. The ratio

$$R = \frac{F_e}{F_y} = \frac{mS_a}{mS_{ay}} = \frac{S_a}{S_{ay}} \tag{4.11}$$

where $S_{ay}$ is the $S_a$ value to cause yield, is a nondimensional parameter often called the strength reduction factor or strength ratio. This definition is commonly used for SDOF oscillators. Although related to the $R$-factor or behavior factor, $q$, used in some code design provisions, the strength ratio differs from those used in code provisions because it is defined here relative to the linear elastic response and yield strength, without addressing over-strength, and risk and safety considerations as required in design applications.

An alternative normalization compares the peak displacement response, $u_{max}$, to the peak displacement of an elastic system of the same period, $S_d$, having the same initial stiffness and period of vibration (Figure 4.17a). Thus,

$$C = \frac{u_{max}}{S_d} = \frac{u_{max}/u_y}{S_d/u_y} = \frac{\mu}{\omega^2 S_d/\omega^2 u_y} = \frac{\mu}{S_a/S_{ay}} = \frac{\mu}{R} \tag{4.12}$$

As discussed in Section 4.8.6, the same oscillator under different ground motions will display widely varying displacement and peak displacement response. The variability in peak displacement response persists even when the ground motions are characterized by the same value of $S_a$, or conversely by the same value of $R$, for systems responding nonlinearly ($R > 1$). Thus, one typically identifies statistics (mean, median, percentiles) for $C$ given the same value of $R$, designated by the term $C_R$. For the same reasons, there is similar variability in the values of $R$ or $S_a$ that are required for different records to produce the same value of ductility. Statistics of $C$ obtained for the same $\mu$ are designated by $C_\mu$.

The degree of nonlinear response is a function of the strength of the oscillator relative to the strength required for elastic response ($R$), and also depends on the details of the excitation, period of the oscillator, hysteretic model, and damping. Clearly, for relatively weak excitations, the system will respond linearly. Imagine that for a particular ground motion the system requires a strength, $F_e$, to stay in the linear range (Figure 4.17). If a lower strength is provided, $F_y = F_e/R < F_e$ (i.e., $R > 1$) the system will yield and therefore respond nonlinearly ($\mu > 1$).

### 4.8.5 Trends in inelastic response

The peak displacement response of a yielding oscillator may be higher or lower than the response of an elastic oscillator of the same period and viscous damping for any individual record. Overall, though, it has been observed (Veletsos and Newmark, 1960) that the average (or median) peak displacement over many ground motions of a moderate/long-period elastic-plastic system tends to be bounded by that of its linear elastic counterpart (having the same initial stiffness and damping). This observation, illustrated in Figure 4.18, has come to be known as the "equal displacement rule." For such systems, changes in yield strength have little effect on the expected peak displacement. Thus, strengths may be substantially less than those required for elastic response ($R > 1$) without incurring an increase in displacement demand. Based on similar triangles, the reduction in strength ($R = F_e/F_y$) is expected to equal the ratio of displacements $\mu = u_{max}/u_y$. Expressed in terms of $C$, we expect $C = 1$ for these systems.

As the natural period of vibration of the oscillator reduces, each impulse (of the sequence of impulses that constitute a ground motion record) has relatively larger duration. In the limit, as $T$ approaches 0, each ground motion impulse is experienced as a sustained load, for which any reduction in strength relative to that required for elastic response (i.e. $R > 1$) results in very large values of $\mu$ and $C$. Thus, short-period nonlinear systems experience $u_{max} > u_e$, which is termed "displacement amplification" and is associated with $\mu > R$ and $C > 1$.

The transition between the above extremes occurs gradually, beginning for short-period systems at a period that corresponds approximately to $T_g$, the corner period at the intersection of the so-called constant acceleration and constant velocity portions of the spectrum

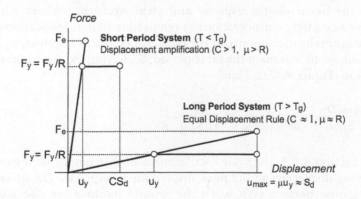

*Figure 4.18* Schematic illustration comparing response of short-period and long-period oscillators. (Note terms apply separately to the short- and long-period systems.)

(Figure 3.17). These observations apply to elastic-plastic systems, and also for more complex systems having no in-cycle strength degradation or significant cyclic strength degradation, as described by the hysteresis of, for example, Figure 4.2, with post-yield stiffnesses in the range of approximately 0%–10% of the initial stiffness. In all cases, systems subjected to ordinary ground motions (i.e., without long duration or directivity characteristics) are considered.

Figure 4.19 illustrates the variation of the mean value of $C_R$ as a function of period for different values of $R$ for SDOF systems subjected to 20 ordinary (far-field) ground motions recorded on Site Class B versus Site Class D sites.

Departures from these trends may be expected for systems with unusual hysteretic behavior (systems with significant stiffness and/or strength degradation or having bilinear elastic response), systems with negative post-yield stiffness, and systems subjected to near-fault motions having forward directivity effects. Figure 4.20 illustrates the effects of changes in the hysteretic model. Figure 4.21 illustrates the effect of changing from ordinary ground motions to near-fault motions containing forward directivity effects. The changes generally

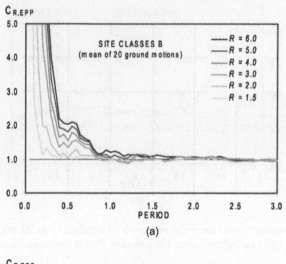

(a)

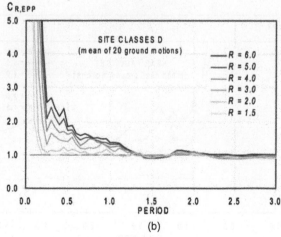

(b)

*Figure 4.19* Effect of site class on $C_R$ determined for SDOF oscillators having bilinear (elasto-plastic) hysteretic response. (From FEMA-440, 2005.) (a) Mean results for 20 site class B motions, (b) mean results for 20 site class D motions.

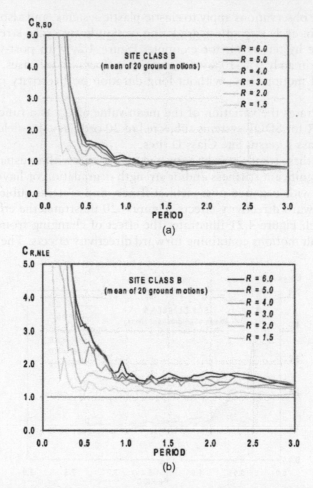

(a)

(b)

*Figure 4.20* Effect of hysteretic model on mean response of oscillators to 20 site-class B motions. (From FEMA-440, 2005.) (a) Stiffness-degrading model, (b) nonlinear elastic model.

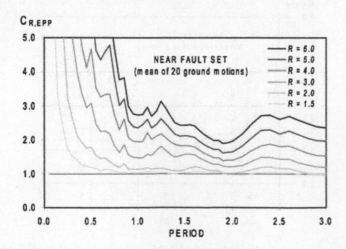

*Figure 4.21* Effect of near-fault motions on response of oscillators having bilinear (elasto-plastic) hysteretic response. (From FEMA-440, 2005.)

increase peak displacements (higher $C_R$) for a given period and $R$ factor, particularly for short-period oscillators $(T < T_g)$. Systems with negative post-yield stiffness are discussed in Section 4.11.

## 4.8.6 Variability in inelastic response as seen with incremental dynamic analysis

Nonlinearity in the load–displacement relation results in variability in response to ground motions for a given $S_a$ or $R$ value. This is easily seen through Incremental Dynamic Analysis (IDA, Vamvatsikos and Cornell, 2002), an analytical approach in which statistics are determined for the peak response of an oscillator subjected to different ground motion records over its entire range of behavior (elastic through inelastic). To do so, multiple ground motions at a number of different intensities are required. Due to limitations in the catalog of ground motions observed, natural records of a given intensity usually are not available; therefore, the amplitude of existing records is scaled. Simply put, one multiplies the acceleration values of a given record by an appropriate constant, the scale factor, to reach the required value of intensity, as represented by the value of $S_a$ or $R$. This is a generally accepted practice for values of the scale factor not too far from unity (i.e., 0.25–4.0) but it becomes controversial for values beyond this range as the estimates of response become increasingly conservative, or biased (Luco and Bazzurro, 2007).

To perform IDA, one only needs to take one record at a time and incrementally scale it at constant or variable steps of the intensity measure (IM), that is, $S_a$ or $R$, for SDOF systems. At each step of each record, a single nonlinear dynamic analysis is performed and the responses of interest, termed engineering demand parameters (EDPs), are recorded. Typically, one starts from a low $S_a$ value where the structure behaves elastically, and stops when a sufficiently large peak displacement has been reached. In cases where in-cycle degradation is present, nominally infinite displacement is deemed to be reached when a well-executed dynamic analysis cannot converge due to the system reaching zero lateral strength, which is considered to be the point at which global instability is reached.

Selecting constant $S_a$ steps is often the simplest but also the most computationally wasteful approach to IDA. Due to the large variability in inelastic response for a given value of intensity, specific values of the EDP may be encountered at widely varying $S_a$ levels for each record. Thus, some ground motions will require twice or thrice the number of dynamic analyses as others. Instead, using variable $S_a$ steps estimated via the hunt & fill procedure of Vamvatsikos and Cornell (2002, 2004) offers consistent accuracy at a predefined computational cost.

The results of IDA initially appear as distinct points, one for each dynamic analysis, in the response (EDP) versus intensity (or IM) plane, as observed in Figure 4.22a for a $T = 1$ s system having no cyclic degradation and a complex backbone with in-cycle degradation similar to Figure 4.5a. Linear or spline interpolation (Vamvatsikos and Cornell, 2004) is employed to generate continuous IDA curves, one for each individual ground motion record, shown in Figure 4.22b. The variability offered by such curves is actually one of the eye-opening features of IDA, visually representing the probabilistic nature of seismic loading and the differences introduced by (seemingly) similar ground motions. Response is linear for $R < 1$; as $R$ increases the peak ductility demands generally increase. Collapse is considered to occur where any further increase in $R$ results in disproportionally large (practically infinite) increases in peak response. Thus, at the highest intensity level that each IDA curve can reach, one can identify the characteristic flatline representing global collapse, caused by the negative stiffness drop of the backbone to zero strength due to in-cycle strength degradation (Figure 4.5a).

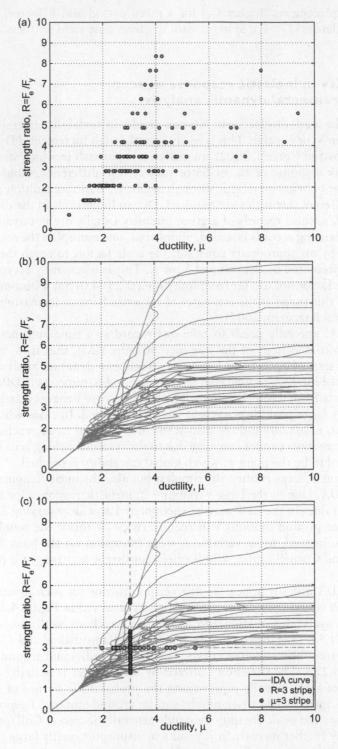

*Figure 4.22* Determination of IDA at constant steps of the IM for a $T = 1$ s SDOF system with in-cycle degradation: (a) the resulting IM-EDP points for 30 ground motion records, (b) the 30 IDA curves derived by interpolating the IM-EDP points, and (c) horizontal stripe of 30 points at $R = 3$ and vertical stripe at $\mu = 3$.

With reference to the earlier discussion on $C_\mu$ and $C_R$ (Section 4.8.4), Figure 4.22c shows a vertical stripe of points, one per record, at a given value of $\mu = 3$, together with a horizontal stripe at a given value of $R = 3$. The first would be employed to determine statistics of $C_\mu$ and the latter to determine statistics of $C_R$. Clearly, the two would be quite different. Still, it is not too difficult to appreciate that both stripes appear to be centered on the point of $(\mu, R) = (3, 3)$. For this level of ductility and value of the strength ratio, this moderate period system is still mainly deforming along the plastic plateau. Therefore, we expect that the equal displacement rule holds and the median value of $R$ given $\mu = 3$ is equal to 3, and also equal to the median of $\mu$ given $R = 3$. Still, the points along any of the stripes are not symmetrically distributed around their medians. Instead, there is a tendency to have outliers toward large $R$ or large $\mu$ values, along the vertical or the horizontal, respectively. This is indicative of a "right-skewed" distribution (or a distribution with a "fat" right tail) and it means that the respective mean will also be to the right of (i.e., larger than) the median. This gives rise to the use of the lognormal model to describe the statistical distribution of IM given EDP or EDP given IM, discussed in Chapter 13.

This characteristic skewness in the distributions of IM and EDP also has another more subtle implication. While each single point on an IDA curve is unambiguous, regardless of whether one determines it for a given $\mu$ or $R$ (barring some switchbacks observed in some IDAs that can produce multiple $\mu$ points, an issue discussed later), this does not hold for the means obtained from multiple ground motion records. Let $E[x]$ denote the mean (or expectation) of $x$. Then, when one estimates $E[C_R] = E[\mu]/R$ from a horizontal $R = 3$ stripe and compares it to $E[C_\mu] = \mu \cdot E[1/R]$ for a vertical $\mu = 3$ stripe, the two will not match (see also Miranda, 2001), despite the (approximate) equal displacement rule. This issue is further exacerbated when one considers published relationships derived via regression for the mean (or generally statistics) of $R$ parameterized by $\mu$ and $T$, of $C_R$ given $R$ and $T$, or of $C_\mu$ given $\mu$ and $T$, collectively termed $R$–$\mu$–$T$ relationships. While the mean $R$ can be transformed to the mean $C_\mu$, simply by dividing by the given parameter of $\mu$, one should never solve $C_\mu$ to find $C_R$ or vice-versa. However, as Vamvatsikos and Cornell (2006) observed, due to the monotonicity of these functions with respect to $\mu$ or $R$ at any specified period, one may solve for the median $C_\mu$ in terms of the median $C_R$ and vice versa.

Overall, IDA is a computationally expensive procedure requiring multiple numerical analyses and extensive post-processing to achieve results similar to Figure 4.22. Still, such results are an indispensable tool in understanding the behavior of inelastic systems and help develop intuition on the significance of different system parameters. Especially for SDOF systems, modern computers can turn the estimation of IDA into a matter of minutes. Tools to automate this process are now available online both in a programming environment (Vamvatsikos, 2005) and as standalone software (Lignos, 2010).

## 4.9 INELASTIC RESPONSE SPECTRA

As in the case of elastic response spectra, inelastic response spectra plot the maximum value of a response parameter for the response of an SDOF oscillator over a range of vibration periods, for a particular hysteretic model and damping. Inelastic spectra may be generated for specific ground motion records, under some specified criteria that are associated with the level of inelasticity (Sections 4.9.1 and 4.9.2). Alternatively, smoothed spectra may be shown, representing trends or curves for use in design. Inelastic response spectra may be plotted as a function of period, frequency, or other parameters (such as yield displacement), as described in Section 4.9.3.

### 4.9.1 Constant ductility iterations

An iterative approach is required to determine the oscillator parameters associated with a specified ductility demand. Iterations on strength may be done for a given stiffness (or initial period)—the yield strength, $F_y$, may be adjusted iteratively until the ductility demand obtained in response to a ground motion record matches the target value of ductility ($\mu$) within a specified tolerance. Inel et al. (2002) present an efficient algorithm for determining constant ductility spectra, which is implemented in the USEE software package (2001). The basic idea is represented in Figure 4.23, which illustrates an approach that involves iterations on yield displacement (or equivalently, yield strength) for each oscillator period of interest.

Figure 4.24 presents constant ductility spectra determined for a particular ground motion record and specified ductility levels.

Any solution approach should address the possibility that multiple $R$ values may result in the same ductility demand (Figure 4.25a) and that discontinuities may exist in the ductility response (Figure 4.25b)—indicating that there may be multiple solutions to choose from or perhaps no solution for a precise value of ductility demand. Thus, some criteria must be established for defining the strength to use.

In order to better understand the results of Figure 4.25a, note that the results plotted are for a constant period ($T = 0.20\,\text{s}$). Thus, the results all pertain to oscillators having the same initial (elastic) stiffness. Therefore, the different values of $C_y$ represent different values of yield displacement (where $u_y = F_y/k = C_y W/k = C_y mg/k = C_y g/\omega^2 = C_y g(T/2\pi)^2$). Consequently, the peak displacements vary ($u_{max} = \mu u_y$) for the different $\mu = 1.4$ oscillators plotted in Figure 4.25a.

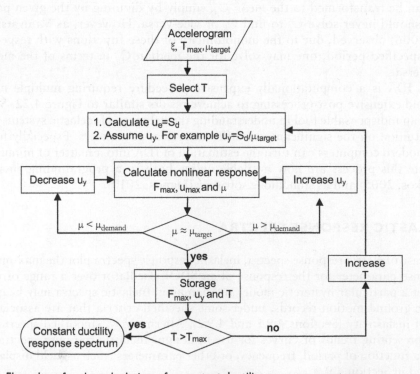

*Figure 4.23* Flow chart for the calculation of a constant ductility response spectrum.

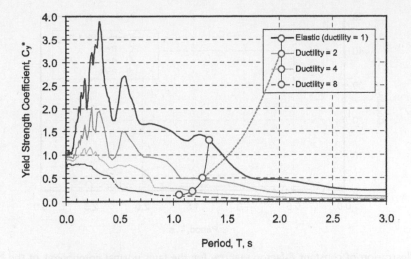

Figure 4.24 Illustration of constant ductility response spectra for the fault normal component of the Newhall LA County Fire Station record of the Northridge Earthquake of January 17, 1994, for bilinear hysteretic response, 5% damping, and 5% post-yield stiffness. Lower yield strengths are associated with higher ductility levels.

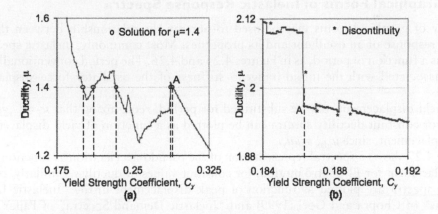

Figure 4.25 The strength-ductility relationship for a bilinear oscillator responding to the NE component of the 1987 Whittier Narrows Earthquake recorded at the Mt. Wilson-Caltech station: (a) different strengths are associated with a ductility of 1.4, for an oscillator period of 0.20 s, and (b) a discontinuity in the strength-ductility relationship, in this case for $\mu = 2$ and a period of 0.15 s. (Adapted from Inel et al., 2002.)

## 4.9.2 Types of Inelastic Response Spectra

As shown in the previous section, in order to generate inelastic spectra, constraints are required to impose $F_y < F_e$. For example, inelastic response can be determined for oscillators having a fixed strength (values of $F_y = C_y W = C_y mg$ are specified), a fixed strength ratio ($R$), or a fixed ductility ($\mu$). Various response parameters (or EDPs) such as peak displacement, displacement ductility, or displacement ratio ($C$) may be plotted. Where inelastic response is determined for constant $R$ values, oscillator strengths vary in accordance with the elastic demands ($F_y = F_e/R = S_a m/R$)—the example plotted in Figure 4.26 illustrates that $\mu$ becomes quite large for short-period oscillators that have $R > 1$.

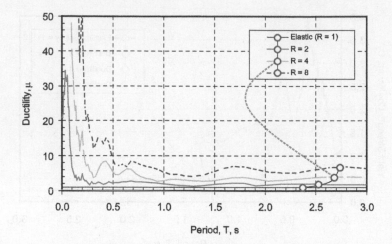

Figure 4.26 Illustration of constant R-factor spectra, for the fault normal component of the Newhall LA County Fire Station record of the Northridge Earthquake of January 17, 1994, for bilinear hysteretic response, 5% damping, and 5% post-yield stiffness.

### 4.9.3 Graphical Forms of Inelastic Response Spectra

A variety of graphical forms may be used to portray the relationship between the peak inelastic response of an oscillator and its properties. Most commonly, inelastic spectra are plotted as a function of period, as in Figures 4.24 and 4.26. The period conventionally refers to that associated with the initial (tangent) stiffness of the oscillator force–displacement response.

The yield displacement may be substituted for period, recognizing that $u_y = C_y g(T/2\pi)^2$. Results for constant ductility spectra can be plotted as a function of yield displacement or peak displacement, since $u_{max} = \mu u_y$.

Figure 4.27 shows a spectral representation of the pseudo-acceleration as function of relative displacement for El Centro motion, for constant values of ductility. Similarly, constant ductility spectra are plotted as function of peak displacement in the "Inelastic Demand Diagrams" of Chopra and Goel (1999) and "Inelastic Demand Spectra" of Fajfar (1999). These spectra resemble the acceleration-displacement response spectra of the capacity spectrum method (Freeman, 2004) as shown in Figure 4.28, but utilize $R$–$\mu$–$T$ relationships for determining the isoductile curves rather than equivalent linearization (used in the capacity spectrum method). In these representations, periods are constant along any line radiating from the origin[2] and may be determined based on the intersection of the radial line and the elastic spectrum.

Note that the definitions of $S_v \equiv \omega S_d$ and $S_a \equiv \omega^2 S_d$ for linear conditions are in conflict with the use of $S_a$ by some investigators to define the strength of an inelastic oscillator (according to $V_y = S_a m$). We reserve pseudo-spectral acceleration ($S_a$) to describe linear response and use the nondimensional yield strength coefficient, $C_y$ ($= V_y/W$), to describe the normalized strength of an oscillator that may yield.

Yield Point Spectra (YPS, Aschheim and Black, 2000) are similar to the previously mentioned "Inelastic Spectra," differing only in that the isoductile strengths are plotted as

---

[2] When plotted on logarithmic axes, periods are constant along the parallel lines.

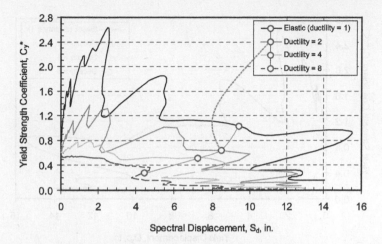

Figure 4.27 Illustration of constant ductility inelastic spectra, for the fault normal component of the Newhall LA County Fire Station record of the Northridge Earthquake of January 17, 1994, for bilinear hysteretic response, 5% damping, and 5% post-yield stiffness. (Note crossing isoductile curves result from interpolation between specified periods.)

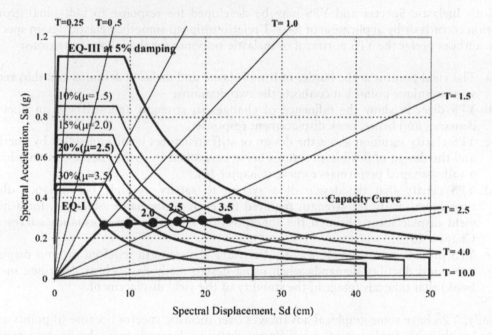

Figure 4.28 Capacity spectrum. (Adapted from Freeman, 2004.)

a function of the yield displacement $(u_y)$ rather than the peak displacement $(u_{max} = \mu u_y)$. Figure 4.29 shows the same data of Figure 4.27 plotted as a function of yield displacement. When plotted using linear axes, periods are constant along any line radiating from the origin.[3] While useful for design, YPS have also been applied to the seismic retrofitting of existing buildings (Thermou et al., 2012).

---

[3] When plotted on logarithmic axes, periods are constant along the parallel lines.

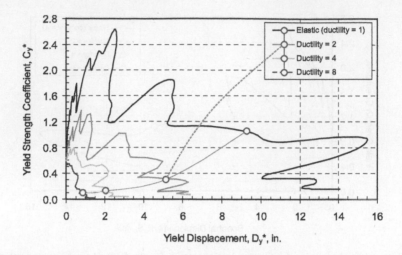

*Figure 4.29* Illustration of constant ductility inelastic spectra, for the fault normal component of the Newhall LA County Fire Station record of the Northridge Earthquake of January 17, 1994, for bilinear hysteretic response, 5% damping, and 5% post-yield stiffness.

Both Inelastic Spectra and YPS may be developed for response to individual ground motion records or by application of $R-\mu-T$ relationships to smoothed elastic design spectra. The authors prefer the YPS portrayal of inelastic response for the following reasons:

a. The yield point is at the border of linear elastic and inelastic domains and thus represents a unique point that connects the two domains;

b. YPS directly show the influence of changes in strength and stiffness on ductility demand, and hence, peak displacement response;

c. YPS clarify intuition about the design of stiff structures being controlled by ductility and the design of flexible structures being controlled by drift, via the construction of a valley-shaped performance curve (Chapter 10).

d. YPS clarify that the design of a system to satisfy multiple performance objectives must satisfy the strength required for each performance objective, for a given yield displacement, through the superposition of multiple performance curves (see Chapter 10);

e. YPS provide a very simple way to determine the strength required to limit displacement and ductility demands when using design methods (such as described in this book) that take advantage of the stability of the yield displacement.

Finally, YPS have some graphical advantages over inelastic spectra because (i) points associated with peak displacements for different ductility demands may obscure one another (such as due to the equal displacement rule, e.g., Figure 4.27), whereas yield displacements are inherently better separated, providing greater graphical clarity (e.g., Figure 4.29), and (ii) there is no difficulty in estimating the required yield strength where kinematic hardening is present, since the yield strength is plotted.

Of course, it is convenient to have a distinct yield point to reference the underlying nonlinear hysteretic relationship, but all that is needed is a unique way to reference the hysteretic curve. For example, a curvilinear hysteretic relationship could be referenced by a unique point that serves as a surrogate for a true yield point and which might be generated by a particular graphical or mathematical approach.

## 4.10 PREDICTIVE RELATIONSHIPS AND DESIGN SPECTRA

### 4.10.1 Development of R–μ–T relationships

Simple estimates of inelastic design response spectra can be obtained in several ways. While the ideal approach would be to conduct a probabilistic seismic hazard analysis in which attenuation relationships are used directly to derive inelastic spectra associated with the desired hazard level (see Chapter 13), more commonly existing R–μ–T relationships are applied to smoothed elastic design response spectra appropriate for the hazard level of interest. Inelastic spectra derived in this way make it possible to determine the strength required to limit ductility and/or peak displacement response to a desired level. The following sections identify some of the most useful relationships, after first discussing the fundamental observations of Newmark and Hall.

Given that the peak displacement response of an inelastic system differs for each ground motion record for a given R (as illustrated in Figures 4.24 and 4.26), researchers have taken different approaches to describing the overall trends, commonly called R–μ–T relationships. We note a few methodological issues below.

Perhaps, the simplest approach is to determine the R values associated with isoductile (i.e., fixed μ) responses for a suite of ground motion records, over a range of periods. A curve is fitted to these R values, resulting in an R–μ–T relationship. While this approach has been used often, two issues arise with this approach. The first is that the required design strength, $F_y$, is determined as $F_e/R$, and hence what is needed is an estimate of the term $1/R$ rather than 1/(estimate of R) as discussed in Section 4.8.6. A simple numerical experiment[4] shows that the latter leads to a slightly smaller strength $F_y$. The second is that a few unusually large responses can be expected (e.g., Figure 4.22), so the application of a single R value to a suite of records (having the same nominal intensity) will result in some unusually large responses that will lead to a larger than expected mean ductility demand. A better approach would be to determine an R value, which when employed across the suite of records, results in the desired mean ductility demand (for an oscillator having period, T, and specified hysteretic behavior and damping). This result is achieved by defining Equation 4.12 in terms of $C_R$, for which the displacement ratios $u_{max}/S_d$ are determined for each ground motion and period for a fixed R. These R–μ–T (or more accurately, R–$C_R$–T) relationships can be solved to determine the R associated with a desired mean ductility demand, since

$$\mu = \frac{u_{max}}{u_y} = \frac{u_{max}}{S_d}\frac{S_d}{u_y} = C_R R \tag{4.13}$$

for a given period. Equation 4.13 makes use of the relationship $R = F_e/F_y = S_d/u_y$, which can be determined considering similar triangles (e.g., in Figure 4.18). This approach also avoids the problems inherent in iterations to establish the strength associated with a desired ductility (Section 4.9.1).

### 4.10.2 Newmark–Hall

Of historical interest are the relationships derived from the study of nonlinear response of oscillators having elasto-plastic hysteretic relationships. The Newmark and Hall observations[5]

---

[4] For R values of 2, 3, and 4, the mean of $1/R$ is 0.361, which is 8% greater than 1/(mean of R).

[5] Presented here in terms of period rather than frequency, as originally described; note that period limits are approximate.

(reported in 1969 and summarized in 1982) identify fundamental period-dependent relationships between strength and ductility as follows:

1. Long period systems ($T > 0.5$ s): the maximum displacement of a long period system is equal to the maximum instantaneous change in ground displacement. This results in the so-called "equal displacement rule" for which

$$R = \mu \tag{4.14}$$

2. Intermediate period systems ($0.5$ s $> T > 0.125$ s): the work done to deform the inelastic system to a ductility $\mu$ is equal to the energy required to deform the elastic system to a displacement $S_d$. Equating these quantities results in the so-called "equal energy" rule, where

$$\frac{1}{2} k S_d^2 = \frac{1}{2} k u_y^2 + (u_{max} - u_y) k u_y \rightarrow R = \frac{S_d}{u_y} = \sqrt{2\mu - 1} \tag{4.15}$$

3. Short-period systems ($0.125$ s $> T > 0.0303$ s): the work to deform the system to a ductility $\mu$ is equal to that associated with an instantaneous ground acceleration moving the mass through $u_{max}$, leading to

$$R = \frac{2\mu - 1}{\mu} \tag{4.16}$$

4. Very short-period systems ($0.0303$ s $> T$): the strength of the oscillator must be preserved to avoid very large ductility demands ($R = 1$).

Based on their observations, Newmark and Hall suggested a procedure to construct inelastic response spectra. Their procedure is described below, with reference to Figure 4.30:

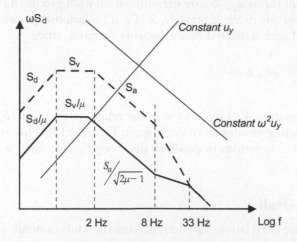

*Figure 4.30* Newmark–Hall inelastic design spectrum, shown on a tripartite logarithmic plot. Because spectral displacement, pseudo-velocity, and pseudo-acceleration are related by $\omega$ for elastic response ($S_d = S_v/\omega = S_a/\omega^2$), all three quantities can be related to frequency on a single plot, known as a tripartite spectrum. Lines plotted above are contours of constant displacement (labeled $u_y$) or constant pseudo-acceleration (labeled $\omega^2 u_y$); any horizontal line represents constant pseudo-velocity ($\omega u_y$). Because $u_y$ represents the limit of linear response, it may be used in place of $S_d$.

1. For long period systems (frequencies smaller than 2 Hz), the peak displacement response is approximately equal to the elastic displacement, $S_d$, independent of the ductility demand $\mu$.

$$S_d(\omega,\xi) = u_{max} = \frac{u_y}{\mu}$$

2. For intermediate period systems (frequencies between 2 and 8 Hz), the equal energy rule applies. Energy required to deform the linear and nonlinear systems to their peak displacements is equal.

$$u_y = \frac{S_d}{\sqrt{2\mu-1}} \rightarrow u_y\omega^2 = \frac{S_a}{\sqrt{2\mu-1}}$$

The Newmark–Hall inelastic design spectrum is presented with respect to the yield displacement in Figure 4.30—$\omega u_y$ and $\omega^2 u_y$ are defined for the case of an elasto-plastic oscillator. Using the tripartite logarithmic representation (e.g., Figure 4.30), the inelastic spectrum (solid curve) is derived by dividing the elastic spectral values according to the following:

1. Plot the elastic design spectrum (dashed curve).
2. Reduce the spectral displacement ($S_d$), in the region of constant $S_d$, by the desired ductility. Also reduce spectral velocity ($S_v = \omega S_d$) in the constant $S_v$ region by the desired ductility.
3. In the constant spectral acceleration region ($S_a = \omega^2 S_d$), reduce the elastic spectrum by $\sqrt{2\mu-1}$.
4. Draw the line of constant acceleration equal to the peak ground acceleration for frequencies exceeding 33 Hz, and draw a straight line to transition between this curve (at 33 Hz) and the previous curve (drawn in the constant spectral acceleration region).

### 4.10.3 FEMA-440 R–μ–T relationship

A very large number of SDOF analyses were used to determine $R$–$\mu$–$T$ relationships in FEMA-440 (2005). In the recommended relationship, $C$ is separated into components $C_1$ and $C_2$, where $C = C_1 \cdot C_2$. The first component, $C_1$, addresses short-period displacement amplification, while $C_2$ addresses cyclic degradation of stiffness and strength. The peak displacement, $u_{max}$, is estimated as $u_{max} = C_1 \cdot C_2 \cdot S_d(T)$. The coefficients $C_1$ and $C_2$ depend on $R$, $T$, and site class as follows:

$$C_1 = 1 + \frac{R-1}{aT^2} \tag{4.17}$$

$$C_2 = 1 + \frac{1}{800}\left(\frac{R-1}{T}\right)^2 \tag{4.18}$$

where $a$ is a parameter that varies with NEHRP Site Class (BSSC, 2000)— $a$ is taken equal to 130 for Site Classes A and B, 90 for Site Class C, and 60 for Site Class D.

The coefficient $C_2$ is taken equal to unity (i) for noncyclically degrading systems and (ii) for cyclically degrading systems having periods greater than 0.7 s. According to FEMA 440, the value of $T$ used in Equations 4.17 and 4.18 need not be taken less than 0.2 s.

### 4.10.4 Cuesta et al. $R-\mu-T/T_g$ relationship

Cuesta et al. (2003) demonstrated a simplification to the Vidic et al. (1994) model that provided a good fit for stiffness-degrading systems. The relationship is indexed by the characteristic period of the ground motion, $T_g$, where $T_g$ is the corner period at the intersection of the constant acceleration and constant velocity portions of a smoothed design spectrum, or is given by Equation 3.35. The relationship is as follows:

$$R = \begin{cases} c_1(\mu-1)\dfrac{T}{T_g}+1 & \dfrac{T}{T_g} \leq 1 \\[3mm] c_1(\mu-1)+1 & \dfrac{T}{T_g} > 1 \end{cases} \qquad (4.19)$$

where $c_1 = 1.3$ for systems with limited stiffness degradation and 1.0 for systems with substantial stiffness degradation.

The relationship was developed considering SDOF systems having a large range of periods ($0 < T \leq 3\,\text{s}$), mass-proportional viscous damping ($2 \leq \xi \leq 10\%$), and post-yield stiffness ($0 \leq \alpha \leq 10\%$), for ground motions selected to span a range of characteristic periods ($0.17 \leq T_g \leq 2.00\,\text{s}$). The ground motion records included both far-fault motions and a limited number of near-fault motions, recorded at various orientations relative to the fault strike. Note that the above relationship is not recommended for use at soft soil sites, which may generate nearly harmonic motion.

### 4.10.5 SPO2IDA

Vamvatsikos and Cornell (2006) developed a tool to estimate response statistics (e.g., the IM given EDP and EDP given IM distributions shown in Figure 4.22) for SDOF oscillators having a moderately pinching or peak-oriented hysteresis without cyclic degradation (as in Figures 4.5a, 4.8c, or 4.9e) and a range of backbone curves (bilinear to quadrilinear, including portions with negative post-yield stiffness). The complex tool is available as a spreadsheet for download.[6] Input consists of a backbone curve, specified in normalized form, and a smoothed elastic response spectrum. Output consists of the estimated 50% (median), 16%, and 84% fractiles of ductility response $\mu$ as a function of $R$ (e.g., Figure 4.31). As discussed in Section 4.8.6, due to the monotonic relationship of $\mu$ and $R$ on any IDA curve, the 16%, 50%, and 84% fractiles of $\mu$ given $R$ directly map to the 84%, 50%, and 16% fractiles of $R$ given $\mu$, allowing an easy conversion from one to the other.

This tool has been used as the basis to estimate yield frequency spectra (described in Chapter 13), which consider the mean annual frequency of exceeding a ductility demand, for a given normalized backbone curve and seismic hazard.

### 4.10.6 Flag-shaped models

Farrow and Kurama (2004) determined coefficients to use with the Nassar and Krawinkler (1991) $R-\mu-T$ model to represent the response of bilinear, bilinear elastic–, and flag-shaped models. Values of ductility were determined for specified values of $R$.

For the results summarized here, (i) oscillator yield strength was established by applying an $R$ factor to the average of the elastic response spectra of the suite of ground motions,

---

[6] http://users.ntua.gr/divamva/software/spo2ida-allt.xls.

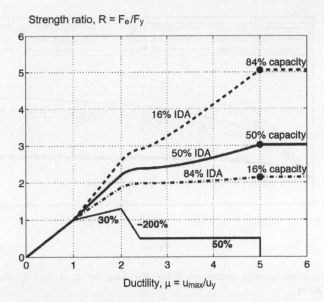

Figure 4.31 Example of SPO2IDA result. (From Vamvatsikos and Cornell, 2006.)

at the relevant period; (ii) the bilinear and bilinear elastic models had post-yield stiffness set equal to 10% of the initial stiffness, and (iii) the flag-shaped model (Figure 4.13c) was established by setting the yield strength and initial stiffness of the bilinear model (Figure 4.13b) equal to 1/3 of the strength and initial stiffness of the bilinear elastic model (Figure 4.13a).

The Nassar and Krawinkler model states

$$R = \left[ c(\mu - 1) + 1 \right]^{1/c}$$

(4.20)

where $c$ varies with period as a function of coefficients $a$ and $b$:

$$c = \frac{T^a}{T^a + 1} + \frac{b}{T}$$

(4.21)

These coefficients, determined by Farrow and Kurama for the Los Angeles "survival" Site Class D motions, are given in Table 4.2.

Equation 4.20 is plotted in Figure 4.32, for the three hysteretic models for ductilities of 2, 4, and 6. It is apparent that the bilinear elastic model requires larger strengths (smaller $R$ values) for a given ductility demand. When a relatively small amount of energy dissipation is provided, resulting in the flag-shaped hysteretic response, the isoductile $R$ values move much closer to those of the common bilinear model (having full hysteretic loops). (See also the numerical comparison of Table 4.1.)

Table 4.2 Coefficients determined for Nassar and Krawinkler model

| Hysteretic model | a | b |
| --- | --- | --- |
| Bilinear | 0.42 | 0.59 |
| Flag | 0.94 | 0.79 |
| Bilinear elastic | 1.30 | 0.87 |

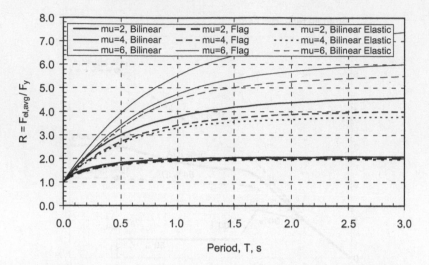

Figure 4.32 Effect of hysteretic relationship on R for ductilities of 2, 4, and 6. (Based on Farrow and Kurama, 2004.)

The preceding describes oscillators with a given stiffness, whose R value has to be adjusted (hence changing the yield displacement) to maintain a given ductility demand. Alternatively, we can look at the change in peak displacement for a given R value and period, due to the change in hysteretic model. Figure 4.33 plots peak displacements for the bilinear elastic and flag-shaped model, normalized by those of the common bilinear model. The increases at shorter periods are much greater than those reported by Christopoulos et al. (2002) and might result from the choice to fit a particular model to the data rather than reflecting an underlying trend toward much larger displacements. At long periods, the bilinear elastic model causes a 30% increase in peak displacement for $R = 6$ and a 20% increase for $R = 4$. The flag-shaped model causes smaller increases, 20% for $R = 6$ and 10% for $R = 4$.

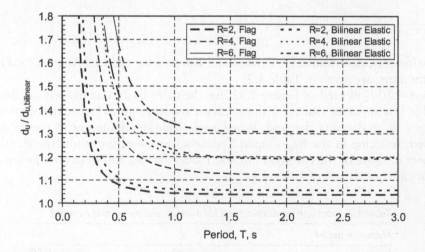

Figure 4.33 Effect of hysteretic relationship on peak displacement response, for R values of 2, 4, and 6, relative to the peak displacement obtained with the bilinear model. (Based on Farrow and Kurama, 2004.).

## 4.11  P-Δ EFFECTS FOR SDOF SYSTEMS

Structural analysis usually considers the geometry of the original (undeformed) configuration when establishing equilibrium. This is recognized to be a good approximation where deformations are small. However, where gravity loads and lateral displacements are significant, so-called second-order effects can have a significant destabilizing effect on the response of structures, increasing lateral displacements, or increasing the lateral strength required to keep these displacements in check.

### 4.11.1  Basic formulation

Geometrical second-order analysis of structures, in contrast to first-order analysis, considers equilibrium in the deformed configuration. In general, tension promotes stability while compression may induce instability. The influence of deformation on the stability of a structure in equilibrium is accounted for by introducing a term known as the geometric stiffness.

For a linear SDOF system subjected to a ground acceleration $\ddot{u}_g$, the elastic second-order equation of motion is:

$$m\ddot{u} + c\dot{u} + (k - k_G)u = -m\ddot{u}_g \qquad (4.22)$$

where $k$ is the first-order elastic stiffness, $k_G$ is the geometric stiffness, $u$ is the displacement of the oscillator relative to the ground, and $\ddot{u}_g$ is the acceleration of the ground.

The simplest way to introduce nonlinearity associated with equilibrium in the deformed configuration is shown in Figure 4.34, where the deformation of an SDOF system is concentrated at an elasto-plastic rotational spring located at the base of a rigid member. For this case, equilibrium of moments in the undeformed configuration (the first-order equilibrium equation) for the linear or pre-yield situation is:

$$Vh = k_r\phi \Rightarrow V = \frac{k_r\phi}{h} = \frac{k_r}{h^2}u = ku$$

where $k_r$ is the stiffness of the spring before yielding, $k$ is the first-order stiffness of the system with respect to $V$ and $u$, and small angles are assumed in setting $u = \phi h$ (Figure 4.34).

The moment equilibrium equation in the deformed configuration (including second-order terms) for the linear or pre-yielding situation is:

$$Vh + Pu = k_r\phi \Rightarrow V = \left(\frac{k_r}{h^2} - \frac{P}{h}\right)u = (k - k_G)u$$

where $k_G$ is the geometric stiffness of the system, $P$ is the vertical load (positive downward), and $h$ is the height of the mass above the base. The yield displacement $u_y$ is the same in the first- and second-order analyses, but the effective stiffness, $k$, and effective period of vibration, $T$, differ in the second-order analysis.

The ratio of the geometric stiffness and the first-order stiffness is called the stability coefficient ($\theta$):

$$\theta = \frac{k_G}{k}$$

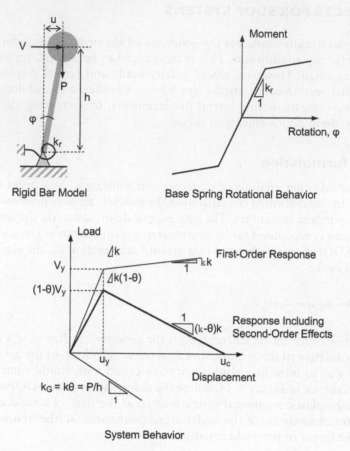

Figure 4.34 Geometrical nonlinearity in SDOF systems.

The stiffness apparent in the second-order analysis, in the linear or pre-yield condition can be expressed as:

$$k(1-\theta)$$

and the effective period of the structure, accounting for P-Δ effects, is:

$$\frac{T}{\sqrt{1-\theta}}$$

where $T$ is the period of the structure based on the first-order stiffness.

For a yielding system, the equation of motion may be solved incrementally in time, with $k$ replaced by the tangent stiffness, $k-k_G$, which varies according to the hysteretic behavior of the nonlinear system. For SDOF systems, the geometric stiffness remains constant, even as yielding occurs.

## 4.11.2 Effective height formulation

We introduce the notion of an effective height to simplify the representation of seismic demands in the presence of P-Δ effects. Without loss of generality, we can consider the vertical load $P$ to be composed of dead $(D)$ and live $(L)$ load components. Then, the geometric stiffness can be represented in terms of dead load alone in conjunction with an effective height, $h_{eff}$, defined as follows (Aschheim and Hernández-Montes, 2003):

$$k_G = \frac{P}{h} = \frac{D+L}{h} = \frac{D}{h_{eff}} = \frac{mg}{h_{eff}} \rightarrow h_{eff} = \frac{D}{D+L}h \tag{4.23}$$

Thus, the conventional representation of Figure 4.34 can be replaced by the normalized representation of Figure 4.35. (Note that the first-order stiffness, $k$, is based on the actual geometry, and is not influenced by $h_{eff}$).

Equation 4.22 can be rewritten in the effective height formulation, considering small deformations $u(t)$, and the presence of a vertical load $P$ ($= mg$):

$$m\ddot{u}(t) + c\dot{u}(t) + f_S(u(t)) - \frac{mg}{h_{eff}} u(t) = -m\ddot{u}_g(t) \tag{4.24}$$

Using this formulation, constant ductility iterations can be used to establish isoductile demands for a given ground motion record, as a function of $h_{eff}$. An example in YPS format is shown in Figure 4.36. Note that the first-order case corresponds to $h_{eff} = \infty$.

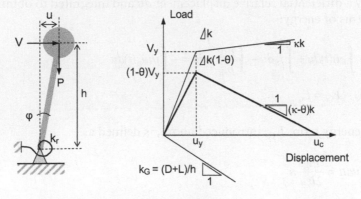

Conventional Formulation

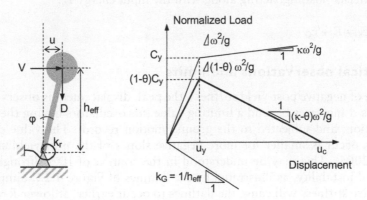

Effective Height Formulation

*Figure 4.35* Conventional and effective height formulations for P-Δ effects.

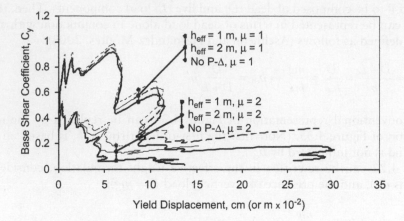

*Figure 4.36* Illustration of constant ductility spectra considering P-Δ effects, for different values of $h_{eff}$, for the 1940 NS El Centro ground motion record.

### 4.11.3 Energy components

As described by Hernández-Montes et al. (2015) the forces expressed in Equation 4.24 can be multiplied by a differential relative displacement *du* and integrated to obtain the equation of motion in terms of energy:

$$\int m\ddot{u}(t)du + \int c\dot{u}(t)du + \int f_S du - \int \frac{mg}{h_{eff}} u du = -\int m\ddot{u}_g(t)du$$

$$E_K + E_D + E_S - E_G = E_i$$

(4.25)

The geometric energy term, $E_G$, introduced above, is defined as:

$$E_G = \int \frac{mg}{h_{eff}} u du = \frac{mg}{2h_{eff}} u^2$$

(4.26)

The terms of Equation 4.25 can be rearranged to show the geometric energy term (or P-δ effect) as an external loading, acting along with the input energy $E_i$:

$$E_K + E_D + E_S = E_i + E_G$$

(4.27)

### 4.11.4 Practical observations and limits

In the presence of negative post-yield stiffness, the peak displacement is observed to increase dramatically as *R* increases beyond a limiting value (for oscillators having the same period, hysteretic relation, and subjected to the ground motion record). The value of *R* at which these increases occur is smaller for more negative slopes of the post-yield stiffness curve (FEMA-440, 2005). This may be understood in the context of IDA through the concept of global lateral instability, as illustrated by the flatlines of Figure 4.22. Simply put, larger (steeper) negative stiffness will cause the flatlines to occur earlier, at lower *R*-values.

Advanced *R–μ–T* relationships such as SPO2IDA (Section 4.10.5), directly incorporate the effect of in-cycle strength degradation. All other simpler *R–μ–T* relationships, discussed herein, do not account for such effects and their applicability needs to be limited before they venture beyond the (maximum strength) capping point (see Figure 4.11). A somewhat simplistic way of doing this is by determining a limiting value of $R$ beyond which global instability is deemed to occur with a high probability. An expression for this limiting value of $R$ was put forward in FEMA-440. Subsequent work has suggested that the FEMA-440 expression for the limiting $R$ value is conservative; Vamvatsikos et al. (2009) recommend the following expression for statistics (mean and percentiles) of the collapse strength ratio, $R_c$:

$$R_c = \left(1 - a_h(\mu_{peak} - 1)\right) R_c^n(a_c, T) + (a_h + 1)\Delta R_c^{pn}(\mu_{eq}, T)$$

where

$$R_c^n(a_c, T) = 1 + \left(1.17 - e^{-\beta_1 T}\right) \, |a_c|^{-\beta_2 + \beta_3 T - \beta_4 \ln T} \tag{4.28}$$

$$\Delta R_c^{pn}(\mu_{eq}, T) = \gamma_1 \left(\mu_{eq} - 1\right)\left[1 - e^{\left(-\gamma_2 (\gamma_3)^{\frac{1}{\mu_{eq}}} T\right)}\right]$$

$$\mu_{eq} = \mu_c + \frac{a_h(\mu_c - 1)}{|a_c|}$$

The basic terms defining the backbone curve are defined in Figure 4.37, while the $\beta_i$ and $\gamma_i$ terms are as defined for different percentiles in Table 4.3. These may be used to provide the mean and the 16%, 50%, and 84% values of $R$ to cause global instability. Typically a user may choose to limit the applicability of simple $R–\mu–T$ relationships by the mean or median $R_c$ value for collapse. A more conservative limit would be supplied by the 16% percentile, at which only 16% of the ground motions have led to global instability. Even better results may be gained by taking advantage of the entire distribution of $R_c$ within a performance-based approach, as discussed in Chapter 13.

Another important issue that appears when employing published $R–\mu–T$ relationships is the need to fit a piecewise linear model to an actual curvilinear backbone. Different concepts have appeared in the past, yet few have been tested rigorously. Trying to fit the piecewise linear model by balancing the area discrepancy vis-à-vis the actual backbone is a popular strategy (e.g., EN, 1998), yet it has been shown to offer low accuracy (De Luca et al., 2013), especially due to the potential mismatch of the initial elastic segment

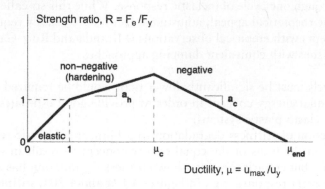

Figure 4.37 Primary curve used to establish $R_c$.

*Table 4.3* Coefficients for determining mean and fractile values of $R_c$

| | Coefficient | 16% | 50% | 84% | Mean |
|---|---|---|---|---|---|
| $R_c^n$ | $\beta_1$ | 0.07 | 0.35 | 1.50 | 0.37 |
| | $\beta_2$ | 1.11 | 1.32 | 1.32 | 1.29 |
| | $\beta_3$ | 0.03 | 0.19 | 0.14 | 0.12 |
| | $\beta_4$ | 0.19 | 0.29 | 0.15 | 0.16 |
| $\Delta R_c^{pn}$ | $\gamma_1$ | 0.77 | 1.20 | 1.90 | 1.20 |
| | $\gamma_2$ | 0.80 | 1.00 | 2.40 | 2.00 |
| | $\gamma_3$ | 5.50 | 2.70 | 0.50 | 1.90 |

and the corresponding period of the systems. For bilinear systems, FEMA 440 (2005) recommends matching the elastic segment to the point of the actual backbone that corresponds to 60% of the maximum strength observed. This is a better strategy that may only fail for systems exhibiting high initial changes in stiffness due to extensive cracking (typical of some detailed reinforced concrete models). Thus, De Luca et al. (2013) recommended fitting to the point of the actual backbone that corresponds to 20% of the maximum strength, while making sure that any subsequent positive or zero stiffness postyield segment captures the maximum strength and ductility without regard for balancing areas above and below the fitted segment. The same authors also offered a number of fitting rules for complex quadrilinear backbones for use with SPO2IDA (Vamvatsikos and Cornell, 2006).

## 4.12 EQUIVALENT LINEARIZATION

Prior to the widespread availability of software to compute inelastic response, estimates of the peak response of an inelastic system were computed using a linear elastic system having reduced stiffness and increased damping (e.g., Jacobsen, 1960). The substitute structure method (Shibata and Sozen, 1976) and more recently, direct displacement-based design (Priestley et al., 2007) use this approach.

Figure 4.38 illustrates energy for a full hysteretic loop that would be dissipated by a linear oscillator (having stiffness equal to the secant stiffness of the nonlinear oscillator at its peak displacement) in which additional damping is provided. The figure illustrates the reduction in spectral response associated with increased damping. In some formulations, effective damping is established to dissipate the same energy (under one cycle of harmonic response) as is dissipated through one cycle of inelastic response. While this so-called "equal energy" approach may have theoretical appeal, adjustment factors are usually required to reconcile equal energy concepts with empirical observations (Miranda and Ruiz-Garcia, 2002).

We note some issues with equivalent damping approaches:

a. Damping levels must be significantly lower than would be required by strict application of the equal energy concept in order to provide good estimates of the peak displacement of elasto-plastic systems.
b. The introduction of stiffness degradation to a bilinear hysteretic relationship would be expected on the basis of the equal energy concept to result in much larger peak displacements, but empirical results show stiffness degradation has a minor effect on peak displacement response (e.g., in Figures 4.19a and 4.20a, stiffness degradation is seen to cause a minor reduction in peak response for many values of $T$ and $R$).

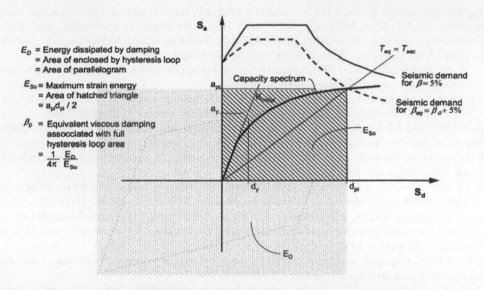

$E_D$ = Energy dissipated by damping
= Area of enclosed by hysteresis loop
= Area of parallelogram

$E_{So}$ = Maximum strain energy
= Area of hatched triangle
= $a_{pi}d_{pi}/2$

$\beta_0$ = Equivalent viscous damping associated with full hysteresis loop area
= $\dfrac{1}{4\pi}\dfrac{E_D}{E_{So}}$

*Figure 4.38* Notion of equivalent linear systems and use of equivalent damping to reduce spectral demand at the period associated with the secant stiffness. (From FEMA-440, 2005.)

c. While small changes in post-yield stiffness have a minor effect on peak response for positive values of post-yield stiffness (e.g., 0%–10% of the initial stiffness), the introduction of a negative post-yield stiffness has a disproportionately large effect on peak displacement response. Yet, such changes in post-yield stiffness have minor effects on the energy dissipated within a cycle and thus, on the equivalent damping level.

An optimization was done to minimize extreme errors in displacement estimates made using equivalent linear systems, resulting in recommendations for effective period and damping in the FEMA-440 (2005) report, in the chapter on the modified acceleration-displacement response spectrum.

Generally speaking, both approaches (displacement modification via $R-\mu-T$ relationships and equivalent linearization) can be made to provide results of similar accuracy. However, due to the complexity and specificity of equivalent linear approaches and the lack of empirical validity of the equal energy concept, the authors prefer the use of $R-\mu-T$ relationships to the use of equivalent linearization. As a final note, the $R-\mu-T$ relationships in Section 4.10 are applicable to oscillators having the specified hysteretic behavior subjected to ordinary ground motions on firm soils, rather than long duration motions or those exhibiting near-field directivity effects. Cases with unusual hysteretic behavior, on soft soil sites, and/or subject to near-fault motions with forward directivity pulses may best be addressed using case-specific results.

## REFERENCES

Aschheim, M., and Black, E. (2000). Yield point spectra for seismic design and rehabilitation, *Earthquake Spectra, Earthquake Engineering Research Institute*, 16(2):317–335.

Aschheim, M. A., and Hernández-Montes, E. (2003). The representation of P-Δ effects using yield point spectra, *Engineering Structures*, 25(11):1387–1396.

Building Seismic Safety Council (BSSC) (2000). NEHRP Recommended Provisions for Seismic Regulations for New Buildings and Other Structures, Part 1, Provisions, and Part 2, Commentary, FEMA 368 and 369, prepared by the Building Seismic Safety Council, published by the Federal Emergency Management Agency, Washington, DC.

Chopra, A. K., and Goel, R. K. (1999). Capacity-demand-diagram methods based on inelastic design spectrum, *Earthquake Spectra*, 15(4):637–656.

Christopoulos, C., Filiatrault, A., and Folz, B. (2002). Seismic response of self-centring hysteretic SDOF systems, *Earthquake Engineering & Structural Dynamics*, 31(5):1131–1150.

Clough, R. W., and Johnston, S. B. (1966). Effect of stiffness degradation on earthquake ductility requirements, *Proceedings of Japan Earthquake Engineering Symposium*, Tokyo, Japan.

Comitè Europèen de Normalisation (1998). Eurocode 8 – Design of Structures for earthquake resistance. EN 1998, CEN, Brussels, 2003.

Cuesta, I., Aschheim, M., and Fajfar, P. (2003). Simplified R-factor relationships for strong ground motions, *Earthquake Spectra*, Earthquake Engineering Research Institute, 19(1):1–21.

De Luca, F., Vamvatsikos, D., and Iervolino, I. (2013). Near-optimal piecewise linear fits of static pushover capacity curves for equivalent SDOF analysis, *Earthquake Engineering and Structural Dynamics*, 42(4):523–543.

Fajfar, P. (1999). Capacity spectrum method based on inelastic demand spectra, *Earthquake Engineering and Structural Dynamics*, 28(9):979–994.

Farrow, K. T., and Kurama, Y. C. (2004). SDOF displacement ductility demands based on smooth ground motion response spectra, *Engineering Structures*, 26(12):1713–1733.

FEMA-440 (2005). Improvement of Nonlinear Static Seismic Analysis Procedures, Applied Technology Council, Project ATC-55, Redwood City, CA.

Freeman, S. A. (2004). Review of the development of the capacity spectrum method, *ISET Journal of Earthquake Technology*, 41(1):1–13.

Hernández-Montes, E., Aschheim, M., and Gil-Martín, L. M. (2015). Energy components in nonlinear dynamic response of SDOF systems, *Journal of Nonlinear Dynamics*, 82:933–945.

Inel, M., Aschheim, M. A., and Abrams, D. P. (2002). An algorithm for computing isoductile response spectra, *Journal of Earthquake Engineering*, 6(3):375–390.

Ibarra, L. F., and H. Krawinkler. (2005). Global Collapse of Frame Structures Under Seismic Excitations John A, Blume Earthquake Engineering Center Report No. TR 152, Department of Civil Engineering, Stanford University and PEER Report 2005/06.

Ibarra, L.F., Medina, R.A., and Krawinkler, H. (2005). Hysteretic models that incorporate strength and stiffness deterioration, *Earthquake Engineering and Structural Dynamics*, 34:1489–1511.

Jacobsen, L. S. (1960). Damping in composite structures, *Proceedings of the 2nd World Conference on Earthquake Engineering*, Japan, Vol. 2, 1029–1044.

Kazantzi, A. K., and Vamvatsikos, D. (2018). The hysteretic energy as a performance measure in analytical studies, *Earthquake Spectra*, 34(2):719–739.

Lignos, D. (2010). Interactive Interface for Dynamic Analysis Procedure. RESSLab École Polytechnique Fédérale de Lausanne, Switzerland. https://zenodo.org/record/1312278#. W0ro5dgzZgc.

Luco, N., and Bazzurro, P. (2007). Does amplitude scaling of ground motion records result in biased nonlinear structural drift responses? *Earthquake Engineering and Structural Dynamics*, 36(13):1813–1835.

Miranda, E. (2001). Estimation of inelastic deformation demands of SDOF systems, *Journal of Structural Engineering, ASCE*, 127(9):1005–1012.

Miranda, E., and Ruiz-Garcia, J. (2002). Evaluation of approximate methods to estimate maximum inelastic displacement demands, *Earthquake Engineering and Structural Dynamics*, 31(3):539–560.

Nassar, A., and Krawinkler, H. (1991). Seismic Demands for SDOF and MDOF Systems, Report 95. John A. Blume Earthquake Engineering Center, Department of Civil Engineering, Stanford University, Stanford, CA.

Newmark, N. M., and Hall, W. J. (1969). Seismic design criteria for nuclear reactor facilities, *Proceedings of the 4th World conference on Earthquake Engineering*, Vol. 4.

Newmark, N. M., and Hall, W. J. (1982). *Earthquake Spectra and Design*. EERI Monograph Series, Earthquake Engineering Research Institute, Oakland, CA.

Otani, S. (1974). SAKE, A Computer Program for Inelastic Response of R/C Frames to Earthquakes, Report UILU-Eng-74-2029, Civil Engineering Studies, University of Illinois at Urbana-Champaign, USA.

Popov, E. P., and Bertero, V. V. (1973). Cyclic loading of steel beams and connections, *Journal of the Structural Division*, 99(6):1189–1204.

Priestley, M. J. N. (1993). Myths and fallacies in earthquake engineering – Conflicts between design and reality, *Bulletin of the New Zealand National Society for Earthquake*, 26(3):329–341.

Priestley, M. J. N., Calvi, G. M., and Kowalsky, M. J. (2007). *Direct Displacement-Based Design of Structures*, IUSS Press, Pavia.

Ramberg, W., and Osgood, W. R. (1943). Description of stress-strain curves by three parameters. Technical Note No. 902, National Advisory Committee For Aeronautics, Washington DC.

Riddell, R., and Newmark, N. M. (1979). Statistical Analysis of The Response of Nonlinear Systems Subjected to Earthquakes, Civil Engineering Studies, Structural Research Series No. 468, University of Illinois at Urbana-Champaign, Urbana, IL, 291p.

Shibata, A., and Sozen, M. A. (1976). Substitute-structure method for seismic design in R/C, *Journal of the Structural Division*, American Society of Civil Engineers, 102:1–18.

Takeda, T., Sozen, M. A., and Nielsen, N. N. (1970). Reinforced concrete response to simulated earthquakes, *Journal of the Structural Division, ASCE*, 96:ST12.

Thermou, G. E., Elnashai, A. S., and Pantazopoulou, S. J. (2012). Retrofit yield spectra-a practical device in seismic rehabilitation. *Earthquake and Structures*, 3(2):141–168.

Uang, C., and Bertero, V. (1988). Use of Energy as a design criterion in earthquake-resistant design, UCB/EERC-88/18, Earthquake Engineering Research Center, University of California, Berkeley, 1988-11, 46 pages.

USEE (2001). Utility software for earthquake engineering report and user's manual M Inel, EM Mretz, EF Black, M Aschheim, DP Abrams - 2001.

Vamvatsikos, D., and Cornell, C. A. (2002). Incremental dynamic analysis, *Earthquake Engineering and Structural Dynamics*, 31(3):491–514.

Vamvatsikos, D., and Cornell, C. A. (2004). Applied incremental dynamic analysis, *Earthquake Spectra*, 20(2):523–553.

Vamvatsikos, D. (2005). *Matlab Bundles for Running and Postprocessing IDA Curves*, National Technical University, Athens. http://users.ntua.gr/divamva/software.html.

Vamvatsikos, D., and Cornell, C. A. (2006). Direct estimation of the seismic demand and capacity of oscillators with multi-linear static pushovers through Incremental Dynamic Analysis, *Earthquake Engineering and Structural Dynamics*, 35(9):1097–1117.

Vamvatsikos, D., Akkar, S. D., and Miranda, E. (2009). Strength reduction factors for the dynamic instability of oscillators with non-trivial backbones, In *Conference on Computational Methods in Structural Dynamics and Earthquake Engineering*, Rhodes, Greece.

Veletsos, A. S., and Newmark, N. M. (1960). Effect of inelastic behavior on the response of simple systems to earthquake motions, *Proceedings of the 2nd World Conference on Earthquake Engineering*, Vol. 2, Tokyo and Kyoto, Japan.

Vidic, T., Fajfar, P., and Fischinger, M. (1994). Consistent inelastic design spectra: Strength and displacement, *Earthquake Engineering and Structural Dynamics*, 23:507–521.

# Chapter 5

# Dynamics of linear and nonlinear MDOF systems

## 5.1 PURPOSE AND OBJECTIVES

This chapter considers the linear and nonlinear dynamic of discrete systems having more than one degree of freedom in the time domain. Background material on the linear response of single-degree-of-freedom (SDOF) systems is provided in Chapter 3, while extensions to address the nonlinear response of SDOF systems are covered in Chapter 4.

## 5.2 LINEAR ELASTIC SYSTEMS

Linear elastic systems are those that experience small deformations (or strains), in which the stiffness is constant (i.e., stress is proportional to strain and force is proportional to displacement). The equations of static equilibrium are established (and solved) using the undeformed geometry of the structure. As a result, the principles of superposition are applicable.

### 5.2.1 Equation of motion of a linear elastic system subjected to applied forces

We develop the equation of motion for the simplest multi-degree-of-freedom (MDOF) system in this section, consisting of two masses having two degrees of freedom, and subjected to applied forces. The same system subjected to motion of the reference frame (base excitation) is addressed in Section 5.3.

Just as for SDOF systems (Section 3.5), the equation of motion of an arbitrary $j$th degree of freedom can be established considering inertial ($f_{Ij}$), damping ($f_{Dj}$), and spring ($f_{Sj}$) forces[1] that resist the applied force $p_j$. With reference to the two degrees of freedom illustrated in Figure 5.1:

$$f_{Ij} + f_{Dj} + f_{Sj} = p_j(t)$$

$$f_{Ij} = m_j \ddot{u}_j(t)$$

$$f_{Sj} = \begin{cases} f_{S1} = k_1 u_1 + k_2(u_1 - u_2) \\ f_{S2} = k_2(u_2 - u_1) \end{cases}$$

$$f_{Dj} = \begin{cases} f_{D1} = c_1 \dot{u}_1 + c_2(\dot{u}_1 - \dot{u}_2) \\ f_{D2} = c_2(\dot{u}_2 - \dot{u}_1) \end{cases} \tag{5.1}$$

---

[1] We have chosen variable names that generally are consistent with those used by Chopra (2017).

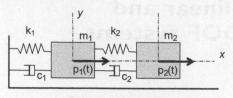

(a) Forced vibration (damped)

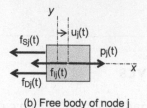

(b) Free body of node j

*Figure 5.1* Excitation of two-degree-of-freedom system by applied forces. (a) Forced vibration and (b) free body diagram of node j.

The spring forces, $f_{Sj}$, are linearly elastic, with springs 1 and 2 having stiffnesses $k_1$ and $k_2$, respectively. The damping force represents the mechanism by which a linear elastic structure dissipates energy; in the most common formulation, the damping force resists the movement (velocity) of one node relative to other nodes.

The matrix expression of dynamic equilibrium of the two-degree-of-freedom system is:

$$\begin{bmatrix} m_1 & 0 \\ 0 & m_2 \end{bmatrix}\begin{bmatrix} \ddot{u}_1 \\ \ddot{u}_2 \end{bmatrix} + \begin{bmatrix} c_1 + c_2 & -c_2 \\ -c_2 & c_1 \end{bmatrix}\begin{bmatrix} \dot{u}_1 \\ \dot{u}_2 \end{bmatrix} + \begin{bmatrix} k_1 + k_2 & -k_2 \\ -k_2 & k_1 \end{bmatrix}\begin{bmatrix} u_1 \\ u_2 \end{bmatrix} = \begin{bmatrix} p_1(t) \\ p_2(t) \end{bmatrix}$$

(5.2)

Using matrix notation, the more general equation of motion for MDOF systems is:

$$\mathbf{M}\ddot{\mathbf{u}} + \mathbf{C}\dot{\mathbf{u}} + \mathbf{K}\mathbf{u} = \mathbf{P}(t)$$

(5.3)

where **M** is the mass matrix, where the $m_{ij}$ entry of **M** is the $(i,j)$ component of the mass matrix and strictly speaking, it is the force at the $i$th degree of freedom due to a unit acceleration of the $j$th degree of freedom. Typically, masses are lumped at each node, resulting in **M** being a diagonal matrix, composed of just the mass $m_{ii}$ at each ($i$th) degree of freedom. Other approaches exist to formulating **M**, such as the consistent mass matrix, which result in mass matrices that include off-diagonal terms. Whatever method is used to establish **M**, the mass matrix inherently must be symmetric.

The structure stiffness matrix, **K**, is a symmetric matrix well known from its use in the direct stiffness method of structural analysis. The $k_{ij}$ entry is the force at the $i$th degree of freedom due to a unit displacement applied at the $j$th degree of freedom. The structure stiffness matrix **K** must be symmetric, as deduced by Maxwell's Reciprocity Theorem, which states that the deflection d$_i$ (at point $i$) due to a force $p_j$ (at point $j$) is equal to the deflection $d_j$ due to a force $p_i$.

The vector of damping forces $\mathbf{F}_D (= \mathbf{C}\dot{\mathbf{u}})$ provides a mechanism to dissipate the energy of vibration over time. The $c_{ij}$ entry of the damping matrix, **C**, is the force at the $i$th degree of freedom corresponding to a unit velocity of the $j$th degree of freedom. Generally **C** will be symmetric; methods to establish the entries of **C** are discussed in Section 5.2.4.

## 5.2.2 Equation of motion of a linear elastic system subjected to base excitation

The previous section considered MDOF systems subjected to discrete forces applied to the lumped masses. This section considers the effect of motion of the reference frame (or base excitation), which could be due to earthquake-induced ground motion. As shown in Figure 5.2, the ground motion is represented by an imposed displacement $u_g(t)$ and its derivatives. The total displacement (or absolute displacement relative to a fixed reference point) of the $j$th degree of freedom is $u_j^t(t)$, while the relative displacement of the $j$th node relative to the ground is $u_j(t)$. Therefore:

$$u_j^t(t) = u_j(t) + u_g(t) \tag{5.4}$$

This is represented for the two degrees of freedom model as:

$$\begin{bmatrix} u_1^t(t) \\ u_2^t(t) \end{bmatrix} = \begin{bmatrix} u_1(t) \\ u_2(t) \end{bmatrix} + u_g(t) \begin{bmatrix} 1 \\ 1 \end{bmatrix} \tag{5.5}$$

and in a more general form using vector notation as:

$$\mathbf{u}^t(t) = \mathbf{u}(t) + u_g(t)\mathbf{1} \tag{5.6}$$

If the applied loads are removed ($\mathbf{P}(t) = 0$), the dynamic equation of equilibrium is simply

$$\mathbf{F}_I + \mathbf{F}_D + \mathbf{F}_S = 0 \tag{5.7}$$

Among these terms, as detailed in Chapter 3, only the inertial force $\mathbf{F}_I$ is dependent on the total displacements or their derivatives. The spring force, $\mathbf{F}_S$, is expressed in terms of the relative displacements (nodal displacements relative to the base of the structure). The damping force, $\mathbf{F}_D$, typically is expressed in terms of the relative velocities of adjacent nodes, but sometimes is expressed in terms of absolute velocities. Assuming the typical relationship, Equation 5.7 can be specialized to the linear elastic response of a viscously damped MDOF system:

$$\mathbf{M}\ddot{\mathbf{u}}^t + \mathbf{C}\dot{\mathbf{u}} + \mathbf{K}\mathbf{u} = 0 \tag{5.8}$$

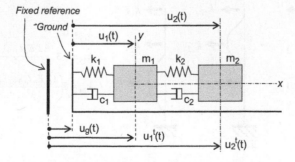

Figure 5.2 Excitation of a two-degree-of-freedom system by ground motion.

Substituting $\mathbf{u}^t$ from Equation 5.6:

$$\mathbf{M\ddot{u}} + \mathbf{C\dot{u}} + \mathbf{Ku} = -\mathbf{M1}\ddot{u}_g(t) = \mathbf{P}_{eff}(t) \tag{5.9}$$

By considering the term $-\mathbf{M1}\ddot{u}_g(t)$ equivalent to $\mathbf{P}_{eff}(t)$, the response of a structure subjected to base motion $u_g(t)$, represented by Figure 5.2, is equivalent to that of a fixed base structure (Figure 5.1) subjected to a corresponding external load $\mathbf{P}_{eff}(t)$.

In a more general formulation, one or more degrees of freedom will have a different direction (with reference to the six Cartesian translations and rotations) or will not be aligned with the ground motion. In such cases, the unit vector 1 (Equation 5.6) can be generalized to the influence vector, $\iota$, and Equation 5.9 is replaced by the more general expression:

$$\mathbf{M\ddot{u}} + \mathbf{C\dot{u}} + \mathbf{Ku} = -\mathbf{M}\iota\ddot{u}_g(t) = \mathbf{P}_{eff}(t) \tag{5.10}$$

The influence vector, $\iota$, represents the influence of a unit displacement $u_g$ on the $j$th degree of freedom. It generally contains 1s and 0s, with a 1 assigned where the degree of freedom is aligned with $u_g$ and a 0 assigned where $u_g$ has no influence on the degree of freedom. For example, for the framed structure of Figure 5.3, the idealized model has $\iota^T = \mathbf{1}^T = [1, 1, 1]$; however, if six degrees of freedom had been considered, such as the horizontal and vertical displacement at each node, then $\iota^T$ would be $[1, 0, 1, 0, 1, 0]$. More generally, where the ground motion component is not aligned with the degrees of freedom, the entries if $\iota$ can be determined by trigonometry.

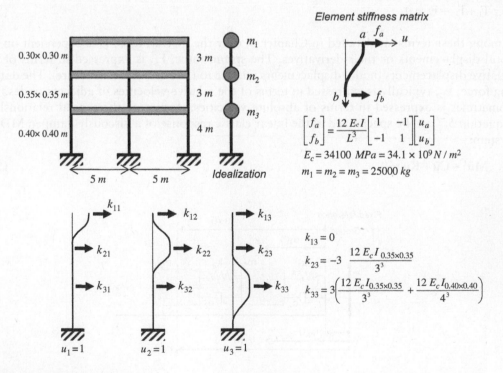

*Figure 5.3* Framed structure.

### 5.2.3 Undamped free vibration and natural modes and frequencies

The undamped free vibration of MDOF systems is governed by the following equation, which can be deduced from Equation 5.3:

$$\mathbf{M}\ddot{u} + \mathbf{K}u = 0 \tag{5.11}$$

This equation represents $n$ homogeneous differential equations that are coupled, where $n$ is the number of degrees of freedom. The solution $u(t)$ to Equation 5.9 depends on two particular solutions, often constrained by the initial conditions $u(0)$ and $\dot{u}(0)$.

As an extension of the SDOF system, consider a possible solution of this form:

$$u(t) = \phi \sin(\omega t + \theta) \tag{5.12}$$

where $\phi$ is a vector representing the displaced shape at the degrees of freedom, $\theta$ is the phase, and $\omega$ is the circular frequency of the solution. Given the second derivative, to be a solution to Equation 5.11 requires:

$$\left.\begin{array}{l} u(t) = \phi \sin(\omega t + \theta) \\ \ddot{u}(t) = -\omega^2 \phi \sin(\omega t + \theta) \end{array}\right\} \rightarrow -\omega^2 \mathbf{M}\phi \sin(\omega t + \theta) + \mathbf{K}\phi \sin(\omega t + \theta) = 0 \tag{5.13}$$

Equation 5.13 can be rewritten as:

$$\left[\mathbf{K} - \omega^2 \mathbf{M}\right]\phi = 0 \tag{5.14}$$

and can be set in the traditional form of the eigenproblem, $\left[A - \lambda I\right]\phi = 0$, by pre-multiplying Equation 5.14 by $\mathbf{M}^{-1}$

$$\left[\mathbf{M}^{-1}\mathbf{K} - \omega^2 \mathbf{I}\right]\phi = 0 \tag{5.15}$$

for which the eigenvalues and eigenvectors are $\omega^2$ and $\phi$, respectively. If the inverse of the matrix $\left[\mathbf{M}^{-1}\mathbf{K} - \omega^2 \mathbf{I}\right]$ exists, then both sides of Equation 5.15 can be pre-multiplied by $\left[\mathbf{M}^{-1}\mathbf{K} - \omega^2 \mathbf{I}\right]^{-1}$, resulting in $\phi = 0$, known as the trivial solution. To find a nontrivial solution requires that the inverse must not exist, i.e., the determinant of the matrix must be equal to zero:

$$\det\left[\mathbf{M}^{-1}\mathbf{K} - \omega^2 \mathbf{I}\right] = 0 \tag{5.16}$$

Equation 5.16 is known as the characteristic equation and it is a polynomial equation whose solution provides the $n$ eigenvalues $\omega_j^2$. For each eigenvalue, the shape of the corresponding eigenvector $\phi_j$ is obtained using Equation 5.15—because the determinant is null, the system of equations is a linearly dependent system. Notice that if $\phi$ satisfies Equation 5.15, $\alpha\phi$ also is a solution, for any scalar $\alpha$. Thus, the eigenvectors are usually presented in one of several normalized forms, such as scaling the vector such that the largest amplitude is equal to one or so that the vector satisfies $\phi^T \mathbf{M}\phi = 1$. In much of the remainder of this book, we assume the mode shape $(\phi_j)$ is normalized to unit amplitude at the roof in order to simplify some mathematical expressions, as illustrated in Example 5.1 (Box 5.1). One of the most popular methods for obtaining the $n$ eigenpairs $(\omega_j, \phi_j)$ is the QR algorithm (Wikipedia, 2017). The eigenpairs are usually ordered from lowest to highest frequency $(\omega_1 < \omega_2 < \dots \omega_n)$.

## BOX 5.1—EXAMPLE 5.1

In this example, an idealized version of a moment frame structure is analyzed. The frame consists of three stories and two bays. Unlike typical moment frames, for simplicity the girders are assumed to be rigid, resulting in a so-called "shear building" in which only three degrees of freedom are considered. Columns are assumed to be axially rigid, shear deformations are neglected, and small deformations are assumed. Lumped masses of 25,000 kg are present at each level. The structure stiffness matrix entries $k_{ij}$ are determined by direct application of the Hermitian shape functions for each member when the structure is subjected to unit displacements at each of the degrees of freedom, one by one.

The mass matrix (lumped mass) and the stiffness matrix are:

$$\mathbf{M} = 25,000 \begin{bmatrix} 1 & 0 & 0 \\ 0 & 1 & 0 \\ 0 & 0 & 1 \end{bmatrix} \text{kg}$$

$$\mathbf{K} = \begin{bmatrix} 3.069 & -3.069 & 0 \\ -3.069 & 8.7547 & -5.6857 \\ 0 & -5.6857 & 9.7777 \end{bmatrix} \times 10^7 \, \text{N/m}$$

Solving the characteristic equation (Equation 5.16), the frequency vector is obtained[2]:

$$\det\left[\mathbf{K} - \omega^2 \mathbf{M}\right] = 0 \rightarrow \omega = \left\{ \begin{array}{c} \omega_1 \\ \omega_2 \\ \omega_3 \end{array} \right\} = \left\{ \begin{array}{c} 18.566 \\ 46.486 \\ 78.325 \end{array} \right\} \text{rad/s}$$

For each $\omega_i$ Equation 5.14 is a linear dependent system that provides information only on the relative amplitudes of the entries of $\phi_j$, because $\alpha\phi_j$ also satisfies the equation.

One way to proceed is to impose that the first entry of $\phi_j$ be equal to 1 and solve for the other two components, using two of the three equations given by Equation 5.16. The remaining equation will be an identity, and can be used to check that the values obtained are correct. The solutions for the three free vibration mode shapes, below, are plotted in Figure 5.4.

$$\phi_1 = \left\{ \begin{array}{c} 1 \\ 0.7192 \\ 0.4586 \end{array} \right\}; \phi_2 = \left\{ \begin{array}{c} 1 \\ -0.7603 \\ -0.9881 \end{array} \right\}; \phi_3 = \left\{ \begin{array}{c} 1 \\ -3.9975 \\ 4.0882 \end{array} \right\}$$

---

[2] The complete structure, composed of 15 beam elements (BEAM3) with six DOF at each node was analyzed with ANSYS, considering lumped masses (MASS21) and flat beams of 0.2 × 5.0 m. The circular frequencies for the first three modes [14.50, 41.03, 74.32] rad/s, are lower than for the shear building assumption because of the flexibility of the beams.

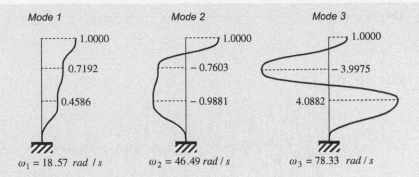

*Figure 5.4* Frequencies and modes of vibration.

One may appreciate the increasing tortuosity of the mode shapes for the higher modes. Specifically, the nodal translations for the first-mode shape are all positive, while for the second mode shape, some values are negative. There are points which have no translation, at the base of the structure for all modes, in the third story for the second mode, and in the second and third stories for the third mode. These points remain fixed in position (stationary) during the response in any mode, even as the amplitude of the modal response changes (both positive and negative) over time.

## BOX 5.2—EXAMPLE 5.1 (*Continued*)

The shear building of Example 5.1 is subjected to the Corralitos ground motion (Figures 3.11 and 3.14). Damping is assumed to be equal to 5% of critical damping for each mode.

$$\mathbf{M} = 25,000 \begin{bmatrix} 1 & 0 & 0 \\ 0 & 1 & 0 \\ 0 & 0 & 1 \end{bmatrix} \text{kg}; \imath = \begin{bmatrix} 1 \\ 1 \\ 1 \end{bmatrix} \Rightarrow \mathbf{s} = \mathbf{M}\imath = 25,000 \begin{bmatrix} 1 \\ 1 \\ 1 \end{bmatrix} \text{kg}$$

$$\Gamma_1 = \frac{\boldsymbol{\phi}_1^{\mathrm{T}} \mathbf{s}}{\boldsymbol{\phi}_1^{\mathrm{T}} \mathbf{M} \boldsymbol{\phi}_1} = 1.26061$$

$$\Gamma_2 = \frac{\boldsymbol{\phi}_2^{\mathrm{T}} \mathbf{s}}{\boldsymbol{\phi}_2^{\mathrm{T}} \mathbf{M} \boldsymbol{\phi}_2} = -0.29299$$

$$\Gamma_3 = \frac{\boldsymbol{\phi}_3^{\mathrm{T}} \mathbf{s}}{\boldsymbol{\phi}_3^{\mathrm{T}} \mathbf{M} \boldsymbol{\phi}_3} = 0.03237$$

To determine the response of the structure to this ground motion, we solve three SDOF systems, one for each frequency ($\omega_1$, $\omega_2$, and $\omega_3$). This results in the three displacement histories $D_1(t)$, $D_2(t)$, and $D_3(t)$ shown in Figure 5.5a, following the same approach used to generate Figure 3.12. The displacement of the structure as a function of time is obtained by superposition using Equation 5.22. Nodal displacement histories are plotted in Figure 5.5b.

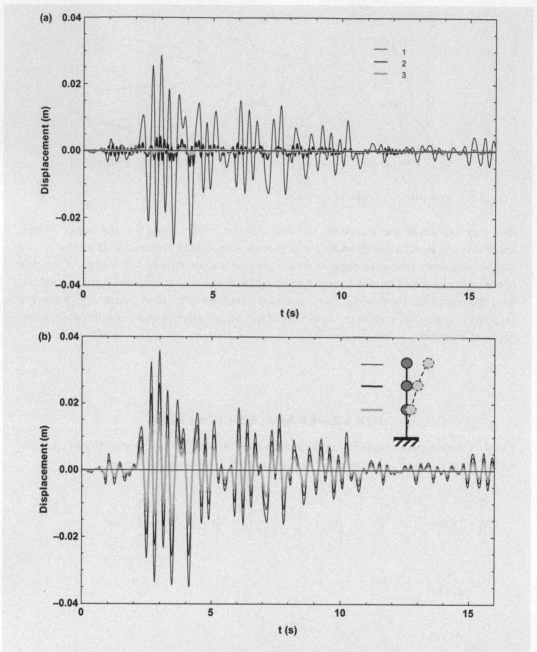

Figure 5.5 Displacement histories for (a) uncoupled SDOF systems at the three natural frequencies (solution of Equation 5.21), and (b) the three nodes of the structure.

The $n$ eigenvectors ($\phi_1...\phi_n$) are the normalized deformed shapes (or mode shapes) associated with the undamped free vibration of the structure. These modes are associated with the $n$ circular vibration frequencies ($\omega_1...\omega_n$, given by the square root of the eigenvalues $\omega_j{}^2$). Since any single mode shape $\phi_j$ is a solution of the undamped free vibration problem (Equation 5.11), any linear combination of the $n$ mode shapes ($\phi_1...\phi_n$) is also a solution of

## BOX 5.3—EXAMPLE 5.1 (Continued)

Let us estimate the peak base shear $V_{b,max}$ and peak roof displacement of the structure of Figure 5.3. The results of MRHA (Example 5.2) include the vector of displacements $u(t)$ plotted in Figure 5.5b. The equivalent static force vector $F(t)$ is obtained using the stiffness matrix $F(t) = Ku(t)$. Horizontal (static) equilibrium establishes that the base shear is the sum of the components of vector $F(t)$, which is given by the dot product of the vector $1^T$ and $F(t)$: $V_b(t) = 1^T F(t)$. Figure 5.8 shows the time history of $V_b(t)$, for which $V_{b,max} = 691,321$ N at $t = 2.98$ s. From Figure 5.5b, the maximum displacement at node 1 is 35.89 mm at $t = 2.98$ s.

Using RSA with the SRSS combination, we note $V_{bj}$, is given by $1^T \cdot s_j$ (having units of mass), while $S_a(T_j)$ is the pseudo-acceleration obtained from the pseudo-acceleration spectra from Figure 3.14 for the frequencies of each of the three modes. Then $V_{bj,max} = V_{bj}S_a(T_j)$.

$$V_{b1} = 1^T s_1 = 1^T \Gamma_1 M \phi_1 = 1.26061 \cdot 2{,}500 \begin{bmatrix} 1 & 1 & 1 \end{bmatrix} \begin{bmatrix} 1 & 0 & 0 \\ 0 & 1 & 0 \\ 0 & 0 & 1 \end{bmatrix} \begin{bmatrix} 1 \\ 0.7192 \\ 0.4586 \end{bmatrix} = 68{,}635.60 \text{ kg}$$

$$V_{b2} = 1^T s_2 = 1^T \Gamma_2 M \phi_2 = 5481.67 \text{ kg}$$

$$V_{b3} = 1^T s_3 = 882.76 \text{ kg}$$

$$S_a(T_1) = 9.7256 \text{ m/s}^2 \rightarrow V_{b1,max} = V_{b1}S_a(T_1) = 667{,}524 \text{ N}$$

$$S_a(T_2) = 12.7206 \text{ m/s}^2 \rightarrow V_{b2,max} = V_{b2}S_a(T_2) = 69{,}730 \text{ N}$$

$$S_a(T_3) = 9.1086 \text{ m/s}^2 \rightarrow V_{b3,max} = V_{b3}S_a(T_3) = 8040 \text{ N}$$

$$V_{b,max} \approx \left( \sum_{j=1}^{3} V_{bj,max}^2 \right)^{1/2} = 671{,}205 \text{ N}$$

The peak displacement obtained by RSA using the SRSS is given by:

$$s_j = \Gamma_j M \phi_j s_j = K d_j$$

$$d_j = K^{-1} s_j \Rightarrow d_1 = \begin{bmatrix} 0.003657 \\ 0.002630 \\ 0.001677 \end{bmatrix} \quad d_2 = \begin{bmatrix} -0.0001356 \\ 0.0001031 \\ 0.0001340 \end{bmatrix} \quad d_2 = \begin{bmatrix} 5.277 \times 10^{-6} \\ -0.000002109 \\ 0.00002157 \end{bmatrix}$$

$$d_{roof,max} \approx \left( \sum_{j=1}^{3} (d_{j,1max})^2 \right)^{1/2} = \left( \sum_{j=1}^{3} (S_a(T_j)d_{j,1})^2 \right)^{1/2} = 0.03561 \text{ m}$$

This example illustrates a more general observation, which is that the higher modes contribute little to the peak roof displacement (0.03561 m is nearly equal to the contribution of the first mode alone, which is 0.003657(9.7256) = 0.03557 m).

the undamped free vibration problem. The mode shape $(\phi_1)$ and modal frequency $(\omega_1)$ associated with the lowest frequency is called the first mode; the corresponding period of vibration is $T_1 = 2\pi/\omega_1$). The next higher frequency is associated with the second mode $(\phi_2, \omega_2)$, and so on.

### 5.2.4 Orthogonality of mode shapes

The eigenpairs (natural frequencies and mode shapes) are characteristics of the structure and are independent of the loading. However, as will be seen in the following sections, the mode shapes are useful for simplifying the solution of the equation of motion for forced vibration (Equation 5.3) and base excitation (Equation 5.10). These solution approaches make use of a very interesting property, the orthogonality of the eigenvectors (mode shapes) with respect to $\mathbf{M}$ and $\mathbf{K}$. Appling Equation 5.14 to two different eigenpairs, $(\omega_j, \boldsymbol{\phi}_j)$ and $(\omega_k, \boldsymbol{\phi}_k)$ and pre-multiplying by the transpose of the other eigenvector yields:

$$\left.\begin{array}{l} \mathbf{K}\boldsymbol{\phi}_j = \omega_j^2 \mathbf{M}\boldsymbol{\phi}_j \\ \mathbf{K}\boldsymbol{\phi}_k = \omega_k^2 \mathbf{M}\boldsymbol{\phi}_k \end{array}\right\} \rightarrow \left.\begin{array}{l} \boldsymbol{\phi}_k^\mathrm{T}\mathbf{K}\boldsymbol{\phi}_j = \omega_j^2 \boldsymbol{\phi}_k^\mathrm{T}\mathbf{M}\boldsymbol{\phi}_j \\ \boldsymbol{\phi}_j^\mathrm{T}\mathbf{K}\boldsymbol{\phi}_k = \omega_k^2 \boldsymbol{\phi}_j^\mathrm{T}\mathbf{M}\boldsymbol{\phi}_k \end{array}\right\} \xrightarrow{\text{Substracting}} 0 = \left(\omega_j^2 - \omega_k^2\right)\boldsymbol{\phi}_k^\mathrm{T}\mathbf{M}\boldsymbol{\phi}_j \quad (5.17)$$

Subtracting one equation from the other in the second step allows for a simplification because, due to the symmetry of $\mathbf{M}$ and $\mathbf{K}$, $\boldsymbol{\phi}_k^\mathrm{T}\mathbf{K}\boldsymbol{\phi}_j = \boldsymbol{\phi}_j^\mathrm{T}\mathbf{K}\boldsymbol{\phi}_k$ and $\boldsymbol{\phi}_k^\mathrm{T}\mathbf{K}\boldsymbol{\phi}_j = \boldsymbol{\phi}_j^\mathrm{T}\mathbf{M}\boldsymbol{\phi}_k$. The last step demonstrates that $\boldsymbol{\phi}_j$ and $\boldsymbol{\phi}_k$, are orthogonal with respect to $\mathbf{M}$ for $j \neq k$. Substituting the equality of the third step in the first equality of the second step shows that $\boldsymbol{\phi}_j$ and $\boldsymbol{\phi}_k$ are also orthogonal with respect to $\mathbf{K}$ for $j \neq k$. The $n$ eigenvectors are orthogonal one by one (with respect to $\mathbf{K}$ and $\mathbf{M}$) and because we have $n$ independent vectors, each containing n components, the $n$ eigenvectors constitute a basis for the $\mathbb{R}^n$ vector space.

### 5.2.5 Modal decomposition of displacement history

The differential equation of the dynamic response of a linear elastic MDOF structure subjected to a horizontal base excitation $\ddot{u}_g$ is given by Equation 5.10:

$$\mathbf{M}\ddot{\mathbf{u}} + \mathbf{C}\dot{\mathbf{u}} + \mathbf{K}\mathbf{u} = -\mathbf{M}\iota\ddot{u}_g(t) = \mathbf{P}_{\text{eff}}(t) = -\mathbf{s}\ddot{u}_g(t) \tag{5.18}$$

In the case that $\mathbf{u}$ is a vector of n components that represent the lateral displacements of the floors relative to the base (as in Example 5.1), $\iota = 1$. More generally, the vector $\iota$ is a column vector having components equal to 1 or 0, and $\mathbf{s}$ is a vector equal to $\mathbf{M}\iota$ and represents the spatial distribution of the effective forces $\mathbf{P}_{\text{eff}}(t)$, whose amplitude is modulated over time by the ground motion. The entries of $\mathbf{s}$ have units of mass.

The displacement vector, $\mathbf{u}$ (having n components) can be expressed as a linear combination of any set of basis vectors (n independent vectors that span the space). In particular, we can use the free vibration mode shapes $(\boldsymbol{\phi}_j)$ as the basis vectors, which have the benefit of orthogonality and orthonormality with respect to $\mathbf{M}$ and $\mathbf{K}$ if properly normalized. Thus, $\mathbf{u}(t)$, can be expressed as a linear combination of spatially invariant mode shapes $(\boldsymbol{\phi}_j)$ weighted by time-varying amplitudes:

$$\mathbf{u}(t) = \sum_{j=1}^{n} \boldsymbol{\phi}_j q_j(t) \tag{5.19}$$

where $q_j$ is termed the $j$th modal coordinate.

## 5.2.6 Modal response history analysis

The expression of the displacement vector in terms of the mode shapes (Equation 5.19) allows the system of $n$ coupled equations represented by Equation 5.18 to be uncoupled into $n$ individual equations expressed in terms of the modal coordinates, provided that the mode shapes $\phi_j$ and $\phi_k$ ($j \neq k$) also are orthogonal with respect to the damping matrix, $\mathbf{C}$. Methods to construct the damping matrix are discussed in Section 5.2.8.

The $n$ uncoupled equations of motion are obtained by substituting Equation 5.19 into Equation 5.18, pre-multiplying by $\phi_j^{\mathrm{T}}$ and applying the properties of orthogonality of the free vibration mode shapes with respect to $\mathbf{M}$, $\mathbf{C}$, and $\mathbf{K}$ to eliminate zero terms. This allows the response of the $j$th mode to ground excitation to be expressed as:

$$\ddot{q}_j(t) + 2\xi_j\omega_j\dot{q}_j(t) + \omega_j^2 q_j(t) = -\Gamma_j \ddot{u}_g(t)$$

where 
$$\omega_j = \sqrt{\frac{\phi_j^{\mathrm{T}}\mathbf{K}\phi_j}{\phi_j^{\mathrm{T}}\mathbf{M}\phi_j}}$$

$$\xi_j = \frac{\phi_j^{\mathrm{T}}\mathbf{C}\phi_j}{2\phi_j^{\mathrm{T}}\mathbf{M}\phi_j\omega_j} \tag{5.20}$$

$$\Gamma_j = \frac{\phi_j^{\mathrm{T}}\mathbf{M}\iota}{\phi_j^{\mathrm{T}}\mathbf{M}\phi_j} = \frac{\phi_j^{\mathrm{T}}\mathbf{s}}{\phi_j^{\mathrm{T}}\mathbf{M}\phi_j}$$

where $q_j$ is the modal coordinate, $\xi_j$ is the damping ratio of the mode $j$, $\omega_j$ is the undamped vibration frequency of mode $j$, and $\Gamma_j$ is the corresponding modal participation factor. Like other modal parameters, $\Gamma_j$ reflects the characteristics of the structure, and is independent of the ground excitation.

In order to compare results with the SDOF system shown in previous chapters a further simplification can be achieved by dividing Equation 5.20 by $\Gamma_j$ and setting $q_j(t) = \Gamma_j D_j(t)$, resulting in:

$$\ddot{D}_j(t) + 2\xi_j\omega_j\dot{D}_j(t) + \omega_j^2 D_j(t) = -\ddot{u}_g(t) \tag{5.21}$$

Equation 5.21 is analogous to the differential equation of motion for a SDOF system (Equation 3.51).

The solution of Equation 5.21 for the $D_j(t)$ corresponding to each mode is the basis of *modal response history analysis* (MRHA). The response of the linear elastic structure is given by the superposition of responses determined independently using each uncoupled modal equation. The response of the system is represented by the vector $\mathbf{u}(t)$, and is obtained as the sum of the modal contributions:

$$\mathbf{u}(t) = \sum_{j=1}^{n}\mathbf{u}_j(t) = \sum_{j=1}^{n}\Gamma_j\phi_j D_j(t) \tag{5.22}$$

## 5.2.7 Modal decomposition of effective force

We can examine the components of the effective force that drive each modal response. To do so, we repeat the process of deducing the SDOF Equation 5.20 from the MDOF Equation 5.18, except now we decompose the effective force, $\mathbf{P}_{\mathrm{eff}}(t) = -\mathbf{s}\ddot{u}_g(t)$, into its components. Taking note of the orthogonality of the mode shapes with respect to $\mathbf{M}$, we choose to decompose $\mathbf{s}$ (the vectorial part of $\mathbf{P}_{eff}(t)$) into components $\mathbf{s}_j = a_j \mathbf{M}\boldsymbol{\phi}_j$, where $a_j$ are unknown scalars to be determined. Thus,

$$\mathbf{s} = \mathbf{M1} = \sum_{j=1}^{n} \mathbf{s}_j = \sum_{j=1}^{n} a_j \mathbf{M}\boldsymbol{\phi}_j \qquad (5.23)$$

Substituting Equations 5.23 and 5.10 into Equation 5.18, and pre-multiplying both sides by $\boldsymbol{\phi}_k^{\mathrm{T}}$ results in

$$\boldsymbol{\phi}_k^{\mathrm{T}} \left( \mathbf{M}\boldsymbol{\phi}_j \ddot{q}_j(t) + \mathbf{C}\boldsymbol{\phi}_j \dot{q}_j(t) + \mathbf{K}\boldsymbol{\phi}_j q_j(t) = -a_j \mathbf{M}\boldsymbol{\phi}_j \ddot{u}_g(t) \right) \qquad (5.24)$$

Orthogonality eliminates terms containing $\boldsymbol{\phi}_k^{\mathrm{T}} \mathbf{M}\boldsymbol{\phi}_j$ for $j \neq k$, indicating that only the $\mathbf{s}_j$ component of $\mathbf{P}_{eff}$ results in a nonzero response in the $j$th mode, and that $a_j$ must equal the participation factor $\Gamma_j$. Thus, $\mathbf{s}_j$ and $\mathbf{P}_{eff,j}$ can be expressed as:

$$\mathbf{s}_j = \Gamma_j \mathbf{M}\boldsymbol{\phi}_j$$

$$\mathbf{P}_{eff,j} = -\mathbf{s}_j \ddot{u}_g(t) = -\Gamma_j \mathbf{M}\boldsymbol{\phi}_j \ddot{u}_g(t) \qquad (5.25)$$

with only $\mathbf{P}_{eff,j}$ causing response in the $j$th mode.

## 5.2.8 Damping of linear elastic systems

As with all real materials, the free vibration portion of the response of a building will diminish over time. While this phenomenon may be physically manifested in the deformation of materials, resistance to motion in air, conversion of vibration to sound and heat, and by friction, damping in structural systems typically is represented as a function of velocity. This is done more for the sake of mathematical simplicity than precision. Errors resulting from the use of viscous damping in place of a more authentic representation of damping are thought to be small and of little significance for low values of damping in most structural systems (typically not exceeding 5%–10% of critical damping).

The viscous damping matrix, $\mathbf{C}$, produces damping forces $f_{Di}(t) = \mathbf{C}\dot{u}$ that resist motion in the direction of the velocity $\dot{u}$ (Figure 5.1). Generally, $\mathbf{C}$ is a symmetric semi-positive definite matrix, which ensures that damping retards response.

Classical damping refers to the case that mode shapes are orthogonal with respect to $\mathbf{C}$. Because the mode shapes $\boldsymbol{\phi}_j$ and $\boldsymbol{\phi}_k$ ($j \neq k$) already are orthogonal with respect to $\mathbf{M}$ and $\mathbf{K}$, the simplest way to construct a classical damping matrix is to choose a linear combination of $\mathbf{M}$ and $\mathbf{K}$, given by $\mathbf{C} = \alpha\mathbf{M} + \beta\mathbf{K}$. This is known as Rayleigh damping.[3] As discussed in Section 5.2.5, the $n$ simultaneous equations representing the dynamic response of a classically damped MDOF system can be uncoupled, allowing the solution obtained as the superposition of $n$ independent SDOF responses. The damping coefficients required for the solution of the $n$ SDOF equations (represented by Equation 5.21) are usually specified as a percentage of critical damping (for each SDOF system) rather than by forming and operating on $\mathbf{C}$ (for the MDOF system). Nevertheless, an understanding of the construction of $\mathbf{C}$ and its implications is important both for the incremental (time-step) solution of the linear elastic MDOF system and for the solution of the response of nonlinear MDOF systems (discussed in Section 5.3).

---

[3] A more general way to construct $\mathbf{C}$ while maintaining modal orthogonality is known as Caughey damping, where $\mathbf{C} = \mathbf{M} \sum_{i=o}^{N-1} a_i (\mathbf{M}^{-1}\mathbf{K})^i$, and $N$ is an arbitrary natural number. Rayleigh damping is a special case, for which $N = 2$.

By operating on the definition of $\xi_j$ in Equation 5.20 by introducing $\mathbf{C} = \alpha\mathbf{M} + \beta\mathbf{K}$, we obtain:

$$\xi_j = \frac{\alpha}{2\omega_j} + \frac{\beta\omega_j}{2} \tag{5.26}$$

The values of $\alpha$ and $\beta$ can be determined to set the modal damping ratios associated with just two frequencies, $\omega_j$ and $\omega_k$, (i.e. $\xi_j$ and $\xi_k$) to desired values. The frequencies $\omega_j$ and $\omega_k$ often selected to be those corresponding to two of the lower modes of vibration (e.g., the first and third modal frequencies) although there is no requirement that the frequencies used in Equation 5.26 match the modal frequencies. Thus, for specified values of $\omega_j$, $\omega_k$, $\xi_j$, and $\xi_k$, Equation 5.26 can be solved for the coefficients $\alpha$ and $\beta$:

$$\alpha = \frac{2\omega_i\omega_j\left(\xi_j\omega_i - \xi_i\omega_j\right)}{\omega_i^2 - \omega_j^2} \quad \text{and} \quad \beta = \frac{2\left(\xi_i\omega_i - \xi_j\omega_j\right)}{\omega_i^2 - \omega_j^2} \tag{5.27}$$

Once the values of $\alpha$ and $\beta$ have been established, the fraction of critical damping at any modal frequency $\omega_i$ can be determined using Equation 5.26.

Figure 5.6 plots the percent of critical damping ($\xi_i$) associated with the mass ($\alpha/2\omega_i$) and stiffness ($\beta\omega_i/2$) contributions of Equation 5.26, as well as their sum, anchored at $\xi_1 = \xi_3 = 0.05$ for $\omega_1 = 2\pi$ rad/s ($T_1 = 2\pi/\omega_1 = 1$ s) and $\omega_3 = 20\pi$ rad/s ($T_3 = 0.1$ s). It is apparent in Figure 5.6 that the mass contribution to damping (associated with $\alpha$) diminishes rapidly with increasing frequency, while the stiffness contribution to damping (associated with $\beta$) increases linearly with frequency. The sum, therefore, results in higher modes being more heavily damped, while modal frequencies between $\omega_1$ and $\omega_3$ have damping less than the specified values of $\xi_1$ and $\xi_3$.

Inspection of $\mathbf{C}$ shows that mass-proportional damping produces forces that retard the velocity of each mass relative to the base. (Of course, no such forces develop at massless degrees of freedom.) Mass-proportional damping has been criticized as lacking physical relevance, since no material is physically present to dissipate vibrational energy (through deformation or friction) between any particular discrete mass and a reference frame fixed to the base. However, reliance on only stiffness proportional damping ($\alpha = 0$) has the drawback that the higher modes are excessively damped and the lower modes may have too little damping (depending on the frequencies at which the damping ratios are specified).

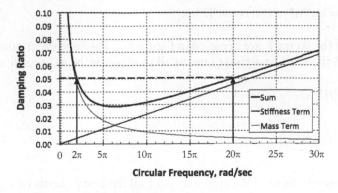

*Figure 5.6* Effective damping as a function of frequency using Rayleigh damping.

An alternative to Rayleigh damping is modal damping, wherein the desired percentage of critical damping of $\xi_j$ is specified for each mode. As described by Clough and Penzien (1993), C is constructed by a summation of terms associated with each mode:

$$C = M \left( \sum_{j=1}^{n} \frac{2\xi_j \omega_j}{M_j} \phi_j \phi_j^T \right) M \tag{5.28}$$

where $M_j$ is the generalized mass $(= \phi_j^T M \phi_j)$ and $\xi_j$ is the fraction of critical damping for the $i$th mode. Each mode contributes to the damping matrix in proportion to $\xi_j$ and the desired damping levels can be set for each mode (unlike the case with Rayleigh damping). The resulting C is fully populated, unlike the case for Rayleigh damping in which only the diagonal entries corresponding to those in K (and M) are populated.

Damping values for bare frame reinforced concrete buildings (i.e., those lacking nonstructural components) are generally on the order of 2%–3% of critical damping, while actual buildings (containing nonstructural components such as partitions, mechanical systems, and cladding) generally have damping in the range of 3%–5% of critical damping. Damping in taller structures (e.g., more than 30 stories) tends to be lower. Thus, damping values of around 4% or 5% may be considered as reasonable for many typical reinforced concrete buildings.

## 5.2.9 Equivalent (statically applied) lateral forces

Engineers are accustomed to think of loading in terms of applied forces (e.g., dead, live, and wind loads). Although the solution to the equation of motion is found in terms of displacements, it is helpful to interpret response in terms of the forces that would be applied quasi-statically to cause the computed displacement response. The static forces required to obtain $u_j(t)$ in the absence of ground motion are given by $F_j(t) = Ku_j(t)$. Forces in individual members can be recovered by multiplying the element stiffness matrices by the relevant nodal displacements contained within u. Note that damping results in additional forces between the nodes (that vary with the velocity of one node relative to another); since each node is in dynamic equilibrium, these velocity-dependent forces influence the response of the structure.

The contribution of the $j$th mode to the displacement of the structure relative to the ground is given by $u_j(t) = \phi_j q_j(t) = \Gamma_j \phi_j D_j(t)$. Using Equations 5.14 and 5.22, the time-varying forces associated with any one mode that would result in the displacement $u_j(t)$ (if applied quasi-statically) are given by:

$$F_j(t) = Ku_j(t) = \omega_j^2 M \Gamma_j \phi_j D_j(t) = \omega_j^2 D_j(t) s_j \tag{5.29}$$

Components $f_{ji}$ of the vector $F_j$ are depicted in Figure 5.7. The base shear due to displacement of each mode and the total base shear (due to all modes) can be expressed as:

$$V_{bj}(t) = \omega_j^2 D_j(t) 1^T s_j \tag{5.30a}$$

$$V_b(t) = \sum_{j=1}^{n} \omega_j^2 D_j(t) 1^T s_j \tag{5.30b}$$

where $1^T s_j$ is the matrix form of the summation of all the components of $s_j$. See Figure 5.7.

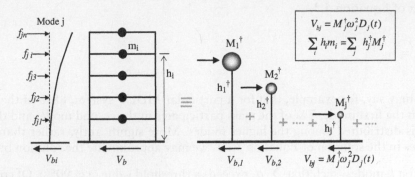

Figure 5.7 Effective modal mass and effective modal height.

## 5.2.10 Effective modal mass

The effective modal mass is a concept that conveys the portion of total mass mobilized within each mode. The effective modal mass is derived based on the relative contribution of each mode to the base shear. The resultant of the modal forces (Figure 5.7) is equal to the modal base shear, under quasi-static conditions. Just as for the modal participation factor, the effective modal mass (and effective modal height, Section 5.2.11) reflect the modal characteristics of the structure (determined by $\mathbf{K}$ and $\mathbf{M}$) and do not represent the degree to which a particular ground motion excites responses in each mode.

For an SDOF system, the base shear is equal to the force (applied quasi-statically) that produces a relative displacement $u(t)$ of the system (Figure 3.15):

$$V_b = ku(t) = m\omega^2 u(t) \tag{5.31}$$

The base shear associated with the $j$th mode of vibration was expressed in Equation 5.30a. Comparing both Equations 5.30a and 5.31, we can define the effective modal mass for the $j$th mode in the context of the $j$th modal contribution to the base shear. Thus, the effective modal mass, $M_j^\dagger$, is defined by the following equivalencies:

$$M_j^\dagger \equiv \mathbf{1}^T \mathbf{s}_j = \mathbf{1}^T \Gamma_j \mathbf{M} \boldsymbol{\phi}_j = \Gamma_j \boldsymbol{\phi}_j^T \mathbf{M} \mathbf{1} \tag{5.32}$$

Moreover, because it can be demonstrated that:

$$\sum_{j=1}^{n} M_j^\dagger = \sum_{i} m_i = \text{total mass of the building} \tag{5.33}$$

where $i$ is the story number (and ranges from 1 to the total number of stories, $n$), we can define the mass participation factor $\alpha_j$ (sometimes called the modal mass participation factor) as

$$\alpha_j = \frac{M_j^\dagger}{\displaystyle\sum_{j=1}^{n} M_j^\dagger} = \frac{\Gamma_j \boldsymbol{\phi}_j^T \mathbf{M} \mathbf{1}}{\mathbf{1}^T \mathbf{M} \mathbf{1}} \tag{5.34}$$

As a result of Equation 5.34,

$$\sum_{j=1}^{n} \alpha_j = 1 \tag{5.35}$$

Thus, we may say, for example, that for a particular MDOF system, 88% of the mass participates in the first mode, 7% of the mass participates in the second mode, and the remaining mass is distributed among the higher modes. More significantly, rather than including all $n$ modes in the solution of Equation 5.22, we may approximate the solution by including only the first $k$ modes, such that $\sum_{j=1}^{k} \alpha_j$ exceeds a threshold value (e.g. 90%). Of course, since the mass participation factors are functions of structural properties ($\mathbf{K}$ and $\mathbf{M}$) and do not reflect the degree to which these modes are excited by the ground motion, inclusion of the first $k$ modes may miss important responses if the ground motion contains significant content at frequencies higher than $\omega_k/2\pi$.

## 5.2.11 Effective modal height

The effective modal height represents the relative contribution of a mode to the overturning moment at the base of a MDOF structure. The product of the resultant and the effective height is equal to the overturning moment, under quasi-static conditions.

We define the effective height $h_j^\dagger$ for the $j$th mode (Figure 5.7), which has an effective mass $M_j^\dagger$, by comparing the overturning moment (at the base) that occurs in the $j$th mode with the overturning moment at the base, $M_b$, of the SDOF system with a mass $m$ at height $h$. For the SDOF system, $M_b$ is:

$$M_b = hV_b = hm\omega^2 u(t) \tag{5.36}$$

Applying Equation 5.29, the product of each component of $\mathbf{F}_j(t)$ and its height is given by $\mathbf{h}^T\mathbf{s}_j$, where $\mathbf{h} = (h_1, h_2, h_3, \dots, h_n)$, where $n$ is the total number of stories. Thus, the overturning moment at the base due to the quasi-static forces for the $j$th mode is:

$$M_{bj}(t) = \omega_j^2 D_j(t)\mathbf{h}^T\mathbf{s}_j \tag{5.37}$$

The expression at the right side of last equality of Equation 5.37 is defined to have the same form as the term at the right side of Equation 5.36. This allows the effective modal height $h_j^\dagger$ to be defined as:

$$h_j^\dagger = \frac{\mathbf{h}^T\mathbf{s}_j}{\mathbf{1}^T\mathbf{s}_j} = \frac{\sum_{i=1}^{N} h_i s_{ji}}{M_j^\dagger} \tag{5.38}$$

Finally, it can be demonstrated that:

$$\sum_{j=1}^{n} h_j^\dagger M_j^\dagger = \sum_{i=1}^{N} h_i m_i \tag{5.39}$$

## 5.2.12 Peak response estimates by response spectrum analysis

MRHA provides information at any time $t$, while structural design is mainly based on peak values of response quantities. Response spectrum analysis (RSA) provides estimates of peak response quantities based on the peak responses of each mode as determined using response spectra. Either the response spectrum for a specific earthquake excitation or a smoothed design response spectrum may be used. In using response spectra, information related to the timing of the individual modal peaks is not available, and so a statistically based combination of the individual modal responses is used instead.

Equation 5.29 expressed the equivalent forces (applied quasi-statically) associated with each modal response over time:

$$F_j(t) = Ku_j(t) = \omega_j^2 M\Gamma_j \phi_j D_j(t) = \omega_j^2 D_j(t)s_j \tag{5.29}$$

The maximum of $|D_j(t)|$ is the spectral displacement, $S_d(T_j)$, and the maximum of $\omega_j^2 |D_j(t)|$ is the pseudo-spectral acceleration, $S_a(T_j)$ (Equation 3.30), where $T_j$ is the period associated with the $j$th mode ($T_j = 2\pi/\omega_j$). Then, the first and last terms of Equation 5.29 can be expressed as:

$$F_{max,j} = S_a(T_j)s_j \tag{5.40}$$

where $F_{max,j}$ has units of force and $s_j$ has units of mass. The principle of superposition is applicable to linear systems; therefore, any response quantity $r_{j,max}$ (e.g., deformation, shear force, or flexural moment at a specific location) obtained by applying the vector $F_{max,j}$ is equal to scalar $S_a(T_j)$ multiplied by the same quantity $r_j$, where $r_j$ is obtained by applying the vector $s_j$.

$$r_{j,max} = r_j S_a(T_j) \tag{5.41}$$

Now, imagine that $r$ is the moment at a certain point, then $r_j(t)$ is the moment at this point due to $F_j(t)$ and $r_{j,max}$ is the maximum value of $|r_j|$ over time. For two different modal forces $F_j(t)$ and $F_k(t)$, the instant $t$ at which $r$ is a maximum is a function of the timing of the modal responses and varies with each ground motion. A upper bound on $r_{max}$ considers each modal contribution to reach its maximum at the same instant, for which $r_{max} = \Sigma r_{j,max}$ over all $j$. A more accurate estimate of $r_{max}$ considers contributions from all the modes at less than their individual maxima. Among the most common combination rules for linear elastic systems are the square root of the sum of squares (SRSS) and complete quadratic combination (CQC) rules.

The SRSS estimate is:

$$r_{max} = \left( \sum_{j=1}^{n} r_{j,max}^2 \right)^{1/2} \tag{5.42}$$

The CQC establishes:

$$r_{max} = \left( \sum_{j=1}^{n} \sum_{k=1}^{n} \rho_{jk} r_{j,max} r_{k,max} \right)^{1/2} \tag{5.43}$$

In the CQC rule each of the $n^2$ terms is the product of the peak responses of the $j$th and $k$th modes (independently) and a correlation coefficient $\rho_{jk}$, which must be defined $a$ $priori$. There are different estimations available for the correlation coefficient (e.g., see Chopra,

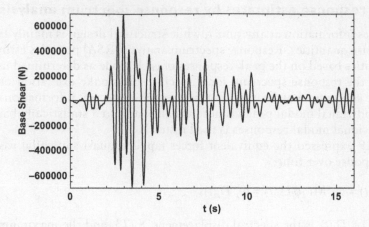

*Figure 5.8* Base shear history.

2017). The CQC and SRSS methods result in nearly the same estimates unless some of the modal frequencies with significant contributions to the response quantity are close to one another. In this case, the CQC estimates will be higher and are considered better estimates.

## 5.3 NONLINEAR SYSTEMS

Nonlinearity in the force–displacement response of structural components and geometric effects (statics in the deformed configuration, also termed second-order or $P$-$\Delta$ effects) are addressed in this section. Nonlinearity associated with large strains is not considered.

### 5.3.1 Equation of motion for nonlinear systems

In the case of linear systems, $\mathbf{F}_S = \mathbf{Ku}$. For nonlinear systems, the loading curve may gradually soften (curvilinear) or may be segmentally linear, such as that shown in Figure 5.9 (extracted from Figure 4.9). For elastic response, unloading proceeds along the loading curve and no energy is dissipated by the material. In the case of inelastic response, at any given displacement the force resisted by the element depends on the history. Thus, for the general case of inelastic response (which must be nonlinear), it is necessary to track the

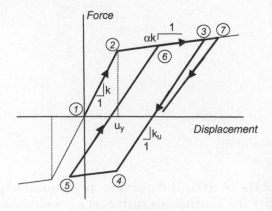

*Figure 5.9* Nonlinear model (bilinear degrading hysteresis).

displacement history. The solution proceeds step-by-step, and for each step the direction of loading determines the stiffness to use for determining the increment in force. For example, at Point 3 in Figure 5.9, the post-yield stiffness $ak$ is used for a positive increment in displacement, while the unloading stiffness $k_u$ is used for a negative increment.

The dependence of the spring force $F_S$ on the displacement history may be expressed as $F_S = F_S(u, \dot{u})$ where $u$ and $\dot{u}$ are function of time, or $F_S(t)$ in a general form:

$$\mathbf{M}\ddot{u}(t) + \mathbf{C}\dot{u}(t) + \mathbf{F}_S(t) = -\mathbf{M}\ddot{u}_g(t) \tag{5.44}$$

In the preceding equation, equilibrium is based on the initial configuration. Geometric nonlinearity considers equilibrium in the deformed configuration. This is also termed second-order or $P$-$\Delta$ effects (see Section 4.11). For linear elastic systems, the forces $\mathbf{F}$ required to maintain static equilibrium for the displacements defined by $\mathbf{u}$ are given by

$$\mathbf{F} = (\mathbf{K} - \mathbf{K}_G)\mathbf{u} \tag{5.45}$$

where $\mathbf{K}$ is the (ordinary, or first-order) stiffness matrix and $\mathbf{K}_G$ is the geometric stiffness matrix. The elements of both matrices, $\mathbf{K}$ and $\mathbf{K}_G$, are obtained from the terms of the stiffness matrices of the individual structural elements, as discussed in courses on structural analysis. Thus, the response of linear elastic systems under seismic excitation is represented by the following equation of motion:

$$\mathbf{M}\ddot{u} + \mathbf{C}\dot{u} + (\mathbf{K} - \mathbf{K}_G)\mathbf{u} = -\mathbf{M}\ddot{u}_g(t) = \mathbf{P}_{eff}(t) \tag{5.46}$$

Considering nonlinear force–displacement relationships where $\mathbf{F}_S(t)$ is used to express $\mathbf{F}_S = \mathbf{F}_S(u, \dot{u})$, the general form of the nonlinear equation of motion is given by:

$$\mathbf{M}\ddot{u}(t) + \mathbf{C}\dot{u}(t) + \mathbf{F}_S(t) - \mathbf{K}_G\mathbf{u}(t) = -\mathbf{M}\ddot{u}_g(t) \tag{5.47}$$

## 5.3.2 Solution by direct integration time history analysis

Methods such as RSA and MRHA that are applicable to linear systems cannot be used with nonlinear systems because superposition requires linearity. As a result, solution of the nonlinear equation of motion is obtained by direct integration in a step-by-step fashion. For each time step, a solution is found based on the state of the system at the conclusion of the previous time step and actions taking place within the time step. (This is described for nonlinear SDOF systems in Section 4.8.) The element stiffness is tracked and whenever changes occur the stiffness must be updated. Thus, the stiffness matrix is reevaluated frequently, although it is considered to be constant within any time step; the duration of the time step may be reduced in order to track changes in stiffness more precisely. Thus, Equation 5.47 is the equation of motion to be solved within any time step.

The incremental form of the equation of motion considers equilibrium between two positions, 1 and 2, in a manner similar to what was done in Section 4.8.1 for the SDOF case:

$$\Delta\mathbf{F}_I(t) + \Delta\mathbf{F}_D(t) + \Delta\mathbf{F}_S(t) + \Delta\mathbf{F}_G(t) = 0$$

where

$$\Delta\mathbf{F}_I(t) = \mathbf{M}\left(\ddot{u}_2^t(t) - \ddot{u}_1^t(t)\right) = \mathbf{M}(\ddot{u}_2(t) + \iota\ddot{u}_{g2}(t) - \ddot{u}_1(t) - \iota\ddot{u}_{g1}(t)) = \mathbf{M}\Delta\ddot{u} + \mathbf{M}\iota\Delta\ddot{u}_g \tag{5.48}$$

$$\Delta\mathbf{F}_D(t) = \mathbf{C}\Delta\dot{u}$$

$$\Delta\mathbf{F}_S(t) = \mathbf{K}\Delta u$$

$$\Delta\mathbf{F}_G(t) = -\mathbf{K}_G\Delta u$$

In the above equations, increments of spring and damping forces, $\Delta F_S$ and $\Delta F_D$, as well as forces arising from second-order effects, $\Delta F_G$, are expressed as functions of the increments of the relative displacements, $\Delta u(t)$, and velocity, $\Delta \dot{u}(t)$.

As mentioned in Chapter 4, several numerical procedures are available for solving this equation. Two of the most common are the constant acceleration method and the linear acceleration method. The latter is developed in Section 4.8.3 and easily can be extended to MDOF systems.

An upper bound to the time step used in the solution is the time step with which the ground acceleration data is specified. Smaller time steps may be needed to accurately represent element stiffnesses. If a constant time step is used throughout the analysis, the direct-integration analysis can be solved several times using decreasing time-step sizes until the step size is small enough that results are minimally affected by the time-step size. Other solution algorithms adjust the time step dynamically, reducing the time-step size when changes in stiffness occur, so that the error within the time step is less than some specified level. The specific techniques used to control the numerical solution procedure will vary with the software package used.

### 5.3.3 Treatment of damping

In nonlinear systems, hysteretic losses associated with the softening and yielding of structural components dissipate vibrational energy, and viscous damping is a less significant form of energy dissipation for inelastic systems (Figure 4.7). Damping within the linear elastic portion of response should be identical to that for linear elastic systems. However, some caution is warranted in the specification of damping in the analytical model to avoid generating unrealistically large damping forces as the structural components undergo inelastic behavior (see the discussion on SDOF systems in Section 4.7). Methods that may be used to obtain more realistic values damping forces vary depending on the options available in the analysis software.

The incremental equation of motion considers equilibrium of each node at the beginning and end of each time step. Damping forces are generated between nodes based on the relative nodal velocities, as specified by the damping matrix, $C$. Damping is needed in general in order to reduce response amplitudes over time, but it is difficult to consider the damping forces within any time step as real forces—we normally are concerned with the forces in structural components that are generated directly through their deformation. Peak member forces may occur at an instant when the relative velocities of adjacent nodes are near zero (possibly indicating maximum deformation, at the instant that a reversal is about to begin, therefore, resulting in relatively small damping forces), but this is not necessarily the case. The presence of unrealistically high damping forces, due to modeling inaccuracies, suggests too severe an opposition to further deformation of the structure and thus an underestimation of member deformations (and possibly member forces, depending on the component force–displacement relationship).

In the case of linear elastic MDOF systems, it was convenient to construct $C$ to comply with the orthogonality relationships. Specifically, if Rayleigh damping is used (Section 5.2.8), whereby $C = \alpha M + \beta K$, then $\phi_j$ and $\phi_k$ also are orthogonal with respect to $C$ for $j \neq k$, the $n$ simultaneous equations of motion that describe the response of the MDOF system can be uncoupled, producing $n$ independent SDOF equations of motion (and leading to the methods of analysis MRHA and RSA). Because the $n$ simultaneous equations of motion that describe the response of the nonlinear MDOF system cannot be uncoupled in this way, they are solved simultaneously in a computationally expensive step-by-step approach. Some advantage remains in specifying damping as a linear combination of $M$ and $K$: because

M and K are essentially sparse matrices (most of their terms are zero),[4] when C is defined as a linear combination of M and K, it also becomes sparse, which allows for the fast solution of the simultaneous equations using sparse-matrix solvers and the efficient storage of the three matrices in memory (Wikipedia, 2019). Hence, Rayleigh damping is often preferred even for nonlinear systems.

Section 5.2.8 discussed drawbacks associated with using Rayleigh damping for linear systems—mainly that the $\beta$ term introduces unrealistically high amounts of damping in higher modes and the $\alpha$ term (mass-proportional damping) lacks physical plausibility. As nonlinearity develops, softening (often attributable to cracking and yielding) causes a reduction in the instantaneous modal frequencies. If $\alpha$ and $\beta$ are established based on linear elastic properties, then the effect of a reduction in instantaneous frequencies with nonlinearity will be an increase in the instantaneous damping (as a percentage of critical damping) for the lowest modes—see Figure 5.6 (Charney, 2008).

Available options for specifying damping vary with the software package used for the dynamic analyses. If linear viscous damping is to be used, then an option may be to require reevaluation of C upon significant changes in stiffness. For example, one may target viscous damping of 4%–5% of critical damping at any instant. Should Rayleigh damping be used to establish C, the coefficients $\alpha$ and $\beta$ can be reevaluated from time to time, using $C = \alpha M + \beta(K_t - K_G)$, where $K_t$ is the current tangent stiffness matrix. While this is an improvement over the use of the initial stiffness matrix, K, a concern is that negative values within $K_t - K_G$ (due to in-cycle strength loss of structural components or $P$-$\Delta$ effects on the system) may generate negative terms within C. The negative terms indicate that some components of C provide a velocity-proportional force that accelerates the system rather than retards it, which is mathematically chaotic. One way to address this concern is to exclude the geometric stiffness term, using only $K_t$ (as suggested by Charney, 2008), and impose a lower limit of zero on entries of C, although doing so will cause the effective damping to depart from the specified values of $\xi_j$ and $\xi_k$.

If Rayleigh damping must be fixed for the duration of the response, one could use Equation 5.27 with reduced modal frequencies to account, in a crude way, for the softening of the system in nonlinear response. Noting that a 50% loss of stiffness would correspond to modal frequencies that are $\sqrt{2}/2 = 70.7\%$ of their original values, one could determine $\alpha$ and $\beta$ for a fixed damping ratio at reduced frequencies (e.g., $0.7\omega_1$ and $0.7\omega_3$). The fixed damping ratio could be set less than 4%–5% to reduce the unrealistically large damping forces that may occur as the system softens during the response.

More generally, modal damping (Equation 5.28) may be preferred for higher accuracy, although it comes at the cost of computational efficiency. C may be recomputed for a desired damping ratio (e.g., 4%–5%) from time to time. If modal damping must be fixed for the duration of the response, the initial, elastic mode shapes may be used to compute C, but with modal frequencies that are reduced to account for nonlinearity in the response.

PEER/ATC 72-1 (2010) provides a summary of damping values derived from shaking table tests conducted on bare structures (lacking partitions, mechanical equipment, cladding, and contents). Damping values typically are inferred from decrements in free vibration response after testing. Based on the summary, we suggest appropriate values for bare reinforced concrete structural systems in Table 5.1. The presence of additional structural members (gravity framing), partitions, mechanical equipment, cladding, and contents, along with soil-structure interaction, will tend to further dampen response. We suggest the values in Table 5.1 be increased by perhaps 2%–3% to account for this increase in effective

---

[4] Only the diagonal terms of M typically are nonzero, while nonzero stiffness appears in K only for the degrees of freedom of nodes that are physically connected (a small percentage of the total).

Table 5.1 Approximate viscous damping value for bare reinforced concrete
structural systems (as a percent of critical damping in the first mode)

| | Damage level | | |
|---|---|---|---|
| None | Slight | Yielding/moderate | Significant |
| 2% | 3% | 5% | 7% |

damping for typical buildings, subject to the engineer's discretion. Where nonlinearity is modeled and adequately represents the energy dissipated through inelastic behavior (i.e., due to material damage), viscous damping should be based on bare frame values associated with minimal damage, modified to address damping contributions of gravity framing and nonstructural components. This is typically the case when using distributed-plasticity fiber elements that can directly account for concrete cracking and the associated inelasticity, thus requiring bare frame viscous damping ratio values on the order of 1%. Most lumped plasticity elements do not show any inelasticity before (nominal) yielding, thus a higher damping value on the order of 4%–5% increases accuracy for serviceability-level limit-states. For larger deformations (and associated damage levels) the damping from hysteresis typically overshadows viscous damping, thus the value of the latter becomes much less important, provided that fictitious large viscous damping values are avoided, as discussed earlier.

### 5.3.4 Inelastic response assessment via nonlinear response history analysis

Just as discussed in Section 4.8.6 for SDOF systems, nonlinearity in the model of an MDOF structure introduces variability in the response to ground motions for a given $S_a$ value. In addition, whereas an SDOF system would display no response variability before the onset of nonlinear behavior (see the unique linear response between $(R, \mu) = 0,0$ and $(1,1)$ in Figure 4.22b) the existence of multiple modes of vibration and the associated variability in the corresponding spectral ordinates that cannot be captured by a single intensity measure (IM) mean that variability in response peaks appears even in the elastic response of an MDOF system. For example, Figure 5.10a shows the pseudo-acceleration spectra of the FEMA P695 far-field set (FEMA, 2009). Even after scaling all records to the same value of $S_a(T_1)$, considerable variability is still visible at other periods, both higher and lower. This, along with characteristics of the ground motion waveform such as the amplitude and sequencing of individual pulses, introduces unavoidable dispersion.

Thus, it is imperative that the assessment of nonlinear MDOF structural response employs multiple dynamic analyses, typically more than an equivalent SDOF model, involving several ground motion records and intensity levels. One of the foremost challenges in this process (either for an MDOF or an SDOF system) is the selection or generation of accelerograms that will be representative of the ground motion that is likely to be encountered at each intensity level. Or, in other words, given an IM and its level that is of interest, what combination of ground motions should one employ to estimate the (distribution of) engineering demand parameter (EDP) response?

Due to limitations in the catalog of available ground motions (see e.g., PEER NGA, 2018; Akkar et al., 2014) natural records of a given intensity cannot usually be found. While generation of artificial ground motions is rapidly becoming a viable approach, amplitude scaling of natural records is typically employed. In simple terms, the acceleration values of a given record are uniformly multiplied by a single scale factor to reach the required value of intensity (Figure 5.10b). This practice is generally accepted for "small" values of the scale

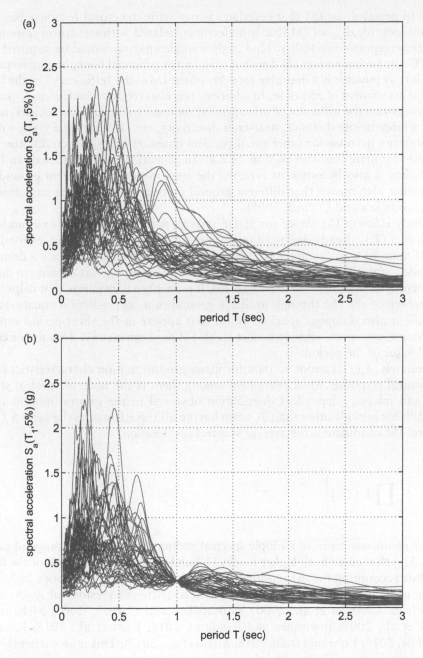

*Figure 5.10* The pseudo-acceleration spectrum of 44 ground motions from the FEMA P695 far-field set (FEMA, 2009): (a) naturally recorded ground motions; (b) scaled to $S_a(1 \text{ s}) = 0.3g$.

factor (i.e., 0.25–4.0), but it becomes controversial for larger values because the estimates of response become increasingly biased (Luco and Bazzurro, 2005). Thus, emphasis has been placed on employing IMs that can offer low values of bias but also of variance.

Theoretically, any scalar that describes the intensity of an accelerogram can be employed as an IM, such as the peak ground acceleration, velocity, or displacement (see Chapter 2) or the spectral acceleration, velocity, or displacement (Chapter 3) at a given period and damping. In practice, an IM that correlates better with structural response has consider-able advantages. Ideally, an IM that is perfectly correlated with structural response would result in zero response variability. Thus, only a single analysis would be required at any of its levels. While such a perfect IM does not exist in any practical form, having response vari-ability as low as possible is a desirable property of the IM—the "efficiency" of the IM allows economical assessment of response. In addition, one also requires "sufficiency," i.e., the IM should incorporate the influence of other ground motion characteristics, such as magnitude, duration, source-to-site distance, near-fault directivity, etc., so that for a given value of the IM one need not account for other site-dependent characteristics that could otherwise bias the response estimate (see also Section 13.2.8). In general, the more efficient an IM is, the more sufficient it also becomes, as bringing the response due to different ground motions closer together also means that different ground motion characteristics stop mattering as much (Vamvatsikos and Cornell, 2005).

A relatively efficient IM choice for SDOF systems is the 5% damped first-mode spectral acceleration, $S_a(T_1)$, already discussed in Section 4.8.6. It retains many of its good qualities for MDOF systems (e.g., Shome et al., 1998), especially when their response is dominated by the first mode at the period $T_1$. Note that the use of 5% damping has nothing to do with the actual damping of the structure to be assessed. It is simply a convention that helps us predict the occurrence of the IM through available ground motion prediction equations (Section 13.3), while it also dampens spurious peaks that appear in the short-period range of the ground motion spectrum, offering a more stable IM performance vis-à-vis minor changes in the actual value of the period.

Unfortunately, $S_a(T_1)$ cannot account for many ground motion characteristics that influ-ence structural response. Most prominent among them is the issue of spectral shape, i.e., the site- and intensity-dependent distribution observed in the ground motion spectra of Figure 5.10b for periods other than $T_1$ when having all records match the same $S_a(T_1)$ value. A far better IM candidate is the average spectral acceleration,

$$\text{Avg}S_a = \left[ \prod_{i=1}^{N} S_a\left(T_{Li}\right) \right]^{\frac{1}{N}} \tag{5.49}$$

that is, the geometric mean of multiple spectral ordinates $T_{Li}$ within a range of periods of, say, $[T_2, 1.5T_1]$ that cover both higher modes $(T_2)$ and an elongated version of the first mode $(1.5T_1)$, thus accounting for the sensitivity of nonlinear behavior to periods higher than $T_1$ due to the general post-yield reduction in stiffness. Different versions of $\text{Avg}S_a$ have been proposed (e.g., Cordova et al., 2000; Vamvatsikos and Cornell, 2005; Mehanny, 2009; Bianchini et al., 2009; Bojórquez and Iervolino, 2011; Eads et al., 2015; Kazantzi and Vamvatsikos, 2015; Kohrangi et al., 2016; Adam et al., 2017). This is an extremely welcome development to improve the validity of assessment results, yet at the same time it is consider-ably more difficult to take advantage of in guiding design, as it lacks the direct connection that $S_a(T_1)$ enjoys with equivalent SDOF quantities, such as the yield base shear. Since an equivalent SDOF proxy is practically given in our design approaches, we do not employ $\text{Avg}S_a$ at present, but keep it as a useful reminder of how assessment can be improved.

Whenever an IM is not sufficient enough to remove the influence of other seismological parameters, their distribution at each IM level should be properly accounted for by appro-priately selecting ground motion records. Thus, for example, if spectral shape, duration, or any other feature that cannot be accounted for by the IM is deemed to be important, the ground motion records used at different intensities should be chosen to reflect the anticipated distribution of those features given the IM. A prime example of a selection method is the conditional spectrum (CS) by Lin et al (2013a,b), whereby the distribution of spectral shape, characterized by the parameter epsilon (Baker and Cornell, 2006) is taken into account. Furthermore, Bradley (2010) has proposed the generalized conditional intensity measure (GCIM) approach to also incorporate other significant characteristics such as the duration and number of significant cycles of the ground motion. While undoubtedly useful, ground motion selection is not a panacea. The growing but nevertheless limited catalog of recorded ground motions limits what can be represented by natural records. Thus, it is often the case that a combination of selection and scaling needs to be employed.

Given that automated tools for the selection of site-specific records are not yet widely available, a good interim rule for general use is to utilize strong records that, when unscaled, can still damage the investigated structure, together with a relatively sufficient IM to allow scaling within reason. For most existing low/mid-rise structures having low-to-moderate ductility capacities that are not subject to near-field motions, this generally means employing $S_a(T_1)$ together with records having naturally high $S_a(T_1)$ values. One such set is the FEMA P695 far-field set (FEMA, 2009). Alternatively, one can employ records that have been pre-selected for a given site or collection of sites, such as the INNOSEIS high/moderate seis-micity sets for Europe (INNOSEIS, 2017). For modern ductile buildings or non-ductile buildings that are not first-mode dominated (e.g., plan-asymmetric or tall structures), or whenever the site of interest is subject to near-field motions, also employing an improved IM such as $AvgS_a$ should be preferred.

Having decided on the IM and the ground motion records, choosing the IM levels, run-ning the analyses, and post-processing the results remains a challenge. Different options to resolve it have given rise to three distinct approaches (Jalayer and Cornell, 2009): cloud analysis, stripe analysis, and incremental dynamic analysis (IDA).

Cloud analysis is perhaps the most intuitive of the three approaches. It is named after the characteristic cloud of results that it creates in the IM versus EDP plane (Figure 5.11a). Each point corresponds to one dynamic analysis performed. Such analyses are run for one or more groups of ground motions, each group either left unscaled or scaled (usually by the same factor). Considerable freedom in selecting and scaling records is thus available to the user. For example, one can employ only scaling using a fixed record set that is scaled to several levels of intensity. This is the case of Figure 5.11a, where two scale factors, 1.0 and 3.0, were applied to the FEMA P695 set. Scaling can also be completely avoided (assuming catalog limitations are not prohibitive) by using multiple separate sets of natural records to represent low, moderate, and high levels of the IM. On the other hand, post-processing the results to assess the distribution of response requires some care. The usual approach is to account for collapse "infinite EDP" points via logistic regression (see Shome and Cornell, 1999) and employ a simple linear regression in the logarithmic space to capture the non-collapsing IM-EDP results (Jalayer and Cornell, 2009). As the latter assumes a constant variability of EDP response given the IM, when experience has shown that variability tends to increase considerably with intensity, it is often advised to perform the latter regression locally, in the region of interest, rather than incorporating all available data points.

Stripe analysis takes its name after the characteristic rows of points aligned in distinct horizontal stripes at different IM-levels (Figure 5.11b). It may involve any number of sets of records, even a different set per stripe, to account for the intensity dependence of ground motion characteristics. Some scaling is needed to ensure that all records within a stripe have the same value of the IM. While this may initially seem as a weakness, especially if one subscribes to the view that even minor scaling should be avoided at all costs, it is also a great strength as it streamlines post-processing. Assuming that a large number of stripes have been employed, the distribution of EDP given IM is directly represented at each stripe

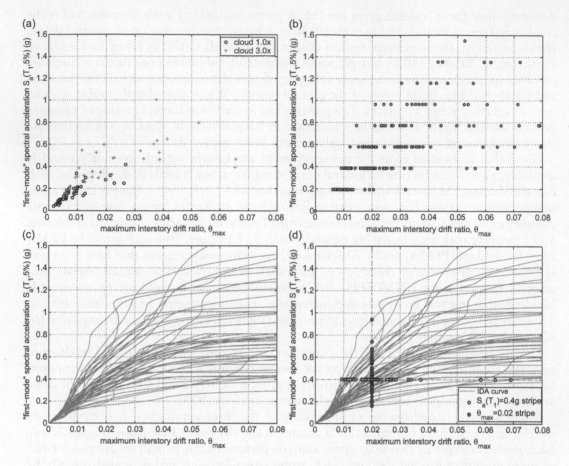

*Figure 5.11* Determination of the maximum interstory drift response for a 12-story reinforced concrete moment-resistant frame building through nonlinear response history analysis using 44 ground motions (FEMA, 2009): (a) two clouds of IM-EDP points formed by scaling each motion by 1.0 and 3.0; (b) eight stripes of IM-EDP points at constant steps of the IM, showing only one non-collapsing point at the highest level; (c) the 44 IDA curves derived by interpolating the stripes of IM-EDP points; (d) horizontal stripe of 44 points at $S_a(T_1) = 0.4g$ and vertical stripe at $\theta_{max} = 0.02$.

level by the pertinent EDP results extracted from the pertinent analyses. In other words, any statistical quantity of EDP given the IM (mean, 16/50/84 percentile, standard deviation, see Chapter 13) can be estimated directly from the corresponding EDP values without any need for regression, thus considerably simplifying post-processing. On the other hand, employing a large number of stripes is an expensive proposition. Figure 5.11b shows the results of 8 stripes times 44 records for a total of $8 \times 44 = 352$ nonlinear response history analyses, which can take anywhere from several hours to several days to perform depending on the complexity of the model and the computational power available. Thus, for reasons of economy, we often only perform analyses at one or at most two stripes near the IM level of interest, this for example being the code design level value of the IM. When two stripes are employed, in a so-called *double-stripe* analysis, we borrow a trick from cloud analysis and perform linear regression in logarithmic space to estimate the distribution of the EDP given the IM both within and outside the two stripes. When only one stripe is available, in a so-called *single-stripe* analysis, we instead employ a linear regression in linear coordinates with an intercept of zero, essentially accepting that the equal displacement rule holds (see Chapter 4). These two options form the mainstay of our approach to practical performance assessment, described in detail in Section 13.4.2.

IDA (Vamvatsikos and Cornell, 2002) employs a single set of ground motion records, each of which is scaled to several levels of intensity, typically until collapse is reached. Such analyses can be arranged in stripes of IM, offering practically identical results to a (multiple) stripe analysis. Still, it is always advantageous to employ nonmatching IM-levels algorithmically chosen to offer a fixed number of nonlinear dynamic analyses for each record (Section 4.8.6), thus achieving a more economical representation of response from elastic to global dynamic instability. Then, interpolation of the results pertaining to each record is employed to generate continuous IDA curves (Figure 5.11c), clearly showing the elastic response at lower intensities, the increase in variability as damage spreads through the structure, terminating in a horizontal flatline when global collapse is reached. These can be interpolated to provide vertical or horizontal stripes at arbitrary IM or EDP values (Figure 5.11d) and derive any statistics of interest (Vamvatsikos and Cornell, 2004).

In practice, any of the above approaches can be used to derive more-or-less equivalent results. The use of different approaches may become important only in the sense of scaling versus selecting ground motion records, as well as the availability of software tools to perform the often challenging post-processing required by each. Chapter 13 discusses in detail the use of IDA, single stripe and double stripe analysis results for performance assessment, while Chapter 19 offers a number of case studies with details on the applications of each approach. In the end, selecting the exact approach to use is a choice that depends on the desired accuracy, the capabilities and experience of the analyst, and his/her understanding of the problem at hand.

## REFERENCES

Adam, C., Kampenhuber, D., Ibarra, L. F., and Tsantaki, S. (2017). Optimal spectral acceleration-based intensity measure for seismic collapse assessment of P-delta vulnerable frame structures, *Journal of Earthquake Engineering*, 21(7):1189–1195.

Akkar, S., Sandıkkaya, M. A., Şenyurt, M., Azari Sisi, A., Ay, B. Ö., Traversa, P., Douglas, J., Cotton, F., Luzi, L., Hernandez, B., and Godey, S. (2014). Reference database for seismic ground-motion in Europe (RESORCE), *Bulletin of Earthquake Engineering*, 12(1):311–339.

Baker, J. W., and Cornell, C. A. (2006). Spectral shape, epsilon and record selection, *Earthquake Engineering & Structural Dynamics*, 35(9):1077–1095.

Bianchini, M., Diotallevi, P., and Baker, J. W. (2009). Prediction of inelastic structural response using an average of spectral accelerations, In *10th International Conference on Structural Safety and Reliability (ICOSSAR09)*, Osaka, Japan.

Bojórquez, E., and Iervolino, I. (2011). Spectral shape proxies and nonlinear structural response, *Soil Dynamics and Earthquake Engineering*, 31(7):996–1008.

Bradley, B. A. (2010). A generalized conditional intensity measure approach and holistic ground-motion selection, *Earthquake Engineering & Structural Dynamics*, 39(12):1321–1342.

Charney, F. A. (2008). Unintended consequences of modeling damping in structures, *Journal of Structural Engineering*, American Society of Civil Engineers, 134(4):581–592.

Chopra, A. K. (2017). *Dynamics of Structures*, 5th edition, Pearson, London.

Clough, R. W., and Penzien, J. (1993). *Dynamics of Structures*, 2nd edition, McGraw-Hill, New York, NY.

Cordova, P. P., Deierlein, G. G., Mehanny, S. S., and Cornell, C. A. (2000). Development of a two-parameter seismic intensity measure and probabilistic assessment procedure. In *2nd US–Japan Workshop on Performance-based Earthquake Engineering Methodology for RC Building Structures*, Sapporo, Hokkaido, 2000.

Eads, L., Miranda, E., and Lignos, D. G. (2015). Average spectral acceleration as an intensity measure for collapse risk assessment, *Earthquake Engineering & Structural Dynamics*, 44(12):2057–2073.

FEMA P695 (2009). Far field ground motion set. http://users.ntua.gr/divamva/resourcesRCbook/FEMA-P695-FFset.zip.

INNOSEIS (2017). Ground motion sets for moderate and high seismicity European sites. http://innoseis.ntua.gr/deliverables.php?deliverable=records.

Jalayer, F., and Cornell, C. A. (2009). Alternative non-linear demand estimation methods for probability-based seismic assessments, *Earthquake Engineering & Structural Dynamics*, 38(8):951–972.

Kazantzi, A. K., and Vamvatsikos, D. (2015). Intensity measure selection for vulnerability studies of building classes, *Earthquake Engineering and Structural Dynamics*, 44(15):2677–2694.

Kohrangi, M., Bazzurro, P., and Vamvatsikos, D. (2016). Vector and scalar IMs in structural response estimation: Part I - Hazard Analysis, *Earthquake Spectra*, 32(3):1507–1524.

Lin, T., Haselton, C. B., and Baker, J. W. (2013a). Conditional spectrum-based ground motion selection. Part I: Hazard consistency for risk-based assessments, *Earthquake Engineering & Structural Dynamics*, 42(12):1847–1865.

Lin, T., Haselton, C. B., and Baker, J. W. (2013b). Conditional spectrum-based ground motion selection. Part II: Intensity-based assessments and evaluation of alternative target spectra, *Earthquake Engineering & Structural Dynamics*, 42(12):1867–1884.

Luco, N., and Cornell, C. A. (2007). Structure-specific scalar intensity measures for near-source and ordinary earthquake ground motions, *Earthquake Spectra*, 23(2):357–392.

Mehanny, S. S. F. (2009). A broad-range power-law form scalar-based seismic intensity measure, *Engineering Structures*, 31(7):1354–1368.

PEER/ATC 72-1 (2010). Pacific Earthquake Engineering Research Center/Applied Technology Council, Modeling and Acceptance Criteria for Seismic Design and Analysis of Tall Buildings, Report Number PEER/ATC 72-1, Applied Technology Council, Redwood City, CA.

PEER NGA (2018). Pacific Earthquake Engineering Research Center—Next Generation Attentuation Relationships. https://peer.berkeley.edu/research/nga-west-2.

Shome, N., Cornell, C. A. (1998). Normalization and scaling accelerograms for nonlinear structural analysis. *Proceedings of the 6th US National Conference on Earthquake Engineering*, Seattle, WA.

Shome, N., Cornell, C. A., Bazzurro, P., and Carballo, J. E. (1998). Earthquakes, records, and non-linear responses, *Earthquake Spectra*, 14(3):469–500.

Vamvatsikos, D., and Cornell, C. A. (2002). Incremental dynamic analysis. *Earthquake Engineering and Structural Dynamics*, 31(3): 491–514.

Vamvatsikos, D., and Cornell, C. A. (2004). Applied incremental dynamic analysis. *Earthquake Spectra*, 20(2): 523–553.

Vamvatsikos, D., and Cornell, C. A. (2005). Developing efficient scalar and vector intensity measures for IDA capacity estimation by incorporating elastic spectral shape information, *Earthquake Engineering & Structural Dynamics*, 34:1573–1600.

Wikipedia (2017). QR algorithm. https://en.wikipedia.org/wiki/QR_algorithm. Last modified December 8, 2017.

Wikipedia (2019). Sparse matrix. https://en.wikipedia.org/wiki/Sparse_matrix. Last modified March 12, 2019.

# Chapter 6

# Characterization of dynamic response using Principal Components Analysis

## 6.1 PURPOSE AND OBJECTIVES

Principal Components Analysis (PCA) provides a simple approach to visualize the complicated dynamic response of a multi-degree-of-freedom system in terms of its principal modal responses. Where the masses are distributed uniformly over the height, the principal modes coincide with the elastic modes for linear response. For nonlinear response, the modes indicate the nature of the plastic mechanism that develops, providing a useful validation that the intended mechanism has developed. Extensions allow applications to nonuniform mass distributions. The theory and applications of PCA is presented in this chapter, as viewed from a structural dynamics perspective.

## 6.2 INTRODUCTION

PCA is a technique used in statistics for the analysis of multivariate data. When applied to dynamic response data, PCA determines an orthonormal set of basis vectors organized such that the first basis vector represents the predominant "mode" of response over time, the second represents the predominant mode of response with the first mode removed, and so on, such that the entire data set may be represented as a linear combination of the basis vectors (modes) over time. The method is applicable to time series data of all types (e.g., displacements, forces, etc.) that may be obtained from structures responding linearly or nonlinearly. Thus, PCA may be used to quickly characterize predominant response characteristics, providing rapid insight into behavior contained within the mountains of data that can be generated in nonlinear dynamic analysis.

When applied to the displacement response of linear systems that have uniform values of mass over the height of the structure, the PCA mode shapes (or principal components) coincide with the conventional mode shapes determined from mass and stiffness matrices (Section 5.2.3). For nonuniform mass distributions, the PCA mode shapes computed for linear elastic response are related to the linear elastic mode shapes, as described in Section 6.3. As structural response becomes increasingly nonlinear, the empirically determined PCA mode shapes generally will deviate from those obtained for linear elastic response, and will depend on properties of both the structure and the excitation. The PCA mode shapes are eigenvectors computed from the response data; the corresponding eigenvalues provide a quantitative measure of the degree to which the response is represented by each PCA mode shape. Thus, it is possible to identify the "first mode" of a structure responding nonlinearly and to quantify the degree to which response is represented by this mode.

## 6.3 THEORY

PCA was first developed by Pearson (1901) and was later independently developed by Hotelling (1933). This section summarizes the theoretical basis for the technique using the nomenclature of structural engineering; further information may be found in Aschheim et al. (2002).

Consider a vector of response data $\mathbf{v}$ that could represent, for example, the displacement relative to the ground at $n$ discrete locations of a multi-degree-of-freedom system at an instant of time. There are $t$ observations of the $n \times 1$ vector $\mathbf{v}$ over the time interval. The $t$ observations may consist of the entire response history or may be a subset of the response history. In a moving window analysis, the interval is advanced incrementally over the data set in a series of analyses, allowing changes in the principal components to be seen over the response history.

The orthonormal basis for the $n$-dimensional space that contains $\mathbf{v}$ is given by the identity matrix, $\mathbf{I}$. It is obvious that the displacement response, $\mathbf{v}$, at any instant of time represents a linear combination of the unit basis vectors (i.e., $\mathbf{v} = \mathbf{Iv}$).

The deviation of $\mathbf{v}$ from its mean over $t$ observations, $\bar{\mathbf{v}}$, may be expressed in terms of a new orthonormal basis, $\mathbf{B}$:

$$\mathbf{v} - \bar{\mathbf{v}} = \mathbf{Bu} \tag{6.1}$$

where $\mathbf{u}$ represents the displacements relative to their means in terms of the basis vectors contained in $\mathbf{B}$. Since $\mathbf{B}$ is orthonormal, $\mathbf{B}^T\mathbf{B} = \mathbf{I}$. Therefore, $\mathbf{B}^T = \mathbf{B}^{-1}$, and pre-multiplying Equation 6.1 by $\mathbf{B}^T$ gives:

$$\mathbf{u} = \mathbf{B}^T(\mathbf{v} - \bar{\mathbf{v}}) \tag{6.2}$$

The mean of $\mathbf{u}$ is the $n \times 1$ vector $\mathbf{0}$.

Let the covariance of the $n \times 1$ vector $\mathbf{v}$ be represented by the $n \times n$ covariance matrix $\mathbf{C_v}$. The $(i, j)$ element of $\mathbf{C_v}$ is the covariance between the displacements at the $i$th and $j$th degrees of freedom, $v_i$ and $v_j$, over the $t$ observations, given by:

$$\text{cov}(v_i, v_j) = \frac{1}{t}\sum_{k=1}^{t}(v_{i,k} - \bar{v}_i)(v_{j,k} - \bar{v}_j) \tag{6.3}$$

where $\bar{v}_i$ and $\bar{v}_j$ are the means of $v_i$ and $v_j$, respectively, over the $t$ observations. Because $\text{cov}(v_i, v_j)$ is computed using real-valued data, $\mathbf{C_v}$ is real, and because $\text{cov}(v_i, v_j) = \text{cov}(v_j, v_i)$, $\mathbf{C_v}$ is symmetric.

Because $\mathbf{C_v}$ is real and symmetric, it can be diagonalized by an orthogonal matrix $\mathbf{\Phi}$:

$$\mathbf{\Lambda} = \mathbf{\Phi}^T\mathbf{C_v}\mathbf{\Phi} \tag{6.4}$$

where $\mathbf{\Phi}$ consists of the orthonormal eigenvectors of $\mathbf{C_v}$ and $\mathbf{\Lambda}$ is a diagonal matrix containing the eigenvalues $\lambda_i$ of $\mathbf{C_v}$. By convention, the eigenvectors of $\mathbf{C_v}$ are arranged in sequence so that their corresponding eigenvalues are in descending order ($\lambda_1 \geq \lambda_2 \geq \ldots \lambda_n$).

By operating with standard identities on Equation 6.2, the covariance of $\mathbf{u}$, $\mathbf{C_u}$, can be expressed in terms of $\mathbf{C_v}$:

$$\mathbf{C_u} = \mathbf{B}^T\mathbf{C_v}\mathbf{B} \tag{6.5}$$

The parallel structures of Equations 6.4 and 6.5 indicate that $C_u$ will be a diagonal matrix consisting of the eigenvalues of $C_v$ if the new basis $B$ is selected to be the set of orthonormal eigenvectors of $C_v$.

Because $C_u$ is diagonal, $\text{cov}(u_i, u_j) = 0$ for $i \neq j$, leading to the result that the displacements $u_i$ and $u_j$ (expressed in terms of the orthonormal basis $B = \Phi$) are uncorrelated for $i \neq j$. Similarly, each displacement $u_i$ has variance $\text{var}(u_i) = \text{cov}(u_i, u_i) = \lambda_i$.

The orthonormal eigenvectors of $C_v$ ($= B = \Phi$) are the principal components (or PCA mode shapes) of the displacement response $v$. This means that $\phi_1$ is oriented so that it maximizes the variance (i.e., $\text{var}(u_i)$ is a maximum), $\phi_2$ maximizes the variance with $\phi_1$ removed, and so on). Also, the variance in the displacement response "explained" by each principal component is proportional to the eigenvalue associated with the principal component, $\lambda_i$.

The PCA mode shapes are found empirically to correspond to the elastic mode shapes for structures with uniformly distributed mass responding within their linear elastic domain. To address the case of nonuniform mass, the free vibration eigenproblem of elastic dynamics, given by $K\phi_{el} = \lambda M\phi_{el}$, may be transformed to a standard form (Bathe, 1982) by introducing a Cholesky factorization of the mass matrix given by $M = LL^T$. The entries of $L$, $L_{ij}$, are zero for $i > j$. This factorization allows the elastic dynamics eigenproblem to be restated as $\tilde{K}\tilde{\phi} = \lambda\tilde{\phi}$, where $\tilde{K} = L^{-1}KL^{-T}$. The transformed elastic mode shapes, given by $\tilde{\phi} = L^T\phi_{el}$, are orthogonal to one another while satisfying the orthogonality relationships required of the elastic mode shapes:

$$\tilde{\phi}_i^T \tilde{\phi}_j = \phi_{el,i}^T LL^T \phi_{el,j} = \phi_{el,i}^T M \phi_{el,j} = \delta_{ij} \tag{6.6}$$

where $\delta_{ij} = 1$ for $i = j$, and $\delta_{ij} = 0$ otherwise.

Because the elastic modal responses are independent and, in general, are uncorrelated with one another, the displacements $\tilde{u}_i$ and $\tilde{u}_j$ expressed in terms of the transformed mode shapes $L^T\Phi_{el}$ will be uncorrelated (for closely spaced elastic response data of sufficient sample size). Therefore, the off-diagonal terms in $C_{\tilde{u}}$ will tend to zero as $t$ increases. If the principal components are determined for transformed displacements $\tilde{v} = L^T v$ then $C_{\tilde{u}}$ will be a diagonal matrix, provided that the displacements $\tilde{u}$ are expressed in terms of the set of orthonormal basis vectors that are the eigenvectors of $C_{\tilde{v}}$. The PCA mode shapes obtained for $L^T v$ exhibit the orthogonality required of the elastic mode shapes (Equation 5.17), and, in the limit, coincide with the transformed elastic mode shapes given by $L^T\Phi_{el}$.

Two special cases are of interest:

1. If $M$ is diagonal with uniform masses, $M = mI = LL^T$, requiring that $L = L^T = \sqrt{m}I$. Because this reduces $L^T v$ and $L^T\Phi_{el}$ to scalar operations on $v$ and $\Phi_{el}$, respectively, the PCA mode shapes obtained on $v$ will coincide with the elastic mode shapes after normalization.
2. If $M$ is diagonal with entries $m_i$, $M = LL^T$ requires that $L = L^T$ with the $i$th diagonal entry of $L$ given by $\sqrt{m_i}$. In this case, the PCA mode shapes obtained on $L^T v$ will coincide with the transformed elastic mode shapes given by $L^T\Phi_{el}$. Thus, the elastic mode shapes are given by $\Phi_{el} = \left(L^T\right)^{-1}\tilde{\Phi} = L^{-1}\tilde{\Phi}$, provided that sufficient response data $v$ is available and response is linear.

## 6.4 APPLICATION TO DISPLACEMENT RESPONSE

To illustrate the use of PCA, the eight-story moment-resistant frame whose capacity curve is plotted in Figure 7.4 was subjected to the Corralitos record of the 1989 Loma Prieta

Earthquake. Lumped masses of 10 tons were applied at each level, resulting in a first-mode period of 0.639 s based on uncracked properties. Viscous damping of 3% was applied, based on the current (tangent) stiffness matrix. The Corralitos record was applied at three intensities: 1% of recorded intensity, corresponding to uncracked (linear elastic) response, 100% of recorded intensity, and 300% of recorded intensity.

The principal components of the displacement response were computed, in this case using the free shareware statistical package "R." (See box for command syntax). Elastic and PCA mode shapes are compared for different record intensities in Figure 6.1. The proportion of variance represented by each of the first three PCA modes is given in Table 6.1.

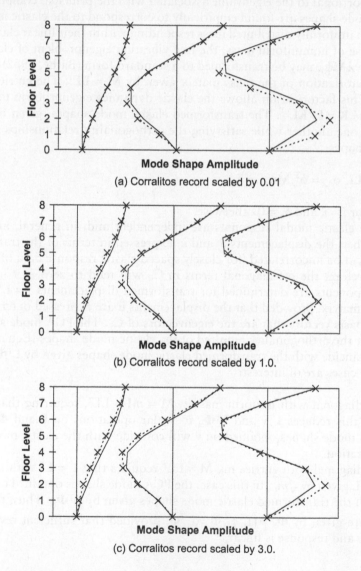

(a) Corralitos record scaled by 0.01

(b) Corralitos record scaled by 1.0.

(c) Corralitos record scaled by 3.0.

*Figure 6.1* Elastic and PCA mode shapes for different intensities of response, for an eight-story reinforced concrete moment-resistant frame. Solid lines indicate the first, second, and third elastic mode shapes as determined for initial, uncracked properties. Dashed lines show the first, second, and third PCA mode shapes. (a) Low-amplitude elastic response, (b) Corralitos record, and (c) Corralitos record scaled by 3.0.

*Table 6.1* Variance in displacement response represented by first three PCA modes for Corralitos record

| | Scale factor applied to Corralitos record | | | | | |
|---|---|---|---|---|---|---|
| | 0.01 | | 1.0 | | 3.0 | |
| PCA mode | Proportion of variance (%) | Cumulative proportion of variance (%) | Proportion of variance | Cumulative proportion of variance (%) | Proportion of variance | Cumulative proportion of variance (%) |
| 1 | 99.9714 | 99.9714 | 99.7813 | 99.7813 | 99.1374 | 99.1374 |
| 2 | 0.0283 | 99.9997 | 0.2146 | 99.9959 | 0.8459 | 99.9833 |
| 3 | 0.0003 | 100.0000 | 0.0037 | 99.9996 | 0.0146 | 99.9979 |

The scale factors of 0.01, 1, and 3 applied to the record resulted in peak roof displacements of 0.0038, 0.2468, and 0.8705 m, respectively. The associated peak roof drift ratios (0.016, 1.03, and 3.63%) suggest uncracked elastic behavior, cracked elastic behavior, and a ductility demand of about 2, based on comparison with Figure 7.4

## BOX 6.1    COMPUTING PRINCIPAL COMPONENTS OF RESPONSE USING R

1. Obtain and install the R software program (shareware).
2. Prepare a text file containing the displacement response history at the nodes of interest. It is suggested that the top of each column of data be labeled with the node ID, the first column consist of the time at which the displacements occurred, and the second and subsequent columns consist of the displacements at each floor, resulting in a matrix of dimension $t+1$ by $n+1$.

   If the analyses were done using Seismostruct, such a data table can be obtained using the global response parameters/structural displacements tab of the post-processor, selecting one node per floor and viewing the data values. A fixed node should be selected as the base node. This data can be selected (Control-A) and copied (Control-C) to a new text file in Notepad. This file should be saved as a *.txt file for subsequent use.

3. Run R. Within the R graphical user interface environment:
   a. Change the directory to that containing the displacement response history text file, using the menu commands File-Change dir.
   b. Import the displacement response history data using the command

```
> v<-as.matrix(read.table("file.txt", header=TRUE, row.names=1))
```

   Note that for data that lacks headers or time data, the values specified for "header" and "row.names" may need to be altered.
   c. Where nonuniform masses are present, define the diagonal matrix L containing the square roots of the floor masses:

```
> L<-diag(c(m_1,m_2,m_3 ... m_n)^0.5,n,n)
```

   where $m_i$ = mass of floor i

d. Determine the principal components (eigenvectors) and variances (eigenvalues). For uniform masses, use the command

```
> eigen(cov(v))
```

For nonuniform masses, use

```
> eigen(cov(v%*%L))
```

Note: if v has dimension *n* by *t*, use `t(v)` in place of `v` and `t(v)%*%L` in place of `v%*%L` above.

4. For those unfamiliar with R, it is suggested that the results displayed on screen be copied and pasted into a spreadsheet program for subsequent manipulation (e.g., normalizing the mode shapes to unit value at the roof) and plotting. The proportion of response represented by the *i*th mode is given by $\lambda_i / \sum \lambda$.

Figure 6.2 illustrates that the first PCA mode shape is effectively identical to the first elastic mode shape for uncracked (linear elastic) response. The small deviation of the higher mode PCA shapes from the elastic mode shapes is attributed to the very poor representation of higher modes in the response (Table 6.1). At higher record intensities, nonlinearity in member response leads the PCA mode shapes to diverge somewhat from the uncracked elastic mode shapes. Nevertheless, the first PCA mode shape still represents over 99% of the variance in the displacement response.

It should be noted that the eight-story frame was designed to have strong column weak beam behavior. For frames that develop story sway mechanisms, the variance represented by the first PCA mode tends to be lower, and the PCA mode shapes reflect the sway mechanism. To illustrate this, the reinforcement of the lowest story columns of the eight-story frame was reduced to induce weak story behavior, clearly evident in the PCA mode shapes of Figure 6.2. The proportion of variance in the displacement response represented by the first PCA mode was reduced to 97.9%.

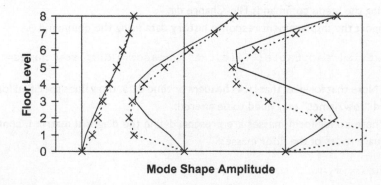

**Mode Shape Amplitude**

*Figure 6.2* Elastic and PCA mode shapes for weak-story behavior. Solid lines indicate the first, second, and third elastic mode shapes as determined for initial, uncracked properties. Dashed lines show the first, second, and third PCA mode shapes.

## 6.5 PCA MODE SHAPES OF VARIOUS RESPONSE QUANTITIES

This section compares PCA mode shapes computed for various response quantities, using a reinforced concrete moment-resistant frame and a wall as examples. Response quantities considered are displacements, interstory drifts, story overturning moments, story shears, and the floor level forces required to equilibrate the story shears in a static analysis.

Both the moment frame and wall models are six stories in height; story heights are 3.66 m (12 ft) on center. Discrete masses (lumped at the floor levels) were uniform in value such that the first-, second-, and third-mode periods of the uncracked structures were equal to 0.823 s, 0.268 s, and 0.125 s for the frame and 0.524 s, 0.083 s, and 0.029 s for the wall, respectively. Both the frame and wall structure were driven to significant levels of inelastic response. The structures were modeled in DRAIN-2DX (Prakash et al., 1993) using a fiber element (Element 15). Results of interest were collected using scripts developed using the AWK programming language. PCA was done using R.

The moment frame and shear wall models were subjected to N-S component of the Takatori record from the 1995 Hyogo-ken Nambu (Kobe) Earthquake, resulting in peak roof drift ratios of 2.55% and 2.34% (559 mm and 513 mm), respectively. This corresponds to displacement ductilities of about 3.4 and 6.5, respectively, relative to the yield displacements observed in a first-mode pushover analysis, indicating that each system experienced significant inelastic response.

Figure 6.3 compares the first three PCA mode shapes and Table 6.2 compares the proportion of variance represented by each PCA mode for different response quantities. It can be observed that a very large proportion of the variance in displacement response is represented by the first PCA mode, and that the proportion of variance represented by the first mode generally decreases in subsequent rows of the table. Note that story shears for the shear wall have much greater higher mode contributions than in the moment frame, while the interstory drifts for the moment frame have much greater higher mode contributions than for the shear wall.

At any instant in time, any response quantity of interest can be represented as a linear combination of the PCA mode shapes (determined for that response quantity):

$$\mathbf{v}(t) = \mathbf{\Phi}\mathbf{u}(t) + \overline{\mathbf{v}} = \sum_{i=1}^{n} \phi_i u_i(t) + \overline{\mathbf{v}} \tag{6.7}$$

Because the PCA modes are an orthonormal basis for the space, selected such that the first PCA mode is the best fit to the response quantity (over time) and successive modes are the best fits to the data with the preceding modal contributions removed, the response quantity over time can be approximated using just one or several of the mode shapes:

$$\mathbf{v}(t) \approx \sum_{i=1}^{j<n} \phi_i u_i(t) + \overline{\mathbf{v}} \tag{6.8}$$

where $j$ is selected based on the desired accuracy and the proportion of variance represented by the modes included in the summation. Note that the mean of the response quantity, $\overline{\mathbf{v}}$, is often close to 0 for elastic response and even for moderate levels of inelastic response. Clearly, a good approximation for the displacement response can be obtained for $j = 1$, while at the other extreme, good estimates of floor forces may require $j = 2$ or 3. Physically, this signifies that higher mode vibrations have very small displacement amplitudes and comparatively larger force amplitudes, relative to the amplitudes of the first-mode vibrations.

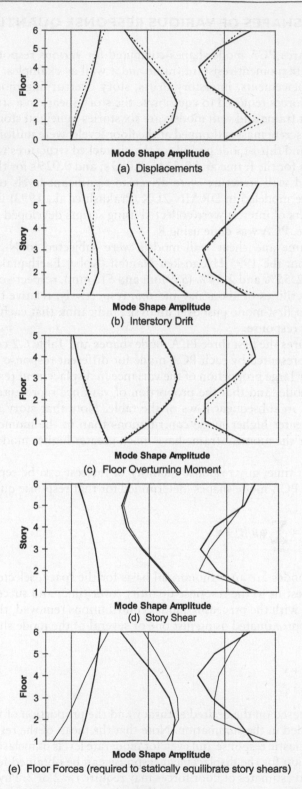

**Mode Shape Amplitude**

(a)  Displacements

**Mode Shape Amplitude**

(b)  Interstory Drift

**Mode Shape Amplitude**

(c)  Floor Overturning Moment

**Mode Shape Amplitude**

(d)  Story Shear

**Mode Shape Amplitude**

(e)  Floor Forces (required to statically equilibrate story shears)

*Figure 6.3* PCA mode shapes (first, second, and third) determined for six-story moment frame (solid line) and six-story shear wall (dashed line). Mode shapes are normalized to unit length.

Table 6.2 Variance represented by first three PCA modes for six-story frame and wall structures subjected to the N-S Takatori record of the 1995 Kobe (Hyogo-ken Nambu) Earthquake

| Response quantity | Mode | Moment frame (%) | Wall (%) |
|---|---|---|---|
| Floor displacements | first | 99.4957 | 99.931 |
| | second | 0.4644 | 0.067 |
| | third | 0.0369 | 0.002 |
| Interstory drifts | first | 95.314 | 99.615 |
| | second | 3.826 | 0.352 |
| | third | 0.750 | 0.031 |
| Floor overturning moment | first | 99.579 | 98.165 |
| | second | 0.404 | 1.788 |
| | third | 0.016 | 0.041 |
| Story shear | first | 98.038 | 90.025 |
| | second | 1.747 | 9.032 |
| | third | 0.178 | 0.779 |
| Floor forces (required to statically equilibrate story shears) | first | 82.480 | 53.393 |
| | second | 12.859 | 32.674 |
| | third | 3.346 | 10.907 |

The similarity of the PCA mode shapes for the moment frame and wall models (Figure 6.3) for both floor overturning moment and story shear is intriguing, and suggests the possibility that good estimates of these and possibly other quantities might be obtained using a simplified analytical approach.

## 6.6 MODAL INTERACTIONS

Equation 6.8 represents displacement response very efficiently using a reduced basis consisting of the first $j$ PCA mode shapes. Recognizing the orthonormality of these mode shapes $\left( \phi_i^T \phi_j = \phi_j^T \phi_i = \delta_{ij} \right)$, modal amplitudes $u_i(t)$ can be determined by premultiplying Equation 6.7 by: $\phi_j^T$

$$u_i(t) = \phi_i^T \left( v(t) - \bar{v} \right) \qquad (6.9)$$

Given displacement response data $v(t)$ and the first two PCA mode shapes computed for this data results in the two-mode estimate:

$$v(t) = \phi_1^T u_1(t) + \phi_2^T u_2(t) + \bar{v} \qquad (6.10)$$

Interaction between the modes can be seen graphically by plotting $u_1$ and $u_2$ as a parametric function of $t$. This interaction is plotted as the solid curvilinear trace in each panel of Figure 6.4. The first panel shows low amplitude (linear elastic) response. Inelastic response corresponding to beam-hinging behavior or the development of a weak story mechanism in the first or sixth story is shown in the remaining panels. Elastic properties of all frames were identical. Differences in the curvilinear traces indicate clearly that modal interaction varies for the different systems.

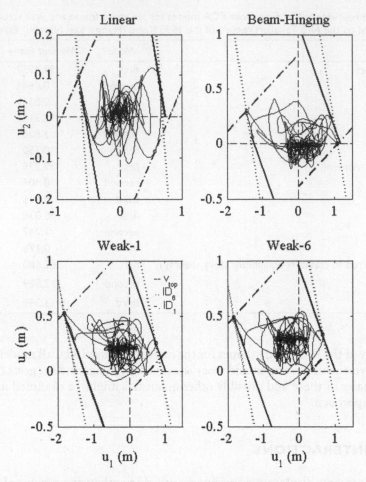

*Figure 6.4* Interaction of displacement amplitudes $u_1$ and $u_2$ for the first two principal components of the displacement response for variants of a 12-story steel moment-resistant frame. Roof displacement (solid line), interstory drift at the sixth floor (dashed line), and interstory drift at the first floor (dash-dot line). (From Cuesta and Aschheim, 2002.)

These modal interaction plots can be used directly to evaluate peak displacements at any floor level as well as interstory drifts, based on the first two modes. Mathematical extensions can be used to consider interactions among three or more modes.

For two modes, the displacement of floor $j$ is estimated as $v_j(t) = \phi_{j,1}(t)u_1(t) + \phi_{j,2}u_2(t)$, where $\phi_{j,i}$ is the amplitude of the $i$th mode at the $j$th floor level. Therefore, $v_j$ is a straight line in the space $u_1$ versus $u_2$. The slope of this line, $\partial u_2/\partial u_1 = -\phi_{i,1}/\phi_{i,2}$ is a function of terms associated with the mode shapes alone. The $u_1$ and $u_2$ intercepts of such a line indicate the accompanying value of the floor displacement $v_j$, which is a constant value for any specific line drawn with the proper slope. Thus, the maximum observed value of $v_j$ is given by the line farthest from the origin that just intersects the domain of interaction between $u_1(t)$ and $u_2(t)$. Displacements at different floors are indicated by lines having different slopes; therefore, maximum floor displacements intersect the interaction "surface" at different points, as shown in Figure 6.4. It is clear from the figure that different floors reach their maximum displacements at different times, and that some floor displacements have greater sensitivity to $u_2$ while others depend almost solely on $u_1$.

An estimate of interstory drift for the $j$th story based on the contributions of two modes is

$$IDI_j(t) = \gamma_{j,1}u_1(t) + \gamma_{j,2}u_2(t) \tag{6.11}$$

where the contribution of the $i$th mode to drift in the $j$th story is determined from the PCA displacement mode shapes as $\gamma_{j,i} = \phi_{j,i} - \phi_{j-1,i}$. The parallel structures of Equations 6.10 and 6.11 indicate that the expression for constant interstory drift plots as a line having slope $-\gamma_{j,1}/\gamma_{j,2}$ on the modal interaction plots of Figure 6.4: the slope is determined by properties of the mode shapes alone. The largest interstory drift that occurs for a given excitation can be found by maximizing the function over the domain of interaction between $u_1$ and $u_2$.

A *priori* estimates of floor displacements and interstory drifts require estimation of the bounding interaction surface, as well as estimates of the mode shapes and peak modal response amplitudes. It is not clear whether the bounding surface is best represented by an ellipse (equivalent to a square root of the sum of squares combination), a rectangle (equivalent to absolute sums), or another shape. Similar questions could be investigated for the PCA mode shapes determined for other response quantities.

## 6.7 COMPARISON OF ELASTIC AND PCA MODE SHAPES

This section investigates similarities between PCA mode shapes and the elastic mode shapes and quantities derived from these shapes as a function of response intensity and structural system type. PCA mode shapes were computed for the moment frame and wall models of Section 6.5. In this case, the Takatori record was used, scaled by 0.20 and 1.00. For the moment frame, this resulted in peak roof drift ratios of 0.307% and 2.55%, corresponding to displacement ductility demands of about 0.41 and 3.4 (based on the yield displacements observed in a first-mode pushover analysis), respectively. For the wall, this resulted in peak roof drift ratios of 0.326% and 2.34%, corresponding to displacement ductility demands of about 0.41 and 2.91, respectively. These responses correspond to significant cracking in the case of the 0.20 scale factor and significant yielding in the case of the 1.00 scale factor.

A *priori* estimates of the height-wise distribution of various response quantities were determined on the basis of the uncracked elastic mode shapes as follows: interstory drifts for each mode shape were determined by subtracting modal amplitudes for the floors immediately above and below the story of interest; floor forces were calculated as proportional to the floor mass and modal amplitude at the floor of interest, story shears were calculated as the sum of the floor forces above the story of interest, and overturning moments at a floor were calculated based on the floor forces above.

The PCA mode shapes determined for the Takatori record scaled by 0.20 and 1.00 are compared to the *a priori* estimates in Figures 6.5 (for the moment frame) and 6.6 (for the shear wall). The influence of cracking and yielding on the PCA mode shapes is plainly evident. The first PCA mode shapes for story shear and overturning moment, and to a lesser extent, floor displacement, are close to the *a priori* estimates for both structural systems. However, first-mode estimates of shear in typical shear walls are inadequate due to the significant presence of the second and higher modes (see Table 6.2), which are not estimated as well *a priori*. And while reasonable estimates of floor overturning moments in moment-resistant frames are shown here for a regular moment frame that develops a beam-hinging mechanism, substantial departures occurred where weak story mechanisms developed (FEMA-440, 2005).

Haselton and Deierlein (2007), in the context of the ATC-63 project, showed that very different mechanisms can form for particular distributions of strength and stiffness (as a result of higher mode contributions to local demands) during dynamic response—current

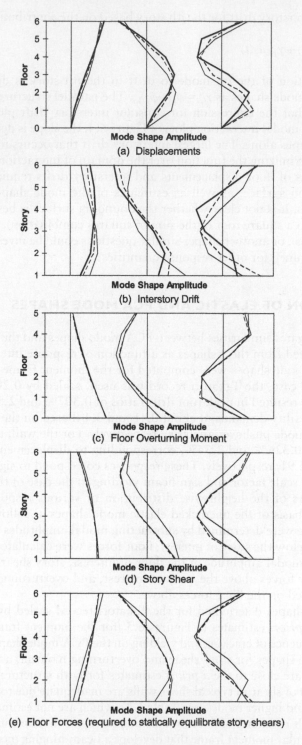

*Figure 6.5* Mode shapes (first, second, and third) determined for six-story moment frame: those derived from elastic mode shapes (solid line) and from PCA of response data obtained from nonlinear response history analysis to the Takatori record are shown (dashed lines). Results for the Takatori record scaled by 0.2 are shown with long dashes while those for a scale factor of 1.0 are shown by short dashes. Mode shapes are normalized to unit length.

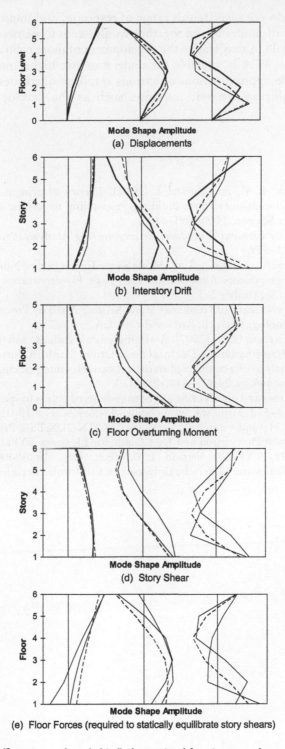

(a) Displacements

(b) Interstory Drift

(c) Floor Overturning Moment

(d) Story Shear

(e) Floor Forces (required to statically equilibrate story shears)

*Figure 6.6* Mode shapes (first, second, and third) determined for six-story shear wall: those derived from elastic mode shapes (solid line) and from PCA of response data obtained from nonlinear response history analysis to the Takatori record are shown (dashed lines). Results for the Takatori record scaled by 0.2 are shown with long dashes while those for a scale factor of 1.0 are shown by short dashes. Mode shapes are normalized to unit length.

pushover procedures do not identify this range of response, and improvements considering various combinations of modes acting together would seem to be necessary. We hope that the insight afforded by PCA may lead to the development of more robust methods of analysis for design. At present, PCA is valuable for understanding higher mode interactions, the relevance of first-mode approximations of various response quantities, and for identifying important behavior during dynamic response, such as the development of weak story mechanisms.

## REFERENCES

Aschheim, M. A., Black, E. F., and Cuesta, I. (2002). Theory of principal components analysis and applications to multistory frame buildings responding to seismic excitation, *Engineering Structures*, Elsevier Science, 24(8):1091–1103.

Bathe, K. J. (1982). *Finite Element Procedures in Engineering Analysis*, Prentice-Hall, Englewood Cliffs (NJ), pp. 573–579.

Cuesta, I., and Aschheim, M. A. (2002). Time-Mode Shape Interaction of Nonlinear MDOF Systems Using Principal Components Analysis, *5th European Conference on Structural Dynamics*, Munich, Germany, September 2–5.

FEMA 440 (2005). Improvement of Nonlinear Static Seismic Analysis Procedures, Report FEMA–440, Applied Technology Council, Redwood City, CA.

Haselton, C. B., and Deierlein, G. G. (2007). Assessing seismic collapse safety of modern reinforced concrete moment frame buildings. *Doctoral dissertation*, Stanford University.

Hotelling, H. (1933). Analysis of a complex of statistical variables into principal components, *Journal of Educational Psychology*, 24:417–441, 498–520.

Pearson, K. (1901). On lines and planes of closest fit to systems of points in space (PDF), *Philosophical Magazine*, 2(6):559–572. http://stat.smmu.edu.cn/history/pearson1901.pdf.

Prakash, V., Powell, G. H., and Campbell, S. (1993). DRAIN-2DX Base Program Description and User Guide - Element Description and User Guide for Elements TYPE01, TYPE02, TYPE04, TYPE06, TYPE09, TYPE15- Version 1.10, Tech. Rep. UCB/SEMM-93/18, Structural Engineering Mechanics and Materials, University of California, Berkeley, December 1993.

# Chapter 7

# Equivalent SDOF systems and nonlinear static (pushover) analysis

## 7.1 PURPOSE AND OBJECTIVES

Experience indicates that good approximations of displacement response can be obtained using so-called "equivalent" single-degree-of-freedom (ESDOF) representations of multistory systems. For this reason, ESDOF systems are relied upon for the displacement-based design approaches described in this book. Subsequent chapters illustrate the use of an ESDOF system in conjunction with an estimate of the yield displacement of the multi-degree-of-freedom (MDOF) system to establish the lateral strength required to limit peak displacement and system ductility demands to acceptable values.

## 7.2 INTRODUCTION

Chapter 6 on Principal Components Analysis (PCA) illustrates the high proportion of variance represented by the first PCA mode for displacement response (greater than 99% in the examples of Chapter 6) and the similarity of this mode shape (given by $\mathbf{L}^{-T}\boldsymbol{\phi}_{PCA,1}$ for nonuniform mass) to the first elastic mode shape. This chapter provides the theoretical basis for establishing an ESDOF approximation of the MDOF system, illustrates its use in estimating displacement response using nonlinear static (pushover) analysis, presents an energy-based formulation of pushover analysis that is particularly useful for representing higher modes, and concludes with a discussion of the difficulties of estimating other response quantities.

Note that the ESDOF system is not mathematically equivalent to the MDOF model. While the adjectives "approximate," "analogous," "surrogate," or perhaps even "index" may be a more accurate description of the ESDOF system, the term "equivalent" is retained here for consistency with common usage. Unfortunately, the reference to equivalency may imply to some that ESDOF systems can be used to recover all response quantities of interest, which is a false proposition.

## 7.3 THEORETICAL DERIVATION OF CONVENTIONAL ESDOF SYSTEM

The derivation of the ESDOF system presumes that a single mode (or shape vector), $\boldsymbol{\phi}_i$, can be used to characterize the response of the multistory building. For generality, the ESDOF system is derived for the $i$th mode, which can be any normalized displacement profile. Approximating the displaced shape to be proportional to $\boldsymbol{\phi}_i$, we can express the lateral displacements of the floors relative to the ground as

$$\mathbf{u}(t) = \phi_i u_{\text{roof}}(t) \tag{7.1}$$

where $\phi_i$ is normalized so that the value at the roof level equals 1 (unity). A comparison to Equation 5.19 affirms that only a single mode is considered in the ESDOF formulation (Equation 7.1).

Equation 5.9, representing the equation of motion of a multistory building, is repeated below. For simplicity, only the lateral degrees of freedom in a two-dimensional plane are represented.

$$\mathbf{M\ddot{u}}(t) + \mathbf{C\dot{u}}(t) + \mathbf{Q}(t) = -\mathbf{M1}\ddot{u}_g(t) \tag{5.9}$$

Terms in Equation 5.9 are defined conventionally, with $\mathbf{M}$ = diagonal matrix representing discrete masses lumped at each floor of the building, $\mathbf{C}$ = damping matrix of the building system, $\mathbf{Q}(t)$ = a vector representing the forces resisting relative displacements at the current instant of response (which simplifies to $\mathbf{Ku}(t)$ in the case of linear elastic response), $\mathbf{u}(t)$ = vector of displacements at the floor levels relative to the displacement at the base of the structure, $\mathbf{\ddot{u}}(t)$ = vector of lateral accelerations of the floors relative to the base of the structure, $\mathbf{1}$ = an $n \times 1$ vector having each component equal to 1, and $\ddot{u}_g(t)$ = acceleration of the ground at the base of the structure.

Substituting Equation 7.1 in Equation 5.9 yields:

$$\mathbf{M}\phi_i \ddot{u}_{\text{roof}} + \mathbf{C}\phi_i \dot{u}_{\text{roof}} + \mathbf{Q} = -\mathbf{M1}\ddot{u}_g \tag{7.2}$$

In the following, the equation of motion is specialized to represent the response of the ESDOF system. Parameters of the ESDOF system are indicated with an asterisk. Defining the displacement of the ESDOF system, $D^*$, as

$$D^*(t) \equiv \frac{u_{\text{roof}}(t)}{\Gamma_i} \tag{7.3}$$

allows the displacement response to be represented as

$$\mathbf{u}(t) = \phi_i u_{\text{roof}}(t) = \phi_i \Gamma_i D^*(t) \tag{7.4}$$

where $\Gamma_i$ is the modal participation factor for the $i$th mode, defined previously in Equation 5.20 and repeated here for ease of reference:

$$\Gamma_i = \frac{\phi_i^{\mathrm{T}} \mathbf{M1}}{\phi_i^{\mathrm{T}} \mathbf{M}\phi_i} \tag{5.20}$$

By pre-multiplying Equation 7.2 by $\phi_i^{\mathrm{T}}$ and substituting for $u_{\text{roof}}(t)$ using Equation 7.3, the following differential equation of motion of the ESDOF system is obtained:

$$\ddot{D}^*(t) + \xi_i^* \dot{D}^*(t) + Q_i^*(t) = -\ddot{u}_g(t) \tag{7.5}$$

with new terms defined as:

$$\xi_i^* = \frac{\phi_i^{\mathrm{T}} \mathbf{C}\phi_i}{\phi_i^{\mathrm{T}} \mathbf{M}\phi_i} \tag{7.6}$$

$$Q_i^*(t) = \frac{\phi_i^T Q(t)}{\Gamma_i \phi_i^T M \phi_i} \tag{7.7}$$

In the preceding, $\phi_i$ is a displacement profile (not necessarily an elastic mode shape), normalized to have unit amplitude at the roof level.

Comparison with Equation 3.28 or Equation 4.1 shows that Equation 7.5 represents the response of a single degree-of-freedom (SDOF) system. For linear elastic response, the peak response, $D_{max}^*$ is given by $S_d(T^*)$. For nonlinear response, the response to specific ground motions can be computed, or appropriate $R-\mu-T$ relationships can be used to estimate peak displacement response given knowledge of $S_d(T^*)$, as discussed in Chapter 4. Doing so requires evaluation of $Q_i^*$, which is explained in Section 7.4.

Concepts of effective modal mass and effective height, defined in Sections 5.2.10 and 5.2.11, respectively, also have relevance for ESDOF systems. An asterisk is used for the ESDOF system to emphasize the possibility of nonlinear behavior. Thus, with reference to Equations 5.32 and 5.38, we have for the MDOF system:

$$M_i^* = \Gamma_i \phi_i^T M 1$$

$$h_i^* = \frac{h^T s_i}{1^T s_i} \tag{7.8}$$

where $s_i$ and $h$ are defined in Sections 5.2.7 and 5.2.11, respectively, and $\phi_i$ need not be an elastic mode shape.

Figure 7.1 illustrates forces acting on the MDOF system in one mode, the location of the resultant of these forces at $h_i^*$, and the mass $M_i^*$ associated with response of the MDOF system in this mode. Lumping the mass $M_i^*$ at height $h_i^*$ maintains the correct overturning moment at the base of the structure. The displacement at $h_i^*$ is $D^*$, while the roof displacement is given as the product of $\Gamma_i$ and $D^*$ (Equation 7.3).

## 7.4 NONLINEAR STATIC (PUSHOVER) ANALYSIS

Nonlinear static analysis, also known as "pushover" analysis, is used to establish the load–displacement response of the system as it undergoes nonlinear response under a prescribed distribution of lateral forces. Pushover analysis is used both to characterize the nonlinear response of the system and to determine $Q_i^*$ in the equation of motion of the ESDOF system.

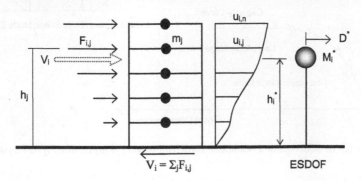

Figure 7.1  Forces acting in the $i$th mode on the MDOF structure—the resultant lateral force $V_i$ acts on mass $M_i^*$ at height $h_i^*$.

In the analysis, lateral forces $F_{ij}$ are applied at each floor level in proportion to $m_j\phi_{ij}$ (in vector form, $\mathbf{F} \propto \mathbf{M}\phi_i$). The nonlinear solution is obtained in a step-by-step displacement-controlled analysis, for which the increment in $\mathbf{F}$ at each step of the analysis is determined to cause a desired increment in the displacement of a designated control node. Results of the analysis are usually displayed as a plot of base shear versus roof displacement. The resulting curve is termed a "capacity curve," which typically is curvilinear (or piecewise linear), as illustrated in Figure 7.2.

If $\phi_i$ is selected to be the $i$th elastic mode shape, this force pattern results in displacements proportional to $\phi_i$ within the linear domain (as shown in Section 5.2.7). For $i = 1$, the displacement pattern typically will not deviate greatly from the first-mode pattern, even as nonlinearity develops.[1] While the force pattern used in a first-mode pushover analysis may be exactly that required to equilibrate a first-mode displacement pattern, some departure from the first-mode pattern may be acceptable, such as when using a displacement pattern representative of the displacement profile that develops during nonlinear response (e.g., one possibility would be to use $\mathbf{L}^{-T}\phi_{PCA,1}$). Since orthogonality is not required in the derivation of the ESDOF system (because only one mode is used—Equation 7.4), where the pushover forces are not proportional to $\mathbf{M}\phi_i$, other modes are present and influence the forces and deformations of the system within the elastic domain. (Of course, even where the pushover forces are proportional to $\mathbf{M}\phi_i$, the onset of nonlinearity changes $\mathbf{K}$ and hence $\phi_i$, indicating that other modes enter into the pushover analysis as nonlinearity develops.) Pushover analysis is best understood as an approximate way to characterize the response of the structure as inelasticity develops.

The capacity curve obtained in a pushover analysis depends on the pattern of lateral forces that is applied. Various codes and guidelines allow or recommend the use of different lateral force patterns. For example, EC8 allows the applied lateral forces to be proportional to $\mathbf{M}\phi_i$ or simply to $\mathbf{M1}$.

The peak displacement response of the ESDOF system is based on the solution of Equation 7.5. Rather than compute $\xi_i^*$ and $Q_i^*$ explicitly (Equations 7.6 and 7.7), we instead use the capacity curve of the MDOF system to establish the force–displacement curve of the ESDOF system, and assume a value of $\xi_i^*$ appropriate to the structural system and material. The peak displacement of the ESDOF system, $D^*$, is mapped to and from that of the MDOF system using Equation 7.3.

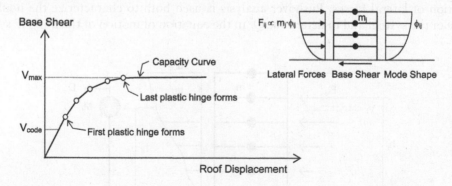

*Figure 7.2* Capacity curve obtained in a nonlinear static (pushover) analysis of a MDOF system.

---

[1] In contrast, applying a displacement pattern in the pushover analysis may require forces that are grossly inconsistent with those developed during nonlinear dynamic response.

Figure 7.3a shows an idealized capacity curve, consisting of a bilinear curve that has been fitted to the computed capacity curve of the MDOF system. Figure 7.3b shows the ESDOF force–displacement curves. The ESDOF curve may be obtained by dividing the MDOF displacements, forces, and modal mass by $\Gamma_i$. This mapping between the MDOF and ESDOF systems is developed in the following.

Dividing the MDOF modal mass $M_i^*$ by $\Gamma_i$ results in the ESDOF mass $M^*$, defined as:

$$M^* = \phi_i^T \mathbf{M1} \tag{7.9}$$

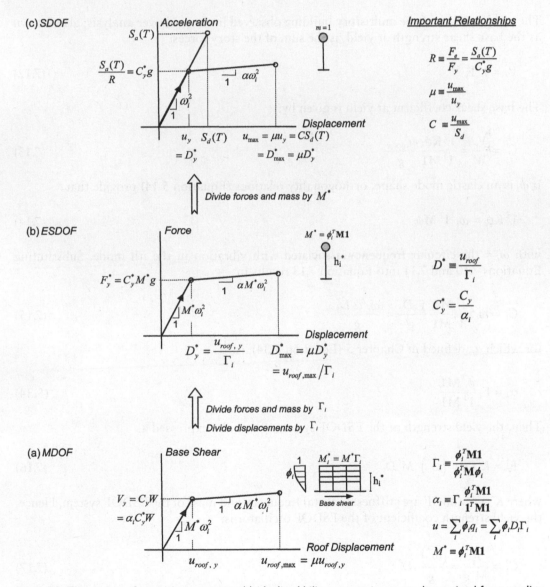

Figure 7.3 The notion of equivalent systems: (a) idealized bilinear capacity curve determined from nonlinear static (pushover) analysis; (b) bilinear curve for ESDOF system having mass $M^*$; (c) bilinear curve for system having unit mass.

Equation 7.3 may be restated to express the yield displacement of the ESDOF system relative to the roof displacement at yield of the multistory system as:

$$D_y^* = \frac{u_{roof,y}}{\Gamma_i} \tag{7.10}$$

The bilinear curve (fitted to the capacity curve) represents a case when yielding occurs at a sharply defined point. Projecting from the elastic domain to this yield point, the vector of lateral forces at this point, $\mathbf{F}_y$, can be expressed as:

$$\mathbf{F}_y = \mathbf{K}\mathbf{u}_y = \mathbf{K}\phi_i u_{roof,y} \tag{7.11}$$

The yield strength of the multistory building observed in the pushover analysis, also known as the base shear strength at yield, is the sum of the story forces:

$$V_y = \mathbf{1}^T \mathbf{F}_y \tag{7.12}$$

The base shear coefficient at yield is given by:

$$C_y = \frac{V_y}{W} = \frac{\mathbf{1}^T \mathbf{K}\phi_i}{\mathbf{1}^T \mathbf{M}\mathbf{1}} \frac{u_{roof,y}}{g} \tag{7.13}$$

If $\phi_i$ is an elastic mode shape, orthogonality relations (Equation 5.14) provide that

$$\mathbf{1}^T \mathbf{K}\phi_i = \omega_i^2 \mathbf{1}^T \mathbf{M}\phi_i \tag{7.14}$$

with $\omega_i$ = the circular frequency associated with vibration in the $i$th mode. Substituting Equations 7.10 and 7.14 into Equation 7.13 results in

$$C_y = \omega_i^2 \frac{\mathbf{1}^T \mathbf{M}\phi_i}{\mathbf{1}^T \mathbf{M}\mathbf{1}} \frac{\Gamma_i D_y^*}{g} = \frac{\omega_i^2 \alpha_i D_y^*}{g} \tag{7.15}$$

for which $\alpha_i$, defined in Chapter 5 (Equation 5.34), is:

$$\alpha_i = \Gamma_i \frac{\phi_i^T \mathbf{M}\mathbf{1}}{\mathbf{1}^T \mathbf{M}\mathbf{1}} \tag{5.34}$$

Then, the yield strength of the ESDOF system, $F_y^*$, can be expressed as

$$F_y^* = K^* D_y^* = \left(\omega^*\right)^2 M^* D_y^* \tag{7.16}$$

where $K^*$, $\omega^*$, and $M^*$ are stiffness, natural frequency and mass of the ESDOF system. Hence, the yield strength coefficient of the ESDOF oscillator is:

$$C_y^* = \frac{F_y^*}{M^* g} = \frac{\left(\omega^*\right)^2}{g} D_y^* \tag{7.17}$$

Because response histories and peak response amplitudes for a given excitation depend strongly on the natural frequency of the system, the period of vibration (or natural frequency) of the $i$th ESDOF system must match the $i$th period of vibration (or natural frequency) of the MDOF system.[2] Setting the circular frequency $\omega^*$ equal to the $i$th circular frequency, $\omega_i$ results in

$$C_y^* = \frac{\omega_i^2}{g} D_y^* = \left(\frac{2\pi}{T_i}\right)^2 \frac{D_y^*}{g} \qquad (7.18)$$

where $T_i$ = the natural period of vibration of the $i$th mode. Equation 7.18 ensures that the natural period of the ESDOF system matches the $i$th natural period of vibration of the MDOF system regardless of whether the shape vector corresponds to an elastic mode or not. A comparison of Equations 7.15 and 7.18 establishes that:

$$C_y^* = \frac{C_y}{\alpha_i} \qquad (7.19)$$

Thus, the yield strength of the ESDOF system is given by $V_y^* = C_y^* W^* = V_y / \Gamma_i$, representing the notion that the yield strength coefficient $\left(C_y^*\right)$ associated with the mass $M^*$ that participates in the $i$th mode can be related to a smaller yield strength coefficient $(C_y)$ that is associated with the total mass of the structure. The similarity to Equation 7.10, which states $D_y^* = u_{\text{roof,y}} / \Gamma_i$, should be noted.

Figure 7.3c illustrates a further reduction of the ESDOF system to a massless SDOF system, obtained by dividing forces and mass by $M^*$. Thus, the response is plotted on the coordinates of spectral (pseudo) acceleration and spectral displacement for elastic systems, with yield point defined by $C_y^* g$ and $D_y^*$ for inelastic systems.

If $\phi_i$ is an elastic mode shape, it ceases to be orthogonal with respect to $\mathbf{K}$ and $\mathbf{M}$ as nonlinearity develops, due to changes in $\mathbf{K}$. (Of course, orthogonality is absent if the shape vector, $\phi_i$, is not chosen to be an elastic mode.) Thus, Equations 7.18 and 7.19 must be considered to be approximate expressions for $C_y^*$.

## 7.5 DISPLACEMENT ESTIMATES

The displacement response of the ESDOF system (Equation 7.5) is the basis for estimating the displacement response of the MDOF system. Rather than explicitly evaluating Equations 7.6 and 7.7, we can transform the capacity curve of the MDOF system to establish the properties of the ESDOF system. Thus, $T^* = T_1$, $\xi^*$ is taken as a value typical of the structural material, perhaps 3% or 5% for reinforced concrete, and the hysteretic response of the oscillator may be taken as representative of reinforced concrete structures (e.g., stiffness degrading) with yield strength given by $V_y^* = C_y^* M^* g$. The equation of motion can be solved for $D^*(t)$ for a specific ground motion, given by $\ddot{u}_g(t)$. Alternatively, where elastic response spectra are available, the peak response of the ESDOF system is given by $S_d(T^*)$ in the linear elastic domain and can be estimated by application of $R$–$\mu$–$T$ relationships in the inelastic domain. The peak displacement, $\mu^* \cdot D_y^*$, can be estimated, using yield point spectra (YPS) or yield frequency spectra to determine the peak ductility response $\mu^*$ (given $C_y^*$ and $D_y^*$).

If the MDOF system truly responded in only a single mode, at any time, $u_{\text{roof}}(t) = \Gamma_i \cdot u^*(t)$. As an approximation, we consider the displacement ductility demands to be identical for both systems, for the mode of response considered

---

[2] Not all ESDOF formulations impose this constraint.

$$\mu^* = \frac{u_{max}^*}{u_y^*} = \mu = \frac{u_{max}}{u_y} \qquad (7.20)$$

where "max" refers to the larger of the maximum and absolute value of the minimum displacement response.

First-mode ESDOF systems tend to slightly underestimate the peak roof displacement for elastic response (attributable to the contribution of higher modes) and to slightly overestimate the peak roof displacement as $\mu^*$ becomes larger (e.g., Cuesta and Aschheim, 2002). Higher modes are associated with relatively small displacements occurring at higher frequencies, in effect representing small perturbations about the slowly moving first-mode displacement.[3]

The trend is nicely illustrated in work by Chopra et al. (2003), who compared the ESDOF estimates of peak displacement response of steel moment-resistant frames of varying heights (3, 6, 9, 12, 15, and 18 stories) to numerous ground motion records to the peak displacement responses computed for the MDOF systems to the same ground motion records. The ratio of the two, denoted $\left(u_r^*\right)_{SDF}$ was calculated for the response of each frame to each ground motion. Median values of this ratio ranged from approximately 0.85–1.2, gradually increasing with ductility. The ESDOF systems tended to slightly overestimate peak roof displacements as ductilities increased above approximately 2.

Thus, the first-mode ESDOF system generally provides a sufficiently accurate representation of displacement response for use in preliminary design, and one that is biased to be slightly greater than the MDOF response for moderate and higher displacement ductility demands.

## BOX 7.1    EXAMPLE ILLUSTRATING THE USE OF PUSHOVER ANALYSIS TO ESTIMATE ROOF DISPLACEMENT RESPONSE

The peak displacement response of a 12-story steel moment-resistant frame is estimated in this example. A steel frame is used because it avoids the complications of changes in stiffness associated with cracking of reinforced concrete members. The frame was subjected to the 1940 NS El Centro record, and is described in more detail by Aschheim et al. (2002). The capacity curve obtained in a first-mode pushover analysis of the beam-hinging frame is shown in Figure 7.4; a bilinear approximation indicates a yield displacement of approximately 0.353 m. Figure 7.5 compares the computed roof displacement response to the estimate $D^* \cdot \Gamma_1$ determined using an ESDOF system. The ESDOF estimate is generally in phase with and of similar amplitude to the computed roof displacement history.

## 7.6 REPRESENTATION OF CRACKING AND CRACK CLOSURE IN MODELS; GEOMETRIC SIMILARITY

One can appreciate that simple bilinear load-deformation models, often used to model steel structures, do not represent pre-yield stiffness reductions associated with crack development. If crack progression is neglected in both MDOF and ESDOF models, the ESDOF models can track the displacement history well, as illustrated in Figure 7.5 for steel moment frame structures.

---

[3] Of course, the higher modes could induce damage that affects first-mode response, such as the shear failure of a wall.

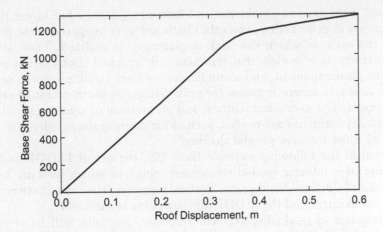

*Figure 7.4* Capacity curve for beam-hinging version of a 12-story steel moment-resistant frame, obtained using the first elastic mode shape.

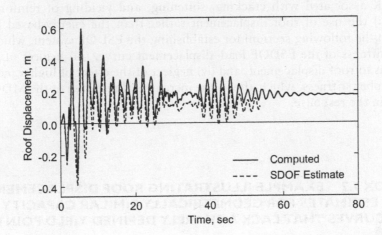

*Figure 7.5* Roof displacement history and estimate based on ESDOF response, for a beam-hinging 12-story steel moment-resistant frame.

The response of "well-behaved" reinforced concrete members typically is more complex than that of steel members, due first to flexural cracking and then softening of the concrete and yielding and strain hardening of the reinforcement.[4] Compressive axial force (e.g., in columns and walls) increases the cracking moment, but also allows the uncracked stiffness to resume after cracking, if only for a small portion of the elastic domain. This complexity of response carries over to the MDOF system under both quasi-static and dynamic response. If this behavior is to be represented in the model, the stiffness of every member must be reassessed often to capture changes during the response. Of course, the current stiffness directly affects the dynamic response computed for any time step, and may be influenced by deformations induced by higher modes.

A limited study suggests that peak displacement response is significantly affected by the uncracked stiffness and progression of cracking for peak displacements less than about

---

[4] "Poorly-behaved" members, such as over-reinforced flexural members and those experiencing shear or bond failure, present other issues.

twice the yield displacement (Aschheim and Browning, 2008). For larger displacements, crack propagation in prior cycles causes the elastic stiffness to approach the cracked section stiffness for the cycle in which the peak displacement is realized. Thus, simple bilinear EDOF load-deformation models that represent only cracked elastic behavior, yielding at the nominal or plastic moment, and strain hardening after yielding seem to be adequate for estimating peak displacement response for peak displacements in excess of about twice the yield displacement. The uncracked stiffness and progression of cracking should be modeled where high-fidelity estimates are needed, such as for meeting the smaller drifts that may be associated with more frequent ground shaking.

As illustrated in the following example (Box 7.2), the use of ESDOF models does not require the use of a bilinear load–displacement model in which inelastic behavior commences at a sharply defined yield point. Rather, geometric similarity between the ESDOF load–displacement curve and the MDOF capacity curve is sufficient.

ESDOF estimates of peak displacement response generally will be close to but not coincide with the displacement response of the MDOF system. The ESDOF system embodies some simplifications and approximations, related to (i) the use of an invariant shape vector for establishing the lateral forces throughout the pushover analysis, despite changes in **K** associated with cracking, softening, and yielding of reinforced concrete members, (ii) the use of roof displacement rather than the energy-based displacement (discussed in the following section) for establishing the ESDOF system, which affects the post-yield stiffness of the ESDOF load–displacement curve, (iii) neglect of higher mode contributions to roof displacement, and (iv) neglect of the effect of higher mode deformations on member stiffness, which affects the incremental response of the MDOF system at any instant in the response.

### BOX 7.2    EXAMPLE ILLUSTRATING ROOF DISPLACEMENT ESTIMATES FOR GEOMETRICALLY SIMILAR CAPACITY CURVES THAT LACK A SHARPLY DEFINED YIELD POINT

The capacity curve for a six-story beam-hinging moment-resistant reinforced concrete frame is shown in Figure 7.6a. The frame was modeled in Drain-2DX (Prakash et al., 1993) using a fiber beam-column element (Element 15). Rayleigh damping was set within the MDOF model to 5% of critical damping in the first and third modes. The first elastic mode shape, based on uncracked behavior, was used as the shape vector in the pushover analysis; lateral forces used to generate the capacity curve were proportional to the amplitude of the first mode and mass at each floor level. A bilinear curve was fitted to the capacity curve, resulting in a yield displacement $u_{roof,y} = 5.5$ in., base shear strength at yield of 173 k, and yield strength coefficient $C_y = 0.321$. Modal parameters $\Gamma_1$ and $\alpha_1$ were determined to equal 1.32 and 0.80, respectively. The initial period, based on uncracked behavior, was 0.732 s.

Based on the properties of the MDOF system, the ESDOF system should have yield strength coefficient $C_y^* = C_y \alpha_1 = 0.321/0.80 \cdot 0.402$ and $D_y^* = u_y/\Gamma_1 = 5.5/1.32 = 4.17$ in. Making use of the relationships $k = V_y^*/D_y^*$ and $V_y^* = C_y^* M^* g'$ the period associated with the secant stiffness at the yield point is given by:

$$T_{yp}^* = 2\pi \sqrt{\frac{m}{k}} = 2\pi \sqrt{\frac{D_y^*}{C_y^* g}} \tag{7.21}$$

Using the preceding values, the ESDOF system should have a secant stiffness period of $T_{yp}^* = 1.03$ s.

To capture the gross degradation of stiffness due to cracking, the ESDOF system was modeled using the same fiber element in Drain 2DX. A single member was modeled. The ESDOF model was established by (i) adjusting the length of the member until the yield point of a bilinear curve fitted to the capacity curve determined for the ESDOF system was equal to the target value of $D_y^* = 4.17$ in.; (ii) determining the required lateral mass as equal to $M^* = V_y^* / C_y^* g = 0.224$ k·s²/in., where $V_y^* (= 34.8$ k) is the base shear at yield corresponding to the yield point of the bilinear curve fitted to the ESDOF capacity curve, and (iii) establishing Rayleigh damping parameters to obtain 5% damping based on the initial, uncracked stiffness. Rayleigh damping for the ESDOF system was established having mass and stiffness proportional damping in the same amounts (percent of critical damping) as for the first mode of the MDOF system. Although not explicit, this procedure (for determining $M^*$) ensures that $T_{yp}^*$ equals the target value of 1.03 s, and avoids having to adjust the parameters of the fiber element to achieve a particular value of $V_y^*$. Gravity load was applied in an amount that corresponds to a representative value of axial load ratio $P/f_c' A_g$ in the MDOF model, recognizing that column flexibility contributes to the order of 1/3 of the flexibility of the frame, with beams (having zero axial load) contributing the remaining 2/3. The capacity curve for the ESDOF system is plotted in Figure 7.6b.

The initial period of the ESDOF system was determined to be 0.669 s. Because the period is inversely proportional to the square root of the stiffness, $T_{yp}^*$, can be related to the initial period, $T_{init}$, considering the initial stiffness, $k_{init}$, and the yield point secant stiffness, $k_{yp}$:

$$T_{yp}^* = T_{init} \sqrt{\frac{k_{init}}{k_{yp}}} \tag{7.22}$$

Referring to Figure 7.6b, $k_{init} \approx 19.7$ k/in. and $k_{yp} = 34.8$ k/4.17 in. = 8.35 k/in., allowing $T_{yp}^*$ to be estimated as $0.669(19.7/8.35)^{0.5} = 1.03$ s, which matches the expected value of $T_{yp}^*$ calculated based on the properties of the MDOF system (Equation 7.21).

The quasi-static load–displacement response of the ESDOF system is compared with the capacity curve obtained from pushover analysis of the MDOF system in Figure 7.6c. The MDOF capacity curve was mapped to ESDOF coordinates by dividing the base shear by $\Gamma_1 M^* (= \alpha_1 W/g)$ and dividing the roof displacement by $\Gamma_1$. Recall that an arbitrary value of yield strength was accepted for the ESDOF system, based on the parameters used to specify the properties of the fiber element, and $M^*$ was selected to achieve the correct base shear coefficient and period (rather than being calculated as $M^* = \phi^T \mathbf{M1}$). For this reason, the ESDOF capacity curve in Figure 7.6c was normalized by dividing the base shear by $M^* = V_y^*/C_y^* g = 0.295$ k·s²/in. Due to these normalizations, the ordinate has units of acceleration.[5]

It is evident in Figure 7.6c that the ESDOF curve obtained by the procedure described above provides a close approximation of the MDOF capacity curve. This match was obtained with relatively minor and simple adjustments to the parameters defining the fiber element model.

---

[5] The authors find the conventional term for this axis, "Spectral Acceleration," to be a misnomer, preferring to retain the classical meaning that this term and its more formal version, "Spectral Pseudo-Acceleration," have—the product of the peak displacement of a linear elastic SDOF oscillator, $S_d$, and the square of its circular frequency, $\omega^2$.

Although the gravity load and fiber model parameters could be refined to obtain an even better match, no such adjustments were made in this case.

The dynamic response of the MDOF and ESDOF models was computed for several earthquake records. Roof displacement estimates were determined by multiplying the ESDOF model displacement by $\Gamma_1$, and are compared to the roof displacement histories computed for the MDOF model in Figure 7.7.

Figure 7.7a compares computed and estimated roof displacement response for the classic NS ground motion recorded at El Centro in the 1940 Imperial Valley Earthquake. The peak displacement response of about 5 in. corresponds approximately to yield in the capacity curve of Figure 7.7a, and indicates the system has gone through substantial softening associated with cracking and perhaps some yielding of reinforcement. The estimated roof displacement response matches the computed roof displacement about as well as for the steel moment frame cases of Figure 7.5.

Figure 7.7b compares computed and estimated roof displacement response for a near-fault ground motion, the NS component of the Takatori record from the 1995 Hyogo-ken Nambu earthquake. Peak displacement response corresponds to about five times the yield displacement observed in the pushover analysis. In both cases, peak displacements of the ESDOF and MDOF systems are generally in phase and have similar values.

The ESDOF and MDOF load–displacement curves are nearly perfectly similar in a geometric sense (Figure 7.6c). This reflects the use of fiber elements in both models, having identical material properties and fiber configurations, and preloaded by gravity load in a representative manner. Only a small difference in the roof displacement response is evident.

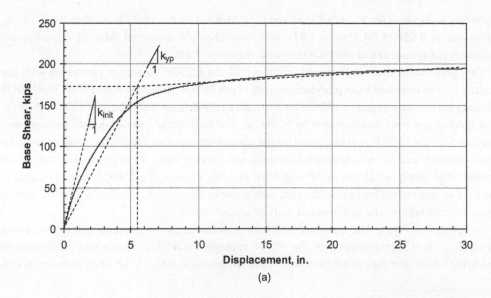

Figure 7.6 Capacity curve determined for a six-story reinforced concrete moment-resistant frame model: (a) MDOF system; (b) ESDOF system; (c) comparison of normalized MDOF (solid line) and ESDOF (dashed line) capacity curves.

(Continued)

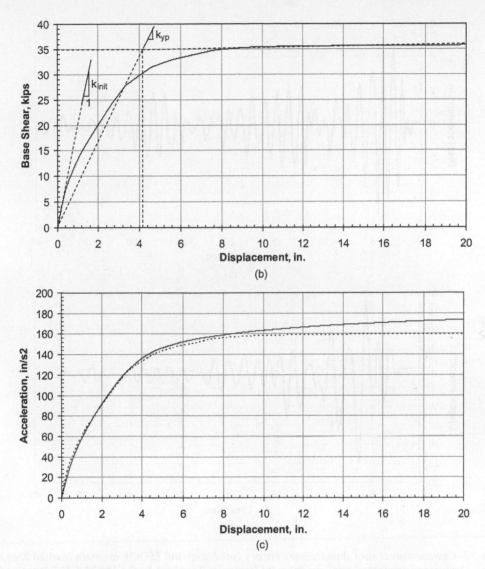

*Figure 7.6* (CONTINUED) Capacity curve determined for a six-story reinforced concrete moment-resistant frame model: (a) MDOF system; (b) ESDOF system; (c) comparison of normalized MDOF (solid line) and ESDOF (dashed line) capacity curves.

## 7.7 ENERGY-BASED PUSHOVER

One may wonder why so much emphasis is given to the roof displacement when deriving the ESDOF system. As will become evident in this section, the movement of the roof node is not always representative of the pattern of deformation that develops over the height of the system. This section develops an alternative pushover method (Hernández-Montes et al. 2004) in which the ESDOF displacement coordinate is computed as an "energy-based" displacement, as an alternative to the conventional definition of $D^* = u_{\text{roof}}/\Gamma_1$. The main point emphasized in this section is that the work done (or energy absorbed) by a multistory building during the quasi-static pushover (as well as the response to a ground motion) is directly related to the work done by the ESDOF system.

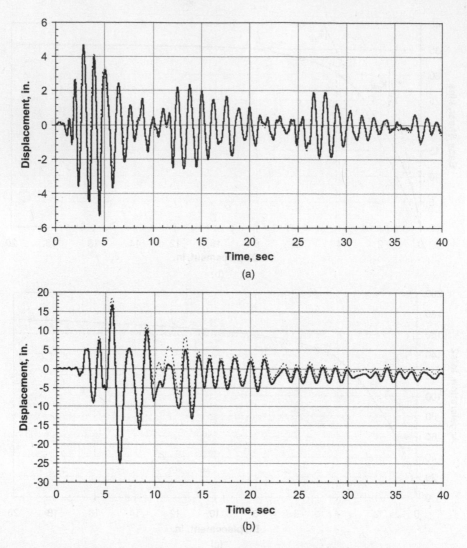

*Figure* 7.7 Comparison of roof displacement history (solid line) and ESDOF estimate (dashed line) for a six-story reinforced concrete moment frame. (a) Response to the 1940 NS El Centro record; (b) response to NS Takatori record of the 1995 Hyogo-Ken Nambu Earthquake.

At its core, the capacity curve of a structure represents the resistance to lateral forces that develops as lateral displacements increase. The capacity curve has great value in characterizing the load-deformation response of the MDOF system. The ESDOF system derived from the capacity curve provides a useful estimate of the peak displacement response and degree of nonlinearity that may develop in the MDOF system, when response is in the first or predominant "mode." Of course, the onset of nonlinearity causes changes in modal properties and invalidates modal superposition.

Because floor displacements over the height of the building generally increase dispro-portionately as response becomes increasingly nonlinear, one cannot rigorously justify the use of the displacement at any one location for the abscissa of the capacity curve, because the apparent post-yield stiffness of the capacity curve will depend on the location selected. Rather than relying on the displacement at a single location (e.g., the roof), the energy absorbed by the structure due to all of the individual floor forces going through their respective displacements is used to determine an energy-based displacement. The energy-based displacement is used for establishing the capacity curves of the MDOF and associated equivalent SDOF systems. Because the energy-based capacity curves do not have the displacement reversals observed in some higher mode pushover curves, they can be used to establish the ESDOF systems and associated response amplitudes in multimode analysis procedures (e.g., along the lines of Chopra and Goel, 2002). However, we have yet to identify a single or multimode method that provides reliable estimates of response quantities other than floor displacements (e.g., Tjhin et al., 2006).

We begin the derivation of the energy-based displacement with the vector form of the strain energy (or absorbed energy), obtained by expanding the third term of Equation 4.3:

$$E_S = \int \mathbf{F}_S^T \cdot d\mathbf{u} \tag{7.23}$$

The strain energy is composed of the recoverable elastic strain energy and expended strain energy associated with energy dissipated by the hysteretic response of the structural components (Figure 4.4).

Focusing initially on response within the elastic domain, as noted in the development of Equation 5.29, the quasi-static force associated with the $i$th mode is $\mathbf{F}_i(t)$. The strain force, $\mathbf{F}_S(t)$, may be represented as the sum of its modal components, $F_i(t)$:

$$\mathbf{F}_S(t) = \sum_i \mathbf{F}_i(t) = \sum_i \omega_i{}^2 \mathbf{M}\phi_i \Gamma_i D_i(t) \tag{7.24}$$

where $D_i(t) =$ is the displacement response to $\ddot{u}_g(t)$, see Equation 5.21. Due to the orthogonality of modes with respect to $\mathbf{K}$ (i.e., $\phi_i^T \mathbf{K} \phi_j \neq 0$ for $i = j$ and 0 otherwise) the force $\mathbf{F}_i$ does work only for displacements in the $i$th mode. The work done by this force on the other modal displacements is zero.

In the elastic domain,[6] the strain energy associated with the static force $\mathbf{F}_i$ going through an elastic displacement from 0 to $\mathbf{u}_i$ may be computed by substituting Equation 7.24 for $\mathbf{F}_i$ and Equation 5.22 for $\mathbf{u}_i$:

$$E_{Si} = \frac{1}{2}\mathbf{F}_i^T \cdot \mathbf{u}_i = \frac{1}{2}\omega_i{}^2 \phi_i^T \mathbf{M}\phi_i \Gamma_i{}^2 D_i{}^2(t) \tag{7.25}$$

Noting that the base shear associated with the $i$th mode pushover within the elastic domain is

$$V_{b,i} = \mathbf{F}_i^T \cdot \mathbf{1} = \omega_i{}^2 \Gamma_i \phi_i^T \mathbf{M} \mathbf{1} D_i(t) \tag{7.26}$$

and using the definition of $\Gamma_i$, we can express the strain energy of Equation 7.25 as

$$E_{Si} = \frac{1}{2} V_{b,i} \cdot D_i(t) \tag{7.27}$$

within the elastic domain.

---

[6] In elastic domain $F_S = \mathbf{K}\mathbf{u}$ so: $\int F_S \, du = \int Ku \, du = ku^2/2 = Fu/2$ or in matrix form: $\int \mathbf{F}_S^T \, d\mathbf{u} = \int \mathbf{u}^T \mathbf{K} \, d\mathbf{u} = \mathbf{u}^T \mathbf{K}\mathbf{u}/2 = \mathbf{F}_S^T \cdot \mathbf{u}/2$.

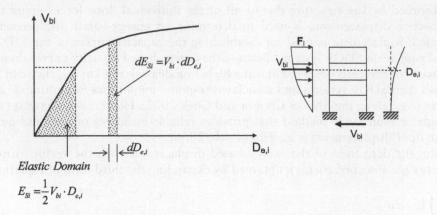

*Figure 7.8* Extension of $D_{e,i}$ definition from elastic domain to inelastic domain for response in the *i*th mode.

Equation 7.27 can be interpreted graphically as the area beneath the curve in a plot of $V_{b,i}$ with respect to $D_i$, in the elastic domain (Figure 7.8). Therefore, we define the energy-based displacement for the *i*th mode, $D_{e,i}$, to be equal to $2E_{Si}/V_{b,i}$ in order to obtain $D_{e,i} = D_i$ in the elastic domain. More generally, for both elastic and inelastic response, the work done by $V_{b,i}$ in a differential displacement $dD_{e,i}$ is $dE_{Si}$:

$$dE_{Si} = V_{b,i} \cdot dD_{e,i} \tag{7.28}$$

which necessarily is equal to the work done by the static force $\mathbf{F}_i$ in a differential displacement of the structure in this mode. Using an incremental formulation, the terms $\Delta E_{Si}$ and $V_{b,i}$ can be computed based on the forces $\mathbf{F}_i$ applied in each step of the pushover analysis. Then, the corresponding increment in the energy-based displacement, $\Delta D_{e,i}$, may be calculated as

$$\Delta D_{e,i} = \frac{\Delta E_{Si}}{V_{b,i}} \tag{7.29}$$

The energy-based displacement $D_{e,i}$ corresponding to the base shear is determined by summation. Equation 7.29 is consistent with Equation 7.27 in the elastic domain.

The possible influence of changes in the deformed shape from static forces associated with modes other than the *i*th mode is neglected in this formulation, because orthogonality of the load vector and the elastic mode shapes is assumed, as described earlier.

As with conventional pushover approaches, the mapping of the ordinate can be obtained by solving Equation 7.26 for the term $\omega_i^2 D_i(t)$:

$$\omega_i^2 D_i(t) = \frac{V_{b,i}}{\Gamma_i \phi_i^T \mathbf{M1}} \tag{7.30}$$

In this case, the values plotted on the ordinate of the representation are determined as before, as:

$$\frac{V_{b,i}}{\Gamma_i} \frac{\mathbf{1}^T \mathbf{M1}}{\phi_i^T \mathbf{M1}} = \frac{V_{b,i}}{\alpha_i} \tag{7.31}$$

## BOX 7.3   ENERGY-BASED PUSHOVER ANALYSIS EXAMPLE

To illustrate pushover analysis without the complexities of cracking and softening characteristic of reinforced concrete, a three-story steel moment-resisting frame is used in this example. The frame was designed for Los Angeles as part of the SAC project, reported in FEMA-355C, 2000. The building has four bays in the north-south direction and six bays in the east-west direction. The building is 120 ft (36.58 m) by 180 ft (54.86 m) in plan and 39 ft (11.89 m) in elevation, with a 2 ft (0.61 m) extension from the perimeter column lines to the edge of the building. Typical floor-to-floor heights are 13 ft (3.96 m). Figure 7.9 shows an elevation of the north-south lateral force-resisting system. All connections within the first three bays of Figure 7.9 are moment resisting. The interior bays consist of frames with simple connections. The last column line of Figure 7.9 utilizes "dummy" columns in order to model $P$-$\Delta$ effects associated with the gravity framing (although $P$-$\Delta$ effects were not modeled in the analyses reported herein). Gravity loads also are shown in Figure 7.9. Modeling assumptions follow the "M1 model" of the frame (FEMA-355C, 2000), for which beam and column framing is located along the member centerlines and panel zone deformations at the beam-column joints are not represented.

Three nonlinear static (pushover) analyses were done, one for each of the first three elastic mode shapes (with $\mathbf{F} \propto \mathbf{M}\phi_n$). Lateral forces were applied at each floor level in proportion to the three modes of the frame. Figure 7.10 presents the capacity curves for modes 1, 2, and 3 as plots of base shear versus roof displacement. Results were computed both with SAP2000NL (CSI, 2000) and Drain-2DX Version 1.10 (Prakash et al., 1993).

The capacity curves for the first- and second-mode pushover analyses of Figure 7.10 display softening behavior. The third mode capacity curve reverses direction after yielding. Clearly, within the linear elastic domain, the displacements over the height of the structure will remain proportional to the force distribution in the pushover analysis. This indicates that the displacements at any level may be used interchangeably for plotting the capacity curve for linear elastic response.

The displaced shapes corresponding to the points indicated by letters A–E on the third mode capacity curve of Figure 7.10 are shown in Figure 7.11. The nonproportional increases in floor displacements, including the reversal of the roof displacement are apparent and illustrate the arbitrariness of the selection of the roof displacement for this conventional three-story structure, for the third-mode pushover.

Capacity curves are plotted with respect to the energy-based displacement $D_{e,n}$ (determined using Equation 7.29) and are compared with those plotted with respect to the roof displacement in Figure 7.12. Figure 7.12a shows that energy-based capacity curve nearly coincides with the conventional capacity curve for the first-mode pushover. Figure 7.12b illustrates some differences in the post-yield portions of the energy-based and conventional capacity curves for the second-mode pushover. Figure 7.12c illustrates that the reversal in the conventional capacity curve is rectified when the energy-based approach is used. The energy-based and conventional capacity curves are identical in the linear elastic domain.

Peak displacement responses for each mode can be determined by superimposing the ESDOF capacity curves for each modal pushover analysis (Figure 7.12) on YPS representations of demands. Figure 7.13 plots these capacity curves together with YPS for a scaled El Centro ground motion that was used in the nonlinear response history analysis of the MDOF frame. In Figure 7.13a, the only YPS curve shown is the one that passes through the effective yield point of the first-mode SDOF system, which is the curve for a constant ductility of 1.68; thus, the peak displacement of the first-mode SDOF system is 1.68 times the effective yield displacement of the SDOF system. Because the second and third modes are determined to respond elastically according to this procedure, the peak displacements are at the intersection of the elastic spectral curve and the second- and third-mode capacity curves, as shown in Figure 7.13b. Using just the first-mode result, the peak roof displacement is estimated to be $u_{\text{roof}} = \Gamma_1(\mu D_y^*) = 228$ mm. This corresponds well with the peak observed in the dynamic analysis using this scaled record, which was 249 mm.

Note that the capacity curves of Figure 7.12 are plotted in terms of spectral acceleration; the associated strength is given by $S_a W^* = S_a M^* g$. The corresponding displacement is sometimes referred to as a spectral displacement. The "performance point" of Figure 7.13 simply is the expected peak displacement response of the ESDOF system.

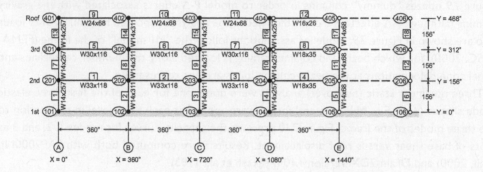

| Gravity Load | Value |
|---|---|
| Floor dead load for weight calculations | 96 psf (4.60 kN/m²) |
| Floor dead load for mass calculations | 83 psf (3.98 kN/m²) |
| Roof dead load excluding penthouse | 83 psf (3.98 kN/m²) |
| Penthouse dead load | 116 psf (5.56 kN/m²) |
| Reduced live load per floor and for roof | 20 psf (0.96 kN/m²) |

*Figure 7.9* Elevation, framing members, and element and nodal numbering for M1 model of SAC three-story steel frame.

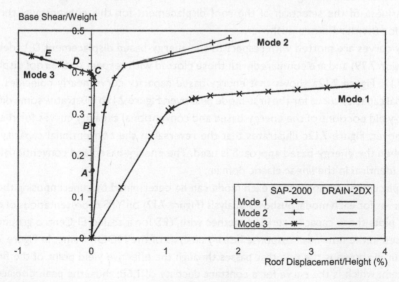

*Figure 7.10* Capacity curves obtained from DRAIN-2DX and SAP2000NL.

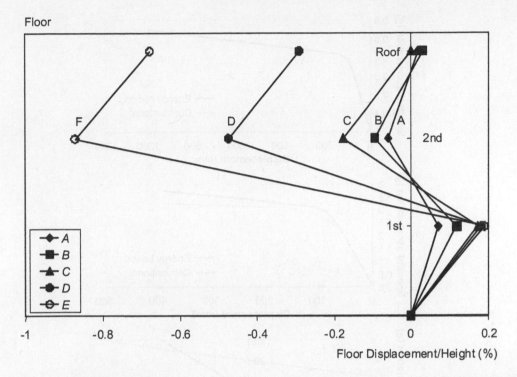

*Figure 7.11* Displacement profiles during the third-mode pushover analysis (locations identified in Figure 7.10).

## 7.8 CHALLENGES FACED IN ESTIMATING OTHER RESPONSE QUANTITIES

Where only a single mode exists, such as a one-story building idealized using a lumped mass model, an "equivalent" SDOF model truly is equivalent to the detailed model of the complete structure and such a model will be fully capable of providing accurate values of all response quantities. As the number of stories increases, the accuracy of the estimates provided using ESDOF models decreases, particularly for response quantities other than floor displacements (FEMA-440, 2005; Valley et al., 2010). Precise guidelines defining the limits of applicability of ESDOF models have not been formulated.

There are several ways to understand this observation:

1. Results from PCA presented in Chapter 6 clearly indicate that the proportion of variance represented by the first mode is highest for floor displacements, and lower for other response quantities, indicating that higher modes have a greater contribution to the other response quantities.
2. Simple expressions are available to relate floor displacements and floor forces to spectral response quantities for elastic response. With reference to Equations 3.30 and 5.22,[7] the vector of peak displacements due to the $i$th mode, $\mathbf{u}_{\max,i}$, is given by

---

[7] The analogy between Equations 3.29 and 5.21 is also worth noting.

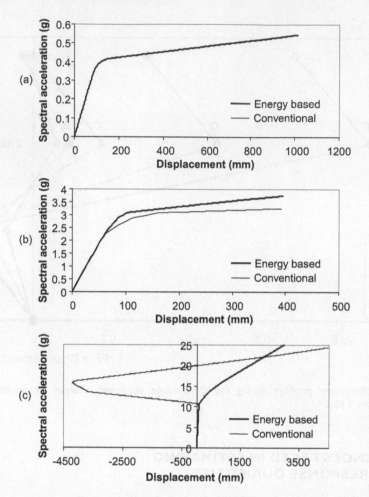

*Figure 7.12* Capacity curve obtained from first-mode pushover analysis—conventional approach compared with energy-based pushover. (a) First, (b) second, and (c) third modes.

$$\mathbf{u}_{\max,i} = \Gamma_i S_d(T_i)\phi_i \tag{7.32}$$

where $S_d(T_i)$ is the spectral displacement associated with the period $T_i$, $\Gamma_i$ is the modal participation factor, and $\phi_i$ is the mode shape for the $i$th mode. Higher mode contributions to displacements typically are minor because both $\Gamma_i$ and $S_d$ typically are much smaller for the second and higher modes, relative to their first-mode values.

In contrast, the vector of lateral forces, $\mathbf{F}_{\max,i}$ required to develop the peak displacements $\mathbf{u}_{i,\max}$ under static loading was expressed by Equation 5.40

$$\mathbf{F}_{\max,i} = S_a(T_i)\mathbf{s}_i = \Gamma_i S_a(T_i)\mathbf{M}\phi_i \tag{5.40}$$

where $\mathbf{M}$ is the mass matrix and $S_a(T_i)$ is the spectral pseudo-acceleration associated with period $T_i$. Because the shape of the response spectrum will often result in substantial higher mode spectral pseudo-accelerations relative to the first-mode value,

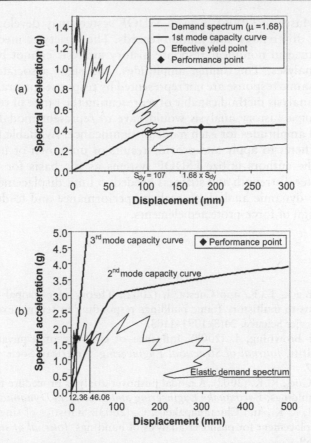

*Figure 7.13* Determination of modal displacement demands using YPS for (a) first-mode response and (b) second- and third-mode responses.

the lateral forces typically have more significant contributions from the higher modes than do the displacements.

3. Interstory drifts over time can be represented as the summation of interstory drifts associated with each mode shape[8] over time. As noted above, the second and higher modes have small contributions to displacement (Equation 7.32). However, because the higher mode shapes have greater tortuosity than the first mode, the higher modes have a greater contribution to interstory drifts than to floor displacements.

In addition, the tortuosity of the higher mode shapes suggests these mode shapes have significant contributions to curvature over the height of the building. Since bending moments are associated with curvature, significant contributions to bending moments may be expected. Following conventional beam theory, similar extensions can be made to shear and applied load, considering that successive derivatives of displacement are related to slope, curvature (and bending moment), shear, and applied load. The tortuosity of the higher mode shapes leads directly to their contributions to these other response quantities.

Nonlinear dynamic response history analyses represent the complicated interaction of multiple modes over time acting on structural components that may respond inelastically.

---

[8] The term "mode shapes" can refer to the elastic mode shapes or those determined by PCA of displacement response data.

As illustrated by Haselton (2006), a given MDOF system may develop different collapse mechanisms under different ground motion records. The variety of mechanisms and range of responses obtained in nonlinear response history analysis cannot be represented with single pushover analyses. The timing, amplitudes, and signs associated with the modal components of dynamic response are not represented by response spectra. It would seem that a nonlinear static analysis method capable of representing the range of responses obtainable by nonlinear response history analysis would have to represent modal interactions using different signs and amplitudes for each mode of significance. Available multimode analysis procedures are difficult to apply and provide results of unknown or uncertain reliability. For this reason, the authors utilize ESDOF systems as the basis for preliminary design (to establish the lateral strength and stiffness needed to limit displacement response), while utilizing nonlinear dynamic analysis to evaluate performance and to determine the forces needed for the design of force-protected elements.

## REFERENCES

Aschheim, M. A., Black, E. F., and Cuesta, I. (2002). Theory of principal components analysis and applications to multistory frame buildings responding to seismic excitation, *Engineering Structures*, Elsevier Science, 24(8):1091–1103.

Aschheim, M., and Browning, J. (2008). Influence of cracking on equivalent-SDOF estimates of RC frame drift, *Journal of Structural Engineering*, American Society of Civil Engineers, 134(3):511–517.

Chopra, A. K., and Goel, R. K. (2002). A modal pushover analysis procedure for estimating seismic demands for buildings, *Earthquake Engineering and Structural Dynamics*, 31:561–582.

Chopra, A. K., Goel, R. K., and Chintanapakdee, C. (2003). Statistics of single-degree-of-freedom estimate of displacement for pushover analysis of buildings. *Journal of structural engineering*, 129(4), 459–469.

CSI (2000). *SAP2000NL, Computers and Structures*, Computers and Structures, Incorporated, Walnut Creek, CA.

Cuesta, I., and Aschheim, M. A. (2002). Peak displacements and interstory drifts of nonlinear MDOF systems using principal components analysis, *7th US National Conference on Earthquake Engineering*, Boston, July 21–25.

FEMA-355C (2000). State of Art Report on Systems Performance of Steel Moment Frames Subject to Earthquake Ground Shaking, prepared by the SAC Joint Venture for the Federal Emergency Management Agency, Washington, DC.

FEMA-440 (2005). Improvement of Nonlinear Static Seismic Analysis Procedures. Applied Technology Council, Project ATC-55, Redwood City, CA.

Haselton, C. B. (2006). Assessing Seismic Collapse Safety of Modern Reinforced Concrete Moment Frame Buildings. *Doctoral dissertation*, Stanford University.

Hernández-Montes, E., Kwon, O., and Aschheim, M. (2004). An energy-based formulation for first and multiple-mode nonlinear static (pushover) analysis, *Journal of Earthquake Engineering*, 8(1):69–88.

Prakash, V., Powell, G. H., and Campbell, S. (1993). Drain 2DX Base Program Description and User Guide Version 1.10, Report No. UCB/SEMM-93/17, University of California, Berkeley.

Tjhin, T., Aschheim, M., and Hernández-Montes, E. (2006). Observations on the reliability of alternative multiple-mode pushover analysis methods, *Journal of Structural Engineering*, 132(3):471–477.

Valley, M., Aschheim, M., Comartin, C., Holmes, W., Krawinkler, H., and Sinclair, M. (2010). Applicability of Nonlinear Multiple-Degree-of-Freedom Modeling for Design-Supporting Documentation, Grant/Contract Reports (NISTGCR)-10-917-9.

# Section III

# Essential Concepts of Earthquake-Resistant Design

# Chapter 8

# Principles of earthquake-resistant design

## 8.1 PURPOSE AND OBJECTIVES

This chapter articulates general principles relied upon in the design of earthquake-resistant structures.

## 8.2 SPECIFIC PRINCIPLES

### 8.2.1 Ductile structural systems can be designed for reduced forces

Figure 8.1 shows the force–displacement response of an SDOF oscillator along with a 5% damped elastic response spectrum that represents the seismic demand. The linear elastic oscillator subjected to the ground motion has a complicated response over time that includes many cycles of response; the diagram of Figure 8.1 is a simplification for which the peak displacement (the maximum of the absolute value of the displacement of the oscillator mass relative to the base) over the response history is given by $S_d$ and the applied lateral force required to develop this displacement under quasi-static conditions is $V_e = mS_a$. The oscillator has stiffness $k = mS_a/S_d$, which is equivalent to the expression $S_a = \omega^2 S_d$, where $\omega^2 = k/m$.

If the oscillator yields at a base shear force $V_y = V_e/R$, where $R$ is the strength ratio of the elastic demand and base shear at yield (similar in concept but not identical to the building code strength reduction factor) its peak displacement $u_{max}$ will depend on the particular characteristics (or signature) of the ground motion and the hysteretic properties of the oscillator. If the ground acceleration in one direction was sustained for a long time, an oscillator having

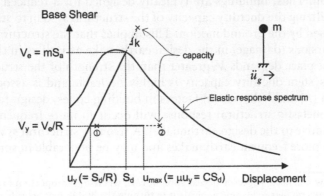

*Figure 8.1* Elastic response spectrum, and peak response of elastic and yielding SDOF oscillators.

strength $V_y < V_e$ (or $R > 1$) would develop large displacements relative to the ground. However, the ground motion is experienced as a series of short-duration impulses that alternate in direction, which serves to limit the peak displacement response of the yielding oscillator.

We define a displacement amplification ratio (also known as an inelastic displacement ratio, see e.g., Miranda, 2001), $C$, and ductility, $\mu$, as

$$C = \frac{u_{\max}}{S_d} \qquad \mu = \frac{u_{\max}}{u_y} \tag{8.1}$$

Then, the expected peak displacement is given by $u_{\max} = CS_d = \mu u_y$, where $C$ and $\mu$ can be described by an $R$–$\mu$–$T$ relationship, with reference to the period $T$ $(=2\pi/\omega)$, based on the elastic stiffness $k$ and strength ratio $R$ that characterize the oscillator. Moderate and long-period structures ($T > T_g$, where $T_g$ = characteristic period of the ground motion, see Section 3.7.5) have $u_{\max}$ generally bounded by $S_d$ (known as the equal displacement rule, described in Section 4.8.5), which is expressed by $\mu = R$ and $C = 1$. For shorter periods, $u_{\max}$ tends to exceed $S_d$ (corresponding to $\mu > R$ and $C > 1$); in this region, $u_{\max}/S_d$ increases with increasing $R$ and decreasing $T$. Both of the above statements are statistical observations of the mean trends (over multiple ground motions) of the peak displacement response of inelastic elasto-plastic hysteretic systems without cyclic or in-cycle degradation (see Chapter 4) rather than precise descriptions of response to be expected for any single ground motion. Also, moderate- and long-period systems that, for example, have entered a negative post-yield stiffness range (e.g., when nearing their global collapse range) may exhibit much larger displacement response than the equal displacement rule would suggest. Still, when concerned with design, one should not focus on a specific ground motion, while recognizing that an elasto-plastic non-degrading hysteresis is a valid description of many systems of interest.

A brittle system having strength $V_y$, represented by point ①, would suffer failure in this earthquake. Taking advantage of the alternating short-duration acceleration pulses of an earthquake ground motion, one may design a structure to yield as long as it has sufficient deformation (or ductility) capacity. Frequent reversals of the seismic excitation will limit ductility demands, but yielding will induce structural damage. Special details may be required to ensure the system has a ductility capacity (represented by point ②) in excess of the ductility demand.

Spectral accelerations ($S_a$) in high seismic regions can approach 2g or more. As a point of reference, consider that a pseudo-spectral acceleration of 1g represents a lateral force comparable to the weight of the building.[1] It is difficult to fathom a structural system that is strong enough to, in effect, support a building turned on its side, cantilevering horizontally out from its foundation. Thus, buildings are typically designed for a reduced lateral force with the intention of utilizing the ductility capacity of the structural system to sustain the reversed cyclic loading caused by the ground motion. This implies that the structural system will sustain inelastic excursions (damage) in the design earthquake as well as in the more frequent, smaller events that place demands $V_e$ greater than the strength of the structural system, $V_y$. While structural system ductility capacity is highly desirable and is associated with larger strength reduction ($R$) and behavior ($q$) factors in building codes, design for higher $R$ values also ensures that inelastic structural response will occur in more frequent earthquakes (of lesser intensity, relative to the design earthquake). A stronger structural system would ensure elastic response in more frequent earthquakes and may be preferable in some cases.

---

[1] More precisely, one would need a spectral acceleration of $g/\alpha_1$, where $\alpha_1$ = proportion of mass participating in the first mode, to represent elastic loading equivalent to turning the building on its side and supporting it as a cantilever from its foundation.

Of course, there is uncertainty in both ground motion and system capacity predictions. A brittle system designed precisely to have strength $V_y = V_e = mS_a$ will fail if the actual ground motion is stronger, or the structural system strength weaker, than considered in design. Similarly, a ductile system designed to reach its ductility capacity at a given earthquake level will fail for any stronger ground motion. Ductility in effect allows uncertainty in ground motion intensity to be converted to uncertainty in the degree of inelastic response. Having "sufficient" reserve displacement capacity provides the ability to handle some of the uncertainties that pervade the design process. Chapter 13 addresses the handling of uncertainties and how one can quantify the level of "sufficient" reserve capacity in relation to the risk of structural failure that one is willing to accept.

## 8.2.2 Energy dissipation is not an objective (but decoupling response from input is)

There is much discussion in the literature referring to the "fullness" of hysteretic loops and the significance of energy dissipation, and this plays into the notion of effective damping in so-called "equivalent linear" systems. There is no doubt that energy dissipation occurs in inelastic response. However, peak displacement response is not heavily influenced by energy dissipation or the fullness of the hysteretic loops. For example, as discussed in Chapter 4, bilinear elastic systems (e.g., rocking posttensioned walls) display nearly the same trends in peak displacement response as their yielding counterparts (Priestley, 1993; Kazantzi and Vamvatsikos, 2018). Rather, what seems to matter is the elastic stiffness (which determines the period and corresponding peak displacement response for long-period oscillators according to the equal displacement rule) and the place (position) that the unloading stiffness sends the oscillator in the absence of further excitation. The change of stiffness brought on by yielding (or rocking) decouples the oscillator from the input. Energy dissipation is an indication of structural damage, rather than being a primary objective in itself.

## 8.2.3 Deformation demands must be accommodated

Inelastic response is expected in most buildings located in higher seismic zones. The distribution of strength to the components of the structural system determines whether a single mechanism is dominant over a large suite of possible ground motions, or whether different mechanisms may develop (e.g., Figure 8.2) under different excitations. Whatever the deformation pattern, sufficient deformation capacity must be provided, and component plastic hinge rotation demands generally will be smaller when the mechanism mobilizes a larger number of plastic hinges. Where weak-story mechanisms form, the increased strength demands and ratcheting that occur with $P$-$\Delta$ effects can drive increasingly large local deformation demands and lead to collapse. System ductility capacity is enhanced in moment-resistant frames if deformation demands are distributed over a larger number of plastic hinges (over many, if not all, stories). While this provides for greater resistance to collapse, damage at more locations increases the cost of repair or the likelihood that replacement will be more cost effective than repair. The demands in so-called "force-protected" components will be determined to some extent by the strengths of the plastic hinges, and the strengths provided at these locations can influence the extent to which force-protected components suffer brittle failures.

## 8.2.4 Choice of structural system impacts performance

Broadly speaking, stiff structural systems (e.g., shear walls) are associated with small interstory drifts but high floor accelerations, while flexible structural systems (e.g., moment frames) are

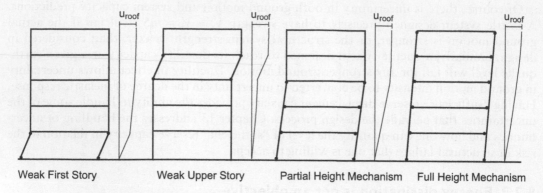

Weak First Story        Weak Upper Story        Partial Height Mechanism    Full Height Mechanism

*Figure 8.2* For a given peak roof displacement, story deformation and plastic hinge rotation demands are smaller when mechanisms are more spatially distributed and mobilize more plastic hinges. Deformed shapes include some elastic deformation (not shown).

associated with much larger interstory drifts but smaller floor accelerations. Thus, shear wall systems are beneficial for limiting damage to partitions, cladding, and other nonstructural components and equipment that can tolerate only small interstory drifts, but the relatively high floor accelerations in these systems may be a problem for equipment that is sensitive to acceleration and may cause sliding and toppling of contents. In contrast, moment frame systems are useful for limiting damage to acceleration-sensitive equipment, but the high interstory drifts in these systems may result in significant damage to nonstructural partitions, cladding, and drift-sensitive equipment such as escalators. FEMA (2012) discusses potentially vulnerable components.

While there are technologies (e.g., supplemental dampers or base isolation) for reducing floor accelerations, such technologies generally are more costly to implement in reinforced concrete construction compared with lighter structural systems (e.g., steel). Thus, where performance objectives constrain both floor accelerations and interstory drifts, the engineer may wish to evaluate more flexible shear wall systems (using more walls of shorter plan length and lower reinforcing ratios) or stiffer moment frame systems (using sections of greater depth and greater reinforcing ratios).

## 8.2.5 Use complete, straightforward, and redundant load paths

Although we tend to lump masses at floor levels for convenience in design, the mass is distributed throughout the building, and is either a part of the structural system or is attached to it. Lateral forces are generated as the distributed mass is accelerated by the ground motion. The engineer must conceive of and provide a complete load path to carry the forces generated between the distributed mass and the base of the building. Figure 8.3 illustrates the role of floor diaphragms, chords, collectors, and vertical elements of the lateral force-resisting system (LFRS) in carrying these forces to the foundation. Any oversights in design or construction (e.g., premature bar curtailment, poor splices, poor joint detailing, weak stories, and poor-quality concrete) are focal points for damage during the response to strong shaking; a reliable load path allows inelastic deformation to concentrate at the locations detailed for ductile response.

The load path should be viewed from both force and deformation perspectives. Load paths that require complicated deformation patterns to mobilize the forces required for static equilibrium (e.g., discontinued shear walls and plan torsion, illustrated in Figure 8.4) are associated with more complex dynamic response and may not work as envisaged.

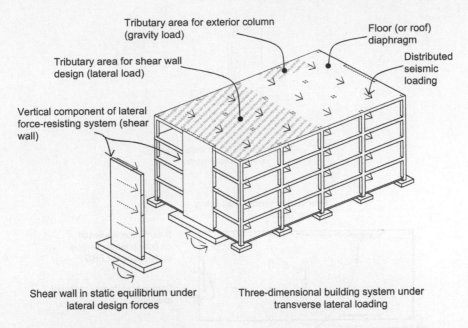

Tributary area for exterior column
(gravity load)

Floor (or roof)
diaphragm

Tributary area for shear wall
design (lateral load)

Distributed
seismic
loading

Vertical component of lateral
force-resisting system (shear
wall)

Shear wall in static equilibrium under
lateral design forces

Three-dimensional building system under
transverse lateral loading

*Figure 8.3* Forces generated during earthquake ground shaking are transmitted to the supporting soils by a load path that includes the vertical components of the LFRS and the foundation. Every structural component and subassembly is in equilibrium.

Generally, simple and straightforward load paths are preferred, are easier to design and detail, and leverage many years of experience in the design of earthquake-resistant systems.

As applied to LFRSs, redundancy refers to having a multiplicity of vertical components, such that failures within any one vertical component do not severely affect the lateral response of the system. The capacity to redistribute forces after unanticipated or premature failures within the vertical components of the LFRS is desirable, and should be considered in the development of the schematic design. In general, redundancy under seismic motion is a concept that is tightly woven with structural reliability under uncertain load and material properties. The target is to make the structure resistant to the effects of such uncertainties (Vamvatsikos, 2015), either by tightly controlling them where possible (e.g., by enforcing strict quality control of a few critical elements), or by spreading the uncertainty to more components (preferably independent in terms of capacity), each of which is less critical for the overall system safety (Wen and Song, 2003). Thus, having fewer vertical components participating in a LFRS does not necessarily indicate lower reliability, but it certainly means that one should be more careful about how they are designed and constructed. In contrast, having more vertical elements can be beneficial to the system's survivability in other scenarios, such as where localized loads of an accidental nature may occur (e.g., explosions or impact).

## 8.2.6 Avoid brittle failures using capacity design principles

Design for reduced forces is contingent on the system having ductility (the ability of the system to sustain resistance to lateral forces as displacements increase beyond the yield point). In reinforced concrete shear walls and moment frames, the ductility of the system is derived from the inelastic response of critical locations in flexure. This occurs at so-called "plastic hinges," which are regions of beams, columns, and walls where flexural yielding is dominated by yielding of longitudinal reinforcement (rather than crushing of concrete).

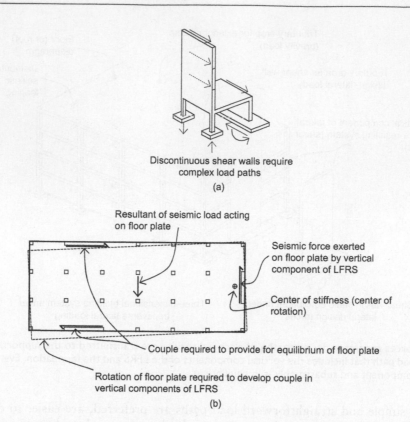

Discontinuous shear walls require
complex load paths

(a)

Resultant of seismic load acting
on floor plate

Seismic force exerted
on floor plate by vertical
component of LFRS

Center of stiffness (center of
rotation)

Couple required to provide for equilibrium of floor plate

Rotation of floor plate required to develop couple in
vertical components of LFRS

(b)

*Figure 8.4* Avoid complex and indirect load paths in the LFRS. (a) Example of complex load paths, and
(b) rotation of floor plate required to mobilize lateral resistance.

The ductility of these hinges is enhanced through ductile detailing requirements, which
typically include more stringent transverse reinforcement requirements along with limits on
longitudinal reinforcement ratios, member slenderness, and materials.

Capacity design principles make use of the concept that a chain is only as strong as its
weakest link (discussed in more detail in Chapter 11)—therefore, if the weak links are made
ductile, the forces developed in the stronger, brittle links, will be resisted without causing
a brittle failure. Thus, the strengths of the plastic hinges that are mobilized in the plastic
mechanism limit the forces that develop throughout the structural system. Although most
codes do not require the explicit use of a plastic mechanism analysis or pushover analysis
in design, their underlying philosophy aims to provide sufficient strength to resist axial
load, shear, and torsion without brittle failure, and to provide plastic hinge zones that are
detailed for ductile behavior. Additional detailing considerations locate lap splices away
from regions of high stress and provide for anchorage (development) of reinforcement.

Capacity design principles are especially effective in components that are isolated by plastic hinges, such as for shear in beams of moment frames that develop plastic hinges at their
ends. As described in the next section, higher modes can influence the demands in force-
protected components.

## 8.2.7 Incorporate higher mode effects

Capacity design principles are used to determine the strengths of the ductile, yielding portions
of the structure as it deforms in a mechanism, typically associated with a quasi-first-mode

deformation pattern. The "chain is as strong as its weakest link" metaphor applies to these first-mode-type deformations. However, higher modes of vibration are present—one may imagine masses located at each link of the chain; relative motion of these masses can induce forces that exceed the strength of the one ductile link. Similarly, in the MDOF system, higher modes will contribute to the forces in members that are not isolated by plastic hinges. For example, computed shears in a shear wall may be several times the tributary design base shear strength at yield.

In addition to inducing forces in excess of those associated with the intended mechanism, higher modes may contribute significantly to interstory drifts in moment frames, and may also contribute significantly to the response of specific members that are especially excited by a specific higher mode.

Examples where higher modes may contribute significantly to the forces that must be sustained by force-protected components (whose response may be non-ductile) include shears carried over the height of shear walls, shears in floor diaphragms, axial forces in collector elements, shear and bending moment in columns, and forces in foundations and foundation (basement) walls. Codes include prescriptive requirements such as determining minimum required column strengths based on using $1.25f_y$ for determining beam plastic hinge strengths, or applying prescriptive overstrength factors ($\Omega_o$) for determining collector reinforcement. Such factors are mainly meant to further protect brittle members/mechanisms against the uncertain strength of ductile ones, where an unaccounted for overstrength in a ductile link will adversely increase the strength demand of the adjacent "force-protected" brittle member. Presumably, such safety factors may also partially protect against the influence of higher modes, yet the adequacy of these approaches is not obvious, while the use of nonlinear dynamic analyses to determine force demands (without simulating all possible failures) leads to a dependence on modeling assumptions that is not always apparent. Presuming suitable modeling, the required strengths for brittle members can be established to ensure with desired confidence that brittle failures will not occur while the structure develops an acceptable plastic mechanism that is compatible with the design performance targets. These concepts are elaborated upon in Chapters 13 and 18.

The effect of higher modes on interstory drifts is accounted for in preliminary design using the coefficient of distortion, defined in Section 12.6 as the ratio of peak story drift ratio observed in the dynamic analysis and the peak roof drift ratio (which may occur at different times in the analysis).

## 8.2.8  Use recognized LFRSs and detailing provisions

Codes of practice recognize and prescribe details for specified structural systems. These systems have had the benefit of review and oversight over many years, and have been refined where experience in the lab or in actual earthquakes has indicated the need for improvement (e.g., the SAC/FEMA (2000) guidelines developed after unexpected fractures in the 1994 Northridge earthquake). In addition, regulatory agencies are familiar with the review and inspection of these systems, and contractors and trades people are familiar with their construction. We consider it far better to rely on this body of knowledge and experience than to tailor ductile details to conform to a particular pattern of computed demands. Consistency across projects helps to maintain quality in project delivery.

We do not wish to discourage the introduction of a new LFRS or ductile detail, particularly significant benefits are anticipated. Yet, this can be a lengthy process, typically requiring laboratory tests and careful analytical modeling to reach parity with established systems and details. FEMA P695 (FEMA, 2009) illustrates such an approach for determining strength reduction or behavior factors for new systems.

## 8.2.9 Recognize limitations of planar thinking and analysis

Points at the base of a structure are subjected to motion in all six Cartesian coordinates (three translations and three rotations relative to a set of coordinate axes). Normally only the three translations are considered in structural design, and for buildings of ordinary dimensions all points at the base are considered to move coherently. Peak response generally is not aligned with a particular set of coordinate axes.

Where two separate LFRSs are used (each aligned in a principal direction of the building), each generally would be designed separately without concern for their interaction. An example would be a shear wall system in one direction and a moment frame system in the orthogonal direction, or two orthogonal perimeter moment frame systems that do not have any columns shared by both systems.

In some cases, some structural components may be common to the orthogonal LFRSs. Examples include corner columns utilized by orthogonal perimeter moment frame systems, or flanged shear walls (e.g., T- or L-shaped cross sections). Conventional design codes require that the column or wall be designed to simultaneously resist 100% of the action in each principal direction in conjunction with 30% of the action in the orthogonal direction. The reasoning behind this rule is probabilistic, in the sense that the two horizontal ground motion components are not perfectly correlated, thus the peak action in one direction generally does not happen simultaneously with the peak action in the other direction (Rosenblueth and Contreras, 1977).

In the case of moment frames, plastic hinges may develop simultaneously in both perpendicular beams that frame into a corner column (Figure 8.5). Plastic moment demand may be summed vectorially to establish column moment demands. To avoid significant plastic hinges from developing in the columns requires the columns to be proportioned to have a resistance in excess of the biaxial moment demands imposed by the beams. Axial load–moment interaction diagrams for biaxial loading of rectangular columns show a slightly reduced capacity for loading along a 45° axis (compared with loading about the principal axes). Consequently, columns in this configuration should be proportioned with some conservatism relative to an in-plane loading condition (the 6/5 rule in ACI given by Equation 15.6 or the EC-8 requirement given by Equation 16.4), or alternatively, these proportioning requirements can be applied to the biaxial plastic hinging condition. Three-dimensional nonlinear dynamic response history analysis can be useful to validate performance.

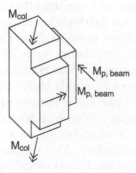

*Figure 8.5* Corner column of reinforced concrete frame subjected to demands from beam plastic hinges in orthogonal frames. (Beam and column shears and column axial forces not shown.)

## 8.2.10 Keep diaphragms elastic and stiff

When floor slabs are called upon to carry in-plane loads, they are referred to as "diaphragms." In-plane forces arise from two sources: (i) acceleration of mass distributed within or directly attached to the floor slabs and (ii) transfer of force between vertical components of the LFRS, which due to differences in their stiffness and/or strength, would displace relative to each other if they were isolated from the slab and free to move independently of each other.

Ordinary reinforced concrete buildings (lacking irregularities) generally have stiff and strong diaphragms that can and should be detailed to provide for elastic response.

Acceptable elastic response may be difficult to achieve where diaphragms contain significant openings, where diaphragms span large distances between vertical components of the LFRS, and where flexible wood or steel deck diaphragms are used on large single-story concrete wall buildings (e.g., single-story tilt-ups and "big box" mass merchandisers). In these cases, special attention is needed to ensure that adequate seismic performance is achieved (e.g., Koliou et al., 2016a,b).

## 8.2.11 Provide for deformation compatibility

The gravity framing must maintain support while subjected to the story drifts associated with the response of the lateral system. While obvious, this point has been overlooked at times, causing failure of the gravity framing and collapse.

Figure 8.6 illustrates schematically the load–displacement response of a shear wall, moment frame, and gravity framing system, where it is assumed that the gravity system is the most flexible. While the moment frame system is the most ductile, the flexibility of the gravity system ensures that it is the most deformable (capable of withstanding the largest displacement). When combined with the shear wall system, the gravity system would remain elastic for the design motion and for the more frequently occurring motions of lower intensity. When combined with the moment frame system, the gravity framing system would undergo some inelastic response for the design ground motion as well as other motions with similarly low annual probabilities of exceedance. Gravity framing systems that have a large deformation capacity improve the ultimate collapse capacity of buildings (see Elkady and Lignos, 2015), because they can maintain vertical load-carrying capacity even after the primary LFRS has exhausted its strength and ductility reserves.

Shear walls are so stiff that the beams and columns used for gravity framing should see little or no inelastic deformation. When using moment frames, economy is achieved

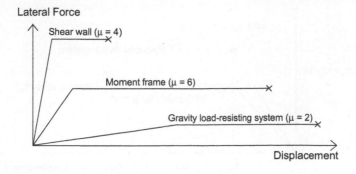

*Figure 8.6* Idealized lateral load response of lateral and gravity load-resisting systems, for assumed ductility capacities. Deformation compatibility requires that the gravity load-resisting system maintain gravity support while subjected to the lateral displacements experienced by the LFRS.

by limiting ductile details as much as possible to the moment frame system—to limit the ductility demands in the gravity framing members, the gravity load-resisting system would be made to be relatively flexible. Stated another way, the moment frame system would be proportioned to have a significantly smaller yield displacement relative to that of the gravity system, which requires making the moment frame beams and columns relatively deep compared with their gravity frame member counterparts. By ensuring that the gravity framing members experience little to no inelastic deformation, the need for ductile details in these members is diminished or eliminated.

## 8.2.12 Eliminate unnecessary mass

For a given structural system (and yield displacement), the base shear strength required to limit ductility demands to an acceptable level is given by $V_y = C_y W$. Thus, reducing unnecessary mass (contained within $W$) allows the base shear strength to be reduced and hence reduces the reinforcement required throughout the LFRS. Reductions in mass can be obtained by using lightweight aggregate in the floor slabs and/or using prestressed concrete floor slabs.

   If only the seismic weight (or reactive mass) is reduced for a given LFRS, the reduction in mass will reduce the fundamental period of the system (reducing $S_d$) while increasing the yield strength coefficient $C_y$ ($=V_y/W$). The effect is to go from Point ① to Point ② in Figure 8.7, which illustrates reductions in peak displacement and ductility responses.

## 8.2.13 Avoid irregularities

Building codes define horizontal and vertical irregularities based on the presence of soft or weak stories, mass discontinuities, offsets in the LFRS, diaphragm discontinuities, and conditions that give rise to plan torsion. Buildings with the most severe irregularities are not allowed in the highest seismic zones. Where irregularities are allowed, they are usually accompanied by penalties in the form of higher design forces (i.e., lower strength reduction or behavior factors) to offset the reduced system ductility capacity of the LFRS due to localization of demands at the points of discontinuity (e.g., De la Llera and Chopra, 1996; Fragiadakis et al., 2006). While such penalties are well intentioned, there is little assurance that the resulting design will achieve the intended performance and low probability of collapse.

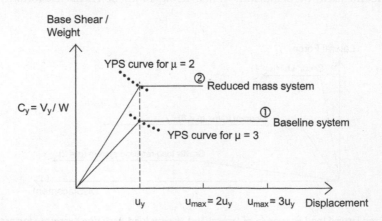

*Figure 8.7* Illustration of reduction in mass for a given LFRS causing a reduction in period, an increase in $C_y$, and hence, a reduction in ductility and drift demands.

Experience indicates that regularity is important to achieving good performance in earthquakes; one can easily make a good structural system (i.e., lacking irregularities) work well. Such systems provide symmetry, have vertical components of the LFRS located at the perimeter to maintain resistance to plan torsion, do not have substantially elongated or irregular plans or significant setbacks in elevation, and do not rely on plan torsion to carry lateral forces.

It may be difficult to make a poorly conceived structural system work adequately. If irregularities cannot be avoided, nonlinear response history analysis should aid in understanding the system response and may be used in design to gain some assurance that the desired performance will be achieved.

### 8.2.14 Anchor nonstructural components to the structure

A significant life safety threat to occupants and passersby is posed by falling hazards. Lights, suspended ceilings, ductwork, parapets, and cladding are among the many nonstructural features that pose falling hazards. To prevent falling hazards and reduce monetary losses, all nonstructural components should be anchored to the structure (FEMA, 2012; Porter et al., 2014).

### 8.2.15 Restrain mechanical equipment and piping

Sliding of heating, ventilation, and air conditioning units, toppling of water heaters, and breakage of pipes may make a facility useless until repairs are complete, and to the extent that liquids or gasses are released, may create additional hazards such as fire. Bracing of pipes and anchorage of mechanical equipment will reduce or eliminate such damage. Dedicated guidelines for the design of such components are being developed (e.g., NIST GCR 17-917-44, 2017).

### 8.2.16 Restrain building contents

Building contents such as furniture (e.g., file cabinets) and items on shelves or in cabinets (e.g., books and bottles in laboratory environments) can slide, tip, and/or topple. Because drift demands are addressed by modern design codes while acceleration demands are not, damage to building contents can present a significant source of monetary losses for modern structures. In addition to the direct physical threat posed by sliding and falling contents, spilling of contents such as laboratory chemicals can release hazardous substances, cause unanticipated chemical reactions, and cause fires. Building contents that pose significant risks should be restrained from sliding and tipping, as recognized by performance assessment guidelines (FEMA, 2012; Porter et al., 2014).

### 8.2.17 Avoid pounding between adjacent structures

Lateral drifts should be accommodated without impact of adjacent structures (pounding), by providing adequate separation between adjacent structures. ASCE-7 drift estimates, determined as $C_d$ times the drifts determined using reduced forces, are recognized to underestimate drifts for the design ground motion. Even larger drifts should be expected for rarer events. Where any doubt exists, floor levels should be matched to those of adjacent buildings, since pounding of floor slabs is preferable to having an adjacent floor slab impacting a column or wall (Karayannis and Favvata, 2005), which can introduce considerable local damage (Kasai and Maison, 1997; Cole et al., 2012) that may compromise the gravity load-carrying system and cause collapse. Where architectural or practical considerations prohibit the use of such measures, vertical members should be protected from impact either by moving them in from the (collision-prone) edge of the floor slab or by introducing additional protective members, such as collision shear walls (Anagnostopoulos and Karamaneas, 2008).

## 8.3 ADDITIONAL CONSIDERATIONS

Structural engineers work together with other professionals in design and construction who may unwittingly introduce changes before or after construction that may negatively impact seismic performance. This includes changes to architectural treatments such as cladding and infill partitions, the introduction of elements contributing mass and possibly stiffness, deviations from the construction drawings, the use of materials stronger or weaker than intended, and issues related to water movement and protection of reinforcing steel from corrosion. Owners should be advised that any changes affecting stiffness, strength, and mass, and their distribution throughout the structure, can have implications on seismic performance. Inspection and quality control during construction also are important for achieving the design intent.

## REFERENCES

Anagnostopoulos, S. A., and Karamaneas, C. E. (2008). Use of collision shear walls to minimize seismic separation and to protect adjacent buildings from collapse due to earthquake-induced pounding, *Earthquake Engineering and Structural Dynamics*, 37:1371–1388.

Cole, G. L., Dhakal, R. P., and Turner, F. M. (2012). Building pounding damage observed in the 2011 Christchurch earthquake. *Earthquake Engineering and Structural Dynamics*, 41:893–913.

De la Llera, J. C., and Chopra, A. K. (1996). Inelastic behavior of asymmetric multistory buildings. *Journal of Structural Engineering*, ASCE, 122(6):597–606.

Elkady, A., and Lignos, D. (2015). Effect of gravity framing on the overstrength and collapse capacity of steel frame buildings with perimeter special moment frames, *Earthquake Engineering & Structural Dynamics*, 44(8):1289–1307.

FEMA (2009). Quantification of Building Seismic Performance Factors, Report No. FEMA P695. Prepared by the Applied Technology Council for the Federal Emergency Management Agency, Washington, DC, 2009.

FEMA (2012). Seismic performance assessment of buildings, FEMA P-58-1. Prepared by the Applied Technology Council for the Federal Emergency Management Agency, Washington, DC, 2012.

Fragiadakis, M., Vamvatsikos, D., and Papadrakakis, M. (2006). Evaluation of the influence of vertical irregularities on the seismic performance of a 9-storey steel frame, *Earthquake Engineering and Structural Dynamics*, 35(12):1489–1509.

Karayannis, C. G., and Favvata, M. J. (2005). Earthquake-induced interaction between adjacent reinforced concrete structures with non-equal heights, *Earthquake Engineering and Structural Dynamics*, 34(1):1–20.

Kasai, K., and Maison, B. F. (1997). Building pounding damage during the 1989 Loma Prieta earthquake. *Engineering Structures*, 19(3):195–207.

Kazantzi, A. K., and Vamvatsikos, D. (2018). The hysteretic energy as a performance measure in analytical studies, *Earthquake Spectra*, 34(2):719–739.

Koliou, M., Filiatrault, A., Kelly, D. J., and Lawson, J. (2016a). Buildings with rigid walls and flexible roof diaphragms. I: Seismic collapse evaluation of existing design provisions, *Journal of Structural Engineering*, ASCE, 142(3):04015166.

Koliou, M., Filiatrault, A., Kelly, D. J., and Lawson, J. (2016b). Buildings with rigid walls and flexible roof diaphragms. II: Evaluation of a new seismic design approach based on distributed diaphragm yielding, *Journal of Structural Engineering*, ASCE, 142(3):04015167.

Miranda, E. (2001). Estimation of inelastic deformation demands of SDOF systems, *Journal of Structural Engineering*, ASCE, 127(9):1005–1012.

NIST GCR 17-917-44 (2017). Seismic Analysis, Design, and Installation of Nonstructural Components and Systems – Background and Recommendations for Future Work, Report No. NIST GCR 17-917-44, National Institute for Standards and Technology, Gaithersburg, MD.

Porter, K., Farokhnia, K., Vamvatsikos, D., and Cho, I. H. (2014). Guidelines for Component-Based Analytical Vulnerability Assessment of Buildings and Nonstructural Elements, GEM Technical Report 2014-13, Global Earthquake Model Foundation, Pavia, Italy. DOI: 10.13117/GEM. VULN-MOD.TR2014.13.

Priestley, M. J. N. (1993). Myths and fallacies in earthquake engineering – Conflicts between design and reality, *Bulletin of the New Zealand National Society for Earthquake*, 26(3):329–341.

Rosenblueth, E., and Contreras, H. (1977). Approximate design for multicomponent earthquakes, *Journal of Engineering Mechanics*, ASCE, 103:895–911.

SAC/FEMA (2000). Recommended Seismic Design Criteria for New Steel Moment-Frame Buildings. Report No. FEMA-350, SAC Joint Venture, Federal Emergency Management Agency, Washington DC.

Vamvatsikos, D. (2015). A view of seismic robustness based on uncertainty, *Proceedings of the 12th International Conference on Applications of Statistics and Probability in Civil Engineering*, ICASP12, Vancouver, Canada.

Wen, Y. K., and Song, S. H. (2003). Structural reliability/redundancy under earthquakes, *Journal of Structural Engineering*, ASCE, 129(1):56–67.

# Chapter 9

# Stability of the yield displacement

## 9.1 PURPOSE AND OBJECTIVES

This chapter examines the kinematics of the yield displacement, clarifying the mechanical basis for using an assumed yield displacement in seismic design. Particular emphasis is placed on the influence of longitudinal reinforcement content on the strength and stiffness of reinforced concrete members. The relative stability of the yield displacement as longitudinal reinforcement content is refined in design iterations makes yield point spectra and yield frequency spectra especially useful in seismic design.

## 9.2 INTRODUCTION

The usual structural engineering task is to determine the physical sizes of members to ensure adequate performance under a given loading or set of load combinations. The strengths of common construction materials are limited to a relatively narrow range, and thus, the strains that accompany the development of key limit states (e.g., yielding, softening, ultimate strength, and so on) are nearly invariant—engineers have little latitude to choose the strain at which concrete fails or the strain at which steel yields. However, the engineer may change the amount of reinforcement (e.g., for common materials, reinforcement ratios can range from around 0.3% to over 1.5% in a beam, a factor of five) and possibly the member dimensions, to achieve the required strength. The limited choice of reinforcing bar materials (grades) limits the strain associated with the yield of the longitudinal reinforcement to a narrow range that is known at the start of the design process. Changes in reinforcement content, however, affect the amount of material present, and hence have very direct and significant influences on stiffness, while having little effect on the displacement at yield (Priestley et al., 1995; Hernández-Montes and Aschheim, 2003). The relative stability of the yield displacement can be seen most easily from a kinematic perspective, and is discussed in this chapter, first for individual members and then for entire lateral force resisting systems.

## 9.3 KINEMATICS OF YIELD—MEMBERS

To begin, consider the design of a steel bar loaded in tension (Figure 9.1). Having selected the grade of steel to use, design for an applied tensile load consists of selecting the cross-sectional area to provide adequate resistance to the applied force. Clearly, the displacement at yield can be found by integrating strain over the length of the bar:

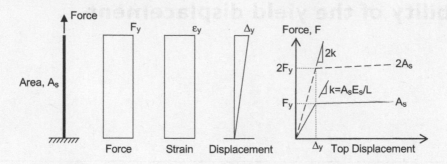

Figure 9.1 An axially loaded steel bar. The yield displacement is a kinematic function of yield strain and length (independent of cross-sectional area), whereas strength and stiffness are proportional to cross-sectional area.

$$u_y = \int \varepsilon_y \, dx \tag{9.1}$$

The Young's modulus ($E_s$, also known as the elastic modulus) of the mild steels used in construction is considered a constant (200 GPa or 29,000 ksi), and is independent of the yield strength of the steel.[1] Then, the yield strain, $\varepsilon_y$, depends on the grade or strength of the steel, and is proportional to yield strength:

$$\varepsilon_y = f_y / E_s \rightarrow u_y = \int \varepsilon_y \, dx = \int_0^L \frac{f_y}{E_s} dx = \frac{f_y}{E_s} L \tag{9.2}$$

Thus, the yield displacement $u_y$ at the tip of an axially loaded bar (Figure 9.1) is a function of the yield strength of the steel material, and therefore is independent of the cross-sectional area $A_s$ selected in the design of the member. The yield displacement is determined kinematically (as a function of strain and geometry), and need not involve consideration of force and stiffness.[2] Note that changes in $A_s$ directly affect the strength ($f_y A_s$) and stiffness ($EA_s/L$) of the bar; changes in stiffness directly affect the dynamic properties (e.g., period of vibration).

These observations are seen most easily in the case of a bar in axial tension, but apply more generally to the deformation patterns present in individual members subjected to bending, shear, and torsion, as well as structural systems subjected to a common deformation pattern, whether composed of reinforced concrete, steel, or other materials, as discussed in more detail in the remainder of this chapter.

Consider a cantilever steel wide-flange beam (Figure 9.2) subject to bending. At the section level, the curvature at yield is given as

$$\phi_y = \frac{\varepsilon_y}{d/2} \tag{9.3}$$

which is a kinematic function of the yield strain and section depth, $d$. For a cantilever configuration, the tip displacement at yield is given as:

---

[1] Concrete differs in that the initial value of Young's modulus ($E_c$) changes with compressive strength, while the strain at ultimate strength is typically estimated as 0.002 for normal strength, normal weight concretes. But crushing of concrete is rarely (if ever) the intended source of inelastic response in ductile reinforced concrete members.

[2] The traditional presentation states $u = FL/EA_s$, which obscures the stability of the yield displacement that is easily seen when integrating the yield strain over the length (Equation 9.2).

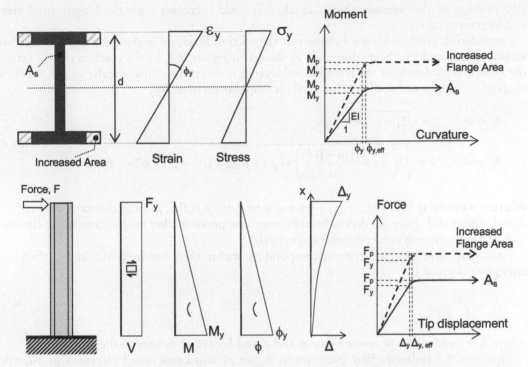

*Figure 9.2* A steel wide flange beam. The yield curvature is a function of the yield strain and section depth. Yield displacement depends on these quantities and the length of the member. Extrapolations to the plastic moment strength can be used to determine the effective curvature or displacement at yield of a bilinear curve fitted to the response.

$$\phi(x) = \phi_y \frac{L-x}{L}$$

$$u_y = \int \left( \int \phi(x)dx \right) dx = \phi_y \left( \frac{x^2}{2} - \frac{x^3}{6L} \right) \Big|_0^L = \phi_y \frac{L^2}{3}$$

(9.4)

Equation 9.4 illustrates the kinematic basis of the yield displacement, meaning that the yield displacement can be found simply as a function of strain and geometry. This simple notion is obscured by traditional presentations that invoke member stiffness and applied forces or moments (e.g., $u = FL^3/3EI$, which can be derived considering that $\phi_y = M_y/EI$).

Figure 9.2 also illustrates the effect of strain penetration over the depth of the section as curvature increases, resulting ultimately in the development of a plastic moment strength, $M_p$, which exceeds the moment at first yield of the extreme fibers, $M_y$. For typical wide flange steel sections, $M_p/M_y$ ($= f_y Z_x/f_y S_x = Z_x/S_x$) $\approx 1.15$–$1.20$. Thus, it is often more useful in practice to characterize an effective yield displacement that is based on the initial elastic stiffness extrapolated to $M_p$. Such modifications can be made without loss of generality of the finding that yield (or effective yield) displacements can be determined kinematically as functions of yield strain and geometry (member depth and length).

The application to reinforced concrete beams is slightly more complex because the depth of the neutral axis varies slightly with the longitudinal steel reinforcement ratio and varies with bending moment (having different values for uncracked and cracked section behavior and varying further as the section approaches the nominal or plastic strength level).

The portion of the member that is cracked at yield increases with the longitudinal steel reinforcement ratio.

Considering cracked elastic behavior at the section level, an under-reinforced beam has strain, stress, and free-body diagrams as shown in Figure 9.3. In the cracked elastic range, the neutral axis depth for the transformed section is determined analytically as $k \cdot d$, where $k$ is given for singly and doubly reinforced rectangular sections by

$$A_s \text{ only:} \quad k = \sqrt{2\rho n + (\rho n)^2} - \rho n$$

$$A_s \text{ and } A_s': \quad k = \sqrt{2\left(\rho n + \frac{\rho'(n-1)d'}{d}\right) + \left(\rho n + \rho'(n-1)\right)^2} - \rho n - \rho'(n-1)$$

(9.5)

where $n$ = modular ratio = $E_s/E_c$, $\rho$ = tension steel ratio = $A_s/bd$, $\rho'$ = compression steel ratio = $A_s'/bd$, and $d$ and $d'$ are the depths from extreme compression fiber to the centroids of the tension and compression reinforcement, respectively.

From first principles, for materials responding within their linear elastic ranges, the yield curvature is given by

$$\phi_y = \frac{\varepsilon_y}{d - kd}$$

(9.6)

where $\varepsilon_y$ = yield strain of reinforcement and $d$ and $kd$ are as defined earlier.

Equation 9.5 indicates that increases in $A_s$ (or $\rho$) will cause small increases in $kd$, and these increases are moderated to a small degree by any increases in $A_s'$ (or $\rho'$). Equation 9.6 indicates that an increase in $kd$ is associated with an increase in yield curvature. Further, as $A_s$ (or $\rho$) increases, greater softening of concrete in compression will occur, leading to slightly larger increases in $kd$ than would be suggested by Equations 9.5 and 9.6 (which are derived for materials in their linear ranges).

At larger curvatures, significant nonlinearity in the concrete stress–strain relation typically develops (Figure 9.4) while strain hardening of the steel reinforcement is expected. As ultimate strength is approached, the presence of additional yielding tension reinforcement requires a larger concrete compression block for equilibrium ($T_s = C_c$); thus, at ultimate, an increase in $A_s$ (or $\rho$) is associated with a larger curvature at ultimate strength.

Figure 9.5 illustrates the analytically determined moment–curvature responses of two cross sections. Although section B has three times the reinforcement area of section A, and nearly three times the strength and stiffness of section A, the yield curvature increases by only about 20%. The uncracked stiffness and cracking moment, $M_{cr}$, are nearly indistinguishable.

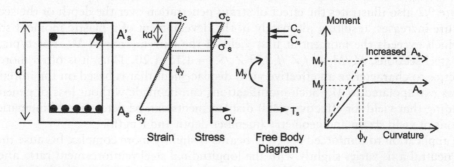

Figure 9.3 A reinforced concrete beam, assuming cracked section elastic behavior until yield. An increase in $A_s$ causes a small increase in neutral axis depth, which results in a slight increase in curvature at yield.

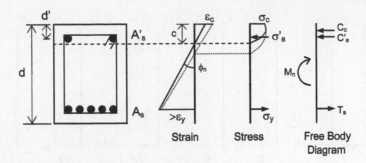

Figure 9.4 Curvature $\phi_n$ associated with nominal (or ultimate) flexural strength. Dotted curves indicate an increase in $A_s$ requires a deeper compression block (larger $c$) to develop sufficient compression to equilibrate the larger $T_s = A_s f_y$. The larger compression block is associated with greater curvature, and hence larger extreme fiber compression strains.

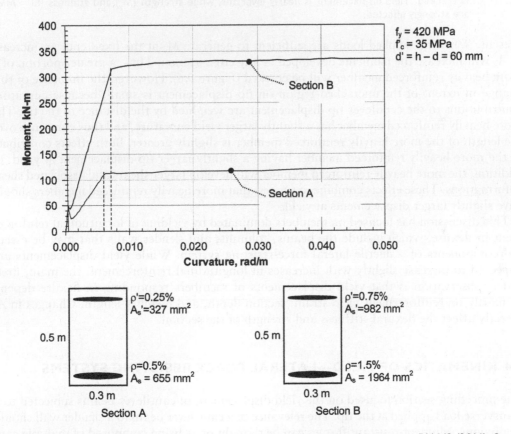

Figure 9.5 Moment–curvature results determined for two sections using the program BIAX2 (2016). Steel area shown by dark ellipses. A nearly threefold increase in strength and stiffness is associated with an increase in yield curvature of about 20%.

While cracks in real members (e.g., beams, columns, and slender walls) occur at discrete locations, for purposes of this discussion it is convenient to divide the member into uncracked and fully cracked regions. The portion of a member subjected to moments less than the cracking moment may be approximated by an uncracked section. As evident in Figure 9.5, the moment required to cause cracking, $M_{cr}$, depends little on the amount of reinforcement

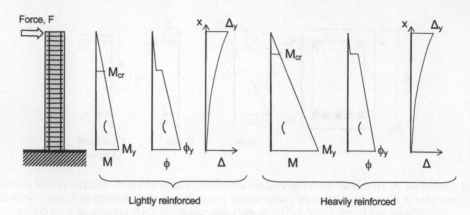

Figure 9.6 Influence of increase in longitudinal reinforcement ratio on the proportion of the member that is cracked. Yield displacement is nearly invariant, while strength ($M_y$) and stiffness ($EI = M/\phi$) are strongly affected.

present. When the applied loads are sufficient to generate $M_y$ at the fixed end, an increase in $A_s$ will increase the moments developed at any cross section. Thus, a greater portion of a more heavily reinforced member will be cracked (Figure 9.6). However, the influence of this change in extent of the uncracked region on tip displacement is small, because curvature contributions to the cantilever tip displacement are weighted by the distance to the tip. The more heavily reinforced member has a slightly larger yield curvature, and thus curvature over the length of the more heavily reinforced member is slightly greater. Both effects contribute to the more heavily reinforced member having a slightly larger tip displacement at yield. In addition, the more heavily reinforced member will develop larger shears and associated shear deformations. These effects combine to suggest that more heavily reinforced members should have slightly larger displacements at yield.

This discussion has focused on members dominated by yielding of longitudinal reinforcement in flexure, which include the beams, columns, and slender walls that may be essential components of a ductile lateral force-resisting system. While yield displacements are expected to increase slightly with increases in longitudinal reinforcement, the main, first-order, observation is that yield displacements of members responding in flexure depend primarily on reinforcement yield strain, section depth, and member length. Changes in $A_s$ directly affect the flexural stiffness and strength of the section.

## 9.4 KINEMATICS OF YIELD—LATERAL FORCE RESISTING SYSTEMS

The preceding section focused on the yield displacement of cantilever beams subjected to a transverse load applied at the tip. The relevance of a cantilever beam to a slender wall should be obvious. Moment-resistant frames can be thought of as being composed of multiple cantilevers, with points of inflection under lateral loading being located at or near midspan (as assumed in the Portal Method and as illustrated in Figure 9.7).

As described in Chapter 6, lateral displacements of multistory multi-degree-of-freedom (MDOF) buildings during elastic and inelastic dynamic response typically are dominated by response in a predominant mode that is similar to the first mode. For this reason, it is useful to characterize the behavior of the system using a first-mode nonlinear static (or pushover) analysis. Such analyses are described in Section 7.4. Results are often portrayed using a "capacity curve," which plots the base shear force as a function of the roof displacement.

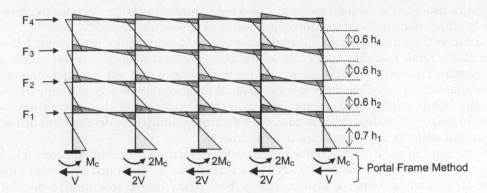

*Figure 9.7* Application of the Portal Frame method to the ASCE 7 lateral force profile associated with the required mechanism yield strength. The relative values of moment and shear forces at the supports and assumed locations of inflection points are shown. (Adapted from Hernández-Montes and Aschheim, 2017a.)

In a first-mode pushover analysis, the applied lateral forces are proportional to the mass and modal amplitude at each floor level. Examples of capacity curves plotted with respect to the roof drift ratio are given in Figure 9.8. Bilinear approximations to these capacity curves are also shown, and were fit by eye. The fitted bilinear curve can be used to establish the yield displacement and yield strength of the MDOF system responding in the first mode.

The capacity curve for the three-story frame of Figure 9.8 having $\rho = A_s/bd = 1.5\%$ was obtained for the beam-hinging frame illustrated in Figure 11.10. The design of this frame was based on beam sections having longitudinal reinforcement ratios of approximately 1.5% (for the top steel at member ends). Capacity curves for two additional designs are also shown. To obtain the $\rho = 0.5\%$ design, the frame was redesigned using just 1/3 of

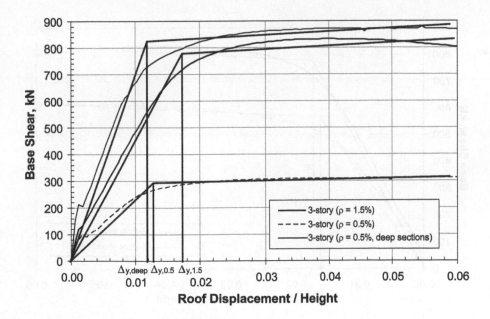

*Figure 9.8* Capacity curves and bilinear idealizations, for three-story reinforced concrete frames, obtained using Seismostruct Version 7.0.6 (2014).

the reinforcement. The original frame was also redesigned using larger sections to carry the same bending moments, with reinforcement ratios of approximately 0.5%.

The curves shown in Figure 9.8 illustrate relatively stiff behavior prior to cracking, followed by an elastic range that shows softening as cracking spreads throughout critical locations in the structure, followed by softening associated with yielding of reinforcement and softening of concrete (at strains exceeding the proportional limit) in the compression zone. Compared with the more lightly reinforced three-story frames, greater softening of the compression zones occurs in the $\rho = 1.5\%$ design; in this case, the softening is sufficient to development of a negative tangent stiffness beginning at around 4% drift.

The strength of the $\rho = 1.5\%$ frame in the pushover analysis was approximately 832 kN, in excess of the design value of 600 kN. This excess may be explained by several aspects of the design and modeling: (i) Reinforcement above that required according to calculation was provided at some critical sections. (ii) Strains in the fiber elements are evaluated at the Gaussian integration points, which are centered along each segment length; yielding at the integration points corresponds to the development of larger moments at the segment ends. (iii) The steel was modeled with an expected yield strength equal to 1.2 times the specified minimum of 420 MPa. (iv) A post-yield modulus equal to 0.5% of the elastic modulus of 200,000 GPa provided some strain hardening. (v) Confined concrete was modeled as having a compressive strength of 1.2 times the specified strength of 35 MPa. Thus, the moments developed at the ends of the members will exceed those estimated in design.

Consistent with the trend for yield curvatures apparent in Figure 9.5, the yield displacement (associated with the bilinear curve fitted to the first-mode capacity curve) for the $\rho = 1.5\%$ case is about 37% greater than that obtained for the $\rho = 0.5\%$ design. The $\rho = 0.5\%$ case has about 37% of the strength and 50% of the stiffness of the $\rho = 1.5\%$ case, based on the bilinear approximations.

Figure 9.9 compares capacity curves for a three-story and eight-story moment-resistant frame. Yield drift ratios are nearly the same, with the yield drift ratio for the eight-story frame being about 6% less than that of the three-story frame. The eight-story frame is

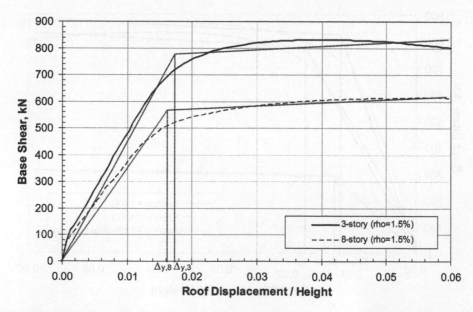

*Figure 9.9* Capacity curves and bilinear idealizations, illustrating a minor effect of the number of stories on yield drift ratio.

composed of cross sections obtained from the $\rho = 1.5\%$ three-story frame, and thus would be expected to have a lateral strength relative to tributary dead weight of about a third of that for the three-story frame. This illustrates a typical pattern that the yield drift ratios tend to reduce slightly as the number of stories increases.

Based on the preceding, one may say that the yield displacement observed in a first-mode pushover analysis is a relatively constant percentage of the height of the structure. Further, for the distribution of mass, strength, and stiffness in a given structure, proportional increases in strength and stiffness, achieved by increasing reinforcement area in all members, will cause the strength and stiffness of the structure to increase while the yield displacement remains relatively constant. While these observations hold in a "first-order" sense, some minor effects are evident from the foregoing discussion:

- for fixed member dimensions, yield drift ratios tend to increase as the amount of reinforcement increases;
- for fixed strengths, achieved with variable section dimensions and reinforcement ratios, yield drift ratios tend to increase as the reinforcement ratio increases; and
- for fixed member dimensions, yield drift ratios tend to decrease as the number of stories increases.

Engineers may have a good sense for the member sizes expected prior to initial proportioning, and thus may simply determine the reinforcement content required to provide sufficient strength for sections of a given depth. In this approach, yield drift may be estimated based on estimated dimensions and reinforcement yield strain (Section 9.5), and will change little as the reinforcement content is determined for the required strength (associated with a given performance objective). Alternatively, the member sizes may be determined based on a target reinforcement ratio and flexural strength. In this approach, more stringent performance objectives will require higher member strengths, achieved by using deeper sections, and the associated yield drift ratio will reduce.

## 9.5 YIELD DRIFT ESTIMATES FOR REINFORCED CONCRETE LATERAL FORCE-RESISTING SYSTEMS

Yield displacements determined in pushover analysis depend on the sources of flexibility that are represented in the structural models employed. At one extreme are software packages that require the user to specify the percentage of gross section stiffness to use for a linear beam-column element, but which readily model shear deformations and flexibility of beam-column joints. Perhaps at the other extreme is software such as Seismostruct that represents axial and flexural deformations in beams and columns by means of fiber elements, but requires significant effort for modeling joints. Codes of practice generally are silent on these modeling issues beyond general statements that the effects of cracking must be considered, while specifying limits on the computed deformations (e.g., interstory drifts). Thus, while yield drifts are relatively stable throughout the design process, simple estimates may deviate from values determined in analysis due to the manner in which sources of flexibility are represented in the analytical models.

To gauge the value of including additional sources of flexibility in analysis models, it is worth gauging their relative contribution. Priestley et al. (2007, p. 162) provides data that suggest that for a typical reinforced concrete moment-resistant frame, beam flexural deformation contributes about 60% of the deformation associated with the yield displacement, while column flexural deformation contributes about 20%, joint deformation about 15%, and shear deformation about 5%. Thus, flexural deformation of beams and columns over their clear spans is expected to contribute about 80% of the deformation. In many cases, analysis models are used in which beam and column elements span between the intersections of beam and column centerlines, in effect substituting additional flexural deformation for deformation occurring within the beam-column joints. Such models may be sufficient for establishing the structural design, depending on the degree of precision needed in meeting the seismic performance objectives.

Chapters 14 and 15 make recommendations for modeling of reinforced concrete members. We note that if conventional approaches are used, where flexural stiffness is taken as a specified fraction of gross stiffness, the influence of reinforcement content on stiffness will not be represented. This creates an anomaly in which stiffness is independent of reinforcement content, and yield displacement increases in proportion to strength.

### 9.5.1 Moment–resistant frames

An estimate of the yield drift ratio of regular moment frames resulting from flexural deformation of beams and columns can be derived from first principles. Because most of the contribution to drift in moment-resistant frames of typical proportions comes from beam deformation, it is convenient to express the yield drift ratio in terms of beam parameters:

$$\frac{u_{\mathrm{roof},y}}{H} = \frac{\phi_y}{6L}\left( \underbrace{(L-d_c)^2}_{\text{beam}} + \underbrace{\frac{\phi_c}{\phi_y}\left(\frac{L}{h}\right)(h-d_b)^2}_{\text{column}} \right) \tag{9.7}$$

where $u_{\mathrm{roof},y}$ = displacement of the roof corresponding to the yield point of a bilinear curve fitted to the capacity curve, $H$ = height of building, $h$ = typical story height, $L$ = typical bay length, $d_b$ = depth of typical beam section, $d_c$ = depth of typical column section, $\phi_y$ = yield curvature of typical beam, and $\phi_c$ = curvature of typical column at a critical section. The terms $L-d_c$ and $L-d_b$ define the clear span of the beams and columns, respectively.

The dependence of yield drift ratio on $\rho$ is reflected in Equation 9.7 because the yield drift ratio depends on $\phi_y$, and $\phi_y$ calculated using Equation 9.6 reflects the dependence of yield curvature on $\rho$, via $kd$ calculated using Equation 9.5. The column contribution, given by the second term of Equation 9.7, is on the order of 20%–30% of the beam contribution, for typical moment frames. This suggests that an estimate for the term $\phi_c/\phi_y$ may be used without significantly affecting the accuracy of the estimated yield drift ratio; a value of around 0.7–0.8 should be adequate in many cases.

Since the accuracy of the estimate obtained using Equation 9.7 depends on the sources of flexibility represented in the mathematical model of the structure, Equation 9.7 may be used to make an initial estimate of yield drift ratio before the frame has been designed in detail. Once available, the results of a pushover analysis can be used to revise the estimate of yield displacement. If needed, the relative influence of potential changes in geometry and reinforcement ratios can be estimated using Equation 9.7. For example, this equation suggests that a 10% reduction in clear span length would reduce the yield drift ratio by about 15%.

Simpler estimates may be derived. If the yield curvature is taken equal to $1.7\varepsilon_y/d_b$, as suggested by Priestley et al. (2007), and the column contribution of Equation 9.7 is taken equal to 30% of the beam contribution, Equation 9.7 may be simplified to

$$\frac{u_{\mathrm{roof},y}}{H} \approx \frac{\varepsilon_y}{3\cdot d_b}L \approx \frac{L}{1,200\cdot d_b} \tag{9.8}$$

for reinforcement having an expected yield strength of around 500 MPa.

Often, yield drift ratios for reinforced concrete moment-resistant frames will be bounded between the estimates given by Equations 9.7 (or 9.8) and approximately 0.5%–0.6%, depending on the specifics of the structural system and the assumptions used in modeling.

## 9.5.2 Cantilever walls

Tjhin et al. (2004) conducted an extensive analytical study of slender reinforced concrete walls having rectangular or barbell cross sections, with varied longitudinal reinforcement amounts and distributions and varied reinforcement yield strengths and concrete compressive strengths. Some results are summarized in Appendix 1. Based on this study, an estimate of the yield drift ratio of the roof of regular (prismatic) walls cantilevered from the base, with uniformly distributed mass and loaded by a first-mode force pattern is

$$\frac{u_{\text{roof,y}}}{H} = 0.3\left(1.8\varepsilon_{\text{y}} + 0.0045\frac{P}{f_c'A_{\text{g}}}\right)\frac{H}{l_{\text{w}}} \tag{9.9}$$

for walls with compressive axial force, $P$ between 0 and 0.2 times $f_c'A_{\text{g}}$, where $f_c'$ = compressive strength of concrete, $A_{\text{g}}$ = gross cross-sectional area of the wall, $\varepsilon_{\text{y}}$ = the yield strain of the steel reinforcement, $H$ = the height of the wall, and $l_{\text{w}}$ = the length of the wall in plan. Given the very light axial load ratios on typical walls, the second term of Equation 9.9 contributes little to the yield drift ratio, and hence a simpler expression is given by

$$\frac{u_{\text{roof,y}}}{H} = \frac{H}{700 \cdot l_{\text{w}}} \tag{9.10}$$

for reinforcement having an expected yield of approximately 500 MPa.

## 9.5.3 Coupled walls

Based on the work of Hernández-Montes and Aschheim (2017b), the displacement at the top of a coupled wall at the effective yield point of the capacity curve, as determined in a first-mode pushover analysis, can be estimated as

$$u_{\text{roof,y}} = 0.68\frac{\varepsilon_{\text{y}}}{D_{\text{cw}}}\frac{H^2}{3} \tag{9.11}$$

where $\varepsilon_{\text{y}}$ is the yield strain of the reinforcement and other terms are as defined in Figure 9.10. This result is applicable to cases in which the coupling beams have an aspect ratio $h_{\text{cb}}/L_{\text{cb}} = 0.7$.

## 9.6 POST-TENSIONED WALLS

For the systems described so far in this chapter, the yield displacement could be assumed to be stable with variations in reinforcement content. This is not the case for post-tensioned walls. For post-tensioned walls, the initial stiffness is based on the cross-sectional area of the gross section; changes in prestressing force cause changes in lateral strength, and the yield displacement can be considered to be proportional to strength. Recommendations for modeling the load–displacement response of post-tensioned walls are provided by Kurama et al. (1999) and Hernández-Montes et al. (2018) and are reviewed in Chapter 19.

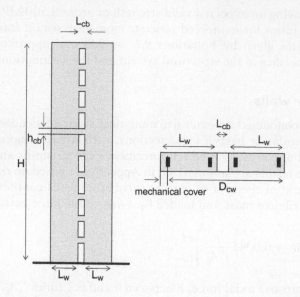

*Figure 9.10* Coupled wall nomenclature.

# REFERENCES

Hernández-Montes, E., and Aschheim, M. (2003). Estimates of the yield curvature for design of reinforced concrete columns, *Magazine of Concrete Research*, 55(4):373–383.

Hernández-Montes, E., and Aschheim, M. (2017a). A seismic design procedure for moment-frame structures, *Journal of Earthquake Engineering*, 26:1–20.

Hernández-Montes, E., and Aschheim, M. (2017b). An estimate of the yield displacement of coupled walls for seismic design, *International Journal of Concrete Structures and Materials*, 11(2):275–284.

Kurama, Y. C., Pessiki, S., Sause, R., and Lu, L.-W. (1999). Seismic behavior and design of unbonded posttensioned precast concrete walls, *PCI Journal*, 44(3):72–89.

Priestley, M. J. N., Seible, F., and Calvi, M. (1995). *Seismic Design and Retrofit of Bridges*, John Wiley & Sons, New York.

Priestley, M. J. N., Calvi, G. M., and Kowalsky, M. J. (2007). *Displacement-Based Seismic Design of Structures*, IUSS Press, Pavia, Italy.

Tjhin, T. N., Aschheim, M., Wallace, J. W. (2004). Yield displacement estimates for displacement-based seismic design of ductile reinforced concrete structural wall buildings, *13th World Conference of Earthquake Engineering*, Vancouver, British Columbia, August 1–5, paper #1035.

Wallace, J. W. (2016). BIAX-2: Analysis of Reinforced Concrete and Reinforced Masonry Sections. The Earthquake Engineering Online Archive NISEE e-Library.

# Chapter 10

# Performance-based seismic design

## 10.1 PURPOSE AND OBJECTIVES

This chapter discusses seismic performance objectives and illustrates how limits on peak displacement and ductility demands constrain the properties (strength and stiffness, period, or yield displacement) of acceptable oscillators. The graphical approach, using yield point spectra, illustrates that the design of stiff systems will tend to be controlled by ductility considerations while the design of flexible systems will tend to be controlled by drift. The required lateral strength may be determined for an estimated yield displacement or period, and one or more performance objectives can be addressed. The treatment of multiple performance objectives in this chapter is extended across the hazard using yield frequency spectra in Chapter 13.

## 10.2 INTRODUCTION

Descriptions of the degree of damage to structural components, nonstructural components, and building contents at various shaking intensities have been articulated in greater detail in recent years, in parallel to the increase in precision in the specification of seismic hazard, the computation of structural demands, and the characterization of component capacities. As a minimum standard, building codes remain focused on the safety of occupants and bystanders in relation to physical harm, and so far have given little attention to the broader exposure of communities to economic and societal disruptions resulting from damage that may render many buildings uninhabitable or prevent their use during repair.

Building code provisions have been developed and refined over many years; many of the current requirements for seismic design in the United States can be traced back to ATC 3-06 (1978) and before. Generally, building code provisions change gradually in response to technical advances, and rapidly in the face of unexpected failures during strong earthquakes. In effect, the built environment is a living laboratory, providing infrequent tests of the adequacy of building code provisions.

Code provisions aim to achieve minimum levels of safety for a large variety of buildings; these buildings are designed, reviewed, constructed, and maintained by people having different degrees of expertise, resources, and interest in seismic safety. The approaches taken by codes are somewhat simplistic (e.g., in the use of $R$ or $q$ factors for design on the basis of elastic analysis results) in their essence, but have been modified by a patchwork of changes that attempt to address particular issues (e.g., torsionally irregular systems, redundancy, and $P$-$\Delta$ effects), while maintaining applicability to a broad variety of buildings. Current building code provisions are prescriptive in that they specify what the designer must do, rather than focusing on the desired performance (i.e., response) under seismic excitation.

Since the prescriptive provisions are used with a broad variety of buildings, we can expect the buildings resulting from these provisions to display a large range of performance.[1] This implies either that the provisions are very conservative (excessively safe and costly to implement) in order to ensure that the worst performing buildings achieve the desired level of safety, or that some proportion of code-compliant buildings will fail to achieve the intended level of safety.

The robust analytical tools that are now available can be used to achieve more uniform performance over the large range of buildings presently designed based on prescriptive requirements and linear elastic analysis, on the one hand, avoiding the cost of needless conservatism, and on the other, reducing the risk of unacceptably poor performance. Doing so requires a greater emphasis on seismic performance criteria.

## 10.3 PERFORMANCE EXPECTATIONS IN BUILDING CODES

The 1933 earthquake in Long Beach, California led the state to institute requirements that mandated seismic design of all new buildings, with more stringent requirements being applied to schools. This and subsequent earthquakes led to significant code development activity, primarily by the Structural Engineers Association of California. Seismic performance objectives were articulated in the *Recommended Lateral Force Requirements and Commentary* (SEAOC, 1960), known informally as the "Blue Book." Structures designed in accordance with the prescriptive provisions of the Blue Book were expected to:

1. Resist minor earthquakes without damage;
2. Resist moderate earthquakes without structural damage, but with some nonstructural damage; and
3. Resist major earthquakes, of the intensity of severity of the strongest experienced in California, without collapse, but with some structural as well as nonstructural damage.

These performance objectives established a frame of reference that continues to guide seismic design nearly six decades later. Refinements in the NEHRP Provisions (BSSC, 2009) clarify that better performance is sought for critical facilities and that reductions in repair costs are intended where they can be achieved with relatively low cost. Specifically, the NEHRP Provisions (BSSC, 2009) state:

> The intent of these provisions is to provide reasonable assurance of seismic performance that will:
>
> 1. Avoid serious injury and life loss,
> 2. Avoid loss of function in critical facilities, and
> 3. Minimize nonstructural repair costs where practical to do so.
>
> These objectives are addressed by seeking to avoid structural collapse in very rare, extreme ground shaking and by seeking to provide reasonable control of damage to

---

[1] This statement is supported in the following quote from PEER 2008/01: "…considering the wide variety of lateral force-resisting systems included in the code over the years (over 80 systems), each controlled by a complex patchwork of prescriptive design requirements and limitations, the large configuration variations allowed for in each system, and the large variation of seismic conditions in the U.S. for which they are designed, it is likely that this methodology if implemented on every system would show large inconsistencies in the code-defined collapse margin."

structural and nonstructural systems that could lead to injury, economic loss, or loss of function for smaller, more frequent ground shaking.

Code provisions have typically focused on design for life safety for a relatively strong shaking level. Use of $R$ factors as large as 8, however, means that structural damage can be expected at more moderate shaking levels, in addition to whatever nonstructural damage may occur. For example, the 1989 Loma Prieta and 1994 Northridge earthquakes, having moment magnitudes ($M_w$) of 6.9 and 6.7, respectively, are considered to be moderate earthquakes. Although they resulted in relatively little life loss, numerous buildings suffered structural damage, and the affected regions suffered substantial economic losses. To prevent significant damage and economic losses in the future would require higher design base shear strengths; achieving this considering the geographic variation in seismic hazard implies the need to explicitly address performance at multiple hazard levels.

## 10.4 MODERN PERFORMANCE OBJECTIVES

Performance-based design refers to the explicit consideration of building performance at several different seismic hazard levels during the design process. Performance may be considered in terms of global parameters (e.g., roof drift), local parameters (e.g., plastic hinge rotation, curvature ductility of cross sections, and strains of steel and concrete materials), and intermediate parameters (e.g., interstory drift, story shear, or peak floor acceleration). Performance-based design resembles limit-states design, with each performance limit attached to a seismic hazard level appropriate for the occupancy of the building. Performance limits may be stated qualitatively (e.g., cracking, yielding, spalling, buckling, fracture, loss of gravity load support), but often must be quantified for purposes of design and evaluation. The coupling of a performance limit and seismic hazard level is referred to as a performance objective. For a given hazard level, better performance is sought for buildings whose function is critical after an earthquake (such as police and fire stations and hospitals) and those containing hazardous substances. While performance objectives are treated discretely in practice, one may imagine performance objectives that are described by continuous functions.

The basic framework for describing performance objectives was developed by the Structural Engineers Association of California (SEAOC Vision 2000 Committee, 1995), and disseminated in the 1999 Blue Book (SEAOC, 1999). Figure 10.1 illustrates these objectives. For example, a basic objective would pertain to a design for a typical new building. Modifications and refinements have been made in more recent guidelines and standards, in some cases focusing on rehabilitation of seismically deficient structures.

In practice, it may be sufficient to consider performance at two hazard levels, sometimes termed "serviceability" and "safety" levels. Serviceability refers to relatively little damage occurring in a frequent event, while safety refers to substantial damage short of collapse in a rare event. Serviceability limits often will be determined by deformation limits associated with cladding or curtain wall elements, or other nonstructural components.

## 10.5 TREATMENT OF PERFORMANCE OBJECTIVES IN DESIGN

Design approaches can be purely deterministic or probabilistic in nature, or may combine deterministic and probabilistic approaches to some degree. Three basic approaches are provided in this book (Chapter 18): a simple approach that provides a deterministic basis for

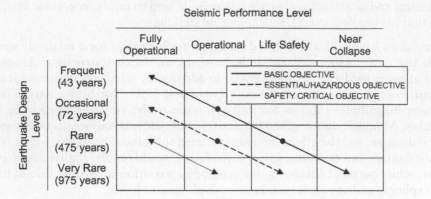

Figure 10.1 Seismic performance objectives.

design at a specified hazard level, which can be used to quickly generate a preliminary design on the basis of interstory drift and system ductility limits, and two more advanced approaches that employ either approximate or detailed hazard information to inform the design approach and essentially allow direct performance-based design, where the mean annual frequency of exceeding specified limit states is the design target.

The deterministic approach is further distinguished from the probabilistic ones in the sense that it can only be employed to target shaking intensities at the design spectrum mean annual frequency of exceedance (e.g., 10% in 50 years). Thus, for example, the deterministic approach would not be used to design for more rare limit-states (e.g., collapse), although inferences may be made about the performance at these limit-states based on the design parameters. In contrast, the two probabilistic design options allow freedom in choosing limit-states and setting performance objectives, subject to later verification by nonlinear dynamic analysis under multiple ground motion recordings.

The design approaches described in this book aim to ensure that the preliminary design can be refined (with further effort) into a system that nominally satisfies performance expectations, with little effort wasted on fruitless design iterations. This is distinct from pure analysis, which can be used to determine the performance of a given design, and differs from procedures that may select from designs contained in a database whose performance was previously evaluated. It is hoped that the design approaches (Chapter 18) will allow the engineer to develop good intuition about performance, while providing a sequence of steps and tools that assures the desired performance is achieved.

## 10.6 CONSIDERATION OF PERFORMANCE OBJECTIVES IN PRELIMINARY DESIGN

An obvious question is, given multiple performance objectives (e.g., the basic objective of Figure 10.1), how can one quickly select among and refine alternatives during schematic design, and for the chosen alternative, determine the required lateral strength and stiffness to use in preliminary design. Each performance objective (representing, for example, limits on interstory drift and system ductility, at a given mean annual frequency of exceedance) constrains the design space. An admissible design region (ADR) represents those combinations of system strength and stiffness that satisfy the performance objectives. ADRs can be determined for each performance objective, and the points common to all ADRs (the intersection of these spaces) constitute the ADR for the design objective.

To consider an example, the (somewhat arbitrary) performance objectives listed in Table 10.1 are applied to determine the ADRs for a single-degree-of-freedom (SDOF) oscillator. This example is of general interest, since it could be an equivalent SDOF system representation of the first-mode response of a multistory building (Chapter 7). The force–displacement response of this SDOF system is represented using a bilinear capacity curve, as shown in Figure 10.2. The yield point of the SDOF capacity curve defines the lateral strength and stiffness of the oscillator. The ductility demands ($\mu = u_{max}/u_y$) that result are due to the intensity of the hazard. Thus, the ADR for a given performance objective consists of the yield points that have acceptable performance for the specified intensity of shaking.

For this example, the hazard curves are defined by smoothed design spectra having the parameters given in Table 10.2. Figure 10.3 plots the elastic and inelastic spectra determined using the FEMA-440 coefficients $C_1$ and $C_2$ (detailed in Section 4.10.3) in the yield point spectra format, for the 10/50 hazard level.

With reference to the top panel of Figure 10.3, ADRs are established by the following logic.

a. Ductility limit: If the system ductility must not exceed 2, the region beneath the $\mu = 2.0$ curve must be eliminated, as shown by the hatched region of Figure 10.3.

b. Drift limit: A simple algebraic process is used to establish the boundary for the drift limit. Assume the height of the oscillator is 12.5 ft (= 150 in.). A drift limit of 4% of the height corresponds to a lateral displacement of 0.04*150 = 6.00 in. For elastic response (assuming $\mu = 1$), the yield displacement is equal to the peak displacement,

Table 10.1 Performance objectives example

| Hazard level[a] | System ductility | Peak drift ratio (%) |
| --- | --- | --- |
| 50/50 | 1 | 1 |
| 10/50 | 2 | 4 |
| 2/50 | 4 | 6 |

[a] Notation refers to probability of exceedance in the stated number of years.

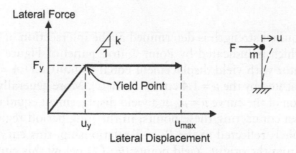

Figure 10.2 Idealized force–displacement response of SDOF oscillator. The peak displacement occurring under dynamic response is illustrated schematically.

Table 10.2 Definition of smoothed design spectra for example

| Hazard level | $S_a(0.2\,s)$ | $S_a(1\,s)$ |
| --- | --- | --- |
| 50/50 | 0.503g | 0.172g |
| 10/50 | 1.104g | 0.465g |
| 2/50 | 1.744g | 0.806g |

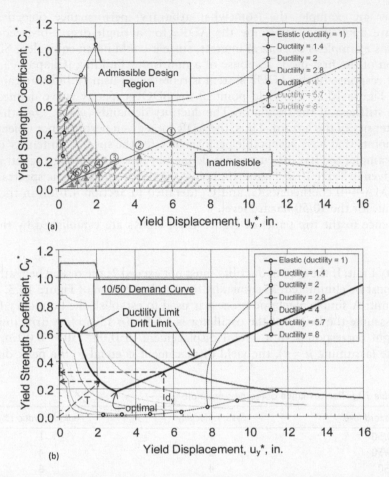

*Figure 10.3* Yield point spectra associated with the 10%/50-year hazard level: (a) mapping performance limits to establish the ADR; (b) demand curve associated with the border of the ADR.

and thus the required strength is determined at the intersection of the $\mu = 1$ curve and $u_y = 6.00$ in., which is indicated by Point ① (top panel of Figure 10.3). By a similar logic, an oscillator with yield displacement equal to $6.00$ in./$1.4 = 4.28$ in. must have the strength indicated by the $\mu = 1.4$ curve (Point ②). More generally, points are plotted at the intersection of the curve $\mu = \mu_0$ at a yield displacement equal to the drift limit/$\mu_0$. A curve is drawn connecting these points; in the long period region where the equal displacement rule is reflected in the $R$–$\mu$–$T$ relationship, this curve will be a straight line, radiating from the origin. Yield points $(u_y, C_y)$ below this curve have larger ductility demands and thus have excessive drift response. Consequently, the region below this curve is ruled out. The drift limit curve may be extended beyond the $\mu = 1$ curve along a line radiating from the origin, representing cases where $R = \mu < 1$ or $F_e > F_y$ and $u_y > S_d$.

The regions below the curves defining the ductility and drift limits result in ductility and drift demands in excess of the performance objective. Thus, ruling out these regions leaves the unshaded area as the ADR. Yield points located within the ADR have acceptable performance. The boundary between admissible and inadmissible regions in Figure 10.3b is labeled as the "10/50 Demand Curve." This valley-shaped curve has a minimum at the intersection of the curves defining the ductility and drift limits.

The yield point defining the change in slope of the bilinear curve of the SDOF oscillator has yield strength $F_y = V_y = C_y \cdot W = C_y \cdot m \cdot g$, yield displacement $u_y$, stiffness $k = F_y/u_y$, and period $T = 2\pi\sqrt{u_y/(C_y \cdot g)}$. Any two of these parameters ($C_y$, $u_y$ and either $k$ or $T$) are sufficient to define the yield point.

If the mechanical properties of the oscillator can be selected optimally, the lowest cost design satisfying the 10/50 performance objectives would have strength (equal to $C_y$ times weight) and yield displacement (or period) in the vicinity of the smallest admissible value of $C_y$. Achieving such a minimum strength design may require changes in materials or geometry (e.g., grade of steel, section depth, or span lengths). Alternatively, different lateral force-resisting systems (e.g., shear wall, braced frame, or moment-resistant frame) may be considered. Architectural constraints may influence the acceptability of such changes.

Alternatively, it may be that the period (or stiffness) or the yield displacement is known or can be estimated easily. Since lines of constant period extend radially from the origin, a line corresponding to the known period can be drawn; its intersection with the 10/50 Demand Curve establishes the required $C_y$ (Figure 10.3b). Similarly, if the yield displacement is known with sufficient precision, the required strength ($C_y W$) can be determined at the intersection with the 10/50 Demand Curve. If lines of constant period or constant yield displacement intersect the left side of the valley-shaped demand curve, ductility limits control while drift limits are easily satisfied. An intersection with the right side of the valley-shaped curve indicates that drift limits control, and that system ductility limits are easily satisfied.

Following a similar process, the 2/50 and 50/50 Demand Curves can be established, using the applicable inelastic spectra (Figure 10.4). Finally, the demand curves obtained for the three performance objectives can be plotted together (Figure 10.5). Each demand curve has the potential to constrain the ADR; the intersection of ADRs obtained for each performance objective determines the composite ADR—i.e., the highest strength at any given yield displacement governs. Inspection of the composite demand curve indicates the particular performance objectives that govern various portions of the composite ADR.

## 10.7 DESIGN VALIDATION AND ITERATION

The intersection of the composite demand curve and a line representing a known (or estimated) period or yield displacement defines the design yield point. As stated earlier, the yield point is defined by two independent variables but is associated with three related quantities: (i) strength ($V_y = C_y W$), (ii) yield displacement ($u_y$), and (iii) stiffness ($k = V_y/u_y$) or period $\left(T = 2\pi\sqrt{u_y/(C_y \cdot g)}\right)$.

In many design scenarios, it will be easiest to provide sufficient strength and verify that the resulting period of vibration matches that associated with the design yield point. If iterations are required, it is worth noting whether the design yield point is controlled by ductility constraints (left side of the valley-shaped curve) or by drift (right side). Clearly, a ductility-controlled system that is more flexible (longer period) than intended will be acceptable, as will be a stiffer (shorter period) system that is controlled by drift. It is plainly evident that systems that are stronger than required will have lower ductility demands and thus, acceptable performance.

The preceding observation is used as an advantage in design based on the stability of the yield displacement. The steps are: (i) enter the YPS with an estimate of the yield displacement,

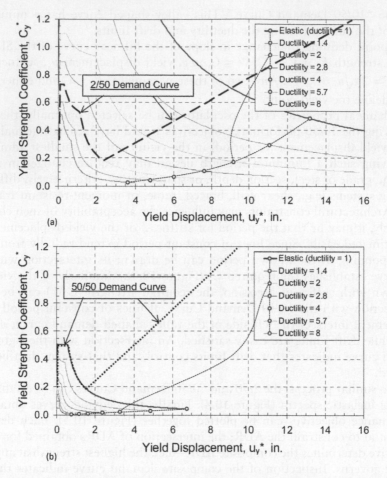

Figure 10.4 Demand curves associated with the (a) 2%/50 year and (b) 50%/50-year hazard levels.

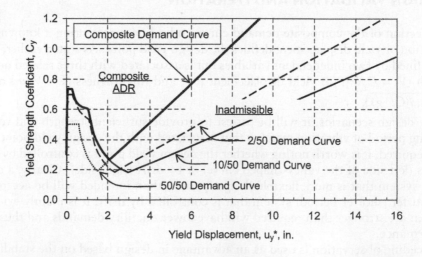

Figure 10.5 Composite demand curve resulting from multiple performance objectives.

(ii) determine the required yield strength coefficient based on the governing performance objective, (iii) proportion the structure to have adequate strength based on a plastic mechanism analysis, and (iv) compute the period of the system using an analytical model of the structural design. If the computed period differs from the value associated with the design values of $u_y$ and $C_y$, then one may deduce that the actual yield displacement is to the left or right of the value that had been assumed in (i). Whether this is acceptable or not depends on whether the design is controlled by drift or ductility. If needed or desired, the design may be revised by repeating steps (i)–(iv), but using a better estimate of the yield displacement, given by: $u_y = \left(\dfrac{T}{2\pi}\right)^2 C_y \cdot g.$

## REFERENCES

ATC 3-06 (1978). Tentative Provisions for the Development of Seismic Regulations for Buildings, Applied Technology Council, Publication ATC 3-06, 505 pp.

BSSC (2009). *NEHRP Recommended Seismic Provisions for New Buildings and Other Structures (FEMA P-750)*, Building Seismic Safety Council, Federal Emergency Management Agency, Washington, DC.

SEAOC (1960). *Recommended Lateral Force Requirements and Commentary*, Seismology Committee, Structural Engineers Association of California, San Francisco, CA.

SEAOC Vision 2000 Committee (1995). Performance-Based Seismic Engineering, Report prepared by Structural Engineers Association of California, Sacramento, CA.

SEAOC (1999). Recommended Lateral Force Requirements & Commentary (Blue Book), Report prepared by Structural Engineers Association of California, Sacramento, CA.

# Chapter 11

# Plastic mechanism analysis

## 11.1 PURPOSE AND OBJECTIVES

Design for ductile response is a well-used strategy for reducing design forces in earthquake-resisting structures (Section 4.2), and typically requires the use of special details (portions of Chapters 15 and 16) to enhance ductility capacity at locations expected to undergo inelastic response (i.e., plastic hinge zones). This chapter describes the use of a kinematically admissible mechanism for determining the locations of the plastic hinges, and a virtual work method of analysis for determining the strengths required at these locations. The base shear strength of the system (at yield) needed to satisfy the performance objectives, is determined in other chapters.

## 11.2 DUCTILE WEAK LINKS

Ductility is provided in specially detailed locations of seismic force-resisting systems. The underlying principle can be explained using the simple analogy of a chain composed of multiple links (Figure 11.1).

It is correctly stated that a chain is only as strong as its weakest link. As long as the weakest link is ductile, the chain will exhibit ductile behavior. The weak link can be thought of as a fuse that limits the force that can be developed in the remaining links (under quasi-static loading). With reference to Figure 11.1, the largest force that can develop in any link, $F_{max}$, is limited to $F_y$. As long as the remaining links are stronger than the weak link, they can be brittle, since they will not experience loads sufficient to make them fail. This principle underlies the proportioning of most earthquake-resistant structural systems.

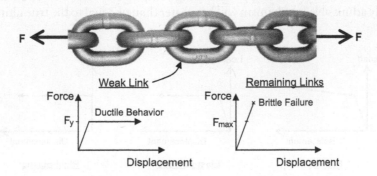

*Figure 11.1* A chain composed of a ductile, weak link and stronger, brittle links exhibits ductile behavior.

A word of caution: for conventional loading (gravity, wind, etc.), an increase in member strength is not harmful. However, for lateral force-resisting systems, an increase in the strength of a ductile component can have the undesirable effect of inducing failures in brittle members that had been intended to remain elastic.

## 11.3 PLASTIC MECHANISM ANALYSIS

Plastic mechanism analysis is a simple method of analysis that can be used to establish the strength of a ductile system (e.g. Neal (1977). The method of analysis is described in this section, while its use in a design context is addressed in Section 11.6.

In essence, plasticity is assumed to occur at discrete locations, termed plastic hinges, which have defined yield (or plastic) strengths and ample rotation capacity. Analysis is simplified by assuming members are rigid until the onset of yielding, at which they are capable of deforming at the yield level an amount that is sufficient to develop the actual mechanism in the real (nonrigid) structure. The assumption of rigid-plastic behavior (Figure 11.2) allows the mechanism strength to be determined using a very simple application of virtual work. Since external work is calculated assuming small deformations, this method of analysis is applicable to ordinary (stiff) structures that develop mechanisms at displacements that are small relative to the dimensions of the structure.

A single hinge is sufficient to develop a collapse mechanism in a statically determinate structure (Figure 11.3a). Additional hinges are required to form a mechanism as the degree of static indeterminacy increases, but the precise number depends on the nature of the mechanism—the plastic mechanism may affect just a portion of the structure (a local mechanism) or the entire structure (a global mechanism). In the case of the fixed-ended beam (Figure 11.3b), symmetry suggests that the midspan plastic hinge can be considered to be two plastic hinges, at either side of the centerline. Structures having a common configuration, but different member strengths or distributions of applied loads, can develop mechanisms requiring different numbers of plastic hinges (compare (d) with (e), and (f) with (g) and (h)).

Of the possible mechanisms for a structure (e.g. Figure 11.3d,e or f–h), the governing mechanism and associated strength for a given loading can be determined using either of two relatively simple approaches. Each involves a trial and error process in which one first assumes the locations of plastic hinges, sufficient in number to develop a kinematically admissible mechanism. Then, either a virtual work analysis is done to establish the collapse strength of the assumed mechanism, or the distribution of internal member forces (e.g., moments, shears, axial forces, and torque) is determined from statics. Identification of the governing mechanism (solution) is based on the following theorems.

- *Upper Bound Theorem:* An applied loading computed on the basis of an assumed (kinematically admissible) mechanism will be greater than or equal to the true ultimate strength.

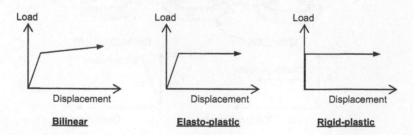

*Figure 11.2* Three types of ductile load-deformation behavior.

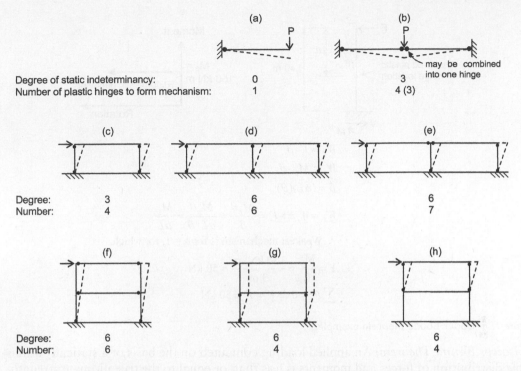

*Figure 11.3* Examples of plastic mechanisms: (a)–(h) illustrate the number of hinges required to form a mechanism and the degree of static indeterminacy.

A kinematically admissible mechanism is one that complies with geometrical constraints associated with rotations (or displacements in the case of axial plasticity) occurring at designated plastic hinges, with rigid members capable only of rigid body rotations and translations.

## BOX 11.1   UPPER BOUND THEOREM EXAMPLE

Figure 11.4 illustrates a reinforced concrete column having uniform plastic moment capacity $M_p = 150$ kN·m over its entire length. Possible locations of hinges are considered parametrically by locating the hinge at $\alpha L$ from the top of the column. Under the constraint of rigid-plastic behavior, the external work consists of the force $F$ moving through a displacement $\delta$. Thus, $W_E = F \cdot \delta$. Internal work consists of a single plastic hinge deforming through a rotation $\theta$; for rigid-plastic behavior this is expressed as $W_I = M_p \cdot \theta$. Setting $W_E = W_I$ allows the force to be solved for in terms of the plastic moment capacity: $F = M_p \cdot \theta / \delta$. For small angles, $\delta = (\alpha \cdot L)\theta$; thus, substitution allows the deformation term, $\theta$, to be eliminated, yielding $F = M_p/(\alpha \cdot L)$. Thus, an infinite number of possible kinematically admissible mechanisms are represented as $\alpha$ varies from 0 to 1. In effect, the upper bound theorem states that the weakest of all possible mechanisms is the governing collapse mechanism. Considering equilibrium of forces in the x-direction, $V = F = M_p/\alpha L$. Thus, the weakest mechanism (smallest $V$) is obtained when the hinge is located at the base of the column ($\alpha = 1$), for which $V = M_p/L = 150$ kN·m/3 m = 50 kN. (Note that this problem also could have been solved by considering a number of discrete locations of the plastic hinge, such as at mid-height ($\alpha = 0.5$) and the base ($\alpha = 1.0$), and determining $V$ separately for each assumed plastic hinge location.)

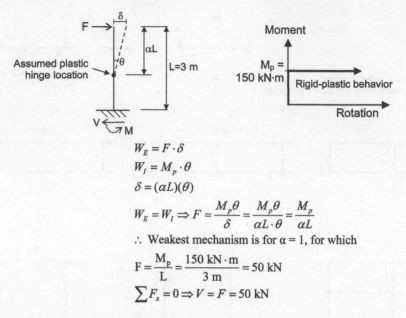

$$W_E = F \cdot \delta$$

$$W_I = M_p \cdot \theta$$

$$\delta = (\alpha L)(\theta)$$

$$W_E = W_I \Rightarrow F = \frac{M_p \theta}{\delta} = \frac{M_p \theta}{\alpha L \cdot \theta} = \frac{M_p}{\alpha L}$$

$\therefore$ Weakest mechanism is for $\alpha = 1$, for which

$$F = \frac{M_p}{L} = \frac{150 \text{ kN} \cdot \text{m}}{3 \text{ m}} = 50 \text{ kN}$$

$$\sum F_x = 0 \Rightarrow V = F = 50 \text{ kN}$$

*Figure 11.4* Upper bound theorem example.

*Lower Bound Theorem:* An applied loading computed on the basis of a statically admissible distribution of forces and moments is less than or equal to the true ultimate strength.

A statically admissible distribution of forces and moments refers to distributions of axial force ($P$), shear ($V$), moment ($M$), and torque ($T$) that are in static equilibrium, such that each is less than its plastic capacity ($P \le P_p$, $V \le V_p$, $M \le M_p$, and $T \le T_p$), where the plastic capacity accounts for interaction where applicable.

### BOX 11.2    LOWER BOUND THEOREM EXAMPLE

The example of Box 11.1 (Figure 11.4) is considered again. A plastic hinge is located parametrically at $\alpha L$ from the top of the column. Shear and moment diagrams are shown, with the heavy line on the moment diagram indicating the location where plasticity ($M_p$) develops. Moments clearly exceed $M_p$ at locations below the plastic hinge; thus, the only statically admissible moment diagram ($M \le M_p$ everywhere) occurs for the plastic hinge located at the base of the column ($\alpha = 1$). Equilibrium ($\sum M = 0$) provides that $F = M_p/L$ while $\sum F_x = 0$ provides that $V = F$. Thus, the mechanism strength is $V = M_p/L = 150$ kN·m/3 m = 50 kN.

*Uniqueness Theorem:* There is a kinematically admissible mechanism which has a statically admissible distribution of forces and moments. This mechanism satisfies the upper and lower bound theorems and is the governing mechanism.

### BOX 11.3    UNIQUENESS THEOREM EXAMPLE

Figure 11.4 illustrates kinematically admissible mechanisms which clearly include some ($\alpha < 1$) that do not result in statically admissible moment diagrams (Figure 11.5). The only mechanism satisfying the Uniqueness Theorem is the one for which $\alpha = 1$.

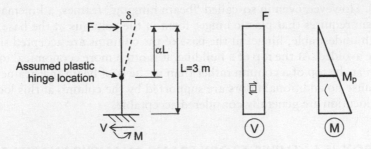

*Figure 11.5* Lower bound theorem example.

In some cases, the plastic hinges that develop for the controlling mechanism may leave portions of the structure statically indeterminate (e.g., Figure 11.3g,h). For this reason, it is usually easier in practice to determine the weakest of the kinematically admissible mechanisms, and then determine the static admissibility of the member internal forces for this case. If the internal forces are statically admissible, the Uniqueness Theorem ensures that the governing mechanism was identified.

Some common types of mechanisms are illustrated in Figure 11.6. Beam plastic hinges are preferable to column plastic hinges, since the failure of beam plastic hinges generally has consequences limited to the supported floor, whereas the development of a column plastic hinge potentially jeopardizes the stability of the floors above. Story sway mechanisms generally lead to very large local plastic hinge rotation demands, often exacerbated by $P$-$\Delta$ effects to the point that a sidesway collapse of the structure occurs. For these reasons, beam-hinging mechanisms (Figure 11.6a) are strongly preferred over story sway mechanisms

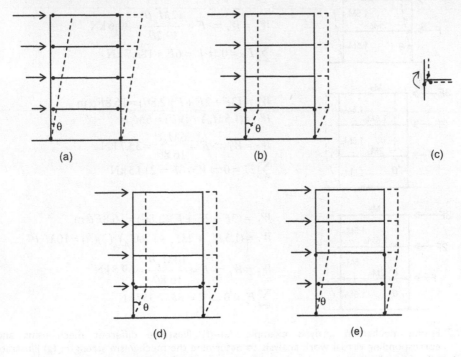

*Figure 11.6* Some common mechanism types: (a) beam-hinging mechanism; (b) weak-story (sway) mechanism; (c) joint mechanism; (d) story sway with joint mechanism; (e) multiple story mechanism.

(Figure 11.6b). However, even in so-called "beam hinging" frames, a kinematically admissible mechanism requires that plastic hinges form in the columns at the base of the frame. Thus, although undesirable, hinges at the base of the columns are accepted simply because they cannot be avoided. At the top of a building, it is often more economical to allow plastic hinges to form at the top of a column rather than in the roof beam that frames into the column, and because no additional floors are supported by the column at this location, plastic hinges at this location are generally considered acceptable.

## BOX 11.4   THREE-STORY FRAME ANALYSIS EXAMPLE

Some of the possible mechanisms of a three-story frame are shown in Figure 11.7. Determine the strengths of the collapse mechanisms shown, and verify that the weakest of these mechanisms is the governing collapse mechanism. Virtual work calculations to establish the base shear strength are shown for the mechanisms identified in the figure. It is evident that (a) is the weakest of the mechanisms considered. Coincidentally, several mechanisms had identical strengths. Internal member forces are determined for case (a) using free-body diagrams of portions of the frame. Appropriate values of plastic moments are shown acting at plastic hinges with sense consistent with the sense of plastic deformation. Beam shears are determined by $\sum M = 0$ about either end of the beam; these shears generate axial forces in the columns. Member moment diagrams are then determined.

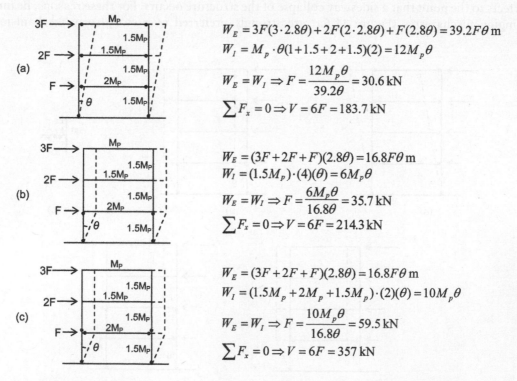

For mechanism (a):
$$W_E = 3F(3 \cdot 2.8\theta) + 2F(2 \cdot 2.8\theta) + F(2.8\theta) = 39.2F\theta \, \text{m}$$
$$W_I = M_p \cdot \theta(1 + 1.5 + 2 + 1.5)(2) = 12M_p\theta$$
$$W_E = W_I \Rightarrow F = \frac{12M_p\theta}{39.2\theta} = 30.6 \, \text{kN}$$
$$\sum F_x = 0 \Rightarrow V = 6F = 183.7 \, \text{kN}$$

For mechanism (b):
$$W_E = (3F + 2F + F)(2.8\theta) = 16.8F\theta \, \text{m}$$
$$W_I = (1.5M_p) \cdot (4)(\theta) = 6M_p\theta$$
$$W_E = W_I \Rightarrow F = \frac{6M_p\theta}{16.8\theta} = 35.7 \, \text{kN}$$
$$\sum F_x = 0 \Rightarrow V = 6F = 214.3 \, \text{kN}$$

For mechanism (c):
$$W_E = (3F + 2F + F)(2.8\theta) = 16.8F\theta \, \text{m}$$
$$W_I = (1.5M_p + 2M_p + 1.5M_p) \cdot (2)(\theta) = 10M_p\theta$$
$$W_E = W_I \Rightarrow F = \frac{10M_p\theta}{16.8\theta} = 59.5 \, \text{kN}$$
$$\sum F_x = 0 \Rightarrow V = 6F = 357 \, \text{kN}$$

*Figure 11.7* Frame mechanism analysis example: (a)–(f) illustrate different mechanisms and the corresponding virtual work analysis to determine the mechanism strength; (g) illustrates the use of free-body diagrams to determine member forces.

*(Continued)*

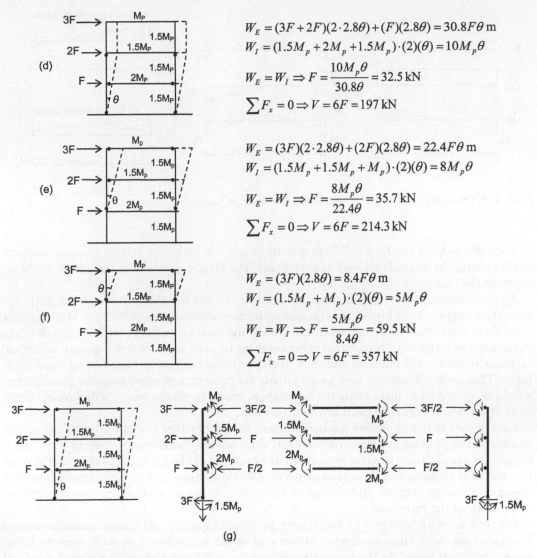

Figure 11.7 (CONTINUED) Frame mechanism analysis example: (a)–(f) illustrate different mechanisms and the corresponding virtual work analysis to determine the mechanism strength; (g) illustrates the use of free-body diagrams to determine member forces.

## 11.4 INTERACTION WITH GRAVITY LOAD

While plastic mechanism analysis can be used to determine the true ultimate strength of systems subjected to only gravity loads,[1] the topic addressed in this section is the influence of gravity loads on the development of plastic mechanisms induced by lateral load.

Figure 11.8 illustrates a beam subjected simultaneously to moments induced by earthquake actions (e.g., as illustrated in Figure 11.7g) and gravity loads applied along the span. Superposition of seismic and gravity loads results in the shear and moment diagrams shown

---

[1] This is done by equating the work done by externally applied gravity loads moving through kinematically admissible displacements to the internal work done by plastic hinge rotations.

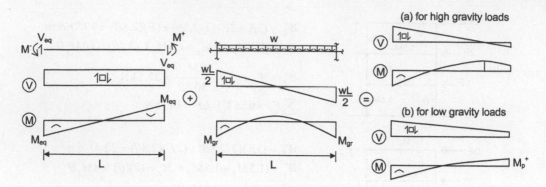

Figure 11.8 Effect of gravity loads on mechanism formation in beams.

at the right side of the figure. Where gravity loads are high, the largest positive moment occurs along the span. For lower gravity loads, the largest positive and negative moments occur at the ends.

A plastic hinge will form if the internal moment reaches the plastic moment strength; the formation of two such hinges is sufficient to form a mechanism. While it is conceptually possible to distribute beam reinforcement to shift the locations where plastic hinges develop, customary practice is to proportion reinforcement to provide sufficient capacity at critical sections located at the midspan and ends, and determine bar cutoffs based on moment envelopes. Thus, where relatively high gravity loads are present, a positive moment plastic hinge would be expected to form along the beam span, whereas plastic hinges will develop at the ends of the beam if gravity loads are relatively low.

For the case with low gravity loads, the plastic hinges at either end of the beam are subject to alternating positive and negative moment, since seismic loads reverse directions. Thus, a hinge that opens due to positive moment in one cycle would be expected to close as the positive moment reduces and then as negative moment develops, and this pattern of reversing moments causing opening and closing of beam plastic hinges would continue throughout most or all of the response.

For the case with high gravity loads, superposition of gravity and seismic moments causes a negative moment hinge to develop at one end of the beam and a positive moment hinge along the beam span. As the seismic loads reverse, the largest positive moment does not develop at the end of the beam and thus is not able to cause a reversal of inelastic deformation at the negative moment hinge. Rather, superposition of the positive gravity load moment and positive seismic moment lead to the formation of a positive moment hinge along the span. The positive moment hinges open under positive moment, but subsequent negative moments are inadequate to cause a reversal of inelastic deformation at the positive moment hinges. Thus, a phenomenon known as *ratcheting* develops, in which each hinge opens more with each cycle of loading. Deformation of the beam continues to increase with each successive cycle, leading to large permanent deformations and perhaps local collapse due to exhaustion of material strain capacity. Such behavior is undesirable and can be avoided by configuring the gravity and lateral force-resisting systems to ensure that the tributary gravity loads do not exceed a critical level.

A peak (maximum or minimum) moment develops where $dM/dx = 0$. Thus, to ensure that peak moments do not develop along the span is equivalent to a constraint that $V = dM/dx \neq 0$ along the span. For the case shown in Figure 11.8, this constraint is satisfied for

$$V_{eq} \geq \frac{wL}{2} \tag{11.1}$$

Considering moment equilibrium about either end of the beam under seismic loading allows $V_{eq}$ to be determined as

$$V_{eq} = \frac{M_p^+ + M_p^-}{L} \tag{11.2}$$

Based on the preceding, in-span hinges can be avoided for

$$w \leq \frac{2}{L^2}\left(M_p^+ + M_p^-\right) \tag{11.3}$$

Thus, if gravity loads are small enough that they do not alter the mechanism that develops under lateral load, the gravity loads do no external work and thus do not influence the lateral strength determined by mechanism analysis. Consequently, as long as the gravity loads satisfy Equation 11.3 (with a sufficient margin) for each beam span, there is no need to consider gravity loads in proportioning the strengths of the plastic hinge regions of the lateral system. This differs from conventional code load combinations that bundle gravity (e.g., dead and live) loads with seismic loads. Load factors applied to gravity loads increase moment demands at locations that also resist seismic loads, thus increasing the lateral strength of the system (for seismic loads). After the design for lateral loads is complete, the beams can be checked to ensure that they have adequate capacity to resist the factored gravity loads (not in combination with seismic).

The presence of gravity loads will cause some plastic hinges to form earlier and others later as lateral load is applied. This is because the gravity loads preload the beam (e.g., inducing negative moments at the ends), and thus hinges will form earlier where moments due to lateral loads add constructively with those due to gravity loads. This will not change the ultimate strength of the mechanism, but will cause the capacity curve to be somewhat rounder than it otherwise would be.

The influence of gravity loads on capacities of members such as columns and walls (due to $P$–$M$ interaction) should be considered when proportioning these members to have the required strength.

## 11.5 REINFORCED CONCRETE LATERAL FORCE-RESISTING SYSTEMS

Recognized reinforced concrete structural systems used to resist lateral forces due to seismic actions are depicted in Figure 11.9. Plastic hinges in reinforced concrete lateral force-resisting systems are all flexural in nature. In contrast, some steel systems make use of members that develop plastic hinges axially (e.g., buckling-restrained braced frames).

It is evident that weak-story mechanisms (e.g. Figure 11.6b,d, and e) are not considered desirable. However, nonlinear dynamic analyses in recent studies (e.g., FEMA P-695, 2009) indicate that code provisions are not adequate to safeguard against weak-story mechanisms; one approach to address this is to proportion column flexural strengths generously compared with that required in Chapter 18 of ACI 318 (2014), discussed in Section 15.6.2.

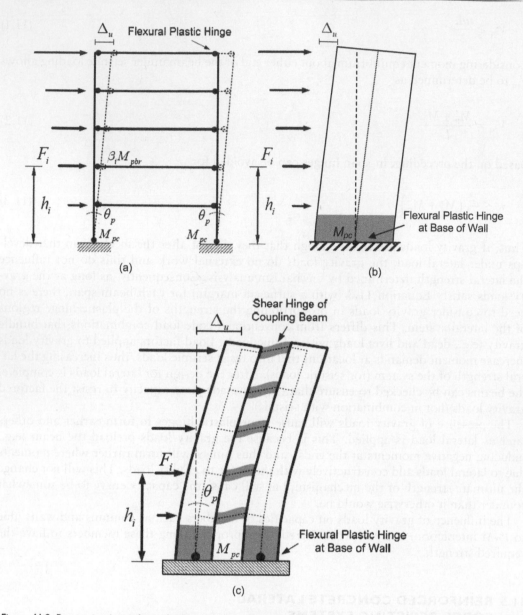

*Figure 11.9* Recognized reinforced concrete lateral force-resisting systems. (a) Moment frame; (b) shear wall; (c) coupled wall system. (From NEHRP Provisions Resource Paper No. 9, 2009.)

## 11.6 DESIGN FOR DESIGNATED MECHANISMS

Much of the effort involved in analysis to determine the governing plastic mechanism can be avoided in design. By proportioning the structure to develop an intended plastic mechanism, we can ensure that the intended mechanism is the one that develops. Given an intended mechanism (e.g., Figure 11.9) and the desired strength (to be discussed in Chapter 12), we can use a virtual work analysis to determine the required strengths of the plastic hinges. Knowledge of the distribution of applied lateral forces and plastic hinge moment strengths is sufficient to establish internal forces (e.g., Figure 11.7g) for most desired mechanisms.

The remaining portions of the frame are designed to remain elastic in the presence of the applied lateral forces and the established plastic hinge moment strengths.

Before illustrating this process, several considerations must be addressed. First, to maintain the simplicity of this design approach, plastic hinges are assumed to be located at the intersections of beam, column, and wall centerlines. If the hinges were considered to be inset from the centerline intersections, higher member shears would develop and a slightly higher collapse mechanism strength would be computed (generally on the order of 5%–10%). This is conservative for system design, but should be kept in mind when designing the elastic (force-controlled) portions of the system, since the forces in the system will be slightly larger than those determined on the basis of centerline dimensions.

Second, for design purposes, we may choose to estimate the plastic hinge strength as the nominal flexural strength, $M_n$, calculated using $1.25f_y$ in place of the specified minimum reinforcement yield strength, $f_y$. This is consistent with what is done in Chapter 18 of ACI 318-14 and reflects to some extent the likelihood that actual material strengths will exceed specified minimums as well as the presence of strain hardening. (This is one component of "overstrength," described in Section 12.5.) $M_p$ computed in this way underestimates the maximum flexural strength that may develop at the plastic hinge, but nevertheless is a useful index for design purposes.

For simplicity, the calculation of $M_p$ in beams can be made neglecting the compression reinforcement, as this usually contributes only a minor amount to strength. (This assumption is conservative for proportioning the lateral force-resisting system, since the presence of compression reinforcement will provide a small increase in strength.) In columns, the influence of gravity loads expected at the time of earthquake shaking on flexural strength should be considered. In many cases, the variation in axial load due to overturning forces associated with lateral response can be neglected at the preliminary design stage. Although overturning axial forces can change the plastic flexural strength, increases in flexural strength on one side of the frame are for the most part counterbalanced by reductions on the other side. These effects can be picked up in nonlinear static and dynamic analyses, and affect the forces to be considered in the design of the elastic portions of the lateral force-resisting system.

Finally, the negative moments due to gravity and seismic loads add together (constructive superposition) at the ends of moment frame beams. This condition may determine the section dimensions and top reinforcement. Chapter 18 of ACI 318 requires that $M_p^+ \geq 0.5M_p^-$ at the critical sections at the ends of the beam. If one assumes that $M_p^+ = 0.5M_p^-$ and that a positive moment hinge forms at one end of the beam and a negative moment hinge at the other, the internal work done by the beam is $1.5M_p^-(\theta)$. Thus, where each beam hinge is assumed to develop $M_p$ in the mechanism analysis, the required negative moment strength is $M_p^- = 1.33M_p$ and the required positive moment strength is $M_p^+ = 0.67M_p$. This approach to proportioning of plastic moment strengths at the beam ends is included in the following example.

## BOX 11.5   THREE-STORY FRAME DESIGN EXAMPLE

The three-story, three-bay frame of Figure 11.10 is to be designed to have a base shear strength of 600 kN for the given lateral force distribution and relative flexural strength distribution (e.g., as discussed in Section 12.8.3). The relative strength distribution is indexed in terms of a reference plastic hinge strength, $M_{pr}$, for the beams, where beam plastic hinges on the first, second, and third floors have relative strengths of $2M_{pr}$, $1.5M_{pr}$, and $1M_{pr}$, respectively.

At the top floor, we choose to have hinges located at the tops of the columns—considering the moment pattern obtained in a portal frame analysis, this results in the plastic hinge strengths at the tops of the intermediate columns being equal to twice the roof beam plastic moment capacities. External and internal work is evaluated to establish the reference hinge strength $M_{pr}$ (Figure 11.10a):

$$W_E = \left(300 \text{ kN}(3) + 200 \text{ kN}(2) + 100 \text{ kN}(1)\right)(3 \text{ m})(\theta) = 4{,}200(\theta) \text{ kN} \cdot \text{m} \tag{11.4}$$

and

$$W_I = \left((1 + 1.5 + 2)(6)(M_{pr}) + 6M_{pc}\right)(\theta) = (27M_{pr} + 6M_{pc})\theta \tag{11.5}$$

We do a separate analysis to establish $M_{pc}$, avoiding a potential weak-story mechanism by increasing the column plastic hinge strength above that associated with a weak-story mechanism (Figure 11.10b). Evaluating external and internal virtual work for the weak-story mechanism results in $M_p = 150$ kN·m, obtained by solving $(3\theta(300+200+100) = \theta(4M_p+4\cdot2M_p))$. In this example, we set the nominal column strength, $M_{nc}$, equal to $1.2M_p$, and evaluating $M_{pc} \approx 1.25M_{nc} = 1.5M_p$ results in $M_{pc} = 225$ kN·m. Substituting this value into Equation 11.5 and setting $W_E = W_I$ allows $M_{pr} = 105.6$ kN·m to be determined.

Equations of equilibrium are used to determine the moment diagram along any column. Considering the pattern of shears in a portal frame analysis, each intermediate column is assumed to carry twice the shear of an end column and has twice the moments of an end column. Thus, for this frame, application of $\Sigma F_x = 0$ results in $V + 2V + 2V + V = 6V = 300 + 200 + 100 = 600$ kN. Therefore, an intermediate column carries $2V = 200$ kN = 1/3 of the applied force. A free-body diagram showing the forces and moments applied to an intermediate column is shown in Figure 11.10c and the corresponding internal moment diagram is given in Figure 11.10d. Because the intermediate column has positive beam plastic moments acting on one side of a joint and negative beam plastic moments acting at the other, the moment diagram for this intermediate column is the same whether the beam plastic moments are partitioned into $M_p^+$ and $M_p^-$ or not. Note that the beam shears shown in Figure 11.10c are determined by considering $\Sigma M$ about either end of the beam shown in the free-body diagram of Figure 11.10e, which results in

$$V = \frac{\sum M_p}{L} = \frac{2M_p}{L} \tag{11.6}$$

For $M_p = 1.5M_{pr}$ and $L = 8$ m, Equation 11.6 yields $V = 39.6$ kN. Similarly, for $M_p = 2M_{pr}$, $V = 52.8$ kN. Shear in the top beam can be determined from the free-body diagram of Figure 11.10f, for which the beam shear, equal to the axial force in an end column, is given by 26.4 kN, which coincides with what would be obtained using Equation 1.2 with $M_p = 1M_{pr}$. This reflects the choice to locate the plastic hinges in the columns at the roof level rather than locating them in the roof beams, a choice that has no consequence on the calculated strength of the plastic mechanism (when the mechanism is calculated assuming the plastic hinges occur at the intersections of the beam and column centerlines).

The beam shears developed by the lateral loads acting on this intermediate column counteract one another, producing no net axial force in the intermediate column in this example. Of course, the design of the columns for flexural strength should consider the presence of axial force due to gravity load. For an end column, the beam shears add from story to story, producing either tension or compression in the column, acting in conjunction with whatever gravity axial force is present.

The kinematic mechanism mobilizes positive moments at one end of a beam and negative moments at the other. For the design of any beam, we require that $M_p^+ + M_p^- = 2M_p$. In addition, Chapter 18 of ACI 318 requires that $M_p^+ \geq \frac{1}{2}M_p^-$. Setting $M_p^+ = \frac{1}{2}M_p^-$ allows preliminary proportioning of the beam. For the first floor beams (for which $M_p = 2M_{pr}$), we have $M_p^+ + M_p^- = (0.5+1)M_p^- = 2M_p = (2)(2M_{pr}) = 4M_{pr} = 4(105.6 \text{ kN} \cdot \text{m})$. This establishes $M_p^- = 281.6 \text{ kN} \cdot \text{m} = 0.2816 \text{ MN} \cdot \text{m}$.

In this example, the plastic moment capacity is estimated as the nominal moment strength, $M_n$, computed using $1.25f_y$ in place of $f_y$. For Grade 420 ($f_y \approx 60.9$ ksi) reinforcement, and assuming a longitudinal reinforcement ratio ($\rho$) of 1.5%, $b = (2/3)d$, and neglecting the contribution of compression steel and slab reinforcement allows preliminary dimensions and reinforcement to be determined. Substituting into

$$M_n = \rho \cdot f_y \cdot b \cdot d^2 \left(1 - 0.59\rho \cdot f_y/f_c'\right) \tag{11.7}$$

results in

$$0.2816 \text{ MN} \cdot \text{m} = 0.015(1.25 \cdot 420 \text{ MPa})(2d/3)(d^2)(1 - 0.59(0.015)(1.25 \cdot 420/35)) \tag{11.8}$$

which may be solved for $d^3 = 0.06185 \text{ m}^3$ or $d = 0.395$ m, from which $b = (2/3)d = 0.264$ m, and $A_s = 0.015(b)(d) = 1.56 \times 10^{-3} \text{ m}^2 = 1,564 \text{ mm}^2$. (3) No. 8 bars provide 1,530 mm², which is close to 1,564 mm², so the height of the trial section was chosen to be 0.450 m, resulting in a depth to the centroid of reinforcement of 0.390 m, which is 1% less than the required depth of 0.395 m, and is close enough for preliminary design. Bottom reinforcement was selected to provide just over half of the top steel area: (2) No. 7 bars provide 774 mm². This establishes the preliminary section for the 1st floor beams, shown in Figure 11.10g.

A similar procedure was used to establish the preliminary section for the second floor beams, using $M_p^+ + M_p^- = 2M_p$, where $M_p = 1.5M_{pr} = 1.5(105.6 \text{ kN} \cdot \text{m})$. The result is the section shown in Figure 11.10h.

If plastic moments are not partitioned into $M_p^+$ and $M_p^-$, the moment diagram for an end column is simply one half of that for an intermediate column (for this example). However, since $M_p^+$ is taken equal to $\frac{1}{2}M_p^-$ and the sum of these moments is equal to $2M_p$, then $M_p^- = (2/3)(2M_p) = 4M_p/3$, and $M_p^+ = 2M_p/3$. Thus, the end column will experience larger moments when the negative beam moments act. The end column can be proportioned for the larger beam moments (associated with the negative moment beam plastic hinges), or can be chosen equal to an intermediate column at this stage in preliminary design.

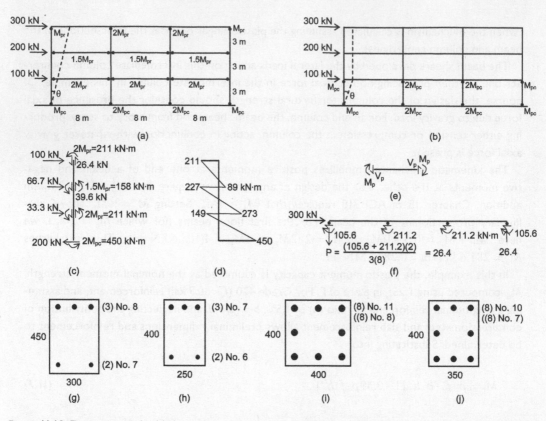

**Figure 11.10** Design example: (a) desired mechanism and distribution of applied forces and plastic strengths; (b) weak-story mechanism; (c) free-body diagram of intermediate column; (d) intermediate column moment diagram; (e) free-body diagram for determining $V_p$ in beams; (f) free-body diagram for determining shear in roof beam; (g) preliminary design of first floor beams; (h) Preliminary design of second floor beams; (i) preliminary design of intermediate column at base (and at roof); (j) preliminary design of end column at base (and at roof). Dimensions in mm.

## 11.7 CONSIDERATION OF MULTI-DEGREE-OF-FREEDOM EFFECTS

The discussion of Section 11.2 considered the chain to be subjected to a gradually increasing (quasi-static) loading. The chain analogy works well for a beam experiencing plastic hinges at either end; the beam shear can be calculated reliably using Equation 11.6, in terms of the plastic hinge strengths at the ends of the beam. However, the chain analogy needs to be extended to address cases where lumped masses may exist at intermediate points along the length of the chain. In such cases, under dynamic response, some links will experience forces that exceed $F_y$, as a result of the forces generated as the masses respond dynamically. This extension of the common weak-link analogy illustrates the effect of higher modes (or "MDOF effects" for nonlinear response) on response quantities. Multi-degree-of-freedom effects cause phenomena such as shear forces in slender multistory shear walls being several times higher than the shear associated with a static mechanism analysis in which a plastic hinge forms at the base of the wall—there are lumped masses at every floor responding dynamically and contributing to the story shears. Consequently, the design approach described in Chapter 18 uses a mechanism analysis to establish the strength of the yielding elements of the lateral force-resisting system, while nonlinear dynamic analysis is relied upon to establish the required strengths of force-protected components, which may be subjected to higher-mode forces, and whose failure is to be avoided.

## REFERENCES

FEMA P-695 (2009). Recommended Methodology for Quantification of Building System Performance and Response Parameters, Report No. FEMA P695A, Prepared for the Federal Emergency Management Agency, Prepared by the Applied Technology Council, Redwood City, CA.

Neal, B. G. (1977). *The Plastic Methods of Structural Analysis*, Chapman & Hall, London.

NEHRP Provisions Resource Paper No. 9 (2009). Seismic Design Using Target Drift, Ductility, and Plastic Mechanisms as Performance Criteria, in Part 3 of the NEHRP recommended seismic provisions for new buildings and other structures (FEMA P-750), Building Seismic Safety Council, Federal Emergency Management Agency, Washington, DC.

# Chapter 12

# Proportioning of earthquake-resistant structural systems

## 12.1 PURPOSE AND OBJECTIVES

This chapter uses generic models for the displacement response of multistory buildings to seismic actions to interpret limits on interstory drift and strength reduction factors for the proportioning of lateral force-resisting systems. This modeling approach is used in the design approaches of Chapter 18. Detailed modeling of systems and components is addressed in Chapters 15–17.

## 12.2 INTRODUCTION

Given that nearly all of the displacement response of multistory buildings responding in the linear and nonlinear domains can be represented by a predominant mode (Chapter 6), a pre-conceived drift profile can be used to approximate the response and to relate constraints on the drift and ductility demands of the multi-degree-of-freedom (MDOF) building to those of its single-degree-of-freedom (SDOF) analogue. Generic drift profiles and the treatment of interstory drift limits and strength reduction ($R$) or ductility modification ($q$) factors in current code provisions are discussed.

## 12.3 GENERIC DRIFT PROFILES

As illustrated in Chapter 6, a large portion of the lateral displacement response of many building systems can be represented using a single, predominant mode, which may differ to some degree from the elastic first mode for nonlinear response. This predominant deformation pattern (or drift profile) is different for moment frame and wall buildings (Figure 12.1).

With reference to Figure 12.1, the following functions provide generic approximations of the typical drift profiles to be expected, in a manner similar to that suggested by Abrams (1985):

$$u_i = \begin{cases} 1 - \left( \dfrac{h_n - h_i}{h_n} \right)^{1.3} & \text{moment frames} \\[2mm] \left( \dfrac{h_i}{h_n} \right) & \text{dual systems and coupled walls} \\[2mm] \left( \dfrac{h_i}{h_n} \right)^{1.7} & \text{slender cantilever shear walls} \end{cases} \qquad (12.1)$$

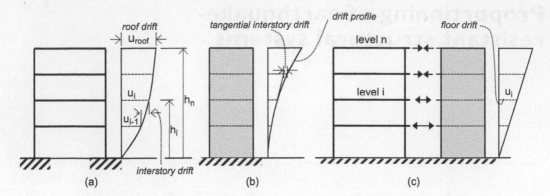

*Figure 12.1* Definitions of various drift quantities, and drift profiles expected for (a) moment frames, (b) slender cantilevered walls, and (c) dual systems, and force transfer between dual system frames and walls.

where $u_i$ is the lateral displacement at the $i$th floor, $h_i$ and $h_n$ are the heights of the $i$th floor and roof, respectively, above the base. The exponents were selected to obtain reasonable shapes, which are plotted in Figure 12.2.

So-called "dual" systems combine moment frames and structural walls. Because the frames and walls are constrained to the same lateral displacement at any floor, internal forces develop to transfer load between the frames and walls. The flexibility of the moment frame drives greater load into the lower stories of the wall, while the flexibility of the wall

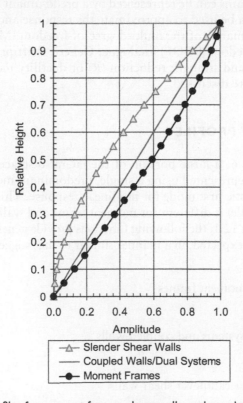

*Figure 12.2* Generic drift profiles for moment frames, shear walls, and coupled walls and dual systems.

drives greater load into upper stories of the moment frame. The result of frame–wall inter-action is a deformation pattern that is somewhat in between the extremes defined by the moment frame and wall systems acting independently, and which often can be approxi-mated by a straight line. The straight-line drift profile is also a reasonable approximation for many coupled wall buildings.

## 12.4 ESTIMATES OF MODAL PARAMETERS FOR PRELIMINARY DESIGN

As described in Chapter 7, the participation factor for the first mode, $\Gamma_1$, and the modal mass coefficient, $\alpha_1$, are needed for mapping between MDOF systems and their equivalent SDOF systems. Defined in Equations 5.20 and 5.34, these unitless parameters are functions of the mass distribution and first-mode shape. Fortunately, they are relatively stable during the design process and can be estimated with reasonable precision early on.

The generic drift profiles of Section 12.3 were used to develop generic estimates of modal parameters for preliminary design (Table 12.1). Mass was assumed to be uniform over the height, except that the roof was assumed to have 80% of the typical floor mass. Note that these are preliminary values, and can be updated once an initial design has been developed and modeled, based on the computed dynamic properties of the model.

Also provided in Table 12.1 are values of $h_1^*/h_n$ used in estimating the height of the resul-tant of the first-mode lateral forces, $h_1^*$ relative to the overall height of the structure, $h_n$. This ratio can be used for determining overturning moments at the base of the structure; more importantly, it is used to adjust the base shear applied when using a lateral force distribution that differs from the first-mode distribution, in order to design the system to have the desired strength in a first-mode pushover analysis (Section 12.8.2).

## 12.5 PROPORTIONING FOR DUCTILE RESPONSE

Figure 12.3 illustrates conceptually the base shear–roof displacement response of a multistory building in a nonlinear static (pushover) analysis; the dots show the sequential formation of plastic hinges and stepwise reduction in lateral stiffness that takes place until a complete mechanism forms (Chapter 7). The mechanism strength, $V_y$, may be several times higher than the base shear strength $V_{code}$ that would be determined according to conventional design provisions. This difference is referred to as "overstrength" and is discussed in Section 12.6.

Table 12.1 Generic estimates of modal parameters for use in preliminary design[a]

| Number of stories | Moment-resistant frames | | | Dual systems and coupled walls | | | Slender cantilevered walls | | |
|---|---|---|---|---|---|---|---|---|---|
| | $\Gamma_1$ | $\alpha_1$ | $h_1^*/h_n$ | $\Gamma_1$ | $\alpha_1$ | $h_1^*/h_n$ | $\Gamma_1$ | $\alpha_1$ | $h_1^*/h_n$ |
| 1 | 1.00 | 1.00 | 1.00 | 1.00 | 1.00 | 1.00 | 1.00 | 1.00 | 1.00 |
| 2 | 1.21 | 0.94 | 0.79 | 1.24 | 0.89 | 0.81 | 1.24 | 0.76 | 0.86 |
| 3 | 1.27 | 0.90 | 0.73 | 1.33 | 0.85 | 0.75 | 1.35 | 0.70 | 0.81 |
| 5 | 1.32 | 0.86 | 0.70 | 1.40 | 0.82 | 0.71 | 1.46 | 0.66 | 0.78 |
| 10 | 1.35 | 0.82 | 0.67 | 1.45 | 0.79 | 0.69 | 1.54 | 0.63 | 0.75 |
| ≥20 | 1.37 | 0.80 | 0.66 | 1.48 | 0.77 | 0.68 | 1.59 | 0.62 | 0.74 |

[a] NEHRP Recommended Seismic Provisions, Part 3 (2009).

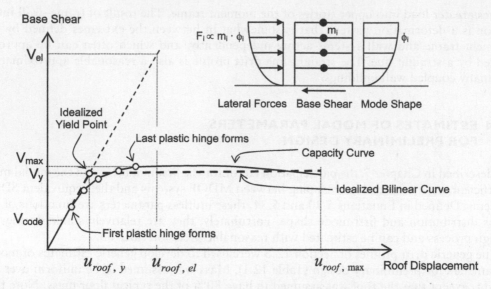

*Figure 12.3* Capacity curve obtained in first-mode pushover analysis, illustrating response of elastic and inelastic systems, mechanism development, associated base shears, and overstrength.

For a relatively rare event, a lateral force-resisting system would have to develop substantial base shear strength, $V_{el}$, if every component were to remain linear elastic during the dynamic response. As mentioned in Chapter 4, the rapidly reversing nature of seismic excitation causes a system with yield strength $V_y$ less than $V_{el}$ to undergo a generally irregular series of reversed cyclic displacements that result in a peak displacement $u_{roof,max}$ (also shown schematically in Figure 12.3). The peak displacement $u_{roof,max}$ is of the same order as $u_{roof,el}$.[1] While the ratio $u_{roof,max}/u_{roof,el}$ varies with each excitation, overall trends for MDOF systems are similar to those for SDOF systems, and these trends are described by $R–\mu–T$ relationships. The basic trade-off in earthquake-resistant structural engineering is that we can provide a system with strength much less than $V_{el}$ if we are willing to tolerate nonlinearity in the response of the lateral force-resisting system. This nonlinearity implies damage to the lateral force-resisting system for nearly all systems, with the possible exception of post-tensioned shear walls, where the nonlinearity is related to a contact problem. This trade-off, allowing a less costly system but one prone to damage during strong shaking, is accepted for nearly all buildings. Thus, in typical cases, the lateral force-resisting system experiences damage in the design ground motion, as well as in some of the lower-intensity ground motions that occur more frequently. One consequence is that the proportioning of MDOF systems must aim to ensure that deformation demands within plastic hinges do not exceed deformation capacities (which often will be achieved by distributing damage throughout the structural system) and that sufficient stiffness is provided so that lateral response is not overtaken by $P$-$\Delta$ effects. Possible damage to the gravity load-resisting system and nonstructural components must also be considered, and will result in restrictions on peak interstory drift that depend on their flexibility and the amount of damage that can be tolerated. These considerations may result in a stiffer and possibly stronger lateral force-resisting system than would be needed when considering only the deformation capacity of the lateral force-resisting system.

---

[1] If the peak seismic load were maintained for a long period of time (e.g., applied statically) $u_{roof,max}$ would approach infinity in the case of an elasto-plastic system, and a very large number in the presence of a small (positive) post-yield stiffness.

Note that in Figure 12.3, the response of the MDOF system simplifies to that of a SDOF system if (i) there is only a single story, or (ii) the lateral forces $F_i$ are made proportional to the mass $m_i$ and mode shape amplitude $\phi_i$ at each floor level $i$ (as shown in Equation 5.29), for which the deformed shape $\mathbf{u}$ is proportional to the mode shape $\phi$ within the initial domain of linear elastic response.

As illustrated in Chapter 6 (Principal Components Analysis (PCA)), displacements in MDOF systems are predominantly in the first mode (an elastic mode or a PCA mode). Thus, a first-mode approximation can be useful for representing displacement response. Mathematical relations allow the displacement response of the MDOF system to be mapped to that of an equivalent SDOF system (Chapter 7).

Yield point spectra (YPS, e.g., Figure 12.4) illustrate the ductility demands $\left(\mu = D_{max}^*/D_y^*\right)$ relative to the yield point associated with an equivalent SDOF (ESDOF) representation of a multistory building. This normalized form of representing seismic demands allows the base shear strength at yield, required to limit roof displacement and system ductility demands to desired values, to be determined on the basis of an ESDOF system through the use of the modal parameters $\alpha_1$ and $\Gamma_1$ (Figure 7.3). Specifically, $D_y^* = u_{roof,y}/\Gamma_1$, and $C_y^* = C_y/\alpha_1$, where $C_y$ is the MDOF system base shear coefficient at yield ($= V_y/W$).

YPS may be used with the conventional assumption that stiffness is a fixed proportion of the gross stiffness (from point ① to point ② in Figure 12.4), or alternatively, that the yield displacement is invariant as reinforcement content changes (from point ① to point ③ in Figure 12.4).

In the conventional approach, an increase in reinforcement content is assumed to increase strength without affecting stiffness, and hence the slope of the load–displacement curve and associated period are constant. For long-period systems, the equal displacement rule requires that the increase in strength causes a reduction in ductility demand but no change in peak displacement (Figure 12.4).

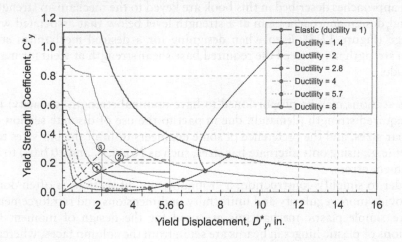

*Figure 12.4* Application of YPS to design illustrating effect of strength on peak displacement response. The curve having yield point ① falls on the $\mu = 2.8$ curve, and thus has a peak displacement of 2.8 (2 in.) = 5.6 in. Point ② illustrates the conventional assumption that stiffness is independent of reinforcement content, which implies that an increase in reinforcement content (and strength) causes a reduction in ductility demand—this example illustrates that the application of the equal displacement rule (for long-period systems) results in no change in peak displacement. Point ③ illustrates the assumption of constant yield displacement, which implies that an increase in reinforcement content affects both strength and stiffness; in this case, the ductility demand reduces to 2, resulting in a peak displacement of 2(2 in.) = 4.0 in.

As discussed in Chapter 9, the assumption that a change in reinforcement content does not affect stiffness conflicts with basic mechanics—since the grade (or strength) of the reinforcement and corresponding yield strain typically are invariant, changes in strength are achieved by changing the amount of material present, and this results in a change in stiffness.[2] Using YPS, one can clearly see how changes in strength and stiffness for a given yield displacement affect ductility demand and peak displacement response.

Yield frequency spectra (YFS, Chapter 13) extend the notion of YPS over the hazard, providing a graphic representation of the relationship of $C_y^*$ to the mean annual frequency of exceeding a ductility demand. In order to retain the simplicity of a two-dimensional plot, ductility-strength-hazard level information is portrayed for a single value of yield displacement, corresponding to that of the applicable equivalent SDOF system.

Using either YPS or YFS, the process of limiting ductility and displacement demands to an acceptable value is greatly simplified relative to the use of conventional, period-based spectra. It is apparent in Figure 12.4 that for any given yield displacement, an increase in strength (① to ③) reduces both ductility and displacement demands. Thus, the design task is to determine the strength that limits ductility and displacement demands to acceptable values (Chapters 15 and 16). To address constraints associated with multiple performance objectives, YPS developed for each hazard level can be used; alternatively, a single YFS can be used, prepared for the estimated value of yield displacement. The estimated yield displacement may be validated or refined after the preliminary design has been completed, in much the same way that the period of vibration is monitored and refined in conventional design approaches.

## 12.6  THE INFLUENCE OF OVERSTRENGTH ON SYSTEM DUCTILITY DEMANDS

The design approaches described in this book are keyed to the mechanism strength, whereas conventional design approaches aim at a strength level below that associated with the first plastic hinge (Figure 12.3). Even when designing for a desired mechanism strength, the mechanism strength may exceed the required base shear strength at yield for many reasons. These include:

1. Cross sections may be proportioned to have nominal strength (capacity) in excess of the required strength (demand), due in part to the use of discrete section dimensions and bar sizes, and the preference of some engineers to limit the bar sizes to a selected subset (e.g., using only alternate bar sizes, such as No. 6, 8, and 10 bars) to reduce field placement errors.
2. In order to simplify construction, a "rounding up" of sections is often done, in order to provide more regularity and uniformity in dimensions and reinforcement.
3. Where simple plastic mechanisms are used for the design of moment frames, the locations of plastic hinges in beams are set in from the column faces, whereas they may have been assumed to be located at the member centerlines to simplify the mechanism analyses). The inset hinges result in shorter spans, and thus the development of higher beam shear forces ($V = \Sigma M_p/L$) and a corresponding increase in the lateral forces associated with mechanism development.

---

[2] Except in the special case of unbonded post-tensioned members, where a higher prestressing force increases strength but not stiffness (Section 19.12).

4. Strain hardening of the reinforcement during large inelastic rotations may result in flexural strengths greater than the plastic moment strengths assumed in mechanism analysis.

5. The gravity framing system may be neglected in the mechanism analysis and the design of the lateral force-resisting system, although it may contribute both strength and stiffness to lateral resistance.

In conventional design approaches, wherein system behavior is linear elastic under the factored design loads, two additional sources of overstrength are present:

6. Expected material yield strengths are typically higher than the nominal values assumed in conventional design.

7. The governing load combination for seismic actions typically includes sources of gravity load. For moment frames, load factors applied to gravity load moments boost the strengths of the plastic hinge zones well above the moments caused by the service-level gravity loads, leaving additional flexural strength available for resisting lateral loads. For slender walls, the governing load combination for seismic actions represents only part of the expected dead load. Since these walls have axial loads well below the balance point, any increase in axial load associated with the full (expected) dead and service-level live loads causes an increase in wall flexural strength.

Of course, any increase in strength also increases the forces that must be carried by force-protected members.

Some of the above sources of overstrength cause an increase in stiffness, whereas other sources do not affect stiffness while increasing strength and yield displacement. For example, using more reinforcement than is required will increase both strength and stiffness. In contrast, actual yield strengths being greater than assumed will increase section strength and member yield displacement, while not affecting stiffness.

Figure 12.4 can also be used to illustrate the effects of overstrength. For example, higher than expected yield strengths affect strength without causing a change in stiffness—the increase in strength ($V_y$) reduces the effective $R$ factor without changing the oscillator period. In the case of long-period systems ($C = 1$), this causes a reduction in the ductility demand, while the peak displacement remains unchanged. For short-period systems ($C > 1$), this source of overstrength causes a reduction in ductility demand due to an increase in yield displacement as well as a reduction in peak displacement. Sources of overstrength that cause an increase in both strength and stiffness lead to reductions in both ductility and peak displacement demands, as illustrated by Point ③ of Figure 12.4.

## 12.6.1 Overstrength and implied system ductility capacities from an American perspective

In conventional practice per ASCE 7 and ACI 318, demands are computed on the basis of elastic analysis, considering defined design load combinations, and each critical section is proportioned such that its reliable capacity exceeds the largest of the factored demands obtained from the applicable load combinations. For example, for beams, $\phi M_n \geq M_u$ where $M_u$ is the factored demand, $\phi$ is a strength reduction factor and $M_n$ is the nominal moment capacity. This notion is also applied to combinations of moment and axial force for critical sections of columns and walls.

Figure 12.3 illustrates that the first plastic hinge to form in a nonlinear static (pushover) analysis develops at a lateral load greater than the design lateral loading ($V_{code}$). In order to

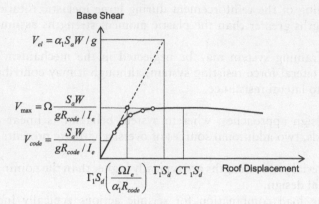

*Figure 12.5* MDOF response in terms of parameters used in American practice. Note: $g$ = acceleration due to gravity and is used where $S_a$ has units of acceleration (where $S_a$ is already normalized by gravity, use $g = 1$).

develop the sequence of plastic hinges needed to form a mechanism, even greater lateral load must be applied, $V_{max}$. Figure 12.5 tailors Figure 12.3 to American practice, and is based on the assumption that the roof displacement response is dominated by the first mode; thus, the base shear required for elastic response is also based on the first mode.

With reference to Figure 12.5, we can define overstrength, $\Omega$, relative to the code design value as

$$\Omega = \frac{V_{max}}{V_{code}} = \frac{V_{max}}{S_a W/g} \frac{R_{code}}{I_e} \qquad (12.2)$$

where $S_a$ = design spectral acceleration, $R_{code}$ is the "R factor" specified in the code (and not equal to $R = V_e/V_y$ as defined in Section 4.8.4), and $I_e$ = importance factor defined by ASCE-7. Note that $\Omega$ defined here differs from the code-specified value of $\Omega_o$, which is used for the conventional design of force-protected members. As can be seen in Table 12.2, larger values of $R_{code}$ are assigned to systems that are provided with details that enhance ductility capacity; these systems are designed for lower lateral forces.

The occupancy categories and corresponding values of the seismic importance factor, $I_e$, are summarized in Table 12.3.

System-level ductility capacities associated with code-specified design requirements have not been explicitly articulated, but can be inferred following the process reported in Resource Paper 9 in Part III of the NEHRP Provisions (BSSC 2010). To begin, with reference to Figure 12.5 we can define $u_{roof,y}$ as the yield displacement associated with the strength of the intended mechanism, $V_{max}$:

*Table 12.2* Seismic design parameters for select reinforced concrete systems given in ASCE/SEI 7-16

| Lateral force-resisting system | Ordinary | | | Intermediate | | | Special | | |
|---|---|---|---|---|---|---|---|---|---|
| | $R_{code}$ | $\Omega_o$ | $C_d$ | $R_{code}$ | $\Omega_o$ | $C_d$ | $R_{code}$ | $\Omega_o$ | $C_d$ |
| Reinforced concrete moment frames | 3 | 3 | 2½ | 5 | 3 | 4½ | 8 | 3 | 5½ |
| Reinforced concrete walls[a] | 5 | 2½ | 4½ | NA | NA | NA | 6 | 2½ | 5 |

[a] Applicable to both cantilever and coupled wall systems, in which a building frame is used to carry gravity loads.

*Table 12.3* ASCE/SEI 7-16 occupancy categories and importance factors

| Occupancy category | Importance factor, $I_e$ | Description (abbreviated) |
|---|---|---|
| I | 1.00 | Buildings that represent a low hazard to human life in the event of failure. |
| II | 1.00 | Ordinary buildings—that is, buildings not listed in Occupancy Categories I, III, and IV. |
| III | 1.25 | Buildings that present a substantial hazard to human life (including buildings with large occupancies and places of assembly), buildings with potential to cause substantial economic impact and/or mass disruption of civilian life (including power stations, water and sewage treatment facilities, and telecommunications centers), and buildings containing hazardous materials. |
| IV | 1.50 | Essential facilities (including hospitals, fire stations, and other public utility facilities required in an emergency, buildings having critical national defense functions, and buildings containing highly toxic substances). |

$$u_{\text{roof,y}} = u_{\text{roof,el}} \frac{V_{\text{max}}}{V_{\text{el}}} = \frac{\Gamma_1 S_d \Omega}{\alpha_1 R_{\text{code}}/I} \tag{12.3}$$

Then, the system-level ductility capacity referenced to $u_{\text{roof,y}}$ is given as

$$\mu = \frac{u_{\text{roof,max}}}{u_{\text{roof,y}}} = \frac{C\alpha_1 R_{\text{code}}/I}{\Omega} \tag{12.4}$$

Equation 12.4 can be used to derive system ductility limits nominally compatible with the code. Consider a long-period structure, for which $C = 1$. Representative values of $\alpha_1$ were taken equal to 0.90 for moment-resistant frames, 0.80 for coupled walls, and 0.65 for cantilever walls. Finally, $\Omega$ was taken equal to the ASCE/SEI 7-16 values of $\Omega_o$ given for each lateral force-resisting system. The resulting ductility capacities, given in Table 12.4, apply to a design basis ground motion (given as 2/3 of the risk-targeted ground motion having a 2% probability of exceedance in 50 years). These values are multiplied by 3/2 to obtain the values tabulated at the risk-targeted 2/50 hazard level. The system ductility limits of Table 12.4 are referenced to $u_{\text{roof,max}}$. For short-period structures (for which $C = C_1 \cdot C_2 > 1$), the explicit consideration of short-period displacement amplification in $R$–$\mu$–$T$ relationships used herein provides better control of system ductility and drift response than is achieved with the ASCE 7 equivalent lateral force (ELF) procedure (because ASCE 7 does not account for short-period displacement amplification). Note that use of the system ductility values of Table 12.4 imply that the code requirements for the proportioning and detailing of these systems are followed.

*Table 12.4* Approximate system ductility values implied in ASCE/SEI 7-16

| Lateral force-resisting system | At 2/3 of the 2%/50 year hazard level | | | At the 2%/50 year hazard level | | |
|---|---|---|---|---|---|---|
| | Ordinary | Intermediate | Special | Ordinary | Intermediate | Special |
| Moment frames | $0.9/I_e$ | $1.5/I_e$ | $2.4/I_e$ | $1.4/I_e$ | $2.25/I_e$ | $3.6/I_e$ |
| Coupled walls[a] | – | – | $1.9/I_e$ | – | – | $2.9/I_e$ |
| Cantilever walls[a] | $1.3/I_e$ | – | $1.6/I_e$ | $2.0/I_e$ | – | $2.4/I_e$ |

[a] In which a building frame system is used to support gravity loads.

## 12.6.2 Overstrength and implied system ductility capacities from a Eurocode perspective

Structural design according to the Eurocodes is based on the use of limit states, which limit values of engineering demand parameters such as stresses, strains, deformations, and drifts experienced by the structure under prescribed loading conditions. Actions on structural elements (axial and shear forces and moments) generally are calculated using an elastic analysis, using load combinations in which the seismic actions are based on an elastic spectrum that is scaled up by $\gamma_I$ to reflect the importance of the structure (Table 12.5) and scaled down to account for system ductility capacity, which is represented in the behavior factor, $q$. System ductility demand is considered at the ultimate limit state (or ULS), generally having 10% probability of exceedance in 50 years (equivalent to a return period of 475 years). Interstory drift is addressed at the serviceability limit state (or SLS), generally having a 10% probability of exceedance in 10 years (equivalent to a return period of 95 years) and represented by scaling the ULS spectrum by a coefficient, $\nu$ (Table 12.5).

The basic value of the behavior factor, $q_o$, is based on the "ductility class" and associated, required, structural detailing; values are tabulated for reinforced concrete systems in Table 12.6. Three ductility classes are recognized in Eurocode 8: high ductility (DCH), medium ductility (DCM), and low ductility (DCL). Higher values of $q_o$ imply greater system ductility capacity and result in lower lateral design forces (assuming that interstory drift at the SLS does not control).

The behavior factor, $q$, is determined as $q = q_o k_w \geq 1.5$, where $q_o$ = basic value of the behavior factor, and $k_w$ reflects the prevailing failure mode (see Section 16.3.2). For frames, $k_w = 1.0$; for walls, $k_w = (1 + \alpha_o)/3 \leq 1.0$ where $\alpha_o$ is the prevailing wall aspect ratio ($\alpha_o = \sum H_i / \sum l_{wi}$), which need not be taken less than 0.5, where $H_i$ is the height and $l_{wi}$ is the length in plan of the $i$th wall.

The overstrength coefficient can be obtained by a pushover analysis (Figure 12.6) as $\alpha_u / \alpha_y$, where $V_{bd}$ is the design value of the shear force at the base. As an alternative to pushover analysis, Eurocode 8 allows the use of prescribed values for the overstrength of regular buildings. For regular multistory, multi-bay buildings, prescribed values of overstrength are: 1.3 for moment frames, 1.2 for coupled walls, 1.1 for cantilevered (uncoupled) walls greater than two in number in each orthogonal direction, and 1.0 where two cantilevered (uncoupled) walls are used in each orthogonal direction.

Approximate system ductility capacities implied within Eurocode 8 can be established following a procedure much like that used in Section 12.6.1. With reference to Figure 12.6,

$$\mu = \frac{u_{\text{roof,peak}}}{u_{\text{roof,y}}} = \frac{CT_1 S_d}{\Gamma_1 S_d \left( \dfrac{\alpha_u V_{bd}}{\alpha_1 S_a W / g} \right)} = \frac{Cq/\gamma_I}{(\alpha_u/\alpha_1)\lambda} \tag{12.5}$$

Table 12.5 Importance classes in Eurocode 8 (EN 1998-1, 2004)

| Importance class | Description | Importance factor $(\gamma_I)$ | Damage limitation reduction factor $(\nu)^a$ |
|---|---|---|---|
| I | Buildings having minor implications for public safety | 0.8 | 0.5 |
| II | Ordinary buildings | 1.0 | 0.5 |
| III | Buildings whose collapse would have significant consequences for public safety, such as schools and places of public assembly[a] | 1.2 | 0.4 |
| IV | Buildings whose integrity is of vital importance for civil protection, such as hospitals, fire stations, and power plants | 1.4 | 0.4 |

[a] Adjusts the ULS motion to the SLS motion.

*Table 12.6* Basic value of behavior factor, $q_o$, for select reinforced concrete systems specified in Eurocode 8 (EN 1998-1, 2004)[a]

| Lateral force-resisting system | Ductility class | | |
|---|---|---|---|
| | DCL | DCM | DCH |
| Moment frames | 1.5 | $3.0\alpha_u/\alpha_y$ | $4.5\alpha_u/\alpha_y$ |
| Coupled walls | 1.5 | $3.0\alpha_u/\alpha_y$ | $4.5\alpha_u/\alpha_y$ |
| Uncoupled (cantilever) walls[b] | 1.5 | 3.0 | $4.0\alpha_u/\alpha_y$ |

[a] In a departure from the Eurocode terminology, we associate the term $\alpha_y$ with first yield and reserve the term $\alpha_1$ for the mass participation factor in the first mode. The term $\alpha_u$ is associated with the formation of a plastic mechanism.

[b] An uncoupled (cantilever) wall system has greater than 65% of the base shear resisted by walls, with more than 50% of the base shear resisted by uncoupled walls.

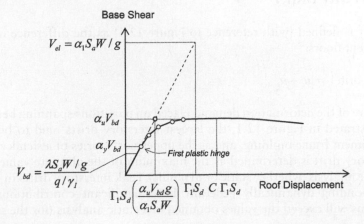

*Figure 12.6* Push-over analysis and overstrength coefficient for $T > T_B$. The term $\alpha_y$ is associated with first yield and $\alpha_u$ is associated with the formation of a plastic mechanism. (We reserve the term $\alpha_1$ for the mass participation factor in the first mode.)

where $u_{roof,y}$ is the yield displacement associated with the mechanism strength $\alpha_u V_{bd}$, $C = 1$ for long-period structures, overstrength ($\alpha_u/\alpha_y$) is taken equal to 1.0–1.3 (as described above), $k_w$ is assumed equal to 1.0 for moment frames and slender structural walls, and $\alpha_y$ is assumed equal to 1.2 for uncoupled (cantilever) walls and 1.3 for coupled walls and moment frames. Representative values of the first-mode mass coefficient, $\alpha_1$, were taken equal to 0.65 for uncoupled (cantilever) walls, 0.80 for coupled walls, and 0.90 for moment frames, while $\lambda$ was taken equal to 0.85, as specified in Eurocode 8 for buildings having three or more stories. The resulting system ductility capacities, provided in Table 12.7, are applicable to regular moment frame buildings and wall buildings in which the structural walls have height at least twice the plan length.

Eurocode 8 scales the design spectrum by the value of $\gamma_I$ accorded to the importance of the structure. The approximate system ductility values given in Table 12.7 are intended for use with an unscaled (elastic) ULS spectrum (typically having 10% probability of exceedance in 50 years).

Table 12.7 was derived considering values reported in Table 5.1 of Eurocode 8, using values of the overstrength coefficient prescribed for regular systems. In the presence of vertical irregularities, the first yield is closer to the structural instability case, leading to smaller values of the overstrength coefficient. Such cases should be validated by nonlinear dynamic analysis.

*Table 12.7* Approximate system ductility values for regular buildings of three or more stories in height for use with the ULS spectrum, implied in Eurocode 8

|  | Ductility class | | |
| --- | --- | --- | --- |
| *Reinforced concrete structural system* | DCL | DCM | DCH |
| Moment frame | $0.9/\gamma_1$ | $2.4/\gamma_1$ | $3.7/\gamma_1$ |
| Coupled wall[a] | $0.9/\gamma_1$ | $2.2/\gamma_1$ | $3.3/\gamma_1$ |
| Uncoupled (cantilever) wall[a] | $0.9/\gamma_1$ | $1.7/\gamma_1$ | $2.5/\gamma_1$ |

[a] The system ductility factors are applicable to wall systems for which the prevailing wall aspect ratio, $\alpha_0$, is greater than 2. An uncoupled (cantilever) wall system has greater than 65% of the base shear resisted by walls, with more than 50% of the base shear resisted by uncoupled walls. Where uncoupled walls are used, more than two are used in each direction.

## 12.7 INTERSTORY DRIFT

Interstory drift is defined (with reference to Figure 12.1) as the difference in the displacements of adjacent floors:

$$(\text{interstory drift})_i = |u_i - u_{i-1}| \tag{12.6}$$

and is a measure of the deformation demand placed on partitions spanning between adjacent floors. As illustrated in Figure 12.1, the largest interstory drifts tend to be at the lowest stories of a moment frame building, and at the uppermost stories of a slender wall building.

Peak interstory drift is determined as the maximum of the absolute values of interstory drift during the analysis (whether static or dynamic). Peak interstory drifts in moment frame buildings responding dynamically generally have significant contributions from higher modes, and thus will exceed the values obtained in a static analysis (for the same peak roof displacement).

Peak interstory drift is often normalized relative to the story height to obtain a nondimensional index of story deformation:

$$(\text{interstory drift ratio})_i = \frac{|u_i - u_{i-1}|_{\max}}{h_i - h_{i-1}} \tag{12.7}$$

Peak interstory drift ratio is a good measure of damage to nonstructural elements such as partitions and mechanical equipment, and is limited in ASCE 7 and Eurocode 8.

Interstory drift is not necessarily a good measure of damage to structural components. For example, as shown in Figure 12.1, large interstory drift near the top of a slender wall includes components associated with rigid body motion of the wall. Tangential interstory drift is a measure of deformation relative to the tangent at the base of the story of interest. This provides an indication of the deformation of structural components that excludes rigid body displacements and rotations of the structural element below the story of interest.

The roof drift ratio is defined as the peak roof displacement normalized by the height of the roof above the base, providing a nondimensional index of building deformation:

$$\text{roof drift ratio} = \frac{u_n}{h_n} = \frac{u_{\text{roof}}}{h_n} \tag{12.8}$$

where $u_n$ is referred to as $u_{\text{roof}}$ for convenience. The ratio of the interstory drift ratio and the roof drift ratio provides a nondimensional measure of the degree to which deformation concentrates within a story. This ratio is also useful in preliminary design for determining a roof drift ratio that will limit interstory drift ratios to acceptable values. The degree to which drift concentrates in particular stories is a function of the number of stories, type of structural system, design requirements (such as the ratio of column and beam flexural strengths at a joint), intensity of inelastic response, and features of the ground motion.

When building code limits on interstory drift are used as a basis to determine a roof displacement limit for design, peak interstory drifts may be approximated using the expressions of Equation 12.1 (plotted in Figure 12.2). Under quasi-static loading, the term $\alpha_{3,\text{stat}}$ may be defined as

$$\alpha_{3,\text{stat}} = \frac{\left(\dfrac{|u_i - u_{i-1}|}{h_i}\right)_{\text{max over } i = 1 \text{ to } n}}{\dfrac{|u_{\text{roof}}|}{h_n}} \tag{12.9}$$

where $u_{\text{roof}}$ is taken at the same step in the static analysis in which the peak interstory drift occurs. Values of $\alpha_{3,\text{stat}}$ determined using the characteristic deformed shapes of Figure 12.2 are summarized in Table 12.8. Once a preliminary design has been developed, a more precise determination can be made for any specific building using a linear or nonlinear static (pushover) analysis.)

Interstory drifts during dynamic response are influenced to varying degrees by higher modes. As described by Qi and Moehle (1991), the coefficient of distortion (COD), represented by $\alpha_{3,\text{COD}}$, is the ratio of the peak interstory drift ratio during dynamic response and the peak roof drift ratio during the response—while these peaks occur during response to a single ground motion, they need not occur at the same instant. Due to the presence of higher modes and the possibility of damage concentrating in particular stories, $\alpha_{3,\text{COD}} > \alpha_{3,\text{stat}}$ for a given lateral force-resisting system and number of stories (greater than one). Generic values that are representative of reported values (e.g., Moehle, 1992; Gupta and Krawinkler, 2000; Aschheim and Maurer, 2007; Ruiz-García and González, 2014) are summarized in Table 12.8.

## 12.7.1 Application of interstory drift limits from an American perspective

Drift limits contained within current building codes provide a useful benchmark for design. This section establishes a roof drift limit that approximately represents the interstory drift limit of ASCE-7. The ASCE-7 interstory drift limits are a function of the structural system

Table 12.8 Generic estimates of interstory drift parameters for use in determining roof drift limits in design

| Number of stories | Moment-resistant frames | | Dual systems and coupled walls | | Slender cantilevered walls | |
|---|---|---|---|---|---|---|
| | $\alpha_{3,\text{stat}}$ | $\alpha_{3,\text{COD}}$ | $\alpha_{3,\text{stat}}$ | $\alpha_{3,\text{COD}}$ | $\alpha_{3,\text{stat}}$ | $\alpha_{3,\text{COD}}$ |
| 1 | 1.0 | 1.0 | 1.0 | 1.0 | 1.0 | 1.0 |
| 2 | 1.2 | 1.3 | 1.0 | 1.0 | 1.2 | 1.3 |
| 3 | 1.3 | 1.4 | 1.0 | 1.1 | 1.3 | 1.4 |
| 5 | 1.3 | 1.5 | 1.1 | 1.2 | 1.4 | 1.5 |
| 10 | 1.3 | 1.7 | 1.1 | 1.3 | 1.5 | 1.6 |
| ≥20 | 1.3 | 1.9 | 1.2 | 1.4 | 1.6 | 1.7 |

*Table 12.9*  ASCE/SEI 7-16 allowable story drift, $\Delta_a$ (for reinforced concrete frame[a] and wall systems)

| Structural system | Occupancy category | | |
|---|---|---|---|
| | I or II | III | IV |
| Structures, four or fewer stories with interior walls, partitions, ceilings, and exterior wall systems that have been designed to accommodate the story drifts. | 0.025 $h_{sx}$ | 0.020 $h_{sx}$ | 0.015 $h_{sx}$ |
| All other structures | 0.020 $h_{sx}$ | 0.015 $h_{sx}$ | 0.010 $h_{sx}$ |

$h_{sx}$ = the story height below Level $x$.

[a]  For moment-resisting frames in the higher seismic design categories (D, E, and F), story drifts are limited to the tabulated values divided by the redundancy factor, $\rho$, where $\rho = 1.3$ except $\rho = 1.0$ may be used for buildings lacking plan irregularities in which at least two bays of moment-resistant framing are present on each side of the structure in each orthogonal direction for any story that carries at least 35% of the base shear.

and occupancy category (or importance) of the building (Table 12.3) and are applied in a prescribed way. The ASCE 7 drift limits applicable to reinforced concrete buildings are summarized in Table 12.9.

The drift limits are applied to drifts determined in both static and dynamic methods of analysis for the design-level ground motion (2/3 of the risk-targeted maximum considered earthquake (MCE)). Particularly for moment-resistant frames, the presence of higher modes and concentration of drifts in particular stories cause the interstory drifts determined during nonlinear dynamic response to be significantly higher than values determined under the ELF method of ASCE 7, at the same level of roof drift or for the same spectral intensity (e.g., $S_a(T_1)$). Because the ASCE 7 drift limits are typically applied to demands calculated according to the ELF method, the ASCE 7 drift limits may be viewed as a requirement to provide a minimum lateral stiffness, rather than as an actual limit on expected interstory drifts.

In the ELF method, story drift demands are determined as the drifts computed under the application of the design-level lateral forces multiplied by the ASCE 7 displacement modification coefficient, $C_d$. (The design lateral forces are 2/3 of the risk-targeted MCE forces divided by $R/I$)[3]. The story drift demands computed in this way are compared with the allowable story drifts of Table 12.9. In addition to ignoring the contributions of higher modes and the concentration of drift during inelastic response, the drifts computed this way are low relative to expectations based on the equal displacement rule and, more generally, relative to expectations embodied in $R$–$\mu$–$T$ relationships.

In the following, a limit on roof drift is derived from the code drift limits. In the ELF method, lateral forces are applied at each floor (in proportion to $w_x h_x^k$) corresponding to a base shear equal to $S_a W/(g R_{code}/I)$.[4] If displacement response is predominantly in the first mode (Figure 12.5), a base shear of $\alpha_1 S_a W/g$ develops for elastic response; the corresponding roof displacement would be $\Gamma_1 S_d$. Since the code load pattern is not a first-mode pattern, the roof displacement under code loading will differ from that obtained in the first mode. Neglecting this difference (as an approximation), the roof displacement determined in the structural analysis for the design lateral forces can be estimated using similar triangles as

---

[3]  Thus, one can view the ASCE-7 drift limits as a requirement for a minimum stiffness at the design-level ground motion, or alternatively, at the serviceability motion having return period corresponding to a shaking intensity of $I/R$ times the design level.)

[4]  The acceleration of gravity, $g$, is used where $S_a$ has units of acceleration. In some cases, $S_a$ already is expressed in terms of $g$, in which case $g$ is taken equal to unity.

$$u_{\text{roof}} = \left( \frac{\dfrac{S_aW}{gR_{\text{code}}/I_e}}{\alpha_1 S_a W / g} \right) \Gamma_1 S_d = \frac{\Gamma_1 S_d}{\alpha_1 R_{\text{code}}/I} \tag{12.10}$$

Based on the drift profiles of Figure 12.2, the tabulated values of $\alpha_{3,\text{stat}}$ (Table 12.8) provide a useful approximation of the ratio of peak interstory drift ratio and roof drift ratio. The maximum (static) interstory drift ratio computed by the ELF may be estimated as

$$IDI_{\max} = \frac{\Gamma_1 S_d}{\alpha_1 R_{\text{code}}/I_e} \frac{\alpha_{3,\text{stat}}}{h_n} C_d \le \frac{\Delta_a}{h_{sx}\rho} \tag{12.11}$$

while the ASCE 7 limit is $\Delta_a/(h_{sx}\rho)$ (see Section 15.5.2.1 for the evaluation of $\rho$). Thus, the peak roof displacement of an elastic system subjected to the ASCE-7 story drift limit, $\Delta_a$, can be estimated as

$$u_{\text{roof,el}} \approx \Gamma_1 S_d \le \frac{\Delta_a}{h_{sx}\rho/h_n} \frac{\alpha_1 R_{\text{code}}/I_e}{\alpha_{3,\text{stat}} C_d} \tag{12.12}$$

To account for short-period displacement amplification in design (which is not done in ASCE 7), the peak roof displacement of the inelastic system, $u_{\text{roof,max}}$ is estimated as $C\Gamma_1 S_d$, and the resulting peak interstory drift ratio is either $\alpha_{3,\text{stat}}(C\Gamma_1 S_d/h_n)$ or $\alpha_{3,\text{COD}}(C\Gamma_1 S_d/h_n)$, depending on whether one wants to estimate the static or dynamic value.

Thus, the peak roof displacement limit to use in design that corresponds approximately to the ASCE 7 drift limits is given by

$$u_{\text{roof,max}} = C\Gamma_1 S_d = C\left(\frac{\Delta_a}{h_{sx}}\right)\left(\frac{R_{\text{code}}/I_e}{C_d\rho}\right)\frac{\alpha_1}{\alpha_{3,\text{stat}}} h_n \tag{12.13}$$

In the above, the two terms in parentheses have values established within ASCE 7, while $\alpha_1$ and $\alpha_{3,\text{stat}}$ can be estimated based on modal properties (Tables 12.7 and 12.10). Although $C = C_1 \cdot C_2 > 1$ for periods less than $T_g$ (Section 4.10), the treatment of drift limits in the code would ignore this (using $C = 1$) and would use a larger tabulated value of allowable drifts for buildings having four or fewer stories (per Table 12.8).

Equation 12.13 would be applied at a hazard level corresponding to 2/3 of the risk-targeted MCE. Within the context of the ASCE-7 and BSSC provisions, demands scale linearly, and thus the interstory drift limit would result in a system ductility limit at the risk-targeted MCE level (having 2% probability of exceedance in 50 years) of approximately

$$u_{\text{roof,max}} = \frac{3}{2} C\left(\frac{\Delta_a}{h_{sx}}\right)\left(\frac{R_{\text{code}}/I_e}{C_d\rho}\right)\frac{\alpha_1}{\alpha_{3,\text{stat}}} h_n \tag{12.14}$$

To get a sense for the significance of Equation 12.14, consider a tall special moment-resistant frame of ordinary importance. Because moment-resistant frames tend to be relatively flexible, we may expect the interstory drift limit to control over the system ductility limit. From Table 12.9, $\Delta_a = 0.020 h_{sx}$, while $u_{\text{roof,y}}$ is estimated to be about (0.5%–0.6%) of the building height (Section 9.5), or approximately $0.0055 h_n$. ASCE 7 provides $(R_{\text{code}}/I_e)/C_d = (8/1)/5.5 = 1.45$. Considering an eight-story building for example, Tables 12.10 and 12.7 suggest $\alpha_1/\alpha_{3,\text{stat}} \approx 0.84/1.34 = 0.63$. Assuming $C = 1$ (which is reasonable for such a tall building) and regularity allowing for $\rho = 1.0$, we obtain $\mu_{\max,\text{drift}} = u_{\text{roof,max}}/u_{\text{roof,y}} = (3/2)(1.0)(0.020)$

*Table 12.10* Limits on interstory drift in Eurocode 8 (EN 1998-1, 2004)

| Condition of nonstructural elements | Limit on $d_r/h$ |
|---|---|
| Brittle nonstructural elements attached to the building framing | $0.005/\nu$ |
| Ductile nonstructural elements attached to the building framing | $0.075/\nu$ |
| Nonstructural elements not interfering with structural deformation, or not present. | $0.010/\nu$ |

$(1.45/1.0)(0.63)(h_n)/(0.0055h_n) = 5.0$. It is interesting that the $\Delta_a$ term is modified by two terms whose product is nearly 1 (i.e., $1.45 \times 0.63 = 0.91$). It is also interesting that even for this moment frame example, the system ductility limit (Table 12.4) of $3.6/I_e = 3.6/1.0 = 3.6$ would control over this drift-based ductility limit.

In the case of structural walls, interstory drifts usually do not control the design—in such stiff systems, higher-mode contributions to drift are small, but generate significant wall shears. However, interstory drifts near the top of slender walls can be significant and may control the design—in such cases, coupled wall systems may be preferable.

By explicitly considering a drift limit in conjunction with an estimated yield displacement, we can use YPS (or YFS) to directly establish the required base shear strength (to limit the roof drift and associated peak interstory drifts), thereby avoiding the iterations that are typically required when using conventional R-factor approaches to obtain a building that satisfies code drift limits.

## 12.7.2  Application of interstory drift limits from a Eurocode perspective

In parallel to Section 12.7.1, this section establishes an expression for a roof drift limit that approximately represents the interstory drift limits contained within Eurocode 8 (EN 1998-1, 2004). In concept, the Eurocode 8 limits are applied at the SLS ground motion, generally defined as that having 10% probability of exceedance in 10 years (equivalent to a return period of 95 years). In practice, the Eurocode 8 limits are specified with respect to the SLS but are applied at the ULS through the use of a scaling parameter, $\nu$. That is, the interstory drift determined under the design spectrum (the ULS spectrum adjusted for the behavior factor ($q$) and importance ($\gamma_1$)) is limited to the value of $d_r$ given in Table 12.10, where $h$ is the story height, and $\nu$ (Table 12.5) is an adjustment to address the higher mean annual frequency of exceedance of the SLS and the importance of the structure (Table 12.5).

Specifically, the Eurocode 8 ULS spectrum is scaled up by $\gamma_1$ (for importance) and down by $q$ for periods greater than $T_B$ (approximately 0.05–0.20 s). Displacements ($d_e$) determined under this scaled ULS spectrum are then scaled up by a term $q_d$ (that is, $d_{s=}q_d \cdot d_e$), where $q_d$ is typically assumed equal to $q$ (although larger values can be used for $T < T_C$ to provide for short-period displacement amplification). The interstory drift ($d_r$) is defined in Eurocode 8 as the difference of the displacements ($d_s$) of adjacent floors. The allowable interstory drift, calculated in this way at the ULS state, is limited to the values given in Table 12.10, which are scaled from the SLS level to the ULS level by the factor $\nu$ (Table 12.5).

Contributions of higher modes and the concentration of drifts in particular stories due to inelastic behavior are neglected when drifts are checked using linear static analysis.

Just as for ASCE 7, due to the neglect of higher-mode contributions and inelastic concentration of interstory drifts, the application of the above drift limits in a static analysis can be viewed as a requirement to provide a minimum lateral stiffness, rather than being an explicit limit on the expected interstory drifts. Therefore, to develop an expression to allow the code drift limits to be used in design (using YPS and YFS), an estimate of the peak roof displacement corresponding to application of the Eurocode 8 interstory drift limits is developed in the following.

The roof drift corresponding to the interstory drift, $d_r$, of Table 12.10 under the ULS spectrum (assuming $q_d = q$) is determined by similar triangles as

$$u_{roof} = \left(\frac{\lambda S_a W/g}{q/\gamma_I}\right)\left(\frac{1}{\alpha_1 S_a W/g}\right)(q)(\Gamma_1 S_d) = \frac{\lambda\gamma_I}{\alpha_1}\Gamma_1 S_d \qquad (12.15)$$

which assumes that differences in the roof drift associated with deviations of the lateral load pattern from the first-mode pattern can be neglected. The corresponding interstory drift, assuming that the drift profiles of Figure 12.1 are applicable, is given as

$$IDI_{max} = \left(\frac{\lambda\gamma_I}{\alpha_1}\Gamma_1 S_d\right)\left(\frac{\alpha_{3,stat}}{h_n}\right) \leq \frac{d_r}{h} \qquad (12.16)$$

Since the peak roof drift may be affected by short-period amplification, $(C = C_1 \cdot C_2 > 1)$, the system ductility limit associated with the Eurocode 8 drift limits is

$$u_{roof, max} = C\Gamma_1 S_d = C\left(\frac{d_r}{h}\right)\left(\frac{1}{\lambda\gamma_I}\right)\left(\frac{\alpha_1}{\alpha_{3,stat}}\right)h_n \qquad (12.17)$$

where $d_r$ is the Eurocode 8 story drift limit $(0.005h$ to $0.010h)/\nu$, $\alpha_{3,stat}$ can be estimated based on modal properties (Table 12.8), and the roof yield displacement, $u_{roof,y}$, can be estimated as a percentage of $h_n$ (Section 9.5), where $h_n$ is the height of the structure.

To get a sense for the significance of Equation 12.17, consider a tall special moment-resistant frame of ordinary importance. Because moment-resistant frames tend to be relatively flexible, we may expect the interstory drift limit to control over the system ductility limit. Assuming nonstructural elements are not rigidly attached, Table 12.10 limits $d_r/h$ to $0.010/\nu = 0.020$ for $\nu = 0.5$. According to Section 9.5, $u_{roof,y}$ is about $(0.5\%-0.6\%)h_n$ or approximately $0.0055h_n$. Considering an eight-story building of ordinary importance $(\gamma_I = 1.0)$, for example, Tables 12.1 and 12.8 suggest $\alpha_1/\alpha_{3,stat} \approx 0.84/1.3 = 0.65$. Thus, using Equation 12.17, we obtain $\mu_{max} = u_{roof,max}/u_{roof,y} = 1.0(0.020)(1/(0.85 \cdot 1.0))(0.65)(h_n)/(0.0055h_n)) = 2.8$. This drift-based ductility limit is less than the structural ductility limit (Table 12.4) of $3.7/\gamma_I = 3.7$ and thus controls, even for a relatively flexible moment frame system.

## 12.8 VERTICAL DISTRIBUTION OF STRENGTH AND STIFFNESS

### 12.8.1 Distribution of base shear over height

Figure 12.3 illustrates the capacity curve obtained in a first-mode static (pushover) analysis. The static application of lateral forces associated with any mode generates a lateral displacement profile that is proportional to the mode shape, within the linear domain. Although displacement response over the height of the building generally is dominated by the first mode, story shears and interstory drifts often have significant contributions from higher modes, especially in the upper stories. Thus, when using an ELF method for design, some higher-mode contributions should be represented in the force profile. The lateral force distributions used in many building codes depart from first-mode distributions in order to introduce additional shear in the upper stories.

For many years the lateral force distribution in the U.S. building codes (e.g., the 1982 Uniform Building Code) followed a simple inverted triangular pattern, supplemented for longer period structures by a force applied at the top, $F_t$. Specifically, $F_t$ is $0.07V_{base}T$ for $T >$ 0.7 s, subject to a cap of $0.25V_{base}$, and is 0 for shorter periods. The lateral force applied to the $i$th floor, $F_i$, is proportional to the weight, $w_i$, and height above the base, $h_i$:

$$F_i = \frac{w_i h_i}{\sum_{j=1}^{n} w_j h_j} \left( V_{base} - F_t \right) \tag{12.18}$$

where $w_i$ and $h_i$ are the reactive weight and height above the base, respectively, of the $i$th floor and $V_{base}$ is the design base shear. As illustrated in Figure 12.7, the presence of $F_t$ increases story shears and overturning moments over the height of the structure. An inverted triangular distribution results for $T \le 0.7$ s.

ASCE 7 addresses higher-mode contributions to story shears and overturning moments by altering the inverted triangular distribution by use of an exponent, $k$, applied to the height term. Thus,

$$F_i = \frac{w_i h_i^k}{\sum_{j=1}^{n} w_j h_j^k} V_{base} \tag{12.19}$$

In the above, $k = 1$ for $T \le 0.5$ s, resulting in an inverted triangular distribution of lateral forces for short-period systems. For $T \ge 2.5$ s, $k = 2$, resulting in a parabolic distribution of lateral forces. For intermediate periods, $k$ is interpolated linearly between 1 and 2.

The "lateral force method of analysis" in Eurocode 8 establishes the base shear force, $F_b$, as:

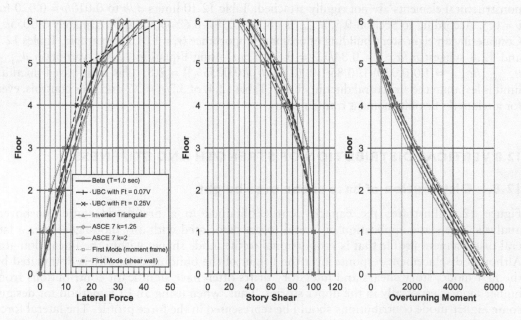

Figure 12.7 Lateral force distributions for the range of UBC, ASCE 7, Eurocode 8 (first mode and inverted triangular), and Beta distribution (for $T = 1.0$ s) applied to a six-story building.

$$F_b = S_d(T_1)m\lambda \tag{12.20}$$

where $S_d(T_1)$ is the ordinate of the pseudo-acceleration design spectrum at the fundamental period of vibration of the building for lateral motion in the direction considered, $T_1$, $m$ is the total mass of the building, and $\lambda$ is a correction factor (taken equal to 0.85 if $T_1 \leq 2T_C$ and the building has more than two stories, or 1 otherwise). $T_C$ is the upper limit of the period of the constant spectral acceleration branch of the Eurocode 8 design spectrum (see Section 16.4).

The "lateral force method of analysis" in Eurocode 8 is applicable to buildings whose response is not significantly affected by contributions from modes of vibrations higher than the fundamental mode in each principal direction, which comply with (a) fundamental periods of vibration in the two main plan directions are smaller than $4T_C$ and 2.0 s, and (b) the structure meets the criteria for regularity in elevation given in Section 4.2.3.3 of Eurocode 8.

A first-mode distribution of lateral forces over the height of the building is given by:

$$F_i = F_b \frac{s_i m_i}{\displaystyle\sum_{j=1}^{n} s_j m_j} \tag{12.21}$$

where $F_i$ is the horizontal force acting on level $i$, $m_i$, and $m_j$ are the masses of the stories $i$ and $j$, respectively, and $s_i$ and $s_j$ are the amplitudes of the fundamental mode at levels $i$ and $j$, respectively.

As an alternative, Eurocode 8 allows the use of an inverted triangular force distribution—using Eurocode notation:

$$F_i = F_b \frac{z_i m_i}{\displaystyle\sum_{j=1}^{m} z_j m_j} \tag{12.22}$$

where $z_i$ and $z_j$ are the heights of the masses $m_i$ and $m_j$ above the level of application of the seismic action.

More recently, Chao et al. (2007) recommended a new lateral force distribution for design, termed the $\beta$ distribution. This lateral force distribution is said to more closely correspond to the peak shears observed in nonlinear dynamic response and to result in a design that develops more uniform interstory drifts over the height of the structure when subjected to earthquake excitation. As presented herein, $\beta_i$ represents the ratio of the story shear, $V_i$, just below level $i$, and the base shear, $V_y$, corresponding to the effective yield strength, as follows:

$$\beta_i = \frac{V_i}{V_y} = \left( \frac{\displaystyle\sum_{j=i}^{n} w_j h_j}{\displaystyle\sum_{j=1}^{n} w_j h_j} \right)^{\frac{\alpha}{T_e^{0.2}}} \tag{12.23}$$

where $V_y$ = effective yield strength, $T_e$ = the effective period of vibration, and $\alpha = 0.75$. Thus, the ELF applied at Level $i$, $F_{i,\beta}$, is given by:

$$F_{i,\beta} = (\beta_i - \beta_{i+1}) \cdot V_y \tag{12.24}$$

In the above, $\beta_{n+1}$ is taken as zero.

Figure 12.7 illustrates the application of these different lateral force distributions to the design of a six-story building having an assumed period of $T = 1.0$ s. The applied lateral forces and resulting story shear and overturning moment distributions over the height of the building illustrate the UBC case for $F_t = 0.07VT$ and the limit of $0.25V$, the ASCE 7 case for $k = 1.25$ and the limit of 2, the Eurocode 8 case for mode shapes representative of slender wall and moment frame buildings, and the Beta distribution for $T = 1.0$ s. An inverted triangular distribution is also shown, representing the UBC distribution for $F_t = 0$, the ASCE 7 distribution for $T < 0.5$ s, and the alternate Eurocode 8 distribution.

Depending on the lateral force distribution, the height of the resultant of the applied lateral forces ranges between 72% and 82% of the height of this example building, which has uniform floor masses and story heights. The resultant moves higher as the lateral force distribution is adjusted to accommodate longer period systems (e.g., an increase in $F_t$ or $k$), to address the greater shear demands in the upper stories associated with higher modes.

## 12.8.2 Modification of base shear

As shown in Chapter 6 on PCA, the roof displacement response typically is dominated by response in mode similar to the first elastic mode. Higher modes (or MDOF effects) generally have a relatively minor contribution to displacement response, assuming no change in mechanism. In the design approaches described in this book, the peak roof displacement is estimated solely on the basis of first-mode response. The "equivalent" SDOF system is derived based on a first-mode nonlinear static (pushover) analysis; the resulting capacity curve is characterized by a yield displacement and yield strength determined for response in the first mode. The proportioning of components of the lateral system, however, must consider the presence of higher modes. The lateral force distributions used in code design procedures increase story shears in the upper stories relative to the first-mode shears (Figure 12.7) and provide a way to address higher modes.

When the distribution of lateral forces for design deviates from the first-mode distribution, there may be a need to adjust the base shear used in design. This is most easily seen by considering the use of a plastic mechanism analysis for the design of the lateral system. The external work may be computed as the product of the resultant lateral force (which equals the design base shear), and its associated displacement. In Figure 12.8, the kinematic mechanism has the same displacement profile for both the moment-resistant frame (in a strong column–weak beam (SCWB) mechanism) and the cantilever wall (hinging at its base). The resultant of the design force distribution goes through a larger displacement, causing an increase in the internal work associated with plastic hinging in these systems, which causes an increase in calculated member strength requirements, relative to that needed for a mechanism that forms under the first-mode forces. For the intended mechanisms for both the moment-resistant frame and the wall, the base shear used in design with a code distribution of lateral forces should be modified relative to that required for the first-mode distribution of forces, in order to preserve the mechanism strength under first-mode forces. For example, in the case of the Beta distribution, the modified base shear is given by $V_y \cdot \left( h_1^* / h_\beta^* \right)$ where $V_y$ is determined as $\alpha_1 C_y^* W$, $h_1^*$ is the height of the resultant of the first-mode forces, and $h_\beta^*$ is the height of the resultant of the $\beta_i$ lateral forces. Estimates of $h_1^*/h_n$ are provided in Table 12.1 for use in preliminary design.

Note that in the case of a weak-story mechanism or a mechanism involving yielding of diagonal braces (Figure 12.8e), the resultant of the code lateral force distribution moves through the same displacement as the resultant of the first-mode forces—in such cases no adjustment in base shear force is indicated.

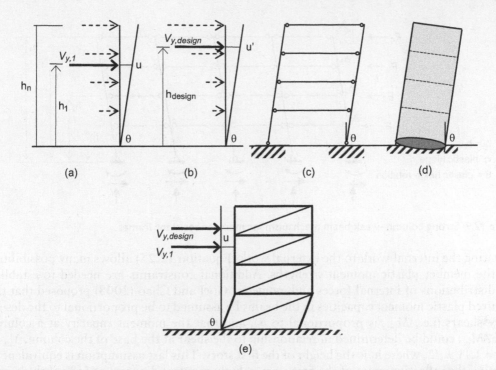

*Figure 12.8* The external work done by different lateral force distributions depends on the location of the resultant and the mechanism that forms. (a) First-mode forces; (b) design forces; (c) moment frame (strong column–weak beam); (d) shear wall; (e) yielding brace (or weak-story frame mechanism).

Also note that in the preceding, we assume that the same mechanism occurs under the lateral design forces and the first-mode forces. This is very likely to be the case, and can be validated by analysis.

## 12.8.3 Design of components based on plastic mechanism analysis

The design base shear of the MDOF system at yield (modified for the height of the resultant) is used to establish the strength of the system at the formation of a mechanism. This approach makes use of plastic design methods (Chapter 11), in contrast to conventional elastic design approaches wherein an elastic distribution of forces and moments is used.

In the case of moment-resisting frames, the SCWB mechanism shown in Figure 12.9 is sought, in contrast to a weak-story plastic mechanism. One may use a virtual work approach to proportion the plastic hinges, equating the internal work of the plastic hinges with the external work of the lateral forces distributed over the height of the structure (having resultant equal to $V_y$):

$$W_{\text{internal}} = W_{\text{external}}$$

$$\text{where} \quad W_{\text{internal}} = \sum_{\text{All the hinges}} M_i \theta_i \tag{12.25}$$

$$W_{\text{external}} = \sum_{\text{Exterior lateral forces}} F_i \Delta_i$$

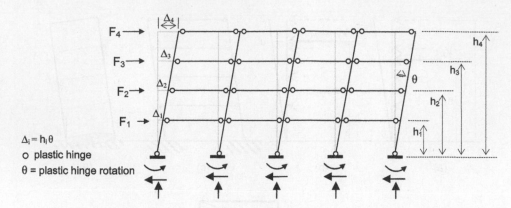

*Figure 12.9* Strong column–weak beam mechanism for moment-resistant frames.

Equating the internal work to the external work (Equation 12.25) allows many possibilities for the member plastic moment strengths. Additional constraints are needed to establish the distributions of internal forces and moments. Goel and Chao (2008) proposed that the required plastic moment capacities in the beams be assumed to be proportional to the design story shears (i.e., $M_{pb,i}$ is proportional to $\beta_i$), and that the moment capacity at a column base, $M_{pc}$, could be determined in relationship to the shear at the base of the column, $V_c$, as $M_{pc} = 1.1 V_c h_1/2$, where $h_1$ is the height of the first story. This last assumption is equivalent to locating the inflection point of the first-story column moment diagram at 55% of the height of the story above the base of the column. In the case of a steel-framed building, the column consists of a prismatic steel shape and the formation of a weak-story mechanism is avoided (Figure 12.10), since a column yielding at the base at a moment $M_C$ has a moment at the top of the column of $0.82M_C$ (which would be in the linear range). In this case, the SCWB mechanism will form preferentially, as it will require less internal work than the weak-story mechanism (provided that the beams are not overly strong).

While appealing, this approach has two drawbacks, which motivated the development of a modified design procedure, described below. The first drawback is that the strength of reinforced concrete columns can be tailored to the moment diagram, so the strength required at the top of the first-story columns to avoid a weak-story mechanism must be stated explicitly (Figure 12.10). The second is that when applying Chao and Goel's approach, the moment diagrams for the columns, associated with a SCWB design mechanism, can be predominantly or entirely in single curvature over one or more stories. This means that larger drifts will occur than otherwise due to accumulation of larger story drifts over multiple stories when developing the design mechanism, and the larger column design moments will require larger member sizes. To obtain a more efficient structural design, we recommend the following (Hernández-Montes and Aschheim, 2017):

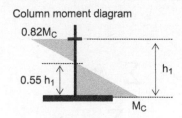

*Figure 12.10* Moment diagrams in a first-story column.

1. For regular framing systems, we rely on the conventional portal frame assumption that end columns resist half of the shear carried by the intermediate columns at any story, for systems having approximately uniform spans. For systems that lack regularity, the engineer should distribute shears in a manner approximately consistent with the distribution observed in linear analysis.

2. Members that develop plastic hinges as part of the mechanism may be proportioned according to conventional code prescriptions. We suggest the use of expected material strengths, with material safety coefficients (or strength reduction factors) set equal to unity.

3. Using the freedom allowed by Equation 12.25, we restrict the locations of inflection points to be near the mid-height of each story (Figure 12.11) in order to keep the moments at the ends of each column segment similar in value. (This allows each column to be proportioned more economically, while the reversal of curvature over the height of the column helps to limit interstory drifts.)

4. Finally, equilibrium of the column at the instant that a SCWB mechanism has formed has beam plastic moments acting as shown in Figure 12.11. In order to assure the formation of the SCWB mechanism shown in Figure 12.9, a hierarchy of strengths must be provided, to ensure the intended mechanism forms at the design yield strength while other locations within the lateral force-resisting system remain nominally elastic. The conventional approach, contained within ACI 318, requires that the sum of column strengths at any beam-column joint exceed 6/5 of the sum of beam hinge strengths (Sections 15.6.2 and 16.6). With this approach, however, the strength at any one column section may fall below that needed to sustain the beam hinging mechanism (e.g., Figure 12.11) since only the sum of column strengths (above and below the beam) is regulated. To avoid a potential weak-story mechanism, we suggest the individual critical column sections be designed for at least 6/5 of the moments required to achieve the SCWB mechanism.

Note that gravity loads do not contribute to external work for the typical mechanisms sought in seismic design, because the vertical displacements at the locations of the applied gravity loads are zero. Thus, the plastic hinges can be designed considering the moments due to lateral load alone. However, the elastic distribution of moments induced by gravity loads may result in nonzero moments at locations of the plastic hinges. Good design practice

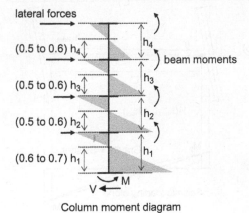

Column moment diagram

*Figure 12.11* Example illustrating use of assumed inflection point locations for determining beam moments in SCWB plastic mechanism analysis. Design column moment strengths are increased above the values plotted to promote the SCWB mechanism.

avoids the development of significant inelasticity at any potential plastic hinge location under the action of only gravity load. This may be achieved by adjusting plastic hinge strengths to more closely match the pattern of gravity demands, and by validating by structural analysis that sufficient capacity is present to resist the gravity load combinations.

The following provides simple adjustments to the beam moments to account for (i) the presence of gravity loads and (ii) the location of plastic hinges being at the ends of the members rather than at the intersections of the member centerlines.

Section 11.4 identifies limitations on gravity loads in order to avoid the development of midspan plastic hinges and "ratcheting" action under reversed cyclic loading. Assuming gravity loads are low enough to preclude ratcheting mechanisms, we can now address the relatively large negative moments that develop near the beam supports under gravity loading. Ordinarily, to provide for good behavior under service loads, near supports such as columns, beams will be provided with negative moment flexural strength that is much larger than the positive moment flexural strength. For example, ACI 318 requires than $M_n^+$ be at least equal to $M_n^-/2$ at the ends of a moment frame beam. This differs from the assumption that the plastic hinges are the ends of a moment frame beam that have the same strength, denoted as $M_p$, for positive and negative bending (e.g., in the analysis illustrated in Figure 12.11).

Thus, if beams are proportioned such that $M_p^+ = M_p^-/2$, plastic hinging at the left and right ends of each beam will develop internal work equal to $1.5M_p^-(\theta)$, which should equate to $2M_p(\theta)$. Consequently, the beam plastic hinges should be proportioned for $M_p^- = 4M_p/3$ and $M_p^+ = 2M_p/3$, where $M_p$ is the beam plastic moment strength assumed in the analysis illustrated in Figure 12.11.

For simplicity, virtual work analyses of kinematically admissible mechanisms will typically be done on the basis of member centerline dimensions. Actual beam plastic hinge locations are inset from the column centerlines (Figure 12.12). If the physical plastic hinge length (in

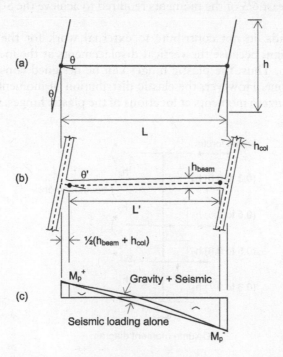

*Figure 12.12* Effect of plastic hinge locations inset from column centerlines: (a) plastic mechanism based on centerline dimensions; (b) plastic mechanism based on plastic hinges set in from column centerlines; (c) beam moment diagram.

contrast to the analytical plastic hinge length discussed in Section 17.3) is taken as $h_{beam}/2$, then the center of the beam plastic hinge is offset from the column centerline a distance of $(h_{beam} + h_{col})/2$, where $h_{beam}$ and $h_{col}$ are the overall depths of the beam and column sections. If this distinction is accounted for, the required beam plastic hinge strengths can be reduced by $(h_{beam} + h_{col})/L$ (in accordance with the moment diagram), resulting in

$$M_p^- = \frac{4}{3} M_p \left( 1 - \frac{h_{beam} + h_{col}}{L} \right)$$ (12.26a)

$$M_p^+ = \frac{2}{3} M_p \left( 1 - \frac{h_{beam} + h_{col}}{L} \right)$$ (12.26b)

where $L$ = the beam span between column centerlines. The inset hinge location also causes the plastic hinge rotation $\theta'$ to be larger than if the hinge were located at the column centerline, with $\theta' = \theta(L/L')$. Plastic hinge rotations may be assessed by nonlinear response history analysis, and if evaluated this way, will include higher-mode contributions to interstory drift.

Finally, two points should be noted: (i) we normally rely on the floor slab cast monolithically with the beams to provide lateral bracing to the beam and its plastic hinge zone; if no slab is present, the stability of the plastic hinge and possible need for lateral bracing should be evaluated, and (ii) it has generally been considered acceptable to replace the roof-level beam plastic hinges with plastic hinges located at the tops of the columns that frame into the roof beams, since hinging of the columns at this location does not jeopardize the stability of multiple floors.

# REFERENCES

Abrams, D. P. (1985). Nonlinear Earthquake Analysis of Concrete Building Structures, Final Report on a Study to the American Society for Engineering Education Postdoctoral Fellowship Program, September 1985.

ASCE/SEI 7-16. *Minimum Design Loads and Associated Criteria for Buildings and Other Structures*, American Society of Civil Engineers, Reston, VA, 2016.

Aschheim, M., and Maurer, E. (2007). Dependency of COD on ground motion intensity and stiffness distribution, *Structural Engineering Mechanics*, 27(4):425–438.

BSSC (2010). Resource Paper 9: Seismic Design using Target Drift, Ductility, and Plastic Mechanism as Performance Criteria, in NEHRP Recommended Seismic Provisions for New Buildings and Other Structures, Building Seismic Safety Council, Federal Emergency Management Agency, Report No. FEMA P-750, published January 28, 2010.

Chao, S.-H., Goel, S. C., and Lee, S.-S. (2007). A seismic design lateral force distribution based on inelastic state of structures, *Earthquake Spectra*, 23(3):547–569.

EN 1998-1 (2004). *Eurocode 8: Design of Structures for Earthquake Resistance – Part 1: General Rules, Seismic Actions and Rules for Buildings*, European Committee for Standardization, Brussels.

Goel, S. C., and Chao, S. (2008). *Performance-Based Plastic Design Earthquake-Resistance Steel Structures*, International Code Council, Washington, D.C.

Gupta, A., and Krawinkler, H. (2000). Estimation of seismic drift demands for frame structures, *Earthquake Engineering and Structural Dynamics*, 29:1287–1305.

Hernández-Montes, E., and Aschheim, M. (2017). A seismic design procedure for moment-frame structures, *Journal of Earthquake Engineering*, 26:1–20, DOI: 10.1080/13632469.2017.1387196.

International Conference of Building Officials (1982). *Uniform Building Code*, Whittier, CA.

Moehle, J. P. (1992) Displacement-based design of RC structures subjected to earthquakes, *Earthquake Spectra*, 8(3):403–428.

NEHRP Recommended Seismic Provisions for New Buildings and Other Structures. 2009 Edition. Resource Paper 9 Seismic design using target drift, ductility, and plastic mechanisms as performance criteria.

Qi, X., and Moehle, J. P. (1991). Displacement Design Approach for Reinforced Concrete Structures Subjected to Earthquakes, Report UCB/EERC-91/02. University of California, Berkeley.

Ruiz-García, J., and González, E. J. (2014). Implementation of displacement coefficient method for seismic assessment of buildings built on soft soil sites, *Engineering Structures*, 59:1–12.

# Chapter 13

# Probabilistic considerations

## 13.1 PURPOSE AND OBJECTIVES

It is a strange cosmic coincidence that the Thirteenth Chapter of this book deals with and hopefully demystifies the (seemingly) "dark art" of probability. This chapter begins with a basic review of probability and statistics and develops the mathematical tools used in Chapters 18 and 19 for seismic performance assessment and design. Section 13.2 provides a basic review of probability and statistics. Probabilistic seismic hazard analysis is presented in Section 13.3. Section 13.4 addresses seismic performance assessment within the framework of performance-based earthquake engineering, and Section 13.5 addresses seismic design to meet desired performance objectives with the required confidence, or margin of safety. These methods are relied upon in the single- and double-stripe assessment approaches introduced in Chapter 5, further described in Chapter 18, and illustrated in the examples of Chapter 19.

## 13.2 PROBABILITY AND STATISTICS FOR SAFETY ASSESSMENT

We review a set of basic concepts and approaches to probability relevant to earthquake engineering, much less than is provided in a complete reference to probability theory and applications (e.g., Benjamin and Cornell, 1970). For our needs, probability can be considered as a rigorous method to quantify what we know about the identified (or known) unknowns of a physical problem. Probability is not some kind of dark art that will magically account for the unidentified (or unknown) unknowns, nor can it substitute for good knowledge and understanding of an engineering problem. Instead, probability may be considered as a tool for carrying defined uncertainty through the components of an engineering problem—it starts where classic, deterministic, engineering models end, and aims to quantify the degree to which our abstract models and calculations match reality. A simple example application is choosing values for partial safety factors (e.g., load and resistance factors) that invariably appear in engineering design equations in order to inject a designated amount of safety against the unknown and the uncertain. Thus, probability becomes essential when we recognize that classical theory is deterministic but reality is not. For example, a deterministic strength of materials approach states that the theoretical tensile yield strength of a steel bar is simply $N_y = Af_y$, where $A$ = cross-sectional area and $f_y$ = material yield stress (Figure 13.1). Thus, a design check that the bar does not yield when subjected to a tensile load of $N$ would ensure that the demand $D$ ($=N$) is less than the capacity $C$ ($=N_y$), or simply $N < N_y$. In reality, several sources of uncertainty may sometimes result in $N > N_y$, despite abiding by the above theoretical result: (i) uncertainty in the material properties, or actual $f_y$; (ii) uncertainty in the bar area, $A$; and (iii) variability in the axial load demand $N$. In each of these sources

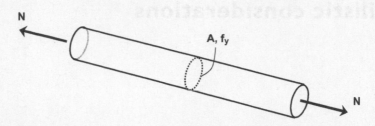

*Figure 13.1* A simple probabilistic problem: a steel bar of uniform cross-section $A$ and yield strength $f_y$ under a tension force $N$.

there happen to be two potential flavors of uncertainty that may give rise to variability: (i) Aleatory variability (from the Latin word alea or dice) is the natural variability that is inherent in a problem and cannot be removed and (ii) epistemic uncertainty (from the Greek επίσταμαι, i.e., to know) is caused by our own incomplete knowledge. The former is irreducible, while the latter can be reduced or (theoretically speaking) even eliminated completely by gaining more knowledge. For example, if we have the bar at hand one may (i) estimate the area of the bar by a single measurement of its diameter, (ii) perform measurements along the length of the bar at frequent intervals to determine the minimum diameter, or (iii) use advanced measurement techniques such as laser-scanning to get an accurate three-dimensional representation of the bar down to the wavelength of light used. Each method increases knowledge and, thus, reduces epistemic uncertainty, but comes with a higher cost of application. This kind of trade-off is not available when considering the case where the bar is not available for testing such as where a nominal reinforcement bar is specified that will be placed within a concrete section. Since the bar is not available for testing, we can only use the (producer-supplied) nominal diameter to estimate its area, and it will be subject to aleatory variability that cannot be alleviated. The same considerations hold for $f_y$, where testing similar bars from the same production run can reduce the uncertainty for a given bar, but probably in the more general case of a random bar from a given steel grade, that may come from any number of producers and batches, testing similar bars will do little to constrain the uncertainty in $f_y$. Similarly, the demand $N$ is subject to aleatory variability due to the nature of the loads themselves (e.g., variable dead, live, and seismic loads), while $N$ is also subject to epistemic uncertainty due to the inaccuracies in the structural model that we use to estimate it.

While the above distinctions are useful, the classification of what is aleatory and what is epistemic is often a function of the model adopted, the situation, or even the current state-of-art (Der Kiureghian and Ditlevsen, 2009). In the above example, depending on whether the bar is available for testing or not, the variability in $A$ may be classified as epistemic or aleatory, respectively. Thus, categorizing the sources of uncertainty in any given problem helps us separate them into those that we *can* control (and whose uncertainty we can perhaps reduce), and those that we *cannot* control. Still, when uncertainty reduction through testing is not of interest, it is much simpler to uniformly refer to all such sources of variability as "uncertainties," without differentiating the source type. This is the approach that we shall adopt henceforth.

As we all very well know, to account for the effect of uncertainties in design we use constants, known as load and resistance partial safety factors, that (i) increase the magnitude of demand $N$ and (ii) reduce the value of capacity $N_y$ to provide a designated level of safety. How one proceeds to back-calculate such safety factors given a required level of safety is a basic application of probability theory, and forms the foundation for this presentation: rather than providing an exhaustive reference for probability theory we instead focus on those properties and definitions that help us understand how and why we approach safety in the way that we do. In other words, we will use probability to do the necessary calculations and retain the needed results and an improved awareness of the contributions of uncertainty to the problem at hand. Bear in mind that probability is not magic—if we cannot identify and model the sources of uncertainty that enter into the important variables (e.g., $N$, $A$, and $f_y$), then we will not be able to achieve the desired safety. On the one hand, identification of the sources of uncertainty rests with the user and his/her understanding and engineering intuition of the problem at hand. There is little that we can offer here, beyond awareness. On the other hand, we shall discuss modeling in detail.

## 13.2.1 Fundamentals of probabilistic modeling

A probabilistic model for sources of uncertainty can be every bit as complex as a finite element (FE) model for a structure. Just like a structural model represents the distribution of stiffness, mass, and strength in the three-dimensional space of the spatial coordinates, a probability model captures the distribution of probability over all possible outcomes for each nondeterministic (i.e., random) variable of the problem and for all (known or identifiable) sources of uncertainty that influence it. To properly do so, one needs to define the following three quantities:

a. *Support*: For each random variable, this is the set of possible outcomes, or, in other words, the range of values whose probability of occurrence is nonzero. For each of the continuous variables $N$, $A$, and $f_y$ of our previous example, this could be $(0, +\infty)$, i.e., the set of all positive real numbers.

b. *Distribution*: For each random variable, how likely/probable/frequent is each value over its support. In other words, how the probability is distributed over this range. If all values are equally probable, one may choose a uniform distribution. In the more likely case where there is preference for some values over others, there are many standard nonuniform distribution models (e.g., normal, lognormal, and exponential) that are discussed in the following.

c. *Dependence*: For every set of random variables we need to decide on whether any are dependent on others or if all are independent; for any one that depends on the other(s), we need to specify the dependency. In the above example, we could easily claim that all three variables are independent. But, if considering both the yield and ultimate strengths, we could equally easily argue for dependence, since knowledge of one helps to inform the likely range of the other—a higher yield strength should be correlated with a higher ultimate strength, and vice versa.

To tackle the above elements, we introduce some mathematical formalism that will make our life easier.

## 13.2.2 Mathematical basis of probability

First, we define the sample space $\Omega$ (the space that contains all possible outcomes). For each variable ($N$, $A$, and $f_y$) in our bar example, when considered individually, the sample space would be $(0, +\infty)$. If we are considering all three parameters together, then $\Omega = (0, +\infty) \times (0, +\infty) \times (0, +\infty)$. In general, $\Omega$ is the Cartesian product of the sample spaces for each individual variable of the problem. Thus, it makes sense to identify them from the very start and appropriately set the scale of the problem (and model) to solve.

The *first axiom of probability* says that the probability of an event (see also formal definition of events below) occurring that encompasses the entire sample space $\Omega$ is 1. Formally:

$$P(\Omega) = 1 \tag{13.1}$$

Practically speaking, this is just a statement that $\Omega$ expresses the entire set of potential outcomes. The value of 1.0 is just a convenient choice to represent absolute surety; a value of 10 or 100 or any other number could have been chosen. However, since a value of 1.0 was chosen, this now represents absolute surety to which all other degrees of lesser surety will be compared to.

Formally, a subset of $\Omega$ is called an event $E$ and it represents a collection of outcomes that is smaller or equal to $\Omega$. The *second axiom of probability* simply says that

$$0 \le P(E) \le 1 \tag{13.2}$$

Implicitly, this assigns a probability of 0 to a null set (i.e., an event of zero outcomes) and informs us that all probabilities of any other event within $\Omega$ will lie between the two extremes of 0 and 1, where 0 and 1 represent the two extremes of absolute surety that something will not or will happen, respectively. Note that the midpoint of 0.5 represents the maximum of unsurety: a 50–50 chance that something will or will not happen, or in other words, a complete lack of information. That would be your regular unbiased coin toss.

The *third and final axiom* tells us how two mutually exclusive events $E_1$ and $E_2$ (events that have no common outcome) are combined together, i.e., the probability that any of them (or both) occurs:

$$P(E_1 \text{ or } E_2) = P(E_1) + P(E_2) \tag{13.3}$$

From the above, a number of common sense attributes for the probability of events can be derived using the rules of logic and set theory. Actually, these rules are our first major tool in performing probabilistic computations. Combining events together and determining their probability is akin to employing the force and moment balance equations to solve statically determinate (isostatic) structural analysis problems. In other words, it is a simple and effective approach, but only for problems of limited complexity. Thus, for example, we can easily derive that

$$P(\text{not } E) = 1 - P(E) \tag{13.4}$$

while in the general case that $E_1$ and $E_2$ are not mutually exclusive we can still claim that

$$P(E_1 \text{ xor } E_2) = P(E_1) + P(E_2) \tag{13.5}$$

where "xor" is the exclusive "or" (i.e., only one of the two happens). In the general case that either or both can happen (where $E_1$ and $E_2$ need not be mutually exclusive), we get

$$P(E_1 \text{ or } E_2) = P(E_1) + P(E_2) - P(E_1 \text{ and } E_2) \tag{13.6}$$

where the last term addresses the lack of mutual exclusiveness when compared to Equation 13.3.

## 13.2.3 Conditional probability

Generally speaking, properly defining sample spaces and the events within them is very important as it sets the scale and size of the problem. Quite often, in the course of solving or modeling a problem, one may need to replace $\Omega$ with a smaller subset of an event $E$, once we have knowledge that $E$ has occurred. This is essentially a rescaling of the problem where our new sample space is reduced from $\Omega$ to $E$ in order to incorporate the fact (or information) that $E$ has happened. Rescaling should always be performed with caution as it changes the probability of all events that are conditioned on $E$ happening (or $E$ being the new sample space, effectively replacing $\Omega$). For example, going back to our bar example, if we receive accurate test information (i.e., precise knowledge) on the diameter of the bar, then we should reduce $\Omega$ from a triple Cartesian product to a double, as the uncertainty on one of the three variables has been eliminated. Formally, the conditional probability that $A$ can occur given that $E$ has (or will) happen is

$$P(A|E) = P(A \text{ and } E)/P(E) \tag{13.7}$$

Obviously, in the case where $P(E) = 0$, then $P(A|E) = 0$ as well.

The concept of conditional probability provides the necessary formalism to define what is meant by independence. Two events $E_1$ and $E_2$ are probabilistically independent if

$$P(E_1|E_2) = P(E_1), \text{ or } P(E_2|E_1) = P(E_2), \text{ or } P(E_1 \text{ and } E_2) = P(E_1) \cdot P(E_2). \tag{13.8}$$

In other words, knowing $E_2$ gives us exactly zero information about $E_1$ and vice versa. Note that this does not require that events $E_1$ and $E_2$ are mutually exclusive, which would translate to $P(E_1|E_2) = 0$, since mutual exclusiveness means that knowing that $E_2$ has happened provides perfect information on $E_1$—that $E_1$ has not happened. Thus, unless we have the trivial case of $P(E_1) = 0$, the two mutually exclusive events are quite dependent.

Conditional probability and mutual exclusivity provide us with our second major tool in modeling and solving probability problems: the theorem of total probability. This theorem tells us that in order to measure, e.g., the weight of a building, which is rather impossible to weigh on a scale, we need only to measure the weights of its individual components and sum them up. We simply have to be careful not to miss any component and not to double-count. In mathematical terms, let complex event $A$ be composed by the mutually exclusive and collectively exhaustive events $E_1, E_2, ..., E_k$. Then:

$$P(A) = \sum_{i=1}^{k} P(A|E_i)P(E_i) \tag{13.9}$$

In other words, the probability of a complex event $A$ can be determined if we can break $A$ into simpler, mutually exclusive, events $E_i$ and determine the probability of $A$ given that each $E_i$ will occur. Thus, if for example one is trying to estimate the probability of a loaf of bread containing any raisins, he or she need only cut the bread to small pieces and estimate the probability that a raisin exists in each. By making the pieces smaller and smaller, one makes the individual probabilities easier and easier to compute, turning a difficult and abstract problem into something manageable. In terms of structural analysis, we can think of the total probability theorem as the compatibility equations. Combined with the "force and moment balance equations" provided by the rules of logic and set theory, they allow us to solve problems of arbitrary complexity.

### 13.2.4 Random variables and univariate distributions

The distribution of probability for a continuous random variable (e.g., N, A, or $f_y$), over its support can be described in three mathematically related ways, termed *distribution functions*. The first is the so-called probability density function (PDF), which describes the density of probability for any potential value of the random variable, in the same way that the density of a material characterizes locally its mass. Just as it makes no sense to discuss the weight of a zero-length piece of a bar (while you can still refer to its density at a specific location), the probability of occurrence of a continuous variable is determined over an interval, say [a, b]. Formally, for a continuous variable x:

$$P(a \leq X \leq b) = \int_a^b f(x)\, dx \tag{13.10}$$

where $f(x)$ is the PDF.

The second approach to describe x is by its cumulative density function (CDF), $F(x)$, which is the probability that x is less than any given value $x_0$:

$$F(x_0) = P(x \leq x_0) = \int_{-\infty}^{x_0} f(x)\, dx \tag{13.11}$$

The third approach is the complementary cumulative distribution function (CCDF) which is simply the probability of x being greater than any given value of $x_0$:

$$G(x_0) = P(x \geq x_0) = 1 - F(x_0) \tag{13.12}$$

Conversely, it becomes obvious that the PDF is the derivative of the CDF, evaluated at any point $x_0$, or, equivalently, the negative (or absolute value) of the derivative of the CCDF:

$$f(x_0) = \left.\frac{dF(x)}{dx}\right|_{x=x_0} = -\left.\frac{dG(x)}{dx}\right|_{x=x_0} = \left|\left.\frac{dG(x)}{dx}\right|_{x=x_0}\right| \tag{13.13}$$

Since it can be difficult to understand and compare distributions expressed as continuous functions, scalars are often used to characterize (i) the central tendency of a distribution and (ii) the dispersion around it.

The most common metric for the central tendency of random variable x is its mean or expected value, represented as $\mu_x$ or $E[x]$:

$$\mu_x = \int_{-\infty}^{+\infty} x f(x)\, dx \tag{13.14}$$

The mean is susceptible to being heavily influenced by extreme values of x. For example, the mean of a discrete variable defined over {1, 2, 3, 4, 100} with 1/5 probability per each outcome is (1 + 2 + 3 + 4 + 100)/5 = 22. In contrast, the mean of equiprobable points {1, 2, 3, 4, 5} is simply (1 + 2 + 3 + 4 + 5)/5 = 3.

A metric of central tendency that is completely insensitive to such outliers is the median $x_{50}$, defined as the value of $x$ above which 50% of the probability lies, or, formally, the value with 50% probability of exceedance:

$$x_{50} = F^{-1}(0.5) \tag{13.15}$$

where $F^{-1}$ denotes the inverse of the CDF. Then, the median estimated over the distributions having equiprobable points {1, 2, 3, 4, 100} or {1, 2, 3, 4, 5} is 3 in both cases.

More generally, one may recover any $p\%$ percentile or fractile value with a corresponding $(1 - p)\%$ probability of exceedance (or $p\%$ probability of non-exceedance):

$$x_p = F^{-1}(p) \tag{13.16}$$

Obviously, the lower $p$ is, the more sensitive $x_p$ is to the lower left "tail" of the distribution, whereas the higher $p$ becomes, the more it is dominated by the right "tail."

To measure the dispersion of a random variable $x$ around its central value, the most common measure is the variance, denoted as $\sigma_x^2$ or Var$[x]$:

$$\sigma_x^2 = \int\limits_{-\infty}^{+\infty} (x - \mu_x)^2 f(x)\, dx \tag{13.17}$$

The square root of the variance, termed the standard deviation $\sigma_x$, is often used instead of the variance, as it has the same units as the mean and the variable itself.

A unitless measure of dispersion is obtained by dividing the standard deviation by the mean to obtain the coefficient of variation (CoV) or $\beta_x$:

$$\beta_x = \frac{\sigma_x}{\mu_x} \tag{13.18}$$

For earthquake engineering problems, variables having $\beta_x < 10\%$ are considered as having very low dispersion and are often considered as deterministic for simplicity—in essence, the trouble of modeling them in a probabilistic problem is often deemed to be unworthy of the effort, considering that ground motion records introduce far higher variability (even for single-degree-of-freedom (SDOF) systems, discussed in Chapter 4).

## 13.2.5 Standard univariate distribution models

In practice, even for variables defined over a continuous interval, one has only a finite number of data points to describe their distributions. For example, for the yield strength of the steel bar, this could be a number of specimens from the same grade of steel and perhaps the same producer, whose strength has been tested. One typically wishes to turn the few data points extracted from the test results into a continuous distribution that will describe the random variable over its entire support. To do so, we typically adopt one of a few simple standardized models that best describe the empirical distribution observed. For our purposes, the most useful models are the uniform, exponential, normal, and lognormal distributions.

The simplest model is the uniform distribution, whereby all points in the support interval of $[a, b]$ receive the same density. Thus, the PDF becomes

$$f(x) = \begin{cases} \dfrac{1}{b-a} & \text{if } x \in [a,b] \\ 0 & \text{otherwise} \end{cases} \tag{13.19}$$

Estimating the CDF is a simple integration operation:

$$F(x) = \begin{cases} 0 & \text{if } x < a \\ \dfrac{x-a}{b-a} & \text{if } x \in [a,b] \\ 1 & \text{if } x > b \end{cases} \tag{13.20}$$

Characteristic plots of the distribution functions appear in Figure 13.2 while the mean, median, and variance appear in Table 13.1.

For cases where the probability density of a random variable $x \geq 0$ drops rapidly as it increases above zero, the exponential distribution is often a useful model:

$$f(x) = \lambda e^{-\lambda x} \tag{13.21}$$

$$F(x) = 1 - e^{-\lambda x} \tag{13.22}$$

$\lambda$ is the so-called mean rate or mean frequency. It is the single parameter that defines the distribution, designating its mean, $1/\lambda$, and standard deviation, also $1/\lambda$. In other words, the exponential has, by definition, a fixed, high CoV of 100%.

The exponential is often used in engineering to describe the interarrival time of so-called memoryless Poisson processes (see Section 13.3.1), and is often used to characterize the occurrence of seismic events on a fault. In other words, lacking more specialized information, we may assume that events (of a given range of magnitude) happen randomly in time with a constant mean rate of $\lambda$. Then, it can be shown that the interarrival time between consecutive events follows the exponential distribution. $T_R = 1/\lambda$ is the mean interarrival time, also known as the mean return period. Although, for a given process, $\lambda$ and $T_R$ may be constants, we prefer to refer to them as *means*. We do so to emphasize that the actual annual frequency and interarrival time between any two consecutive events is not fixed, but instead is highly variable. The overall average over thousands of occurrences is presumed constant (i.e., thousands or millions of years for geological processes).

Another interesting property of an exponential distribution and the corresponding Poisson process appears if we randomly select a subset of all the discrete Poissonian events from a given process with rate $\lambda$, accepting each by a probability of $p$. This is called Poisson filtering and its output is described by another Poisson process whose rate is exactly $p \cdot \lambda$. Thus, for example, if earthquakes occur on a certain fault with a mean annual rate of $\lambda = 0.05\,\text{year}^{-1}$, but only 10% of them are strong enough to cause light damage or worse to our structure, then the mean rate of light damage or worse is $0.1 \cdot 0.05 = 0.005\,\text{year}^{-1}$. Such annualized values of the Poisson rate, $\lambda$, are expressed in units of $\text{year}^{-1}$ and termed mean annual frequency (MAF) or mean annual rate (MAR); they are the basis for all our performance calculations (Sections 13.3–13.5).

A normal (or Gaussian) distribution is arguably the gold standard of random variable models. It is the model for a sum of independent random variables and it possesses a large number of interesting mathematical properties that we shall not dwell upon—interested

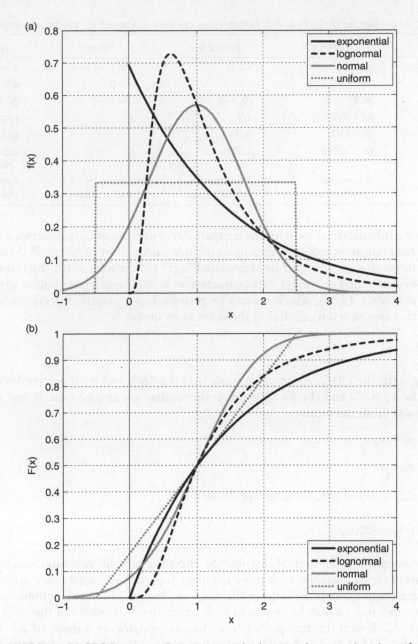

*Figure 13.2* (a) PDF and (b) CDF functions of an exponential, a lognormal, a normal, and a uniform distribution, all having a median of 1.0.

readers are referred to any good probability book (e.g., Benjamin and Cornell, 1970). The normal distribution is used to describe a random variable that may take any real value, positive or negative, with most of the probability concentrated around its mean, while higher or lower values become exponentially less probable with distance from the mean:

$$f(x) = \frac{1}{\sigma\sqrt{2\pi}} \exp\left[-\frac{(x-\mu)^2}{2\sigma^2}\right] \tag{13.23}$$

*Table 13.1* Basic properties of the four distribution types typically employed for earthquake engineering

|  | Uniform | Exponential | Normal | Lognormal |
|---|---|---|---|---|
| Parameters | $a, b$ in $(-\infty, +\infty)$ | $\lambda > 0$ | $\mu$ in $(-\infty, +\infty)$ $\sigma > 0$ | $\mu$ in $(-\infty, +\infty)$ $\sigma > 0$ |
| Support | $[a, b]$ | $[0, +\infty)$ | $(-\infty, +\infty)$ | $(0, +\infty)$ |
| Mean | $(a + b)/2$ | $1/\lambda$ | $\mu$ | $\exp(\mu + \sigma^2/2)$ |
| Median | $(a + b)/2$ | $\ln(2)/\lambda$ | $\mu$ | $\exp(\mu)$ |
| Variance | $(b - a)^2/12$ | $1/\lambda^2$ | $\sigma^2$ | $\exp(2\mu + \sigma^2) \times [\exp(\sigma^2) - 1]$ |
| Properties | Symmetric, uninformative | Monotonic, memoryless | Symmetric all real numbers | Asymmetric, fat right tail |

There is no useful analytical form for the normal CDF, but computational software typically provides a function to compute it. Examples include normcdf in MATLAB® (Mathworks, 2018) and *norm.dist* or *normdist* in Microsoft Excel® (Microsoft, 2010). By inspection, it becomes obvious that Equation 13.23 characterizes a symmetric distribution around the mean $\mu_x = \mu$ (Figure 13.2a), which, as can be proven, has a standard deviation of $\sigma_x = \sigma$. Thanks to this symmetry, its median is the same as its mean:

$$\mu_x = x_{50\%} \tag{13.24}$$

By working with the CDF, one can also show that the 16th and 84th percentiles (or more precisely, the 15.87% and the 84.13% percentile values) are spaced exactly one standard deviation away from the mean:

$$x_{16\%} \approx \mu_x - \sigma_x$$
$$x_{84\%} \approx \mu_x + \sigma_x \tag{13.25}$$

Thus, one may approximate the standard deviation as

$$\sigma_x \approx \frac{1}{2}\left(x_{84\%} - x_{16\%}\right) \tag{13.26}$$

Despite the support of the normal distribution encompassing all real numbers, it is often employed to describe variables with more compact support, as long as they are unimodal (i.e., having a single "bump" of large probability, see Figure 13.2a) and symmetrically distributed. For example, it may be employed to describe the distribution of the yield strength $f_y$ of a grade S235 steel (by European EN standards), employing a mean of $\mu = 300$ MPa (remember that 235 MPa is a characteristic value, not the mean) and a standard deviation of 20 MPa. Although $f_y$ will always be positive, there is a nonzero probability that negative yield strengths appear when a normal distribution is used. Still, for these numerical values, this probability is only $F(0) = 3.7 \cdot 10^{-51}$, which is truly low. Still, some care should be exercised to make sure that the probability of negative yield strengths does not become appreciable, despite the exponentially decreasing tails of the normal.

The lognormal distribution is quite commonly used in earthquake engineering, especially when modeling seismic demand. The lognormal takes its name from the fact that if you take the natural logarithm of a lognormally distributed variable $x$, the result is a normally distributed random variable $\ln(x)$. For this reason, one nearly always uses the parameters of the underlying normal distribution to characterize the lognormal distribution. For example, by employing the mean $\mu_{\ln x}$ and the standard deviation $\sigma_{\ln x}$ of the logarithm of the lognormally distributed variable $x$, the PDF of $x$ is written as:

$$f(x) = \frac{1}{x\sigma_{\ln x}\sqrt{2\pi}} \exp\left[-\frac{(\ln x - \mu_{\ln x})^2}{2\sigma_{\ln x}^2}\right] \tag{13.27}$$

As for the normal distribution, the CDF of the lognormal distribution has no useful analytical expression, but can be computed easily using *logncdf* in MATLAB (Mathworks, 2018) and *lognorm.dist* or *lognormdist* in Microsoft Excel® (Microsoft, 2010).

The lognormal distribution can be used to characterize positive unimodal variables that have a tendency to produce frequent extreme high values. This becomes obvious from Figure 13.2, whereby a "fat" right tail appears, showing the relatively large probability (e.g., vis-à-vis a normal) of encountering extreme positive values. For this reason, this model is commonly used for describing the conditional distribution of peak response demands (especially deformation related) for a given seismic intensity (discussed in Sections 4.8.6 and 5.3.4), since some ground motions will generate significantly larger peak responses than typical for any given intensity. Due to this fat right tail, the mean (which is sensitive to extremes) tends to be pulled to the right, relative to the median which is not sensitive to extreme values (see Table 13.1):

$$\mu_x > x_{50\%} \tag{13.28}$$

At the same time, due to the monotonicity of the lognormal function, percentile values are easily transformed between the lognormal and the underlying normal. Thus, for any $p$ in $[0, 1]$

$$(\ln x)_{p\%} = \ln(x_{p\%}) \tag{13.29}$$

or equivalently, by taking the exponential of both sides:

$$x_{p\%} = \exp\left[(\ln x)_{p\%}\right] \tag{13.30}$$

For example, this means that if one wishes to estimate the median of the lognormal variable $x$, one only need to take the exponential of the underlying normal median, which happens to equal the normal mean:

$$x_{50\%} = \exp\left[(\ln x)_{50\%}\right] = \exp(\mu_{\ln x}) \tag{13.31}$$

Furthermore, one may estimate the lognormal standard deviation via Equation 13.26 as

$$\sigma_{\ln x} \approx \frac{1}{2}\left[\ln(x_{84\%}) - \ln(x_{16\%})\right] \tag{13.32}$$

Also, an interesting result may be derived by estimating the CoV of the lognormal $x$ using the values of the standard deviation and the mean (as appearing in Table 13.1):

$$\beta_x = \frac{\sigma_x}{\mu_x} = \frac{\exp\left(\mu_{\ln x} + 0.5\sigma_{\ln x}^2\right)\sqrt{\exp\left(\sigma_{\ln x}^2 - 1\right)}}{\exp\left(\mu_{\ln x} + 0.5\sigma_{\ln x}^2\right)} = \sqrt{\exp\left(\sigma_{\ln x}^2 - 1\right)} \tag{13.33}$$

If we take the first two terms of the Taylor series expansion for the exponential function, we end up with

$$\beta_x \approx \sqrt{1 + \frac{\left(\sigma_{\ln x}{}^2 - 1\right)}{1!}} = \sigma_{\ln x} \tag{13.34}$$

In other words, the standard deviation of the underlying normal (also called log-standard deviation, or simply dispersion $\beta_x$) is practically the same as the CoV of the lognormal, at least for values of $\sigma_{\ln x} < 0.7$, for which the higher-order terms of the Taylor series are negligible. For this reason, values of $\sigma_{\ln x}$ are often quoted as percentages in practice (e.g., 20% dispersion, meaning $\beta_x \approx \sigma_{\ln x} = 0.2$).

In parallel to the normal distribution, the lognormal distribution is the model for the product of independent random variables, and it possesses a large number of interesting mathematical properties that are discussed in detail in any good probability book (e.g., Benjamin and Cornell, 1970). The lognormal distribution is also our first encounter with a derived distribution, i.e., the distribution of a variable that is a function of another. This will be our basis for a longer discussion on propagating probability (Section 13.2.7). But first, let us turn to the distribution of multiple random variables when viewed as an ensemble.

## 13.2.6 Multivariate probability distributions and correlation

When dealing with a problem having multiple random variables, such as the $N$, $A$, $f_y$ of the bar in tension (Figure 13.1), it is not enough to model the distribution of each one separately. We also need to model their dependence. For example, if $A$ and $f_y$ have a tendency to take small values at the same time, rather than independently taking high and low values (whereby both being small becomes a rare occurrence indeed) then the estimated yield strength, $Af_y$, and the resulting safety will be directly impacted.

A first-order (i.e., simplest possible) view of dependence is described by correlation. Dispensing with the mathematics, the correlation coefficient $p$ simply tells us how often large or small values of two variables tend to appear together. A correlation of +1 implies a linear relationship of the two variables with a positive slope (Figure 13.3a). Positive correlation coefficients of $0.5 \leq p < 1$ imply a strong tendency for coinciding high extremes (and low extremes) of the two variables (e.g., Figure 13.3b). Near-zero correlation practically implies independence (e.g., Figure 13.3c). Negative correlations mean that large values of one variable imply small values of the other (Figure 13.3d,e). A correlation of $-1$ means that the two variables are linearly dependent with a negative slope.

Strictly speaking, correlation and dependence are not the same thing, as there are several cleverly picked cases of almost functionally dependent variables that show zero correlation (e.g., Figure 13.3f). Although the data appearing therein as circularly distributed has a correlation of zero, there is very clear dependence. If we select a value of the variable along, say, the horizontal axis, this clearly limits the range of values for the other variable along the vertical axis. This is not the case for Figure 13.3c, a case where independence and the absence of correlation coincide. The random variables that we are going to deal with in most situations do not have such a peculiar (and nearly nonrandom) structure. Thus, for most practical purposes, correlation and dependence can be taken to be similar (or nearly identical) concepts. They will be treated us such henceforth.

Once their correlation has been defined, one can join the outcomes of separate variables into a single-joint multivariate distribution defined over its support of the Cartesian product of the individual outcome spaces from each constituent variable. For the case of $N$, $A$, and $f_y$, this would be $(0, +\infty) \times (0, +\infty) \times (0, +\infty)$. PDFs and CDFs can be appropriately defined as

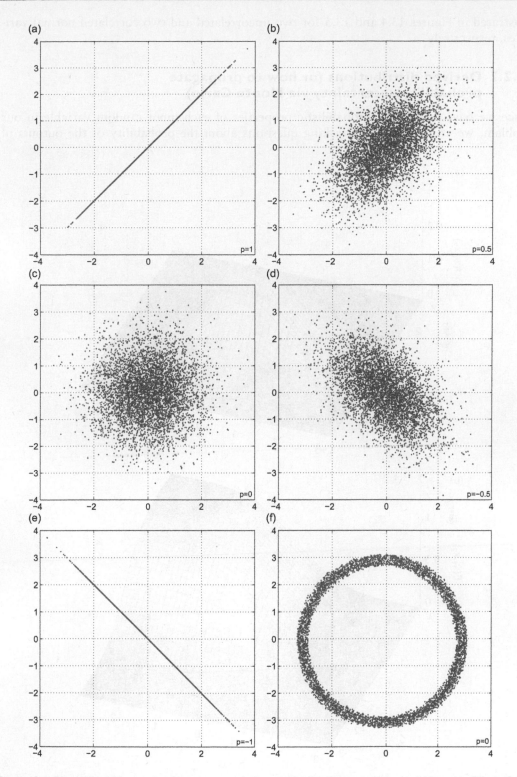

*Figure 13.3* Plots of 1,000 samples from two random variables having correlations of (a) $p = 1$, (b) $p = 0.5$, (c) $p = 0.0$, (d) $p = -0.5$, (e) $p = -1.0$, and (f) an example of dependent variables with zero correlation ($p = 0.0$).

illustrated in Figures 13.4 and 13.5 for two uncorrelated and two correlated normal variables, respectively.

### 13.2.7 Derived distributions (or how to propagate probability/uncertainty via Monte Carlo)

Once we have defined the probabilistic properties of each input random variable in our problem, we are ready to start asking questions about the probability of the outputs of

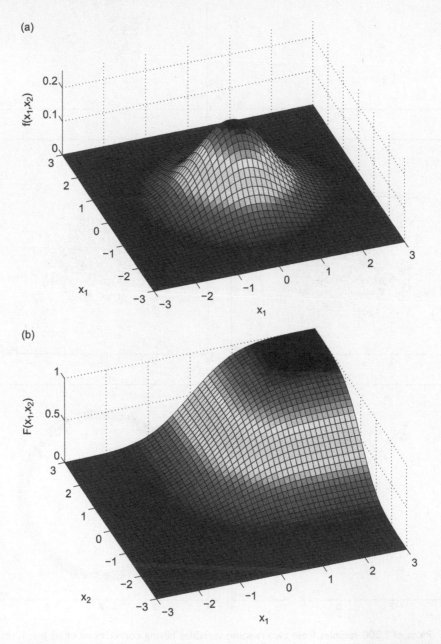

*Figure 13.4* (a) PDF and (b) CDF functions of the joint distribution of two uncorrelated identically distributed normal variables, with a mean of 0.0.

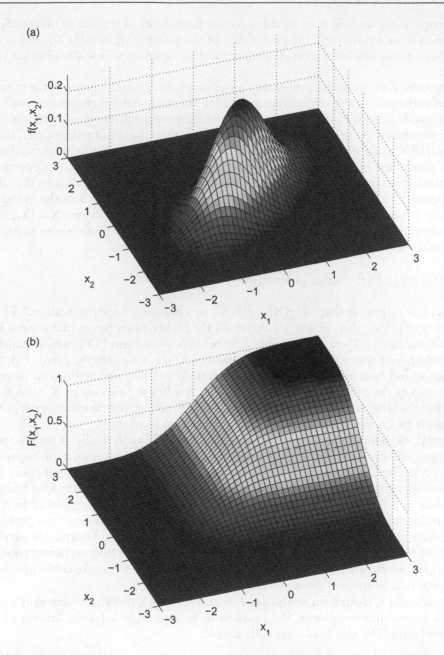

*Figure 13.5* (a) PDF and (b) CDF functions of the joint distribution of two correlated identically distributed normal variables, with a mean of 0.0 and a correlation coefficient of 0.7.

interest. For the bar-in-tension example, the input variables are obviously $N$, $A$, $f_y$. The output of interest is the result of the classic safety verification of demand $D = N$, versus capacity $C = A f_y$. Specifically, we are interested in the probability of failure, or capacity exceedance:

$$P(D > C) = P(N > A \cdot f_y) = P\left(\underbrace{N - A \cdot f_y}_{s_M} > 0\right) = P\left[\underbrace{N/(A \cdot f_y)}_{s_{MR}} > 1\right] \tag{13.35}$$

In this simple case, we only need to check for the probability of the safety margin $S_M$ being greater than 0, or equivalently the probability of the safety margin ratio $S_{MR}$ being greater than 1. Both quantities are essentially output random variables whose distribution we need to derive.

This process of propagating probability (or uncertainty) from the input to the output of a model lies at the core of all performance evaluations in earthquake engineering. Analytically, it is not a simple process by any means, even for the case of the three-variable model describing a bar in tension. At first glance, it seems plausible that one could plug in, e.g., the mean of each variable and estimate the mean of the output $S_M$ or $S_{MR}$. Unfortunately, this is not the case, regardless of the shape of the distributions, or the dependence/independence of the variables (perhaps with the exception of some very carefully crafted cases that have no useful applications in practice). Generally speaking, if a scalar output $Y$ of the model is represented by a function $Y = g(\mathbf{X})$ of the vector of input random parameters $\mathbf{X} = [X_1, ..., X_N]$, then, its expectation (or mean) is $E[Y] = E[g(\mathbf{X})]$. If we consider a Taylor series expansion of $g(\mathbf{X})$ around the mean of all parameters $E[\mathbf{X}]$, it is easy to see that

$$E\big[g(\mathbf{X})\big] = g\big(E[\mathbf{X}]\big) + \text{higher order terms} \tag{13.36}$$

Hence, we can only state that "$E[g(\mathbf{X})] \approx g(E[\mathbf{X}])$ to a *first-order* approximation." This is an elegant way of saying that we are neglecting all the higher-order terms that cannot be estimated without having knowledge of how $g(\mathbf{X})$ behaves away from $E[\mathbf{X}]$, which translates to running additional analyses away from the mean of $\mathbf{X}$ (see, for example, Pinto et al. (2004), Vamvatsikos and Fragiadakis (2010), and Vamvatsikos (2014) on methods to incorporate such terms). On the other hand, note that if $g(\mathbf{X})$ is a linear function of $\mathbf{X}$, then the first-order approximation is exact. Still, even for a linear function, the distribution of $g(\mathbf{X})$ cannot generally be known, only its moments (e.g., mean and variance).

In general, deriving the output distribution of $Y = g(\mathbf{X})$ analytically is only possible in special cases, such as when $g(\cdot)$ is the sum of normal $X_i$ or the product of lognormal $X_i$ ($i = 1, ..., N$). Then, $Y$ is a normal or lognormal variable, respectively, with easily derived parameters (Benjamin and Cornell, 1970), thanks to the first-order assumption being exact for this case of a linear function of the $X_i$ (in the case of a sum of normals) or $\ln X_i$ (for a product of lognormals). Furthermore, we should consider that many practical problems do not even have an analytical expression for $g(\cdot)$, connecting input and output. In earthquake engineering, the output can be any number of response variables (forces, bending moments, deformations, accelerations) estimated from a FE model of the structure subject to the input of a random (i.e., unknown) ground motions.

In all such cases, lacking an analytical expression for $g(\cdot)$, there is a very simple computational solution that reengineers the problem to become fully solvable, known as Monte Carlo simulation. The idea is conceptually simple:

1. Create $K$ equiprobable scenarios based on appropriately *sampling* their joint distribution (i.e., by obtaining sets of values of the input variables to represent their joint distribution).
2. Estimate $K$ sets of output values, one set from each scenario.
3. Determine the joint distribution of the output from the $K$ sets of output values.

Going back to our tensile bar, let $K = 100$ realizations of different scenarios. Sampling the problem input variables is arguably the most difficult part of the problem, but it can be easily performed using appropriate software, such as MATLAB (Mathworks, 2018) or Excel® (Microsoft, 2010). Having the $i = 1...100$ triplets of $\{N_i, A_i, f_{yi}\}$ we simply need to estimate the corresponding $S_{Mi}$ (or $S_{MRi}$) values. These $K$ values represent the distribution of $S_M$ (or $S_{MR}$). If, for example, 25 out of the 100 values have $S_{Mi} > 0$ then the probability of failure is 25%. Obviously, if we chose to run only $K = 10$ scenarios, depending on our luck, we could easily get anything from 1 to 5 out of 10 samples failing, for a probability of failure of 10% or 50%, respectively. In some extreme cases, we could even receive 0 or 10 samples failing! The more samples we use (and the better our sample space coverage is), the better the results, but the higher the computational effort needed to calculate the model output for each. This leads us to a most interesting question: How does one make sense out of samples of observations, rather than infinite populations of them? Linking the mathematics of probability back to the real world means we need to think about statistics.

Statistics is essentially the science of making sense out of limited samples. It is an exceptionally important and equally wide field that in the present context will be treated very simply, giving the shivers to real statisticians. It suffices to say that the larger the variability in any given random variable, the more samples we need in order to avoid underrepresenting it. Similarly, random variables with small (practically unimportant) variability can be represented even by a single sample taken at their central value, i.e., at the mean or median. Statistics provide us with formal methods to evaluate the accuracy achieved in any assessment (or "inference" in statistical parlance) where limited samples are involved. The smaller the sample size used for a variable of given dispersion, the lower the accuracy and the less the "significance" of any observation or comparison that we may try to make. In other words, neglecting to take a large enough sample for influential random variables means that we introduce unnecessary additional uncertainty into the assessment. If this is not smaller than the intrinsic uncertainty of the random variables themselves, then our approach to assessment is unduly influencing the results, muddying the waters to the extent that we become unable to make any useful statements about the problem at hand. Ignoring this issue and proceeding to make statements that are not statistically valid is tantamount to junk science and should be avoided.

## 13.2.8 Modeled versus unmodeled variables and practical treatment

In any model, there may be hundreds or even thousands of random variables; for example, consider the uncertain yield strength, mass, stiffness, etc. of every single member in a building (see Kazantzi et al., 2014). For obvious practical reasons of limiting the number of Monte Carlo simulations and associated analyses needed, one would like to consider only the important parameters and approximately account for the influence of the remaining ones. In mathematical terms, this implies rewriting the model function of $Y = g(\mathbf{X})$ by separating $\mathbf{X}$ into the vector of modeled variables $\mathbf{X}^M$ and the vector of unmodeled ones $\mathbf{X}^U$, and then approximately accounting for the effect of $\mathbf{X}^U$ via a single "model error" parameter $\varepsilon$ that is often taken to be normal or lognormal:

$$g(\mathbf{X}) = g\left(\mathbf{X}^M, \mathbf{X}^U\right) = g\left(\mathbf{X}^M, E[\mathbf{X}^U]\right) + \varepsilon, \quad \text{if } \varepsilon \text{ is normal}$$

$$g(\mathbf{X}) = g\left(\mathbf{X}^M, \mathbf{X}^U\right) = g\left(\mathbf{X}^M, E[\mathbf{X}^U]\right) \cdot \varepsilon, \quad \text{if } \varepsilon \text{ is lognormal}$$

(13.37)

This approach essentially frees us from having to consider thousands of random variables of lesser importance, while at the same time it explicitly provides an error term to recognize that we have neglected them. Conceptually, this is akin to considering a second layer of probability, or, if you wish, an application of probability on probability. Simply put, the modeled variables are employed for "accurate" probabilistic assessment of any responses/

probabilities/quantities of interest. The question then is how accurate the results of such an assessment are given the unmodeled variables. Their existence tells us that our estimates of the mean or variance of $g(\mathbf{X})$ are uncertain themselves, i.e., they are not deterministic but random variables with a distribution of their own. Obviously, now the million dollar question is how one selects the variables to include in $\mathbf{X}^M$ and how one decides on the distribution type for $\varepsilon$ and its parameters, the reply to which depends on the problem at hand as the following examples will attempt to elucidate.

### 13.2.8.1 Examples of modeled versus unmodeled variables

For the bar-in-tension example, assuming good construction quality it would be reasonable to expect that the section area $A$ and the steel yield strength $f_y$ have tight and relatively symmetric distributions around their central values with small dispersion, while the axial force demand $N$ is considerably less certain as it depends on highly variable loads. The problem can thus be considerably simplified by probabilistically modeling only $N$. Given the assumed uncorrelated distributions for $A$ and $f_y$, and the multiplicative (rather than, e.g., exponential) nature of $g(\cdot)$ [see definition of $S_M$ or $S_{MR}$ in Equation 13.35] we do not expect too large or too small output values to be favored. Therefore, a normally distributed model error $\varepsilon$ is a plausible assumption.

For seismic hazard models, to be discussed in Section 13.3, experience from actual recordings of ground motion has shown that larger-than-average extremes may often appear, giving credence to a lognormal distribution for the model error $\varepsilon$. Actually, seismologists have adopted quite complex models to quantify $\varepsilon$ and its effects in their calculations, using elaborate logic trees that carefully account for each "unmodeled" random variable in their estimates; the engineer typically need not concern oneself beyond adopting the mean estimate of the hazard curve that is typically provided for his/her use.

In structural modeling and seismic response assessments, the largest uncertainty tends to be associated with the ground motion itself. In the vast majority of cases, this is the *only* random variable whose distribution needs to be modeled explicitly. Monte Carlo sampling of records is performed by employing a number of natural or artificial accelerograms. How one selects such records to be equiprobable is an evolving and quite advanced discussion that we shall not dwell upon. Interested readers can read, for example, the work of Baker and Cornell (2006), Bradley (2010), Lin et al. (2013), and Kohrangi et al. (2017). The important thing to remember is that record-to-record variability associated with peak responses is so substantial (see Sections 4.8.6 and 5.3.4) that it cannot be accounted for by using only three to seven records. At the very minimum, 15 records from several different events should be employed to avoid underestimating the variability in the output. Using 20 or more records is even better. When nonlinear response history analysis is not an option due to complexity, one can at least employ static pushover approaches with an SDOF approximation (i.e., $R-\mu-T$ relationships) that have explicitly accounted for record-to-record variability. A prime example is SPO2IDA (Section 4.10.5) that we will extensively employ in the development of yield frequency spectra (YFS, Section 13.5.2). Beyond the ground motion, lesser but still nonnegligible uncertainty is introduced by the structural model (especially by what has not been modeled, such as partition walls, staircases, etc.), the approximately known structural properties, and the even less certain thresholds used to define the violation of limit-states (LSs). Accounting for the effect of such neglected sources of uncertainty on structural response is usually done by selecting a lognormal model error $\varepsilon$. This is partly for reasons of mathematical convenience, as the uncertainty in the response due to the ground motion is also approximately lognormally distributed. As explained in Section 13.4.2, the lognormal assumption for the model error distribution helps derive a simple mathematical expression for safety checking. The lognormal assumption is also supported by the observation that structural behavior near collapse is an inherently unstable phenomenon, whereby small changes in the input can often cause large values of response; the lognormal distribution has a relatively fat tail, which helps to address the range of values near collapse.

Beyond physical model variables, there are also sources of uncertainty that may not directly correspond to the properties of an FE model, but they nevertheless contribute variability that cannot be ignored. Perhaps the two most important such sources are the model type, when considered as an imperfect idealization of an actual structure, and the inaccuracies in the analysis method employed, as e.g., when static pushover approaches are used to capture nonlinear dynamic response. Sample size issues (Section 13.2.7) are also another important consideration. As mentioned earlier, reasons of mathematical convenience more than actual results mean that a lognormal model error is typically employed.

### 13.2.8.2 The first-order assumption for model error

Having decided on the distribution to use for the model error, we now have to determine its parameters, i.e., the mean and standard deviation of $\varepsilon$ or $\ln(\varepsilon)$, for a normal or lognormal assumption, respectively. Lacking any further information, the so-called "first-order" assumption is typically made. Specifically, we assume that the central value (mean or median) of $g(\mathbf{X})$ is accurately captured if we run our analyses by employing the central values of the unmodeled input parameters as dictated by Equation 13.37. Equivalently, this translates to the mean of a normal $\varepsilon$ being zero, or the median of a lognormal $\varepsilon$ being 1.0. This is clearly an imperfect assumption (see the discussion on uncertainty propagation in Section 13.2.7 and Equation 13.36), equivalent to the first-order approximation. Essentially, in the interest of practicality, we are assuming that neglecting the randomness of the unmodeled parameters does not change the central value of the estimated output distribution. In statistical parlance, we are saying that despite our neglecting many sources of uncertainty we have still achieved an *unbiased* estimate. Although Equation 13.36 makes clear that this may not be strictly true in mathematical terms, it can still be a fairly accurate assumption for practical purposes given the stipulated small influence of the unmodeled parameters. Note that it is often the case that we may have strong indication of unmodeled variables biasing the results of our assessment (e.g., Vamvatsikos and Fragiadakis, 2010). Even then, incorporating such bias into the distribution of $\varepsilon$ is a risky proposition, as the bias may be severely model dependent, becoming equivalent to a "fudge factor" that we cannot accurately estimate without significant effort, as discussed in the following.

Despite the central value being presumably unbiased, the dispersion of the output due to unmodeled input parameters is surely impacted. In the vast majority of engineering problems, adding more random variables (i.e., noise) to the system means that the uncertainty of the output is increased. Due to the difficulty and the computational burden of properly defining the distribution of unmodeled variables at the input level and trying to propagate it to the output, we tend to adopt one of two simpler options to quantify their effect in terms of the model error dispersion, $\beta_{\varepsilon}$:

a. Directly assume the magnitude of the additional output dispersion $\beta_{\varepsilon}$ due to the unmodeled input parameters.
b. Adopt an approximate analytical model for $g(\mathbf{X})$ and employ it to propagate the uncertainty. This could be a logic tree (such as often used by seismologists) or a Taylor series expansion of Equation 13.36, preferably including at least some higher-order terms. The latter is the basis of the first-order second-moment method (Benjamin and

Cornell, 1970; Pinto et al., 2004) that has already found useful application in seismic assessment of complex structures with good results (e.g., Vamvatsikos and Fragiadakis, 2010).

Option (b) is by far the better one, as it incorporates uncertainty propagation and it allows us to gain at least some intuition on how unmodeled variables influence the results. In addition, it can even be employed in a more expanded approach to waive the first-order assumption and provide us with a more accurate estimate of the central value or even an approximate distribution for $\varepsilon$. Still, it does involve additional computations, making option (a) look considerably more attractive for casual use.

In general, making assumptions for the magnitude of $\beta_\varepsilon$ requires considerable expertise. To use this approach with some modicum of confidence, $\beta_\varepsilon$ should be less than the dispersion contributed by the modeled sources of uncertainty. For example, when assessing the seismic response we would expect that $\beta_\varepsilon$ is smaller than the dispersion in response values associated with record-to-record variability, lest we violate our assumption about what matters and what does not. Still, there are cases where such an obvious limit cannot be complied with, typically due to some important sources of uncertainty being difficult to model due to the lack of data. In such cases, any assumption about the magnitude of $\beta_\varepsilon$ should be considered with care, as it will have a large impact on the output distribution and may eventually become detrimental to the overall accuracy of the assessment. Some guidance on estimating values for $\beta_\varepsilon$, taking into account different sources of uncertainty in structural analysis, can be found in FEMA P-695 (FEMA, 2009) and FEMA P-58-1 (FEMA, 2012a).

### 13.2.8.3 Smeared versus discrete treatment of unmodeled uncertainty

Having adopted the first-order assumption for modeling the error term, and having decided on a value for the additional dispersion of unmodeled variables that we need to account for in our (uncertain) estimates of the model output, there is still one important choice to make: How to incorporate this uncertainty in our assessment. There are two options to consider

a. Smeared uncertainty: Smear the model error generated by unmodeled variables into the distribution estimated from modeled variables to produce a single output distribution, without distinguishing the variance from modeled and unmodeled parameters.
b. Discrete uncertainty: Employ the model error distribution generated by unmodeled variables separately from the output distribution of modeled variables, using it to define multiple possible realizations of the output distribution.

In the first scenario, the output distribution (Figure 13.6a) of the demand parameter of interest (which incorporates the dispersion contributed by the modeled parameters) is augmented by aggregating (or smearing) the additional dispersion from model error (Figure 13.6b). Typically, this means inflating the variability of the output dispersion that was initially estimated by Monte Carlo, something that can be accomplished in any number of ways. In the usual case where the output is modeled as a lognormal variable, we often use a simple square root of the sum of squares (SRSS) rule to combine the dispersion $\beta_{\mathrm{MC}}$ caused by modeled variables and estimated by Monte Carlo with the additional model error dispersion $\beta_\varepsilon$ to estimate the total:

$$\beta_{\mathrm{total}} = \sqrt{\beta_{\mathrm{MC}}^2 + \beta_\varepsilon^2} \tag{13.38}$$

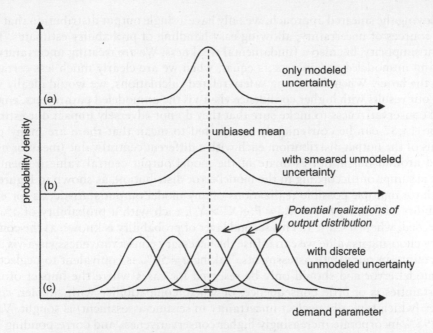

*Figure 13.6* PDFs for the output distribution of a demand parameter determined from model response: (a) incorporating only the uncertainty from modeled sources produces the starting distribution of the output without consideration of model error; (b) the smeared uncertainty approach produces a new output distribution with the same mean but inflated dispersion, while (c) the discrete approach employs multiple possible realizations of the output distribution, each practically a copy of the starting distribution, but with a shifted mean to represent the model error dispersion.

For our tension bar example, let 10% be the allowable probability of failure. If $N$ is a lognormal variable with a median of 150 kN and a dispersion of 0.35, it is obvious that the safety margin ratio $S_{MR} = N/(Af_y)$ of Equation 13.35 will also be lognormally distributed as $A$ and $f_y$ are considered deterministic and equal to their median (or mean) values of, say, 10 cm² and 300 MPa = 30 kN/cm², respectively. If we neglect the uncertainty introduced by the two unmodeled variables, $S_{MR}$ becomes a lognormal with a median of 150 kN/(10 cm²·30 kN/cm²) = 0.5 and a dispersion of $\beta_{MC} = 0.35$. Then, the probability of failure is

$$P(D > C) = P(S_{MR} > 1) = 2.4\%,$$

estimated as "=1 − LOGNORM.DIST(1,ln(0.5),0.35,true)" in Excel®. Let us now assume that unmodeled sources impart an additional $\beta_{add} = 0.2$ dispersion to $S_{MR}$. Then, $S_{MR}$ becomes a lognormal with a median of 0.5 and dispersion of

$$\beta_{total} = \sqrt{(0.35^2 + 0.2^2)} \approx 0.40.$$

Then

$$P(D > C) = P(S_{MR} > 1) = 4.3\%,$$

estimated as "=1 − LOGNORM.DIST(1,ln(0.5),0.40,true)" in Excel®, a clearly higher result due to the increased dispersion but still acceptable vis-à-vis the 10% allowable probability.

By employing the smeared approach, we only have a single output distribution that encompasses all sources of uncertainty, allowing easy handling of probability estimates. There is an inherent simplicity, but also a fundamental weakness: We are treating uncertainty due to modeled and unmodeled parameters as equal, when we are clearly much less certain with respect to the latter. When discussing safety-related calculations, we would ideally want to safeguard our results with higher confidence vis-à-vis the unmodeled parameters, employing additional conservativeness to make sure that they do not adversely impact our estimates.

Equation 13.37 can be conveniently interpreted to mean that there are many possible realizations of the output distribution, each with a different central value (mean or median), distributed around our initial estimate of the model output central value (remember the first-order assumption) according to the model error distribution, as shown in Figure 13.6c. Thus, we have multiple possible realizations of any model output statistic, as for example the probability of failure, $P(Y > Y_0) = P(g(\mathbf{X}) > Y_0)$, each with a probability of $a\%$ of not being exceeded, where $0 < a < 1$. This second layer of probability is known as the confidence $a\%$, and by choosing its value we can infuse the necessary conservativeness vis-à-vis unmodeled uncertainty in our safety assessments. Taking $a = 50\%$ is equivalent to neglecting the effect of model error and should only be reserved for cases where the impact of unmodeled uncertainties is of little consequence, or where compatibility with simpler, code-like approaches (which typically neglect uncertainty in seismic assessment) is sought. Values of $a = 75/90/95\%$ incorporate increasingly higher conservativeness and corresponding safety.

In our tension bar example, let us assume that failure is of high consequence and therefore $a = 90\%$ confidence is required in our assessment. Then, the 90% confidence estimate of the mean of $S_{MR}$ is

$$\mu_{SMR,90\%} = 0.65$$

estimated as "=LOGNORM.INV(0.9,ln(0.5),0.2)" in Excel®. Due to the monotonically increasing relationship between the probability of failure and $S_{MR}$, the $a\%$ percentile of $S_{MR}$ directly produces the $a\%$ percentile estimate of the failure probability. Thus

$$P_{90\%}(D > C) = P(S_{MR,90\%} > 1) = 10.6\%,$$

or "=1 − LOGNORM.DIST(1,ln(0.65),0.35,true)" in Excel®. When compared to the allowable probability of 10%, the bar is now inadequate to guard against a high-consequence failure.

With a bit of trial and error, we can easily show that the smeared approach, in this case, is equivalent to a confidence of $a = 67.6\%$. Following the discrete approach clearly offers an additional knob to help tune safety as we see fit, infusing conservativeness as needed to guard against failure mechanisms of different consequential severity. We are going to take advantage of this in Section 13.4 to set up our framework for performance assessment under the influence of uncertainty.

## 13.3  PROBABILISTIC SEISMIC HAZARD ANALYSIS

### 13.3.1  Occurrence of random events and the Poisson process

When assessing the effects of natural hazards, including earthquakes, the dimension of time becomes indispensable for proper modeling. The reason is that loads coming off natural processes are not an eternal, always-there presence, like the gravity action on dead loads. Instead, they occur intermittently (and randomly) over a period of years, and may be described as outcomes of a probabilistic process. Thus, we need to describe the magnitude (or intensity) of an event together with the corresponding frequency of occurrence. To introduce this dimension of time, engineers and seismologists alike use the so-called Poisson process of events (first discussed in Section 13.2.5). Its main characteristic is that the average rate of events, i.e., the average number of occurrences divided by the period of time in which they were observed, is constant. Since this is taken to hold for any time interval, regardless of how small or large it may be, it is easy to see that it leads to the so-called *memoryless* property that can be roughly described as follows: The occurrence or nonoccurrence of an event within a given period of time has no influence on the future. In other words, if an earthquake happens, it does not hasten or delay the (random) occurrence of the next one. Obviously, this does not hold if one considers the aftershocks of a main event, but it is fairly reasonable when discussing mainshocks themselves. Regardless of evidence that faults may actually behave otherwise, this convention is mathematically and conceptually attractive, as it only requires assessing the single parameter of the Poisson, $\lambda$, also known as the MAF. Thus, even non-Poissonian processes will often be approximated by a Poisson process by averaging out the frequency of events to estimate an equivalent MAF. For seismic hazard purposes, $\lambda$ is expressed in units of year$^{-1}$ and the probability that $k$ events will occur in time $t$ becomes:

$$P(k \text{ events in } t) = \frac{(\lambda t)^k}{k!} e^{-\lambda t}, \, k = 0, 1, \dots \tag{13.39}$$

When discussing earthquakes, where occurrences are rare and decades typically pass between mainshocks of engineering interest (magnitudes of $M > 4.0$ in general), we are usually only interested in the next event. It is not very difficult to see that the probability of having to wait time $t$ to the next event is the same as the probability that there are exactly zero events from now until $t$, thus giving rise to the well-known exponential CCDF of interarrival time between consecutive events:

$$P(\text{next event after time } t) = P(\text{zero events in } [0, t]) = \frac{(\lambda t)^0}{0!} e^{-\lambda t} = e^{-\lambda t}, \tag{13.40}$$

which, as expected, does not depend on any information about previous events, but only on the MAF of $\lambda$, thus being truly memoryless. Compare now Equation 13.40 to the corresponding CDF, or the probability of having at least one event in $[0, t]$ which is one minus the CCDF and leads to the same expression as Equation 13.22:

$$P(\text{one or more events in } [0, t]) = 1 - e^{-\lambda t}. \tag{13.41}$$

As can be observed in Figure 13.7, as the MAF increases, it becomes more and more probable that the next event will arrive shortly. For a given MAF, i.e., rate of occurrence, as we increase the time interval over which an event may be expected, its probability also increases. The catch here is that this probability does not change as we move forward in the future and ask this question again and again, with different starting times but concerning a constant time interval, covering the same number of years. For example, if an $M = 5$ event happens today and right after it we ask the probability that the next one will happen in 1 year, for $\lambda = 0.5$ events/year the answer is $1 - \exp(-0.5 \cdot 1) \approx 0.39$. If after one or more years no such earthquake has happened and we ask the same question for the coming year, then the answer is exactly the same, i.e., 0.39, as the process truly has no memory.

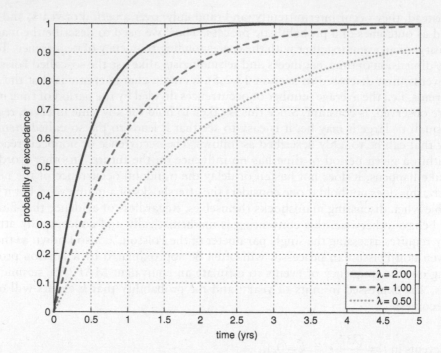

*Figure 13.7* Cumulative distribution function, Equation 13.41, of the exponential distribution for different mean annual rates.

Thus, for a Poissonian assumption, the MAF is all we need to know. We shall use it to describe the occurrence of any event of engineering significance, typically focusing on the exceedance of engineering quantities or LSs of interest. Going back to our tension bar example, let us say that its demand $D = N$ is only driven by seismic events of magnitude $M = 5$ that have a rate of occurrence of $2\,\text{year}^{-1}$. Let us say that we have calculated the probability of failure given an $M = 5$ event, $P(D > C|M = 5)$, then the so-called "filtering property" of the Poisson distribution allows us to "filter" the sequence of $M = 5$ events and only keep those that cause failure, in order to estimate the MAF of tension failure

$$\lambda(D > C|M = 5\,) = P(D > C|M = 5) \cdot \lambda(M = 5), \tag{13.42}$$

Thus, the Poisson model can fully describe the processes of component or structural failures and the assessment of the associated MAF values permeates modern earthquake engineering. Still, most engineers may not be aware of this fact because instead of encountering the (mathematically more tractable) MAF they have interacted with two equivalent forms. First is the mean return period $T_p$,

$$T_p = \frac{1}{\lambda} \tag{13.43}$$

Second is the probability $p$ of having one or more events where $D$ exceeds $C$ (termed exceedance events) over the lifetime $t$ of a structure, as estimated via Equation 13.41 above. If we solve the latter for $\lambda$ we have

$$\lambda = -\frac{\ln(1 - p)}{T} \tag{13.44}$$

Thus, for example, the design basis 10% in 50-year exceedance probability, which will be shortened to 10/50 henceforth, corresponds to setting $p = 0.1$, and $t = 50$ Then,

$$\lambda = -\ln(1-0.1)/50 = 0.0021/\text{year}$$

Through Equation 13.43 this is the same as a return period of

$$T_p = 1/\lambda = 1/0.0021 \approx 475 \text{ years}$$

A similar and equally useful concept is the "annual probability," or the probability $p_1$ of exceedance in $t = 1$ year. If we take a Taylor approximation of the exponential function, valid for small values of $\lambda \ll 1$ (as typical in practical seismic applications), then:

$$p_1 = 1 - \exp(-\lambda) \approx 1 - (1 - \lambda) = \lambda \qquad (13.45)$$

This is a very useful property, as it allows us, for example, to handle MAFs as probabilities and thus employ all the tools of probability presented so far (such as the total probability theorem, as we shall see in the next section) on MAFs as well. Still, it has often led to some confusion in notation, where frequencies and (annual) probabilities are mixed up. To avoid such problems, we shall adopt a purist stance and only talk of frequencies wherever units of time are involved. This makes for a mathematically consistent approach that remains correct even when the value of $\lambda$ is substantial.

## 13.3.2 The seismic hazard integral

The MAF of exceeding any value $x$ of the seismic intensity measure (IM), $\lambda(\text{IM} > x)$ at a given geographical site can be estimated by the application of the total probability theorem, breaking up the calculation by considering all possible events (or scenarios) of given magnitude $M$ and source-to-site distance $R$ that may occur at $i = 1 \dots N$ different faults (or seismic sources) that influence the site of interest (see also Baker, 2008):

$$\lambda(\text{IM} > x) = \sum_{i=1}^{N} v_i \int_{M_{\min}}^{M_{\max,i}} \int_{0}^{R_{\max,i}} P(\text{IM} > x | M, R, \text{site}) f_i(M, R) \, dM \, dR \qquad (13.46)$$

where

- $M_{\min}$ is the minimum magnitude of engineering interest, typically a value of 4.0–4.5 that is chosen to be the same for all sources.
- $M_{\max,i}$ is the maximum magnitude that can be produced by source $i$, as determined by its physical dimensions and seismological characteristics. Simply put, the amount of energy that can be generated by a fault is bounded by the area of rock that can be ruptured along the fault plane.
- $v_i = \lambda(M > M_{\min})$ is the probability of occurrence for an event of magnitude higher than the minimum value $M_{\min}$ at source $i$.
- $R_{\max,i}$ is the maximum possible distance from the site to a rupture of any size on source $i$.
- $f_i(M, R)$ is the joint PDF of magnitude and distance for events that can be produced by source $i$.
- $P(\text{IM} > x | M, R, \text{site})$ is the probability that the seismic intensity IM produced by an event of magnitude $M$ and distance $R$ will exceed the value of $x$, given the site characteristics (e.g., soil type). This is also known as the ground motion prediction model (GMPM) or the ground motion prediction equation (GMPE).

The evaluation of Equation 13.46 essentially requires a seismological model that details all the sources and their characteristics within an area of interest. Such a model provides the basis for probabilistic seismic hazard analysis (PSHA) calculations, in the same way that an FE model is used for structural analysis, and it is every bit as complex. Building such a seismological model is specialized work that requires considerable data and is best left to a team of competent seismologists. In the following, we shall delineate its components and discuss the use of PSHA results.

### 13.3.3 Seismic sources

The seismic source model essentially details all the seismic sources in the area of interest, typically within a radius of 200–500 km from the site, the larger values being used for sites subject to large magnitude subduction zone events (e.g., Chile, Japan, Mexico). The model incorporates their physical properties, such as the actual dimensions of the fault plane, its orientation in space (e.g., dip and rake angles) and its type (e.g., normal, reverse, etc., see Chapter 2) as well as seismological characteristics, such as the maximum magnitude $M_{max}$ that one expects it to produce, and the MAF $v_i$ of events above a minimum magnitude occurring on the fault. In addition, in cases where some or all of the faults cannot be identified (as typical for low seismicity areas), extended areal sources are employed anywhere within which an event can occur.

### 13.3.4 Magnitude–distance distribution

The seismic activity in a region is characterized by the recurrence interval of earthquakes of different magnitudes. The basic idea is that small magnitude earthquakes occur more often. The frequency with which an earthquake of magnitude larger than a given value occurs is known as the frequency of magnitude exceedance $\lambda(M)$. One of the earliest models for this is the Gutenberg–Richter recurrence law. Based on data recollected in Southern California in the early 1940s, Gutenberg and Richter (1944) proposed an empirical recurrence law that provides the MAF of events in a particular zone (or source) exceeding a magnitude $M$:

$$\log \lambda(M) = a - bM \tag{13.47}$$

The coefficients $a$ and $b$ are constants, estimated by regression analysis of the data obtained in a certain region. Therefore, $a$ and $b$ values change from region to region, as shown for example in Figure 13.8.

From Equation 13.47 one can determine the CDF and PDF of magnitude distribution given that an event of magnitude above $M_{min}$ has occurred on a seismic source (Baker, 2008):

$$f(M) = \ln 10 \cdot b \cdot 10^{-b(M-M_{min})}, \quad M > M_{min} \tag{13.48}$$

The GR recurrence law has no upper bound on magnitude. This is not reasonable as the physical size of the source faults provides an inherent limit on the maximum magnitude. If a maximum magnitude $M_{max}$ can be identified, the GR recurrence law can be modified to what is called the bounded Gutenberg–Richter recurrence law. Equation 13.48 becomes:

$$f(M) = \frac{\ln 10 \cdot b \cdot 10^{-b(M-M_{min})}}{1 - 10^{-b(M_{max}-M_{min})}}, \quad M_{max} > M > M_{min} \tag{13.49}$$

Alongside the determination of the magnitude distribution, we also need the distribution of distances. There are many definitions of the distance from site to source, for example, the distance to the hypocenter (or hypocentral distance), the epicentral distance, the closest

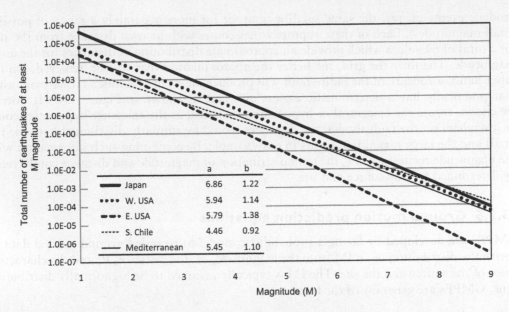

*Figure 13.8* GR recurrence law for shallow earthquakes observed in different seismic zones. (Adapted from Kaila and Narain, 1971).

distance to the fault rupture, and the closest distance to the surface projection of the fault rupture. While magnitudes may be considered to be identically distributed via Equation 13.49 for faults of the same $M_{max}$ in a given area, the distribution of distances depends on the distance metric selected, as well as the magnitude and the source geometry. Since each event of a given magnitude on a fault is associated with the rupture of a certain area, the determination of the joint distribution of magnitude and distance is done by discretizing the source of interest and employing analytical geometry.

If, for example, we consider a fault plane of $80 \text{ km} \times 80\text{km}$ and divide it into a $10 \text{ km} \times 10\text{km}$ grid (Figure 13.9), we have $8 \times 8 = 64$ possible $10 \text{ km} \times 10\text{km}$ ruptures that

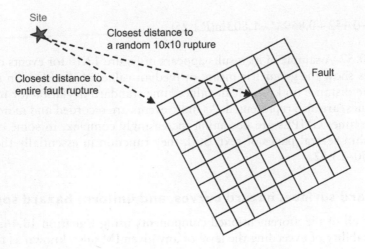

*Figure 13.9* Example of source-to-site closest distances estimated on the surface projection of an $80 \times 80 \text{km}$ fault. The distance to a rupture of the entire fault is practically deterministic, while the distance to a $10 \times 10\text{km}$ rupture varies with the position of the ruptured zone.

produce events of, say, the same small magnitude (or more accurately a range of possible small magnitudes). Each of these rupture zones comes with its own distance from the site, for a total of 64 values which provide an approximate distribution of distances for the given magnitude. The finer the grid, the better the approximation that will be achieved. On the other hand, a rupture of the entire fault will produce an event of a much larger magnitude that practically has a deterministic closest distance, as shown in Figure 13.9. If instead we choose to use the epicentral or hypocentral distance, as the epicenter/hypocenter could be anywhere on the fault shown, the distances would be similarly distributed both for the small and the large magnitude events in our example. By combining such information with the magnitude recurrence law, the joint distribution of magnitude and distance can be readily determined for any source and site.

### 13.3.5 Ground motion prediction equations

GMPEs are developed by fitting a probabilistic model to recorded ground motion data to derive the distribution of an IM from the magnitude, $M$, the distance, $R$, and the characteristics of the source and the site. The IM is typically assumed to be lognormally distributed; thus, GMPEs are generally of the form:

$$\ln \text{IM} = m_{\ln \text{IM}}\left(M, R, \text{source}, \text{site}\right) + \sigma_{\ln \text{IM}}(M, R, \text{source}, \text{site}) \cdot \varepsilon \tag{13.50}$$

where $m_{\ln \text{IM}}$ and $\sigma_{\ln \text{IM}}$ are the conditional mean and standard deviation, respectively, of lnIM, while $\varepsilon$ is the (normalized) lognormally distributed error. Thus, the probability of the IM exceeding a value of $x$ given $M$, $R$, source and site is estimated as:

$$P(\text{IM} > x \mid M, R, \text{source}, \text{site}) = 1 - \Phi\left(\frac{\ln x - \mu_{\ln \text{IM}}}{\sigma_{\ln \text{IM}}}\right) \tag{13.51}$$

Initially, due to the lack of data, GMPEs were simple expressions. For example, Cornell et al. (1979) proposed the following model to predict peak ground acceleration (PGA) (in units of g):

$$m_{\ln \text{PGA}} = -0.152 + 0.859M - 1.803\ln(R + 25) \tag{13.52}$$

with $\sigma_{\ln \text{PGA}} = 0.57$. A sample of its results appears in Figure 13.10 for events of $M = 5.5, 7.0$. The solid lines show the geometric mean, or median value, of the PGA (in units of g) as a function of the distance (in km). Mean values±one standard deviation (or more accurately 16%/84% values) are also represented. As more events are recorded and more data becomes available over time, GMPEs are becoming increasingly complex, in some cases not being able to fit within several pages of text. Still, they function in essentially the same way as Equation 13.50–13.52.

### 13.3.6 Hazard surface, hazard curves, and uniform hazard spectra

By combining all of the aforementioned components using Equation 13.46, one may estimate the probability of exceeding the level of any given IM (also known as the hazard). Of special interest is a comprehensive site hazard representation that may be achieved by estimating the so-called seismic hazard surface, a three-dimensional plot of the MAF of exceeding any level of the elastic spectral acceleration $S_a(T, 5\%)$, spanning the full practical range

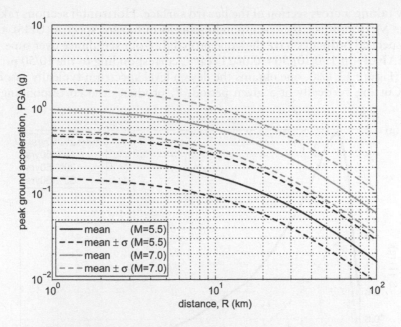

*Figure 13.10* The predicted distribution of the PGA and its variation with distance *R* according to the GMPE of Cornell et al. (1979) for *M* = 5.5 and *M* = 7.0.

of periods (Figure 13.11). This is the true representation of the expected seismic loading (formally, the mean estimate considering epistemic uncertainty) for any given site, in terms of the elastic response of an SDOF oscillator. More familiar two-dimensional plots can be

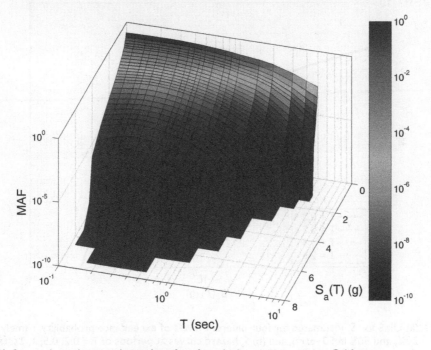

*Figure 13.11* Spectral acceleration hazard surface for a high-seismicity site in California.

produced by taking a cross section of the hazard surface. Horizontal sections taken at given values of the MAF produce the corresponding uniform hazard spectra (UHS), thus called, because all spectral ordinates have the same probability of exceedance over time. For example, at an MAF of $-\ln(1-0.10)/50 = 0.0021\,\text{year}^{-1}$, corresponding to the 10/50 probability of exceedance (Figure 13.12a), one obtains the design level spectrum typically associated with life safety. Cutting vertically at a given period, $T$, produces the corresponding $S_a(T, 5\%)$

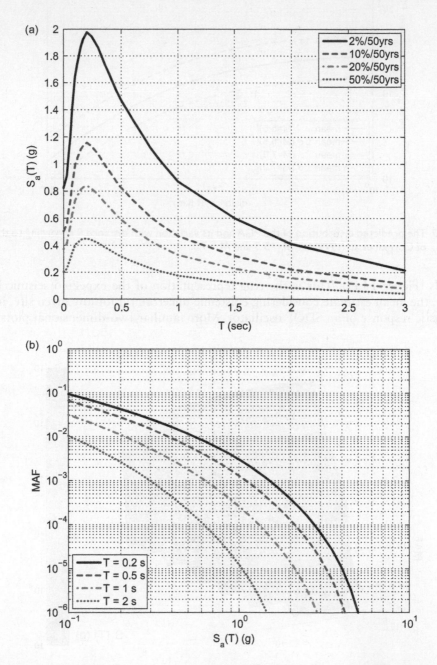

Figure 13.12 (a) UHS for $S_a$ (estimated for four different levels of exceedance probability, namely 2%, 10%, 20%, and 50% in 50 years), and (b) $S_a$ hazard curves at periods of $T = 0.2, 0.5, 1, 2\,\text{s}$, taken from the hazard surface of Figure 13.11.

hazard curve (Figure 13.12b), describing the full range of intensities and their associated MAF that a structure of a given period can experience at the site.

### 13.3.7 Risk-targeted spectra

A weakness of UHS when employed for conventional design applications is that they can only guarantee a uniform seismic hazard *input* (e.g., 10/50 for EN1998 and 2/50 for ASCE/ SEI 7). Due to the seismic hazard surface being ill-reproduced by just a single horizontal cut in Figure 13.11, it actually comes to pass that the output risk (in MAF terms) is much higher than the design hazard MAF. This means that sites having the same design spectrum are not equivalent in terms of the entire hazard surface. See, for example, Figure 13.13, where three European sites may have the same 10/50 design PGA, yet show very different frequencies for lower and higher intensities. This is best characterized by the (absolute value of the) slope $k$ fitted locally in log–log coordinates, showing the different rate of change in the MAF as the intensity increases. Thus, the site with $k = 2.84$ shows more frequent low-intensity events, while the site with $k = 2.16$ is characterized by more frequent high-intensity ones. Obviously, these two sites will not produce the same risk (see also the relevant discussion in Section 13.4).

To harmonize risk among such different sites, Luco et al. (2007) proposed the concept of risk-targeted (or uniform-risk) spectra. These are specially crafted spectra that can theoretically "guarantee" a uniform risk of collapse (or any other targeted LS of interest) regardless of the site. They have only been implemented in U.S. practice so far, targeting a collapse

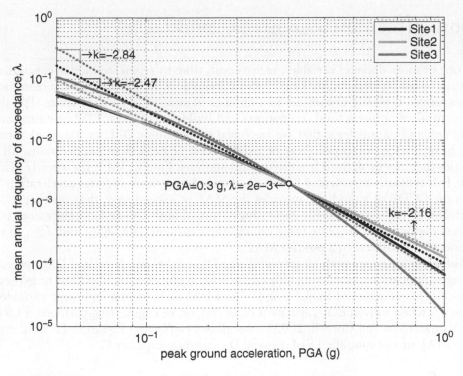

*Figure 13.13* Three different hazard curves of PGA for sites in Europe in log–log coordinates that have the same value of PGA = 0.3 g at $\lambda = 0.0021$ (or a 10/50 level). In terms of EN1998 these three sites would have the same design spectrum (and similar if not same uniform hazard spectrum). The dashed lines are the fitted tangents at the intersection point. (Adapted from Spillatura 2018).

risk of 1% in 50 years, and are being considered for adoption in other parts of the world. In reality, they should not be thought of guaranteeing the targeted level of risk, simply because a single generic distribution is utilized to represent the probability of collapse given the IM (the so-called collapse fragility, see Section 13.4) of all structural systems of the same period regardless of individual system-type characteristics. Still, they very nicely manage to harmonize risk among the sites, achieving a relatively more uniform MAF of collapse, though ultimately an unknown one (Spillatura 2018). Thus, using risk-targeted spectra is not the same as explicitly designing a specific structure to satisfy a given risk target. This will be addressed in Section 13.5.

For our purposes, it is thus important to say that the modifications that had to be made to UHS to produce the risk-targeted spectra make the latter incompatible with the approaches for assessing performance introduced in Section 13.4. Thus, in terms of Chapter 18, where the three design approaches are described, one should only employ risk-targeted spectra (where available) with Design Method A, which does not otherwise account for risk. The other two approaches, termed B and C, explicitly incorporate risk, therefore, they do not need the implicit (and incomplete) incorporation attempted via risk-targeted spectra.

## 13.4 ASSESSMENT OF PERFORMANCE

The theoretical framework that will be employed for assessing seismic performance is based on the Cornell–Krawinkler framing equation, as adopted by the Pacific Earthquake Engineering Research (PEER) Center, (Cornell and Krawinkler, 2000):

$$\lambda(\text{DV}) = \iiint G(\text{DV}|\text{DM})|dG(\text{DM}|\text{EDP})||dG(\text{EDP}|\text{IM})||d\lambda(\text{IM})| \tag{13.53}$$

DV is one or more decision variables, such as cost, time-to-repair, or human casualties, that are meant to enable decision-making by the stakeholders. DM represents the relevant damage measure(s), typically discretized in a number of successive damage states (DS) defined for structural or nonstructural components and building contents. EDP contains the engineering demand parameter(s) that are employed to determine DM, such as peak interstory drifts, beam plastic rotations, shear or axial forces, or peak floor accelerations, all being quantities that can be derived from structural analysis. Finally, IM is the already discussed seismic IM, typically represented by the 5% damped first-mode spectral acceleration $S_a(T_1)$. $G(x)$ is the CCDF of $x$, and $\lambda(y)$ provides the MAF of exceeding $y$, thus making $\lambda(\text{IM})$ the hazard. Due to the monotonically decreasing nature of both $G(\cdot)$ and $\lambda(\cdot)$, their slope is negative, thus requiring an absolute value around their differentials, $dG(\cdot)$ and $d\lambda(\cdot)$, to ensure Equation 13.53 will return a positive MAF.

Defining performance without involving any DV or the closely related DM makes sense for design applications, employing only engineering quantities, such as EDP, to express performance. This is achieved by defining DV and DM to be simple indicator variables that become 1.0 when a given EDP capacity is exceeded, or, in engineering parlance, a LS is violated (Vamvatsikos and Cornell, 2002). This allows employing Equation 13.53 to estimate $\lambda_{\text{LS}}$, the MAF of violating the LS of demand D exceeding capacity C:

$$\lambda_{\text{LS}} = \int G(\text{EDP}|\text{IM})|d\lambda(\text{IM})| \tag{13.54}$$

This can be also be written in the more explicit format of

$$\lambda(D > C) = \int P(D > C|\text{IM})|d\lambda(\text{IM})| \tag{13.55}$$

where $P(D > C|\text{IM})$ is the probability of demand $D$ exceeding capacity $C$ for any given level of the IM. This is the so-called fragility function or curve and its estimation is a central point of assessment, as will be discussed in Section 13.4.2. Essentially, the fragility acts as a weighting function on the integration of the hazard curve in Equation 13.55, signifying which intensity levels (and associated MAFs) matter for the LS of interest. A fragility with high dispersion (e.g., compare building 1 versus building 2 in Figure 13.14) assigns more weight to intensities away from the median of the fragility. Due to the exponentially higher MAF of low intensities relative to higher ones, it is actually the side of the hazard curve left of the fragility median that matters most. Thus, steeper slopes, i.e., $k = 2.84$ in Figure 13.14, indicate a more aggressive hazard compared to progressively milder ones of $k = 2.47$ and $k = 2.16$.

## 13.4.1 Performance objectives

There are seemingly many ways to define the performance of a structure, but very few that are unambiguous in terms of risk. For our purposes, we shall adopt a definition of a performance objective (or target) for design applications that requires a triplet of values: (i) a threshold or capacity value of response, $C$, (ii) a maximum allowable MAF of exceeding this threshold, $\lambda_{\text{PO}}$, and (iii) a desired confidence level of meeting the objective vis-à-vis epistemic

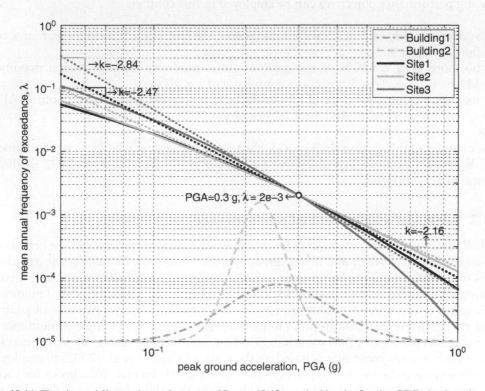

*Figure 13.14* The three different hazard curves of Figure 13.13 overlaid by the fragility PDFs (rather than the typical fragility "curves," i.e., the CDFs) of two different buildings. In general, the steeper their slope $k$ is, the more aggressive the hazard curve. (Adapted from Spillatura 2018).

uncertainty, $x$ in [0.5, 1]. Of interest is the occurrence of LS exceedance, i.e., of demand, $D$, exceeding (i.e., violating), capacity, $C$. Meeting a performance objective is defined to mean that the $x\%$ percentile estimate (due to epistemic uncertainty) of the MAF of LS exceedance should be lower than $\lambda_{PO}$, or

$$\lambda_{x\%}(D > C) < \lambda_{PO} \tag{13.56}$$

Optionally, one may choose to treat epistemic uncertainties by adopting a mean estimate of $\lambda(D > C)$, which is equivalent to prescribing a value of the confidence level $x$ greater than 50%. The actual value of confidence achieved is actually dependent on the distribution of the demand and capacity due to the additional uncertainty, or simply the dispersion of the latter if we make the typical first-order assumption for the influence of uncertainty on demand and capacity (Section 13.2.8). Then, the triplet may be collapsed to a pair of MAF and threshold C values, and Equation 13.56 becomes

$$\bar{\lambda}(D > C) < \lambda_{PO} \tag{13.57}$$

where the bar signifies the mean.

Equations (13.56) and (13.57) are directly based on the performance assessment framework embodied in Equation 13.55 and employ the latter to estimate the MAF of LS exceedance. Furthermore, they provide a conceptual method for verifying whether a design complies with a stated objective. Such a performance objective can be expressed in terms of any EDP of interest, and is related to MAF and confidence requirements that may be as strict as an owner requires, and as lax as the design code allows. For example, any of the following performance objectives can be employed in this context:

- system ductility of $C = 1.5$ with a maximum MAF of $\lambda_{PO} = 10\%$ in 50 year, at a confidence of $x = 60\%$;
- no more than $C = 20\%$ of the columns suffer moderate damage with a maximum MAF of $\lambda_{PO} = 5\%$ in 50 year, at a confidence of $x = 90\%$;
- maximum interstory drift of $C = 0.7\%$ being exceeded with a maximum MAF of $\lambda_{PO} = 10\%$ in 10 years, and a confidence of $x = 75\%$.

The remainder of this section shall provide the tools for assessing whether a structure meets or violates such performance objectives by employing the results of multiple nonlinear response history analyses.

## 13.4.2 Practical assessment of performance

We shall offer two general approaches for assessing structural performance. The first adopts the so-called MAF format, and it is a direct approach based on estimating the associated MAFs of LS exceedance, either in their mean or their $x\%$ confidence values, and directly checking for compliance via Equation 13.57 or Equation 13.56, respectively. At a minimum, this requires two stripes of nonlinear response history analyses, i.e., two sets of analyses run at two different levels of ground motion intensity, each with multiple ground motion records (see Section 5.3.4), plus some numerical integration. The second approach takes an indirect path to assessment that is based on the work of Cornell et al. (2002) to employ an load and resistance factor design (LRFD)-like (AISC, 2003) format that checks for capacity $C$ exceeding demand $D$ to verify safety. This is the so-called demand-capacity factored design (DCFD) format, which requires one or two stripes of analyses, requires no numerical integration, but comes at the cost of a more conservative result.

### 13.4.2.1 MAF format

Let $IM_i$ be $N$ values of increasing IM ($IM_i < IM_{i+1}$) covering the entire hazard curve from the lowest to the highest nonzero MAF available ($N > 50$ recommended). If LS is defined by a specific EDP exceeding a deterministic threshold value of $EDP_C$, then the MAF of violating LS, $\lambda LS$, can be estimated by the sum (or discretized integral):

$$\lambda_{LS} \cong \sum_{i=1}^{N} P\left(EDP > EDP_C \mid IM_i\right) \Delta\lambda(IM_i) \tag{13.58}$$

where $\Delta\lambda(IM_i) = \lambda(IM_i) - \lambda(IM_{i+1}) > 0$, due to the monotonically decreasing hazard, and $P(EDP > EDP_C \mid IM_i)$ is the probability that the EDP demand exceeds capacity $EDP_C$ at the given $IM_i$, also known as the fragility function. Assuming that a stripe has been performed at the given $IM_i$, this probability may be easily estimated by counting the number of points (i.e., dynamic analyses) where EDP has exceeded $EDP_C$, and dividing by the total number of analyses. Formally:

$$P\left(EDP > EDP_C \mid IM_i\right) = \frac{\sum_{j=1}^{N_{rec}} I\left[EDP^j > EDP_C \mid IM_i\right]}{N_{rec}} \tag{13.59}$$

where $N_{rec}$ is the number of ground motion records appearing in stripe $i$, and $EDP_j$, $j = 1 \ldots N_{rec}$ is the EDP value estimated for each record, and $I[\cdot]$ is an index function, which takes the value of 1 when its argument is true and 0 otherwise (Bakalis and Vamvatsikos, 2018).

Although, $N \geq 50$ would be required in Equation 13.58 to achieve an accurate numerical integration, one typically would perform only 5–10 stripes at well-spaced IM levels (Figure 13.15a), estimate the corresponding $P(EDP > EDP_C \mid IM)$ probabilities and employ simple (e.g., linear) interpolation to estimate values of exceedance probability for all $N$ levels of $IM_i$ (Figure 13.15b). The example shown in Figure 13.15a employs 8 stripes of 44 records/analyses each. Note that while only a few points seem to appear for stripes at high IM levels, this is only due to the "missing" points having resulted in global collapse, or an "infinite" EDP. In such cases, one should take care to count such points as "exceedance." Thus, the highest stripe at $S_a = 1.54\,g$ shows one exceedance point and 43 collapsing ones, for a total probability of $P(EDP > 0.04 \mid S_a = 1.54\,g) = 1.0$. Similarly, $P(EDP > 0.04 \mid S_a = 1.36\,g) = 1.0$ for the next lowest stripe, while $P(EDP > 0.04 \mid S_a = 1.16\,g) = 42/44 \approx 0.93$ for the third stripe from the top. These values appear as the top three IM points in Figure 13.15b, together with the corresponding estimates from the other stripes, providing via interpolation the so-called empirical fragility curve for exceedance of a maximum interstory drift of 0.04.

Alternatively, the probability of exceedance may be assumed to follow the lognormal distribution. Then, the points estimated by Equation 13.59 can be fitted by a lognormal CDF (Figure 13.15b) via an appropriate estimate of the corresponding median $IM_{50}$ and dispersion $\beta_{IM}$ as follows:

$$P\left(EDP > EDP_C \mid IM\right) = \Phi\left(\frac{\ln IM - \ln IM_{50}}{\beta_{IM}}\right) \tag{13.60}$$

If enough stripes are available, these two parameters can be evaluated by interpolating the available IM and $P(EDP > EDP_C \mid IM)$ values (i.e., via the empirical fragility curve) to estimate the 16%, 50%, and 84% values of the IM. Then, $IM_{50}$ is obviously the 50%

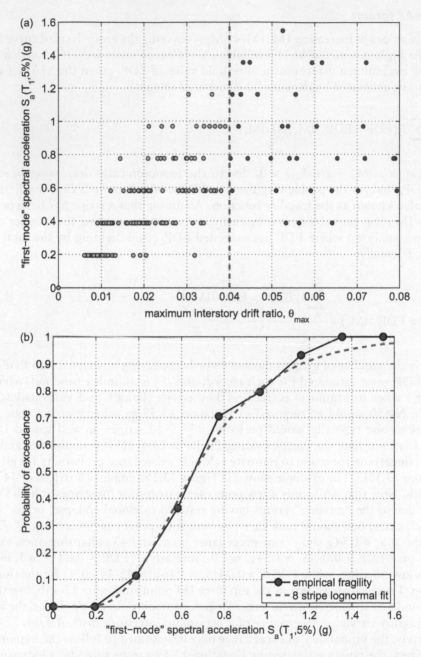

*Figure 13.15* Estimation of the fragility curve corresponding to the exceedance of a maximum interstory drift of $EDP_c = 0.04$ on a 12-story RC frame building using eight stripes: (a) determination of the runs violating the LS for each stripe (in solid black) and (b) the estimated empirical fragility curve and its lognormal fit.

value while the dispersion is estimated from the 16% and 84% values via Equation 13.32. Alternatively, one may employ the formal statistical approach of maximum likelihood estimation to evaluate the parameters of the distribution, available through MATLAB via the *mle* or *fitdist* functions or in Excel® as a standalone spreadsheet included in the FEMA P-58 software tools (FEMA, 2012b). For the case of Figure 13.15b, both the interpolation and the maximum likelihood approach provide practically identical results when using the eight available stripes.

The lognormal approximation comes with the advantage of allowing us to reduce the number of stripes employed to only two. In this special case, one can actually solve a system of two linear equations with two unknowns to determine the needed parameters. If $IM_1$ and $IM_2$ are the low and the high value, respectively, of the IM levels where the stripes have been run, solving the system results in:

$$IM_{50} = IM_2 \exp(K_{P2}\beta_{IM}) = IM_1 \exp(K_{P1}\beta_{IM}) \tag{13.61}$$

$$\beta_{IM} = \frac{\ln(IM_2) - \ln(IM_1)}{K_{P2} - K_{P1}} \tag{13.62}$$

where $P_i$ is the probability of exceedance at intensity $IM_i$ $(i = 1, 2)$. $K_{Pi}$ is the corresponding standard normal variate, $K_{Pi} = \Phi^{-1}(P_i)$, estimated as norminv($P_i$,0,1) in MATLAB and NORM.INV($P_i$,0,1) in Excel®, where obviously "$P_i$" needs to be replaced in the formulas by the proper numerical value each time. Note that $P_i$ cannot equal 0 or 1; otherwise, the relevant $K_{Pi}$ becomes infinite and the system cannot be solved.

This two-stripe approximation is obviously less robust than a multi-stripe approach due to the limited data that it employs. In general, its accuracy is good enough for practical purposes as long as the two stripes are well separated, capturing exceedance probabilities that differ by at least 0.25 probability. An example appears in Figure 13.16 where it becomes clear that the two-stripe approximation can offer good accuracy as long as the two fitted points are well separated along the vertical axis. When they are too close, there is a

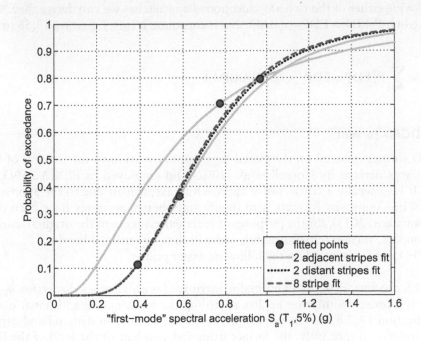

*Figure 13.16* Examples of estimated fragility curves corresponding to the exceedance of a maximum interstory drift of 0.04 on a 12-story RC frame building using two and eight stripes. An accurate approximation can be found using two stripes, as long as they are well separated in terms of probability of exceedance.

nonnegligible chance of a misfit, as shown for the left-most curve in Figure 13.16 where the two fitted points correspond to exceedance probabilities of about 0.7 and 0.8, being only 0.8 − 0.7 = 0.10 apart.

When $EDP_C$ is uncertain, and assuming that the median value, $EDP_{C50}$, of $EDP_C$ is employed, the result of Equation 13.58 is actually the 50% confidence level estimate of $\lambda_{LS}$. This is tantamount to ignoring the influence of uncertainties beyond the ground motion variability. If, instead, an estimate at a confidence level $x\%$ is sought, it can be evaluated by virtue of the monotonically decreasing relationship between $\lambda_{LS}$ and $EDP_C$. Thus, as discussed in Section 13.2.8.3 in the case of discrete treatment of uncertainty, we may estimate the corresponding $\lambda_{LS,x\%}$ value by subtly employing the $(1 − x)\%$ percentile value of $EDP_C$, instead of $EDP_{C50}$:

$$EDP_{C,x\%} = EDP_{C50} \cdot \exp(-K_x \beta_{TU}) \tag{13.63}$$

Alternatively, if the lognormal fit of the 50% estimate of the fragility is already available, it can be directly modified to the account for the increased confidence by appropriately lowering its median IM:

$$IM_{50,x\%} = IM_{50} \cdot \exp(-K_x \beta_{IM,U}) \tag{13.64}$$

where $\beta_{IM,U}$ is the total uncertainty in terms of the IM. As we shall discuss later, an approximation involved in the DCFD format allows us to estimate this from the value of $\beta_{TU}$ that concerns the EDP and appears in Equation 13.63.

By employing either of the two aforementioned approaches we can derive the $x\%$ realization of fragility, $P(EDP > EDP_{C,x\%} | IM_i)$, and incorporate it into Equation 13.58 to establish

$$\lambda_{LS,x\%} \cong \sum_{i=1}^{N} P(EDP > EDP_{C,x\%} | IM_i) \Delta\lambda(IM_i). \tag{13.65}$$

### 13.4.2.2 DCFD format

The DCFD format is based on the closed-form approximation to the integral of Equation 13.55 that was derived by Cornell et al. (2002) and employed in FEMA 350/351 (SAC, 2000a,b). It is generally a conservative approach to assessment due to the approximations involved. While improved formats with significantly better accuracy have been developed (e.g., Vamvatsikos, 2013), for the purposes of practical assessment the original, conservative but less complex, version remains useful, and will thus be outlined herein.

The DCFD format is based on the following assumptions:

i. The LS capacity $EDP_C$ is lognormally distributed with aleatory dispersion $\beta_{CR}$.
ii. The first-order assumption applies regarding the influence of additional uncertainty (see Section 13.2.8.2). The dispersion due to uncertainty in demand and capacity are $\beta_{DU}$ and $\beta_{CU}$, respectively, the former being independent of the level of the IM or the EDP demand. The total uncertainty is thus:

$$\beta_{TU} = \sqrt{\beta_{DU}^2 + \beta_{CU}^2}, \tag{13.66}$$

iii. The hazard curve can be approximated in the intensity range of interest by a power law function:

$$\lambda(IM) \approx k_0 \cdot IM^{-k} \qquad (13.67)$$

This line can be fitted as a straight line in log–log coordinates, and following the observations on Figure 13.14, it is advantageous to employ a biased fit that better captures lower IMs and higher MAFs, following the approach outlined by Vamvatsikos (2014). An example appears in Figure 13.17.

iv. The EDP demand given the IM is lognormally distributed with a median value given by a power law

$$EDP_{50} \approx a \cdot IM^{b} \qquad (13.68)$$

and with a constant (IM and EDP independent) aleatory dispersion of $\beta_{DR}$, typically accounting for record-to-record variability.

The last assumption is perhaps the most troublesome as it essentially implies that no "infinite" EDP values appear, or that collapse is not an issue. This makes the DCFD format applicable only to cases that are not appreciably influenced by collapse.

Fitting Equation 13.68 in general requires some degree of care. When multiple stripes are available, the simplest approach is to employ linear regression (function *LINEST* in Excel®, or *regress* in MATLAB) in logarithmic space to fit a power law relationship between the IM and EDP points. Stripes having more than about 20% collapses must be excluded because their most aggressive records (i.e., those that have caused collapse) cannot be accounted for by this simple regression scheme; thus, the resulting fit may be biased to show lower than

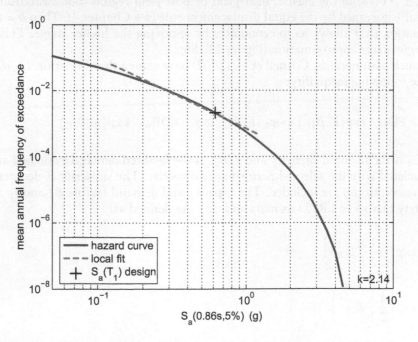

*Figure 13.17* The mean hazard curve for $S_a(0.86s, 5\%)$ at a site in California and a biased power-law approximation in the region of the 475-year intensity level.

actual response values. Stripes with fewer collapses can be employed, although, naturally, the "infinite" EDP points will have to be excluded from the fit. More elaborate regression schemes that separately fit collapsing runs and non-collapsing ones (Jalayer and Cornell, 2009) should be preferred when higher fidelity is required in the near-collapse region.

When only two-stripe analyses are to be performed, for the so-called *double-stripe* approach, one can simply interpolate between the median EDPs of each stripe. The first stripe is run at the intensity corresponding to the performance objective MAF, $\lambda_{PO}$, say $IM_{Po}$, Selecting an appropriate multiplier $c > 1$ to derive the second stripe, $IM' = c \cdot IM_{Po}$, is a bit more difficult. Similar to the MAF format described in the previous section, one would prefer to have a large enough value of $IM'$ that would offer robustness against inherent randomness in the median EDP of each stripe, and this becomes more important as one tries to economize by employing few runs, say 14–17, in each stripe. On the other hand, having a high multiplier $c$ introduces a risk that the stripe will contain too many collapse points and thus become unusable. To guard against this, values of $c = 1.1$–1.3 can be selected, although we would ideally like to use $c = 2$ or larger. So, let $EDP'_{50}$ denote the median demand parameter at the higher $IM'$ level, and $EDP_{Po50}$ the corresponding median demand at $IM_{Po}$. Then the log-slope, $b$, of the median demand parameter curve can be estimated via two stripes as follows:

$$b = \frac{\ln(EDP'_{50}) - \ln(EDP_{Po50})}{\ln(IM'/IM_{Po})} = \frac{\ln(EDP'_{50}) - \ln(EDP_{Po50})}{\ln(c)}, \qquad (13.69)$$

An example with $c = 3$ is illustrated in Figure 13.18a.

Alternatively, one may choose to employ only the single $IM_{Po}$ stripe. This is supported by the fact that deformation-related quantities for moderate and long-period structures (roughly, $T > 0.7\,s$) in the elastic, near-yield or post-yield region (but away from collapse) are generally governed by the equal displacement rule (see Chapter 4). Then $b = 1$ is a valid approximation that allows us to economize by removing the higher stripe. This is the so-called *single-stripe* approximation (Figure 13.18b).

Given such assumptions, Cornell et al. (2002) have shown that Equation 13.56 is equivalent to the following inequality:

$$FC_R \geq FD_{RPo} \cdot \exp(K_x \beta_{TU}) \Leftrightarrow \varphi_R \cdot EDP_{C50} \geq \gamma_R \cdot EDP_{Po50} \cdot \exp(K_x \beta_{TU}) \qquad (13.70)$$

where $FC_R$ is the factored capacity and $FD_{RPo}$ is the factored demand evaluated at the MAF level associated with the selected performance objective. The subscript R denotes that only aleatory uncertainties are included. The capacity and demand factors $\phi_R$ and $\gamma_R$ are similar to the safety factors of LRFD formats and they are defined as:

$$\phi_R = \exp\left[-\frac{1}{2}\frac{k}{b}\beta_{CR}^2\right], \qquad (13.71)$$

$$\gamma_R = \exp\left[\frac{1}{2}\frac{k}{b}\beta_{DR}^2\right]. \qquad (13.72)$$

Figure 13.19 provides an illustrative breakdown of the sources of variability included in the assessment and how they influence the parameters of Equation 13.70.

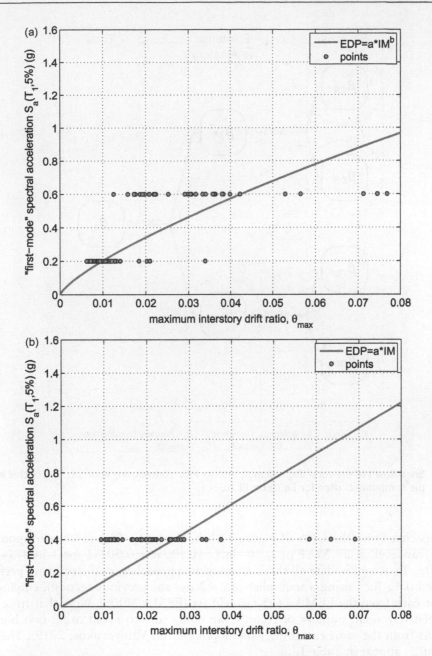

*Figure 13.18* Determination of the power–law relationship between EDP and IM for a 12-story RC frame structure using (a) two stripes (with $IM_{P_0} = 0.2\,g$ and $c = 3$) versus (b) one stripe (with $IM_{P_0} = 0.4\,g$) of dynamic analysis data, each with 44 ground motion records.

## 13.4.3  Example of application

A 12-story reinforced concrete (RC) moment-resistant frame designed for the site hazard of Figure 13.11 is used as an example. Its fundamental period is $T_1 = 2.14\,s$, and we want to ascertain its safety against a $\theta_{max,C} = 0.02$ threshold for a maximum allowable MAF of 10/50 using either 50% or 90% confidence. The total uncertainty for demand and capacity in EDP terms is $\beta_{TU} = 20\%$, while an additional aleatory variability of $\beta_{CR} = 20\%$ is assumed

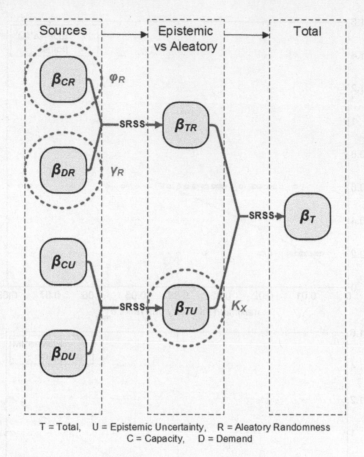

T = Total,    U = Epistemic Uncertainty,    R = Aleatory Randomness
C = Capacity,    D = Demand

*Figure 13.19* Breakdown of the total dispersion $\beta_T$ into its four contributing sources. The circled items are the components used for Equation 13.70.

for the capacity. From the UHS of Figure 13.12b, the $S_a(T_1, 5\%)$ value corresponding to the 10/50 year level, at an MAF of $\lambda_{PO} = -\ln(1 - 0.10)/50 = 0.00211$ (or 1 in 475 years) is $S_{aPo} = 0.20\,g$. To accommodate all three assessment approaches, two stripes are performed, at 0.2g and 0.4g (i.e., using a multiplier of $c = 2$ per the previous section), employing 15 records for each from the FEMA P695 far-field set (FEMA, 2009). Note that since we are using a relatively small number of records, we made sure to avoid using two horizontal components from the same recording (Giannopoulos and Vamvatsikos, 2019). The results in term of $\theta_{max}$ appear in Table 13.2.

### 13.4.3.1 MAF format

For the MAF approach, we need the uncertainty in terms of the IM, rather than the EDP. A rough translation of dispersion values from the EDP ($\beta_{EDP}$) to the IM ($\beta_{IM}$) axis can be effected via the power law approximation of Equation 13.68, which implies that $\beta_{IM} = \beta_{EDP}/b$, at least away from the region of collapse. For our case, $b = 1$ is a reasonable assumption. Still, a further complication in the MAF approach is that there is some numerical complexity in properly accounting for $\beta_{CR}$. A simple option, that is correct only for a level of confidence compatible with the mean value vis-à-vis uncertainties, is to combine this with $\beta_{TU}$ and

Table 13.2 The $\theta_{max}$ results of 15 + 15 dynamic analyses employed for the examples

| Record | 1 | 2 | 3 | 4 | 5 | 6 | 7 | 8 | 9 | 10 | 11 | 12 | 13 | 14 | 15 |
|---|---|---|---|---|---|---|---|---|---|---|---|---|---|---|---|
| $S_a = 0.2\,g$ | 0.0085 | 0.0063 | 0.0071 | 0.0067 | 0.0086 | 0.0070 | 0.0073 | 0.0132 | 0.0091 | 0.0140 | 0.0108 | *0.0203* | 0.0103 | 0.0115 | 0.0106 |
| $S_a = 0.4\,g$ | 0.0166 | 0.0125 | 0.0132 | 0.0164 | 0.0176 | 0.0107 | 0.0118 | *0.0252* | 0.0177 | *0.0691* | *0.0339* | *0.0584* | 0.0186 | *0.0280* | *0.0210* |

Exceedance points for $\theta_{max} = 0.02$ are shown in bold italics.

employ $\sqrt{\beta_{\mathrm{TU}}^2 + \beta_{\mathrm{CR}}^2}/b \approx 0.28$ as the overall dispersion of uncertainty in terms of the IM. This would be overly conservative at confidence levels higher than about 70%. Therefore, we shall assume a reduced value, between the above estimate and $\beta_{\mathrm{TU}}/b = 0.20$, settling on $\beta_{\mathrm{IM,U}} = 0.25$. This may seem less rigorous than desired, but it is important to remember that there is little guidance on selecting dispersion values anyway. Here, we are only making sure that our assumptions in both formats are compatible.

Both stripes appear in the IM–EDP plane of Figure 13.20, showing one exceedance point at 0.20 g and six at 0.40 g. Thus:

$$P_1 = P(\theta_{\max} > 0.02 | S_a = 0.2\ \mathrm{g}) = 1/15 = 0.0667$$

$$P_2 = P(\theta_{\max} > 0.02 | S_a = 0.4\ \mathrm{g}) = 6/15 = 0.4000$$

Thus, the corresponding standard variates are:

$$K_{P1} = \Phi^{-1}(P_1) = -1.5011$$

$$K_{P2} = \Phi^{-1}(P_2) = -0.2533$$

From (13.62), we get the fragility dispersion:

$$\beta_{\mathrm{IM}} = \frac{\ln(\mathrm{IM}_2) - \ln(\mathrm{IM}_1)}{K_{P2} - K_{P1}} = \frac{\ln(0.4\ \mathrm{g}) - \ln(0.2\ \mathrm{g})}{-0.2533 + 1.5011} = 0.5555$$

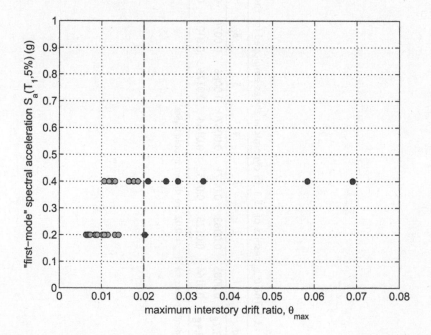

*Figure 13.20* The two stripes of 15 records employed for assessment via the MAF approach, showing the points violating $\theta_{\max,C} = 0.02$ (in solid black).

Then Equation 13.61 provides the fragility IM median:

$$\text{IM}_{50} = \text{IM}_2 \exp(K_{P2}\beta_{\text{IM}}) = 0.4 \text{ g} \exp(-0.2533 \cdot 0.5555) = 0.4605 \text{ g}$$

The resulting fragility appears in Figure 13.21, and thanks to the good vertical separation of the two fitted points we can be confident of a good fit with only two stripes. By numerically integrating with the $T_1 = 2.14$ s hazard curve of Figure 13.22 through Equation 13.58 we get the MAF estimate with a 50% confidence

$$\lambda_{\text{LS}} = 0.61 \cdot 10^{-3} < \lambda_{\text{PO}} = 2.11 \cdot 10^{-3}$$

or equivalently $1 - \exp(-0.61 \cdot 10^{-3} \cdot 50) = 3\%$ in 50 year, which is lower (i.e., rarer and thus better) than the target of 10% in 50 years. For a 90% confidence level we need to employ a fragility with the same dispersion but appropriately reduced median according to Equation 13.64:

$$\text{IM}_{50,0.90} = \text{IM}_{50} \exp(-K_{0.90}\beta_{\text{IM,U}}) = 0.4 \text{ g} \exp(-1.28 \cdot 0.2) = 0.3342 \text{ g}$$

Then, Equation 13.65 provides

$$\lambda_{\text{LS},90\%} = 1.29 \cdot 10^{-3} < \lambda_{\text{PO}} = 2.11 \cdot 10^{-3}$$

or approximately 6% in 50 years, which is again lower than the target of 10% in 50 years. Thus, the structure is safe with a 90% confidence.

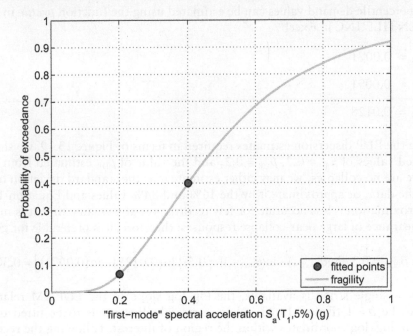

*Figure 13.21* The fitted lognormal fragility function for violating $\theta_{\text{max,C}} = 0.02$ and the fitted points from the two stripes of Figure 13.20.

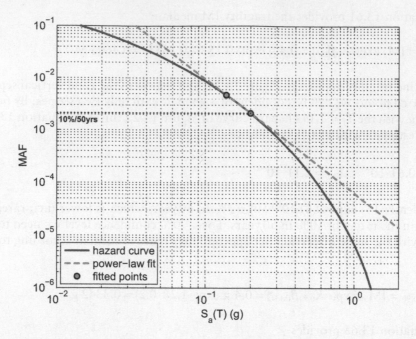

*Figure 13.22* The $S_a(T_1)$ hazard curve for $T_1 = 2.14\,s$, as derived from the hazard surface of Figure 13.11 for a site in California.

### 13.4.3.2 DCFD format—Single stripe

Assessment based on the DCFD format follows the basic steps discussed in Section 13.4.2.2. Only a single stripe is performed at $S_{aPo} = 0.2\,g$, with the results appearing in Table 13.2. The 16/50/84 percentile demand values can be estimated using the function *prctile* in MATLAB or PERCENTILE.INC in Excel®:

$$\theta_{max,16} = 0.0071$$

$$\theta_{max,50} = 0.0091$$

$$\theta_{max,84} = 0.0128$$

Regarding the EDP dispersion estimates required in terms of Figure 13.19, we shall employ the supplied values of $\beta_{TU} = 0.2$, $\beta_{CR} = 0.2$, and the value of $\beta_{DR}$ estimated from the stripe. Since there are no collapses, we may either estimate it as the standard deviation of the logarithm of the data, or approximate it by the 16% and 84% values and Equation 13.32. The latter approximation is less accurate for few records, yet at the same time it is more robust to the appearance of large near-collapse responses; therefore, it is preferable for general use:

$$\beta_{DR} = \beta_{EDP} = 0.5 \cdot [\ln(\theta_{max,84}) - \ln(\theta_{max,16})] = 0.5 \cdot \left[\ln(0.0128) - \ln(0.0071)\right] = 0.30$$

Since only a single stripe is available, the log–log slope of the EDP–IM relationship is assumed to be $b = 1$ (Figure 13.23). Now, a straight line needs to be fitted to the hazard curve in log–log coordinates within the region of interest. Following the recommendations of Vamvatsikos (2014), the biased power law approximation is fitted to pass through the hazard curve points at $IM_{fitA} = S_{aPo} = 0.2\,g$ and $IM_{fitB} = S_{aPo} \cdot \exp(-\beta_{fit}/b)$, where $\beta_{fit}$ is a

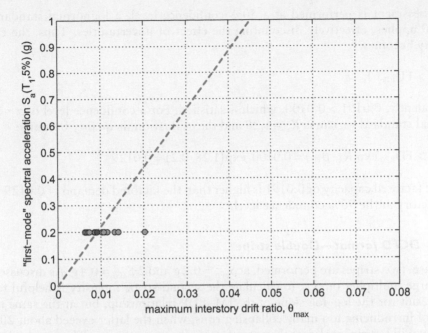

*Figure 13.23* The analyses and the fitted IM–EDP relationship employed for the single-stripe DCFD approach.

dispersion value for the fit that is characteristic of the problem. We can employ either the total dispersion, $\beta_T$, or even just the record-to-record dispersion $\beta_{EDP}$. Both are equally good choices. In our case, we shall take the middle ground and use the SRSS of $\beta_{EDP}$ and $\beta_U$, effectively excluding only $\beta_{CR}$:

$$\beta_{fit} = \sqrt{\beta_{EDP}^2 + \beta_{TU}^2} = \sqrt{0.3^2 + 0.2^2} = 0.36$$

$$IM_{fitB} = S_{aPo} \exp(-\beta_{fit}/b) = 0.2 \text{ g} \cdot \exp(-0.36/1) = 0.14 \text{ gs}$$

The corresponding MAFs for fitting are $\lambda_{fitA} = 2.1072 \cdot 10^{-3}$, corresponding to 10/50 obviously, and $\lambda_{fitB} = 4.6075 \cdot 10^{-3}$, as determined by interpolating the hazard curve, simply by employing linear interpolation in log–log coordinates. Finally, the log–slope of the fit to the hazard curve becomes:

$$k = -\frac{\ln(\lambda_{fitB}) - \ln(\lambda_{fitA})}{\ln(IM_{fitB}) - \ln(IM_{fitA})} = \frac{\ln(4.6075 \cdot 10^{-3}) - \ln(2.1072 \cdot 10^{-3})}{\ln(0.14 \text{ g}) - \ln(0.20 \text{ g})} = 2.18$$

The resulting fit appears as a red dashed line in Figure 13.22. As expected, we overestimate the hazard more or less for any value of $S_a$, thus our approach will result in a somewhat conservative evaluation, typical for the Cornell et al. (2002) approximation.

Finally, factored demand and factored capacity are estimated as:

$$FD_{RPo} = \theta_{max,50} \exp\left(0.5 \cdot k \cdot \beta_{EDP}^2/b\right) = 0.0091 \exp\left(0.5 \cdot 2.18 \cdot 0.23^2/1.00\right) = 0.0100$$

$$FC_R = \theta_{max,C} \exp\left(-0.5 \cdot k \cdot \beta_{CR}^2/b\right) = 0.02 \exp\left(-0.5 \cdot 2.18 \cdot 0.2^2/1.00\right) = 0.0191$$

If the assessment is performed at a 50% confidence level, a lognormal standard variate of $K_x = 0$ applies, effectively discounting the effect of uncertainties. Thus, the evaluation inequality becomes:

$$FC_R > FD_{RPo} \cdot 1,$$

or equivalently, $0.0191 > 0.0100$, which is satisfied. For a confidence level of $x = 90\%$, the lognormal standard variate is $K_x = 1.28$ and the evaluation inequality becomes:

$$FC_R > FD_{RPo} \exp(K_x \cdot \beta_{TU}) = 0.0100 \cdot \exp(1.28 \cdot 0.2) = 0.0129$$

Since the factored capacity of 0.0191 is higher than the factored demand of 0.0129 the result is satisfactory at the 90% confidence level.

### 13.4.3.3 DCFD format—Double stripe

In this case, two stripes are performed, at $S_{aPo} = 0.2\,g$ and $2 \cdot S_{aPo} = 0.4\,g$. As discussed earlier, such a large multiplier on $S_{aPo}$ to calculate the second-stripe intensity is helpful to achieve a good result for the log–log slope of the IM–EDP relationship, but at the same time runs the risk of introducing too many collapsing runs; when the latter exceed about 20% of the runs, they will largely invalidate the second stripe for our purposes. Thus, multipliers on the order of 1.1–1.3 are usually preferable. Since we are checking a relatively low $\theta_{max,C} = 0.02$, we do not expect to have too many collapsing runs. Therefore, to maintain consistency with the MAF format of Section 13.3.3.1, we kept the same stripes of Table 13.2. The 16/50/84 percentile demand values for the first stripe remain the same as in the previous section:

$$\theta_{max,16} = 0.0071$$

$$\theta_{max,50} = 0.0091$$

$$\theta_{max,84} = 0.0128$$

At the higher intensity level, we only need the median, estimated as:

$$\theta'_{max,50} = 0.0177$$

Note how this implies that one could further economize by employing only about half of the analyses in the higher stripe, as the median is always more robust for small sample sizes than the dispersion. For the first stripe, though, this is not an option. Employing the median values of both stripes allows estimating the log–log slope of the median IM–EDP relationship (Figure 13.24) via Equation 13.69:

$$b = \frac{\ln(\theta'_{max,50}) - \ln(\theta_{max,50})}{\ln(2)} = \frac{\ln(0.0128) - \ln(0.0091)}{\ln(2)} = 0.96$$

which, predictably, is very close to 1.0 for a deformation-related quantity away from collapse on a long-period structure.

Just as for the single-stripe approach, we estimate the dispersion to be:

$$\beta_{EDP} = 0.5 \cdot [\ln(\theta_{max,84}) - \ln(\theta_{max,16})] = 0.5 \cdot \left[\ln(0.0128) - \ln(0.0071)\right] = 0.30$$

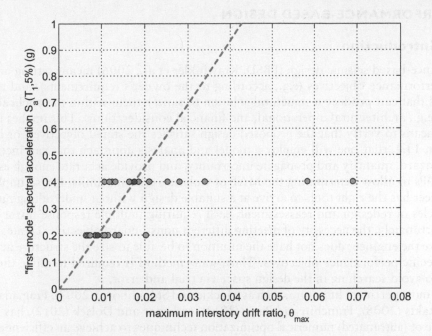

*Figure 13.24* The analyses and the fitted IM–EDP relationship employed for the double-stripe DCFD approach.

Since the value of $b$ and the associated dispersions are practically the same as for the single-stripe analysis, there is no good reason to change the hazard fit. Thus, the same log–slope of $k = 2.18$ is employed, as shown in Figure 13.22.

Finally, factored demand and factored capacity are estimated as:

$$FD_{RPo} = \theta_{max,50} \exp\left(0.5 \cdot k \cdot \beta_{EDP}^2/b\right) = 0.0091 \exp\left(0.5 \cdot 2.18 \cdot 0.23^2/0.96\right) = 0.0100$$

$$FC_R = \theta_{max,C} \exp\left(-0.5 \cdot k \cdot \beta_{CR}^2/b\right) = 0.02 \exp\left(-0.5 \cdot 2.18 \cdot 0.2^2/0.96\right) = 0.0191$$

Obviously, the minor change in $b$ has negligible influence on the above results, thus the assessment leads to identical numbers (at this level of precision) and conclusions. If the assessment is performed at a 50% confidence level, the evaluation inequality becomes:

$$FC_R > FD_{RPo} \cdot 1,$$

or equivalently, $0.0191 > 0.0100$, which is satisfied. For a confidence level of $x = 90\%$, the lognormal standard variate is $K_x = 1.28$ and the evaluation inequality becomes:

$$FC_R > FD_{RPo} \exp(K_x \cdot \beta_{TU}) = 0.0100 \cdot \exp(1.28 \cdot 0.2) = 0.0129$$

which is again satisfactory. Thus, the structure is again found to be safe against exceeding $\theta_{max,C} = 2\%$ at a MAF of 10/50 and a 90% confidence level.

## 13.5 PERFORMANCE-BASED DESIGN

### 13.5.1 Introduction

Performance-based seismic design (PBSD, Krawinkler et al., 2006) means setting any number of performance objectives (e.g., according to the owner's requirements), and applying a method that can produce a structural solution to satisfy them within the applicable constraints (e.g., architectural, operational, and financial considerations). This implies that one has the means to verify that the proposed design satisfies the stated objectives, or in terms of Section 13.4, that one will employ a model and analysis approach that can incorporate the site hazard, quantify and propagate uncertainty, and provide accurate enough estimates of the EDPs to allow determining the MAF of violating relevant limits. It also implies that the engineer has the right tools to arrive at a suitable design without undertaking numerous costly cycles of redesign and reassessment, each requiring multiple response history analyses. Unfortunately, the necessity of meeting different nonstandard objectives creates a situation where the engineer does not have the intuition to be able to size the structure according to the specific performance objectives. Thus, some method is required to guide the design process to avoid searching in the design space via trial and error.

So far, most pertinent literature, such as Mackie and Stojadinovic (2007), Fragiadakis and Papadrakakis (2008), Franchin and Pinto (2012), or Lazar and Dolsek (2012), has focused on the use of (automated) numerical optimization techniques to achieve an efficient exploration of the design space. This is a comprehensive solution approach that requires specialized software and considerable computational power. To lighten such requirements, Sinković et al. (2016) proposed empirical rules to guide a "manual" optimization for RC frames that can reportedly achieve a competitive design for a mid-rise frame within about four steps of design and assessment. Herein, following the work of Vamvatsikos and Aschheim (2016) we adopt a general direct design approach that employs an equivalent SDOF (ESDOF) proxy to produce a viable initial design within a single step.

### 13.5.2 YFS

YFS constitute a visual and factual representation of a structural system's performance that associates the MAF of exceeding any displacement $\delta$ (or ductility $\mu$) value with the equivalent SDOF system yield strength $V_y^*$ (or yield strength coefficient, $C_y^* = V_y^*/W$, where $W$ is the seismic weight). They can be estimated via available Excel® tools (Vamvatsikos and Aschheim, 2018). The calculation of YFS is dependent on the assumption of a stable system property, this being either a constant system yield displacement, $\delta_y$, for the "classical" YFS proposed by Vamvatsikos and Aschheim (2016), or a constant first-mode period, $T$, for the variant termed YFS-T, introduced by Ruiz-Garcia and Miranda (2007). Thus, variations in strength ($C_y^*$), presented on an YFS plot, are related to changes in both vibration period and stiffness, while on a YFS-T plot they denote changes to $\delta_y$ under a constant stiffness. Example YFS plots are presented in Figure 13.25 for an elastic-perfectly-plastic oscillator, where three POs are specified by the "$x$" symbols and curves representing the site hazard convolved with the structural system's fragility are plotted for fixed values of $C_y^*$. The minimum acceptable $C_y^*$ that fulfills all the specified POs can be readily determined and constitutes the strength to be used as a starting point for the performance-based design of the structural system studied. Because YFS determine the $C_y^*$ required for the ESDOF system, mapping to the yield strength of the multi-degree-of-freedom (MDOF) system is obviously needed (Chapter 7).

The formulation of the YFS-based design process is founded on the stability of the value of a single system parameter that essentially remains unchanged with strength. For a given

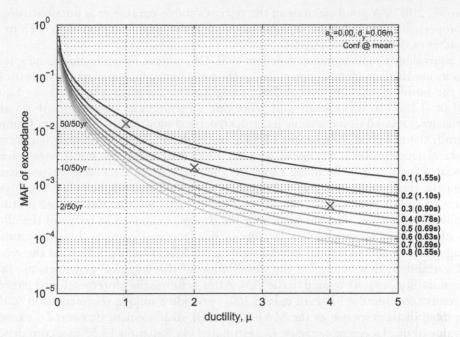

Figure 13.25 An example of (a) YFS estimated for the site hazard of Figure 13.11 showing three performance objectives (indicated by the "x" symbols) overlaid against and ten different realizations of the same elasto-plastic system with normalized yield strength $C_y^*$ ranging in 0.1–1.0. The third objective governs, leading to $C_y^* = 0.36$.

structural configuration of conventional yielding systems, such as moment-resistant frames or concrete shear walls, Chapter 9 discusses in detail the near-insensitivity of $\delta_y$ to changes in the system's strength and stiffness (Figure 13.26a) compared to the less stable fundamental period, $T$, typically employed in code approaches. For rocking systems, though, such as unbonded post-tensioned shear walls (Figure 13.26b), analytical models point instead to the insensitivity of the system's period vis-à-vis changes in strength (e.g., Kurama et al., 1999;

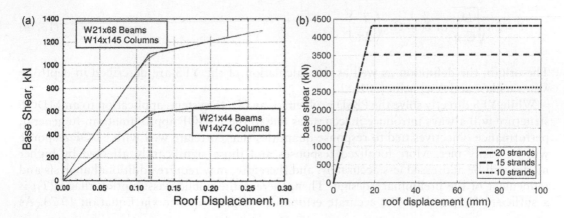

Figure 13.26 Capacity curves determined by first-mode nonlinear static (pushover) analysis of (a) two four-story moment-resistant steel frames with the same story heights and nominal section depths (from Aschheim, 2002) and (b) three different four-story unbonded post-tensioned walls having the same dimensions but different prestressing steel area (see Design Example 6, Section 19.12).

Perez et al., 2007). A good estimate of the system's stable parameter is possible using structural properties that are known *a priori* to the designer, and is exploited by YFS to reduce the number of design/analysis cycles.

The equivalent single-degree-of-freedom (ESDOF) system simplification is adopted similarly to its use in all modern design codes as the basis for approximating the inelastic behavior of the building. A fully probabilistic basis for the design is employed via Equations (13.56) or (13.57) to incorporate and propagate all sources of uncertainty (both aleatory and epistemic) related to the seismic hazard, structural modeling, and analysis framework.

Overall, the essential components of YFS are (i) the site-specific seismic hazard surface for spectral acceleration at a range of periods, (ii) an estimate of the stable parameter of the structural system, (iii) its damping ratio, (iv) the *shape* of its force-deformation backbone (e.g., elastic, elasto-plastic, etc.) and the general *form* of dissipating behavior (e.g., hysteretic system with full or pinched loops versus nonlinear elastic with no dissipation) (v) the set of performance objectives (POs) to be met by the design, expressed in terms of the allowable MAFs of exceeding specific global ductility or displacement limits and (vi) an estimate for the magnitude of the additional uncertainties (i.e., dispersion, $\beta_{TU}$), beyond the record-to-record variability, that encumber the distribution of the response given seismic intensity and the ductility capacity related to the POs. After defining the aforementioned parameters, displacement (or ductility) hazard curves $\lambda(D_C^*)$ provide a unique representation of the system's probabilistic response as the MAF of ESDOF displacement demand $D^*$ exceeding a given value of displacement capacity, $D_C^*$, estimated via Equation 13.58 at a confidence level compatible with the mean

$$\lambda\left(D_C^*\right) \cong \sum_{i=1}^{N} P\left(D^* > D_C^* | \mathrm{IM}_i\right) \Delta\lambda\left(\mathrm{IM}_i\right) \tag{13.73}$$

or similarly via Equation 13.65 as $\lambda_{x\%}(D_C^*)$ to introduce confidence $x\%$.

The main outcome of the YFS framework is the structural strength, normalized as $C_y^*$ that is required for the imposed performance levels to be satisfied. When adopting the assumption of a constant $D_y^*$, $C_y^*$ essentially becomes a direct replacement of the period $T$, while under the assumption of constant $T$, it becomes a replacement of $D_y^*$.

$$T = 2\pi\sqrt{\frac{D_y^*}{C_y^* g}} \text{ or } \delta_y = C_y^* g\left(\frac{T}{2\pi}\right)^2 \tag{13.74}$$

The origin, the definition as well as the calculation of the YFS are described in depth by Vamvatsikos and Aschheim (2016).[1]

While YFS directly solve the PBSD problem of an SDOF system, application to an MDOF structure will always introduce inaccuracies due to the ESDOF approximation. In general, performance objectives tied to response quantities that correlate well with SDOF response will be easily met. More localized responses and those significantly affected by higher modes will be addressed less accurately, and therefore, may require additional analysis and refinement of the preliminary design. Then, there is the implicit assumption that $S_a(T_1)$ is a sufficient IM to allow the accurate estimation of performance via Equation 13.73. As

---

[1] Spreadsheets for the calculation of YFS are available at users.ntua.gr/RCbook.html; specifically, for constant yield displacement, see users.ntua.gr/divamva/RCbook/YFSapp.xls, for constant period, see users.ntua.gr/divamva/RCbook/YFS-Tapp.xls, and for nonlinear elastic constant period systems, see users.ntua.gr/divamva/RCbook/YFS-TNEapp.xls.

discussed in Section 5.3.4, this is not always the case, yet better IMs are not easy to use as they lack the direct connection to the ESDOF yield base shear. Finally, the rapid estimation of YFS relies on the use of approximate regression expressions rather than actual nonlinear response history analyses. These are the so-called $R–\mu–T$ relationships generally connecting the statistics of reduction factor $R$ and ductility $\mu$ for a given period and characteristic hysteretic behavior, as discussed in Chapter 4. These inherently constrain the accuracy of the design $C_y^*$ when based on ground motions that are not representative of site characteristics (due to soft soil, directivity, etc.) or when the hysteretic behavior assumed for their development does not correspond to the actual structure. There are few $R–\mu–T$ expressions that provide the entire distribution of response, limiting us to SPO2IDA for conventional yielding structures (Section 4.10.5, Vamvatsikos and Cornell, 2006) and to the expressions of Bakalis et al. (2019) for rocking systems, especially in the early range of response where they exhibit nonlinear elastic behavior (i.e., a complete or near-complete lack of hysteresis).

In the end, although design according to YFS can eliminate many cycles of iteration, the resulting initial design will not necessarily be perfectly compliant. Some reanalysis and redesign iterations may be required, but the initial design will provide a good starting point, just as a better initial guess will improve the convergence of any iterative method (e.g., Newton–Raphson). Therefore, if strict compliance to the stated performance objectives is desired, YFS design should be followed by a proper performance assessment as per Section 13.4.

## 13.5.3 Example of application

To illustrate the design approach, we shall employ the design example of a bridge pier from Section 1.4. The example consists of the seismic design of a single-column pier of a highway bridge for transverse response. We seek to limit the peak displacement to 2% of the height of the structure (i.e., 0.02(6.25 m) = 0.125 m = 4.92 in.). The pier has a circular cross section and is subjected to a dead load of 6,130 kN (1,378 kips). The diameter of the column is chosen to be 1.3 m (51.2 in.). The steel reinforcement is B-500, having a minimum yield strength $f_{yk} = 500$ MPa (72.5 ksi). The characteristic concrete compressive strength is 30 MPa (4.35 ksi). The effective yield curvature, $\phi_y$, for this circular reinforced concrete column cross section can be estimated (Hernández-Montes and Aschheim, 2003), as

$$\phi_y = \frac{2.3\varepsilon_y}{d}$$

where $\phi_y$ is the yield strain of the reinforcement and $d$ is the depth to the centroid of the extreme tension reinforcement. For this column, $\phi_y = f_y/E_s = 500/200{,}000 = 0.0025$ and $d \approx 1.300 - 0.056 = 1.244$ m (= 49.0 in.). Thus, we estimate $\phi_y = 0.0046$ rad/m.

Obviously, since this is an SDOF system, no real differentiation between the ESDOF and the MDOF needs to be made. The corresponding yield displacement, $u_y = D_y^*$, for a cantilever column subjected to lateral load applied at the end, is estimated as

$$u_y = \phi_y \frac{L^2}{3}$$

or 0.0602 m for this column, having length $L = 6.25$ m (20.5 ft). Thus, the ductility limit associated with a drift limit of 2% of the height is 0.125 m/0.0602 m = 2.08, or approximately 2. We shall set a 60% confidence level and an MAF target of 10% in 50 years. For reasons of comparison with the results of Section 1.4 we selected a California site whose 10/50 UHS approximately matches the spectral acceleration produced by the 1940 NS El

Centro record in the range of 1–2 s (Figure 13.27). An additional dispersion of about 0.20 was assumed. The resulting YFS curves appear in Figure 13.28.

The resulting $C_y^*$ is 0.11. The required base shear strength at yield is given by

$$V = C_y^* \cdot W = 0.11 \cdot W = 152 \text{ kips} = 675 \text{ kN}.$$

A plastic mechanism will form when a plastic hinge forms at the base of the column. Thus, the required strength of the plastic hinge is given by $M = V \cdot h = 3{,}000 \text{ kip} \cdot \text{ft} = 4{,}070 \text{ kN} \cdot \text{m}$. Using the nominal flexural strength calculated using expected material strengths ($1.2 f_y = 72$ ksi = 496 MPa and $1.3 \cdot f_c' = 5{,}200$ psi = 35.9 MPa according to Caltrans) as an approximation of the plastic flexural strength, a 48-in. (1.22-m) diameter cross section with 12 No. 9 bars (12 bars, each 29-mm diameter) has adequate strength to resist these actions under the given axial dead load (1,380 kips = 6,140 kN). This represents a reinforcement ratio of 0.66% (which is acceptable for bridge columns) and compares to 1.26% required when the base shear was determined using YPS in the example of Chapter 1.

One may observe that the required base shear strength was determined as a function of the yield displacement, without having made any reference to the period of vibration. The period is the outcome of the design process, and may be determined as:

$$T = 2\pi \sqrt{\frac{u_y}{C_y^* g}} = 2\pi \sqrt{\frac{2.37 \text{ in.}}{0.11 \cdot 386.1 \text{ in./s}^2}} = 1.48 \text{ s}$$

The results of two approaches are summarized as follows for a side-by-side comparison:

YPS with El Centro 1940 NS: $C_y^* = 0.08$ ($T = 1.74$ s) or $C_y^* = 0.14$ ($T = 1.31$ s), depending on which peak we choose (Figure 13.29). An engineering safety-conscious choice was made to use the higher peak, thus $C_y^* = 0.14$.

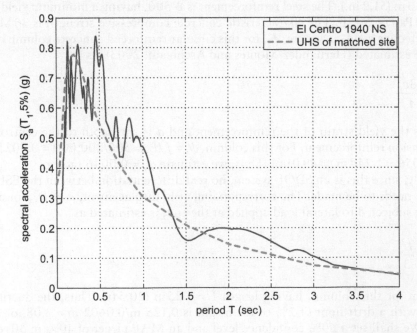

*Figure 13.27* The El Centro 1940 NS component spectrum versus the 10/50 UHS of a site that matches the former in the region of 1–2 s.

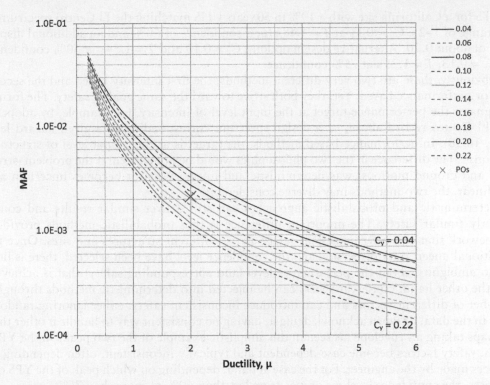

*Figure 13.28* YFS contours for the El Centro site at a 60% confidence.

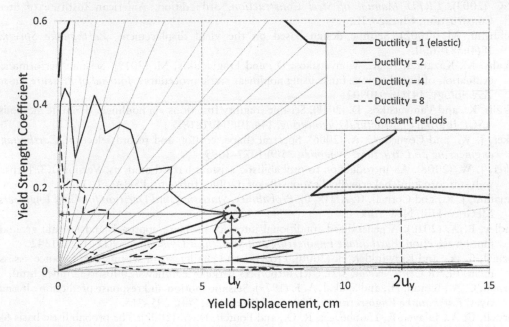

*Figure 13.29* YPS as estimated for the El Centro 1940 NS component. The jagged shape of the spectrum allows two different values for $C_y^*$, namely 0.08 (lower circle) and 0.14 (upper circle).

YFS for a California site with a 10% in 50 years UHS matching the El Centro spectrum in the range of 1–2 s: $C_y^* = 0.11$ at $T = 1.48$ s for a confidence of 60% and an additional dispersion of about 0.20. Alternative design options: $C_y^* = 0.10$ and $T = 1.54$ s at 50% confidence, or $C_y^* = 0.13$, $T = 1.35$ s at 75% confidence.

Obviously, these are two very different methods, the first intensity based and the second risk or performance based. Yet, they both strive toward the same goal of safety. The former designates the performance target at the input level of intensity, for example, by adopting the El Centro record above, or a design spectrum corresponding to a given hazard level (e.g., 10/50). In performance-based methods, the target is at the output level of structural response. The difference in the two philosophies would be nonexistent if the problem (structure and ground motions) was deterministic and linear. When it becomes uncertain and nonlinear, the two methods may diverge considerably.

Deterministic and probabilistic approaches can still deliver similar results and consequently similar safety. The important difference is that probabilistic methods provide a framework that can offer consistent safety across different structures and sites. Once the additional uncertainty dispersion and the confidence level have been selected, there is little or no ambiguity about the performance level (and corresponding safety) that is achieved. On the other hand, conservativeness can be injected into deterministic methods through a number of different vectors and can introduce inconsistent safety, either ignoring randomness in the data, or when acknowledging it, having no consistent way to handle it other than perhaps taking an envelope (as seen in this simplistic example of the two peaks in the YPS). Thus, safety factors become case-dependent and typically inconsistent, often depending on choices made by the engineer. For the case at hand, depending on which peak of the YPS one chooses, the confidence level may range from less than 40% to more than 75%.

## REFERENCES

AISC (2003). *LRFD Manual of Steel Construction*, 3rd edition, American Institute of Steel Construction, Chicago, IL.

Aschheim, M. (2002). Seismic design based on the yield displacement, *Earthquake Spectra*, 18(4):581–600.

Bakalis, K., Kazantzi, A. K., Vamvatsikos, D., and Fragiadakis, M. (2019). Seismic performance evaluation of liquid storage tanks using nonlinear static procedures, *Journal of Pressure Vessel Technology*, 141(1), 010902.

Bakalis, K., and Vamvatsikos, D. (2018). Seismic fragility functions via nonlinear dynamic methods, *ASCE Journal of Structural Engineering*, 144(10):04018181.

Baker, J. W., and Cornell, C. A. (2006). Spectral shape, epsilon and record selection, *Earthquake engineering and structural dynamics*, 35(9):1077–1095.

Baker, J. W. (2008). An introduction to probabilistic seismic hazard analysis. Version 1.3. https://web.stanford.edu/~bakerjw/Publications/Baker_(2008)_Intro_to_PSHA_v1_3.pdf.

Benjamin, J. R., and Cornell, C. A. (1970). *Probability, Statistics, and Decision for Civil Engineers*, McGraw-Hill, New York.

Bradley, B. A. (2010). A generalized conditional intensity measure approach and holistic ground-motion selection, *Earthquake Engineering and Structural Dynamics*, 39(12):1321–1342.

Cornell, C. A., and Krawinkler, H. (2000). Progress and challenges in seismic performance assessment. *PEER Center News*, 3(2). URL http://peer.berkeley.edu/news/2000spring/index.html.

Cornell, C. A., Banon, H., and Shakal, A. F. (1979). Seismic motion and response prediction alternatives. *Earthquake Engineering & Structural Dynamics*, 7(4), 295–315.

Cornell, C. A., Jalayer, F., Hamburger, R. O., and Foutch, D. A. (2002). The probabilistic basis for the 2000 SAC/FEMA steel moment frame guidelines, *ASCE Journal of Structural Engineering*, 128(4):526–533.

Der Kiureghian, A., and Ditlevsen, O. (2009). Aleatory or epistemic? Does it matter? *Structural Safety*, 31:105–112.

FEMA (2009). FEMA P695 Far field ground motion set. http://users.ntua.gr/divamva/RCbook/FEMA-P695-FFset.zip.

FEMA (2012a). Seismic performance assessment of buildings, Volume 1—Methodology, FEMA P-58-1, prepared by the Applied Technology Council for the Federal Emergency Management Agency, Washington, DC.

FEMA (2012b). Seismic performance assessment of buildings, Volume 3—Spreadsheet Tools, FEMA P-58-3.3-3.5, prepared by the Applied Technology Council for the Federal Emergency Management Agency, Washington, DC. http://users.ntua.gr/divamva/RCbook/FragilityCurveFitTool.xlsx.

Fragiadakis, M., and Papadrakakis, M. (2008). Performance-based optimum seismic design of reinforced concrete structures, *Earthquake Engineering and Structural Dynamics*, 37:825–844.

Franchin, P., and Pinto, P. (2012). Method for probabilistic displacement-based design of RC structures, *Journal of Structural Engineering*, 138(5):585–591.

Giannopoulos, D., and Vamvatsikos, D. (2019). Ground motion records for seismic performance assessment: To rotate or not to rotate? *Earthquake Engineering and Structural Dynamics*. DOI: 10.1002/eqe.3090.

Gutenberg, B., and Richter, C. F. (1944). Frequency of earthquakes in California, *Bulletin of the Seismological Society of America*, 34(4):185–188.

Hernández-Montes, E., and Aschheim, M. (2003). Estimates of the yield curvature for design of reinforced concrete columns, *Magazine of Concrete Research*, 55(4):373–383.

Jalayer, F., and Cornell, C. A. (2009). Alternative non-linear demand estimation methods for probability-based seismic assessments, *Earthquake Engineering & Structural Dynamics*, 38(8):951–972.

Kaila, K. L., and Narain, H. (1971). A new approach for preparation of quantitative seismicity maps as applied to Alpide belt-Sunda arc and adjoining areas, *Bulletin of the Seismological Society of America*, 61(5):1275–1291.

Kazantzi, A. K., Vamvatsikos, D., and Lignos, D. G. (2014). Seismic performance of a steel moment-resisting frame subject to strength and ductility uncertainty, *Engineering Structures*, 78:69–77.

Kohrangi, M., Bazzurro, P., Vamvatsikos, D., and Spillatura, A. (2017). Conditional spectrum based ground motion record selection using average spectral acceleration, *Earthquake Engineering and Structural Dynamics*, 46(10):1667–1685.

Krawinkler, H, Zareian, F., Medina, R. A., and Ibarra, L. F. (2006). Decision support for conceptual performance-based design, *Earthquake Engineering and Structural Dynamics*, 35(1):115–133.

Kurama, Y. C., Pessiki, S., Sause, R., and Lu, L.-W. (1999). Seismic behavior and design of unbonded posttensioned precast concrete walls, *PCI Journal*, 44(3):72–89.

Lazar, N., and Dolsek, M. (2012). Risk-based seismic design - An alternative to current standards for earthquake-resistant design of buildings. *Proceedings of the 15th World Conference on Earthquake Engineering*, Lisbon, Portugal.

Lin, T., Haselton, C. B., and Baker, J. W. (2013). Conditional spectrum-based ground motion selection. Part I: Hazard consistency for risk-based assessments, *Earthquake engineering and structural dynamics*, 42(12):1847–1865.

Luco, N., Ellingwood, B. R., Hamburger, R. O., Hooper, J. D., Kimball, J. K., and Kircher, C. A. (2007). Risk-targeted versus current seismic design maps for the conterminous United States. *SEAOC Convention Proceedings*, USA.

Mackie, K. R., and Stojadinovic, B. (2007). Performance-based seismic bridge design for damage and loss limit states, *Earthquake Engineering and Structural Dynamics*, 36:1953–1971.

Mathworks (2018). MATLAB and Statistics Toolbox Release 2018b, The MathWorks, Inc., Natick, MA.

Microsoft (2010). Microsoft Excel version 2010, Redmont, WA.

Perez, F. J., Sause, R., and Pessiki, S. (2007). Analytical and experimental lateral load behavior of unbonded posttensioned precast concrete walls, *Journal of Structural Engineering*, 133(11):1531–1540.

Pinto, P. E., Giannini, R., and Franchin, P. (2004). *Seismic Reliability Analysis of Structures*, IUSS Press, Pavia, Italy.

Ruiz-Garcia J., and Miranda, E. (2007). Probabilistic estimation of maximum inelastic displacement demands for performance-based design, *Earthquake Engineering and Structural Dynamics*, 36:1235–1254.

SAC Joint Venture (2000a). Recommended seismic design criteria for new steel moment-frame buildings, FEMA-350, prepared for the Federal Emergency Management Agency, Washington, DC.

SAC Joint Venture (2000b). Recommended seismic evaluation and upgrade criteria for existing welded steel moment-frame buildings, FEMA-351, prepared for the Federal Emergency Management, Agency, Washington DC.

Sinković, N. L., Brozovič, M., and Dolšek, M. (2016). Risk-based seismic design for collapse safety, *Earthquake Engineering & Structural Dynamics*, 45(9):1451–1471.

Spillatura, A. (2018). From record selection to risk targeted spectra for risk based assessment and design. *PhD Thesis*, University of Pavia, Italy.

Vamvatsikos, D. (2013). Derivation of new SAC/FEMA performance evaluation solutions with second-order hazard approximation, *Earthquake Engineering and Structural Dynamics*, 42(8):1171–1188.

Vamvatsikos, D. (2014). Accurate application and second-order improvement of the SAC/FEMA probabilistic formats for seismic performance assessment, *ASCE Journal of Structural Engineering*, 140(2):04013058.

Vamvatsikos, D., and Aschheim, M. A. (2016). Performance-based seismic design via Yield Frequency Spectra, *Earthquake Engineering and Structural Dynamics*, 45(11):1759–1778.

Vamvatsikos, D., and Aschheim, M. (2018). Yield frequency spectra application tools. http://users.ntua.gr/divamva/RCbook.html.

Vamvatsikos, D., and Cornell, C. A. (2002). Incremental dynamic analysis. *Earthquake Engineering and Structural Dynamics*, 31(3):491–514.

Vamvatsikos, D., and Cornell, C. A. (2006). Direct estimation of the seismic demand and capacity of oscillators with multi-linear static pushovers through Incremental Dynamic Analysis. *Earthquake Engineering and Structural Dynamics*, 35(9):1097–1117.

Vamvatsikos, D., and Fragiadakis, M. (2010). Incremental dynamic analysis for estimating seismic performance uncertainty and sensitivity, *Earthquake Engineering and Structural Dynamics*, 39(2):141–163.

# Chapter 14

# System modeling and analysis considerations

## 14.1 PURPOSE AND OBJECTIVES

This chapter helps the analyst decide how to model the structural system as a whole; the modeling of structural components is addressed in Chapter 17.

## 14.2 USE OF ANALYSIS FOR DESIGN

While we may one day be able to model reinforced concrete members of arbitrary geometry, configured and reinforced in complex ways, and subjected to arbitrary and potentially inelastic combinations of axial load ($P$), shear ($V$), moment ($M$), and torsion ($T$), the best models at present are limited to the relatively simple actions of axial load and moment, with minor amounts of shear, applied in the plane of the lateral force-resisting system. The accuracy of component models affects the accuracy of system response predictions. The most reliable analysis results are limited to a set of structural systems, such as those identified in ASCE-7 or EC-8, which subject components to loading histories for which responses can be modeled reasonably well.

Accuracy can be compromised in many ways—a more obvious case is where small differences in member strengths can affect the locations of inelastic response, such as a chain that has two weak links of similar strength (Figure 14.1). More generally, inaccuracies in any of the parameters defining a component load–deformation model or even the discretization of mass potentially can shift the computed locations and degree of inelastic deformation.

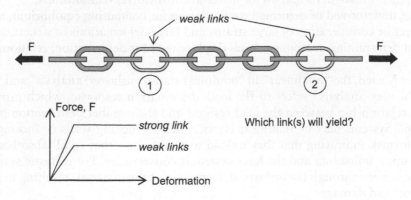

*Figure 14.1* Chain with two weak links. Deformations may take place in either or both of the weak links; for identical elasto-plastic force–deformation relationships the solution is indeterminate.

303

The design methods of Chapter 18 advance design from the linear elastic approaches typical of conventional practice to explicit consideration of distributed inelastic deformation demands through a plastic mechanism analysis, with lateral strength tailored to achieve the desired seismic performance using a simple and direct approach based on the stability of the yield displacement.

Precise performance predictions for frequent motions are hampered by uncertain conditions at the start of the response simulation. The initial state of stress, stiffness, and extent of cracking may be affected by prior earthquake and wind loading, acting on a structure that previously may have been exposed to significant live loads, temperature changes, shrinkage, differential settlement, and creep. Although these will not affect the mechanism strength (as determined by plastic mechanism analysis, assuming that sufficient deformation capacity is available), they make the initial state of stress and initial stiffness characteristics indeterminate. For rare motions, observation of damage in large magnitude events illustrates the complex degradation processes and mechanisms of resistance that can come into play at or near collapse, which can be difficult to predict and model *a priori*.

Thus, we aim for modeling accuracy that is sufficient for design, but cannot claim that the resulting performance predictions will be precise. Nevertheless, the resulting designs should more effectively achieve the stated performance objectives and should do so with greater economy, relative to conventional linear-based design approaches.

Finally, component stiffness must be represented as a function of reinforcement content (rather than taken as a fixed proportion of the gross stiffness) if the stability of the yield displacement is to be evident during the design process. Decisions about whether to represent beam-column joints and the use of rigid-end offsets, and other such modeling decisions, will influence the initial stiffness and yield displacement determined in analysis. The analyst should consider the potential impacts of such disparities on performance, in much the same way that potential errors in computed periods are (or should be) considered using conventional design approaches.

## 14.3 ANALYSIS CONSIDERATIONS

### 14.3.1 Nonlinearities represented in the analysis

Analyses may be classified as linear or nonlinear with respect to three attributes:

1. component load–deformation (or material constitutive) relationships,
2. use of undeformed or deformed configuration for computing equilibrium, and
3. neglect or consideration of large strains and large deformations of structural members when determining component load–displacement (or deformation) relationships.

As commonly used, the "nonlinear" in "nonlinear static (pushover) analysis" and "nonlinear response history analysis" refers to the load–deformation response, which provides for a nonlinear relationship between the load resisted and the member deformation (rotation or translation). Systems may be nonlinear elastic, such as rocking walls or footings (in their idealized forms), indicating that they unload to zero force—that is, all absorbed energy is recovered upon unloading and the force system is conservative. For inelastic systems, some energy is absorbed through the inelastic deformation of the material, resulting in permanent deformation and damage.

For relatively flexible structures, such as moment-resistant frames, it may be critically important to consider equilibrium in the deformed configuration. This gives rise to so-called

"*P-Δ*" effects, a second-order effect (or nonlinearity) in the equations of equilibrium. This second-order effect is generally quite detrimental, causing increases in peak displacements and possibly inducing collapse. *P-Δ* effects generally do not need to be simulated at the component level—typically, consideration of equilibrium in the deformed configuration at the global (or structural model) level is sufficient. Note that there can be second-order effects other than *P-Δ*, including some conditions that may be beneficial.

Finally, although rarely encountered in structural engineering, materials that undergo large strains and structural members that go through large deformations should be modeled considering large deformations. A Lagrange strain formulation can be used to avoid the error induced by the usual first-order (linear) approximations $\sin(\theta) = \tan(\theta) = \theta$ and $\cos(\theta) = 1$.

## 14.3.2 Information required for modeling response

The design approaches described in this book make use of a simple rigid-plastic model of component load–deformation response in the plastic mechanism analyses used for design, along with detailed modeling of component load–deformation responses for use in nonlinear response history analyses. Even for monotonic static (pushover) analyses, some components may unload as nonlinearity develops. Hence, in addition to knowledge of the component backbone curves, additional information including the unloading and reloading slopes and the parameterization of any degradation of strength and stiffness are needed for higher fidelity analyses. Of course, this information also is crucial for any nonlinear response history analyses.

## 14.3.3 Use of equivalent single degree-of-freedom systems

The design methods rely on experience indicating that peak displacements can be accurately estimated using equivalent single-degree-of-freedom (SDOF) systems. This raises two points:

1. Differences may exist in the modeling of damping in the relationships used to estimate the peak displacement response of the equivalent SDOF system and the damping model used in the nonlinear response history analysis of the MDOF model. Ordinarily, hysteretic losses dominate for systems with positive post-yield stiffness, and the precise value of viscous damping is of much less importance.
2. Although equivalent SDOF systems are relied upon conceptually for estimating peak displacement response, nonlinear static (pushover) analyses are not required in the design methods of Chapter 18. Instead, the yield displacement can be inferred based on the plastic mechanism strength and the fundamental period of vibration.

## 14.3.4 Simulated collapse modes and force-protected members

In Methods B and C of Chapter 18, design forces for force-protected members are obtained from the dynamic analyses. Typically, these actions can be modeled as linear elastic—this is acceptable where a high confidence that capacity exceeds elastic demand is desired. Allowing some nonlinearity for these actions (with very limited ductility) will reduce design forces. If inelastic behavior and, ultimately, failure, of force-protected components is not modeled, the computed response will be unrelated to these behaviors. Either inelastic response or failure must be prevented, by design using a suitably high confidence, or the possibility of such behaviors must be allowed for in the component load–deformation relationships.

## 14.4 SPATIAL COMPLEXITY OF MODEL

This section addresses the spatial complexity of the model—that is, in what circumstances must the entire three-dimensional structural system be represented, where would two-dimensional representations be sufficient, and is it necessary to model the gravity framing system? An example illustrating the discretization of a system into elements is provided.

### 14.4.1 Selection of components to represent

The mathematical model of the structural system used in analysis must represent all building components and mass that can significantly affect response, including nonstructural components such as infills that are rigidly connected to the structural system. Where design values (or deformation demands) are determined by nonlinear dynamic analysis (such as for force-protected actions in Methods B and C of Chapter 18), those components have to be represented in the model in order to determine the demands that must be sustained by these components.

### 14.4.2 Choice of two- and three-dimensional models

The entire height of the structural system should be represented in the model. Whether the areal extent of the system has to be represented depends on the presence of irregularities that can induce torsional response (rotation of the floors) in plan and how the lateral stiffness of the floor diaphragm is represented (Figure 14.2a).

As discussed in Appendix 2, plan irregularities may promote eccentricities between the center of mass and the center of stiffness at any level. Where the structural framing system and mass distribution follow a regular, symmetric, pattern, the centers of mass and lateral stiffness coincide. The lack of eccentricity between the centers of mass and stiffness indicates an absence of predicted torsional response. If a rigid diaphragm assumption applies, then a two-dimensional (planar) model of the framing will adequately represent the building response. Both gravity and lateral framing may be modeled (Figure 14.3); framing along each frame line can be linked together in the same plane using pin-ended bars to form a 2D model, in which each level of every frame experiences essentially identical lateral displacements (presuming these members have substantial axial stiffness).

Evaluation of diaphragm stiffness and modeling of diaphragms are addressed in Section 14.5.

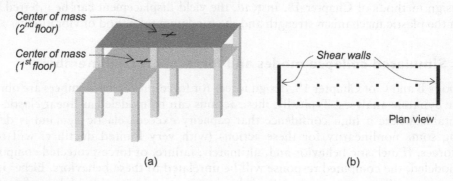

Center of mass
($2^{nd}$ floor)

Center of mass
($1^{st}$ floor)

Shear walls

Plan view

(a)                                                    (b)

*Figure 14.2* The need for three-dimensional models in cases of (a) complicated geometery and (b) diaphragm flexibility.

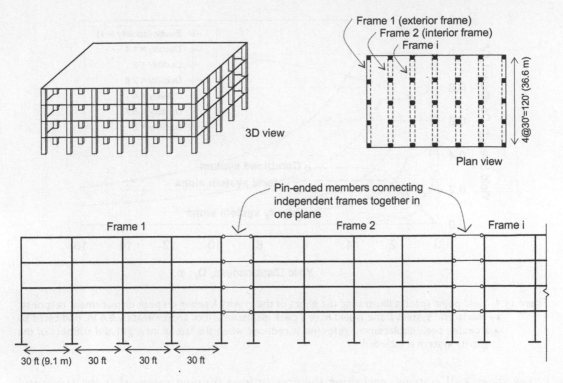

*Figure 14.3* Planar idealization for regular systems with rigid diaphragms.

## 14.4.3 Representation of gravity framing in the model

It is common for engineers to model only the lateral force-resisting system; floor and roof diaphragms are added where three-dimensional models are needed. However, gravity framing can have a beneficial effect on response; its overall influence can be readily seen using yield point spectra. Figure 14.4 shows (conceptually) the response of an elasto-plastic system that represents the lateral force-resisting system alone, with peak displacement of 5.6 in. represented by a circle. The lateral load–displacement behavior of the gravity framing, generally more flexible and weaker than the designated lateral force-resisting system, is also shown. Since these systems work together (in parallel), the combined system has the load–displacement response shown. The peak displacement of the trilinear curve was estimated on the basis of an approximate yield displacement of 2.1 in. and ductility demand of 2.5, resulting in a peak displacement of 5.25 in. The stiffness contributed by the gravity system resulted in a reduction in peak displacement.

Neglecting the stiffness and strength provided by the gravity load-resisting system is conservative for the design of moment-resisting frame systems. However, moment frames, in particular, may experience significant $P$-$\Delta$ effects, and the stiffness contributed by the gravity load system can significantly reduce or perhaps even eliminate the negative post-yield stiffness associated with $P$-$\Delta$ effects acting on the moment framing alone. The gravity framing also can help resist any tendency for weak-story mechanisms to develop, reducing concentrations of interstory drift and, thus, reducing the mean annual frequency of collapse.

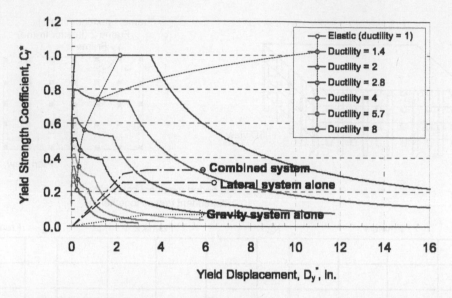

*Figure 14.4* Yield point spectra illustrating the effect of the gravity framing on peak displacement response. The lateral system alone would have a peak displacement of approximately 5.6 in. (indicated by a circle); peak displacement response is reduced when the lateral strength and stiffness of the gravity system is included.

For shear wall systems, neglecting the gravity load framing means that the structural model will not represent frame–wall interaction (also see Figure 17.14), and this will affect the computed wall shears and deformed shape.

### 14.4.4  Use of simplified models

Simpler models may be useful for capturing the salient aspects of response, particularly for preliminary design, for performing parameter studies, and for validating more complex models.

Figure 14.5 illustrates three structural systems on a continuum. In the case of the effectively uncoupled cantilever system (Figure 14.5a), response may be idealized using a so-called "stick" model, using a single element for each story that represents the composite flexural and shear behavior of the walls at each story. In the case of the strong beam–weak column system (which is ill-advised, due to its tendency to form weak-story mechanisms), a "stick" model may also be used (Figure 14.5c)—flexural and shear stiffnesses can be selected to represent the response of each floor relative to the floor below, while strengths can be selected to represent the capacity of each story under weak-story behavior.

For the intermediate case, a so-called "fishbone" model (also termed "generic frame" and "notional frame") may be useful (Nakashima et al., 2002; Lepage, 1997). As shown in Figure 14.5b, a "stick" model is augmented by beams at every floor level extending halfway toward the adjacent column, with a roller supporting each beam at midspan. The model allows plastic hinges to form in the beams or columns, and thus is capable of representing the development of mechanisms involving beam plastic hinges and weak stories.

The above models include a column element (along the "stick"). If column hinging is not anticipated, this column can be used directly as a leaning column for representing P-Δ effects, by simply applying the vertical gravity loads at each floor and instructing the software to

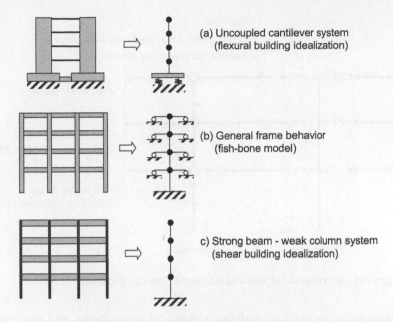

Figure 14.5 Simplified structural models for dynamic response simulation: (a) cantilevered shear walls; (b) moment-resistant frames; (c) weak-story behavior (adapted from FEMA-440, 2005).

include $P$-$\Delta$ effects in the analysis (computing equilibrium in the deformed configuration). Where the axial load can affect formation of plastic hinges in the column elements, an auxiliary leaning column should be added to the model.

These simplified models are useful for understanding lateral response with a minimum number of elements, thereby simplifying data management and reducing computational time. The simplifications come at the cost, however, of not obtaining detailed information on the response of individual components. For example, the development of axial forces in columns due to overturning effects is not represented.

## 14.4.5  Discretization in modeling structural system

Reinforced concrete structures can be considered to be composed of some combination of floor slabs, beams, girders, columns, walls, coupling beams, and interconnections (joints) among these components. Important sources of flexibility should be represented in the model, along with potential modes of failure[1] (due to axial load, moment, shear, torsion, and combinations thereof).

Design requirements influence potential modes of failure and hence locations where inelastic deformations may concentrate. Current design criteria for new moment-resistant frames should preclude or delay joint failure to large deformations, and should restrict inelasticity in the columns to minor flexural yielding. Beam design typically will be governed by the negative moment resistance required at the support; some nominal positive moment resistance will be provided, and gravity loads will be limited to avoid development of plastic hinges along the span. Thus, the discretization shown in Figure 14.6 would be suitable for most two-dimensional analyses (in the plane of the frame). Modeling details, such as

---

[1] *failure* can refer to exceedance of any performance limit.

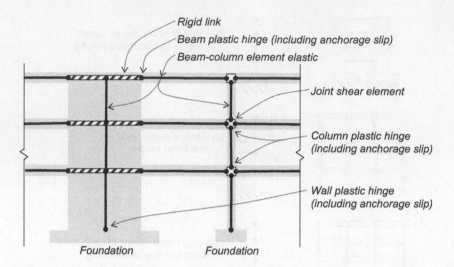

Figure 14.6 Suggested discretization of wall and moment frame components.

the choice of elements to use for modeling structural components and the specification of element properties, are discussed in Sections 17.7 and 17.8.

It should be noted that choices in discretization of the structure and modeling of structural components affect computed deformations, and thus the acceptability of a given design is a function not only of the modeling choices, but also of the modeling basis assumed when the acceptance criteria values were established.

## 14.5 FLOOR AND ROOF DIAPHRAGM CONSIDERATIONS

Floor and roof slabs participate in the lateral force-resisting system. When carrying lateral forces to the vertical components of the lateral force-resisting system, these components are termed "diaphragms." They serve two primary functions in lateral response: (i) they connect the distributed floor mass to the vertical components of the lateral force-resisting system, thereby loading these components as their mass is accelerated, and (ii) they interconnect the vertical components (of the gravity and lateral force-resisting systems), and thereby provide a means for transferring or redistributing load between these components.

Load transfer between the vertical components develops where the vertical elements have different relative stiffnesses, when the relative stiffnesses change during nonlinear response, and where discontinuities in these vertical elements require force transfer to maintain compatibility of deformations. A simple example is frame–wall interaction, where the diaphragm is called upon to transfer forces between the frames and walls acting along a single horizontal line. In addition, any eccentricities between centers of mass and stiffness lead to the development of torsional response, mobilized through diaphragm actions on the vertical components of the lateral force-resisting system.

The extent to which a floor or roof diaphragm deforms under lateral load, relative to deformation of the vertical components of the lateral force-resisting system, allows the diaphragm to be classified as rigid, stiff, or flexible. Diaphragm flexibility causes the floor mass to be less firmly coupled to the lateral system, lengthens periods of vibration, and increases deformations (including deformations of the gravity system). Because mass now participates in modes involving diaphragm response, the mass participating in the first mode may be reduced.

Reinforced concrete diaphragms of ordinary plan dimensions and without significant openings are normally quite stiff in comparison with moment-resistant framing, and would be modeled using a rigid diaphragm approximation (Figure 14.7d). Formally, a rigid diaphragm assumption is permitted for concrete slab diaphragms with span-to-depth ratios of 3 or less in structures that do not have horizontal irregularities (ASCE 7). In the rigid

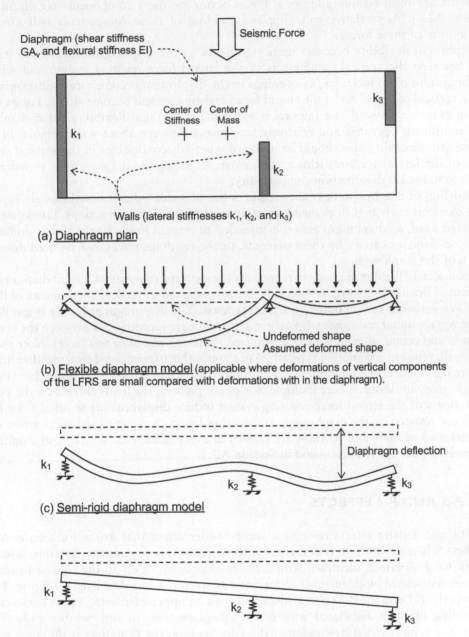

Figure 14.7 Illustration of diaphragm stiffness assumptions for a single-story building (adapted from PEER ATC 72-1): (a) plan; (b) flexible diaphragm; (c) semirigid diaphragm; (d) rigid diaphragm.

diaphragm approximation, the entire floor plate is assumed to move as a rigid body; any eccentricity between the center of mass and center of stiffness induces rotation of the diaphragm (Figure 14.7a). According to the rigid diaphragm approximation, lateral forces tributary to a floor level are distributed to the vertical components of the lateral force-resisting system in proportion to the lateral stiffnesses of these components; any torsional moments are resolved into additional forces acting on these components (as discussed in Section A2.4). Any relative softening and yielding of these components will affect the distribution of these forces.

Diaphragm flexibility becomes more significant as the diaphragm aspect ratio (ratio of span between the vertical components of the lateral force-resisting system and width of diaphragm, in plan) increases, as openings in the diaphragm become more substantial, and as the vertical components of the lateral force-resisting system become stiffer. For example, at one extreme, it would be important to model diaphragm flexibility for a diaphragm with significant openings and relatively long spans between shear walls (Figure 14.7b,c). Diaphragm flexibility also should be modeled where discontinuities in the vertical components of the lateral force-resisting system exist, requiring the diaphragm to transfer large forces to maintain deformation compatibility.

Modeling of diaphragm flexibility requires use of a three-dimensional model. Typically, finite elements such as shell elements are used to represent in-plane actions. Linear elements are often used, as diaphragm action is intended to remain in the linear elastic domain, or nearly so. Stresses carried by these elements, or the resultant forces, can be used directly in design of the diaphragm.

A completely flexible diaphragm represents the extreme opposite of a rigid diaphragm. In the case of flexible diaphragms, lateral forces are assigned to vertical components of the lateral force-resisting system based on tributary mass. The diaphragm generally is too flexible to develop torsional resistance where there are significant eccentricities between the center of stiffness and center of mass (since the required rotations would be too large). Note that the flexible diaphragm assumption typically is not applicable to reinforced concrete diaphragms.

Note that Figure 14.7 illustrates the continuum of rigid to flexible diaphragm behavior for a single-story building. Where multiple stories are present, the loads carried by the vertical components of the lateral force-resisting system induce displacements at other floor levels, which are resisted by the diaphragms at the other floors. A complex situation arises where eccentricities of mass and stiffness are present at every floor, even where rigid diaphragms can be assumed. See the discussion in Section A2.5.

## 14.6 P-Δ AND P-δ EFFECTS

$P$-Delta and $P$-delta effects refer to a second-order effect that arises for axially loaded members when equilibrium is evaluated in the deformed configuration. By convention, the $P$-delta (or $P$-$\delta$) effect, identified with a lower-case letter, refers to the change in internal moment associated with member deformations over the member length (Figure 14.8a), whereas the $P$-Delta (or $P$-$\Delta$) effect, identified with an upper-case letter, refers to the change in bending moment associated with relative displacements of the member ends (Figure 14.8c). For typical structures under earthquake loading, the $P$-$\Delta$ effect is the more significant of the two, owing to the reduction in lateral stiffness and increase in internal bending moments at a given displacement, in cases where $P$ acts to induce compression in the member. Since flexural hinges form at defined strengths, the moments arising due to $P$-$\Delta$ effects reduce the strength available to resist lateral loads. This reduction in strength takes place without causing a change in yield displacement.

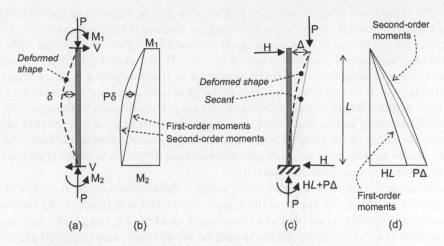

Figure 14.8 Internal bending moments of a column increased by P-δ and P-Δ effects: (a) column deformations give rise to (b) internal P-δ moments, and (c) column tip displacement gives rise to (d) internal P-Δ moments.

Figures 14.8c,d illustrate a cantilever column acted up on by a lateral load, H, and vertical load, P. The ordinary, first-order, moments determined using the underformed configuration, are shown in Figure 14.8d. The internal moment at y from the top is increased by P·δ, where P is positive in compression and δ is the deformation of the member, Figure 14.8b. Furthermore, moments at the secant are increased due to the relative displacements of the member ends by an amount P·Δ·y/h (Figure 14.8d, dotted line). The increase in moment causes a corresponding increase in curvature and deformation. Thus, the second-order moment, considering the deformed configuration including P·δ and P·Δ, is as given in Figure 14.8d.

In contrast to the example of Figure 14.8, the columns of multistory buildings rarely deform in single curvature. For columns deforming in double curvature (Figure 14.9), the displacements relative to the secant are substantially reduced. This reduces the moments due to deformations δ, and for this reason P-δ effects are small and usually are neglected. Where desired, P-δ effects can be represented either by using an element explicitly formulated to consider P-δ effects, or by subdividing the member into shorter segments and using an ordinary element for each segment. As many segments can be used as needed to reduce the deformations δ from the secants to an acceptably small amount. With this strategy, the P-δ analysis of a single member is approximated by a P-Δ analysis of the member divided into multiple smaller segments.

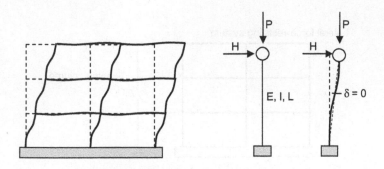

Figure 14.9 P-δ effects are reduced in columns deforming in double curvature.

While $P$-$\delta$ effects are often negligible, $P$-$\Delta$ effects can be significant for relatively flexible and/or heavily loaded structures responding to earthquake ground motions. Moment-resistant frames are recognized as particularly vulnerable to $P$-$\Delta$ effects—not only are the initial stiffness and lateral strength reduced by $P$-$\Delta$ effects, thereby leading to larger displacements, but the reduction in stiffness may lead to the development of a negative post-yield stiffness, which can heavily bias displacement response in one direction and lead to much larger displacements if not collapse. Further, for multistory buildings, $P$-$\Delta$ effects can induce weak-story mechanisms over one or several stories in a structure that otherwise would have formed a beam-hinging mechanism. This concentration of deformation in one or several stories greatly increases inelastic deformation demands in some components and reduces the overall system ductility capacity.

Where structural systems are represented using a three-dimensional model, $P$-$\Delta$ effects can be represented simply by applying the expected gravity loads tributary to each node.

Where gravity framing is included in a two-dimensional model, the gravity framing tributary to the lateral force-resisting system should be represented, and expected gravity loads tributary to each node are applied.

Where two-dimensional models are used without including gravity framing, $P$-$\Delta$ effects can be represented using a so-called "leaning column." Tributary gravity loads are applied to each node of the lateral force-resisting system, and the remaining gravity load at each floor level is applied to a node of the leaning column at that level (Figure 14.10). The leaning column may be composed of axially stiff members that are pinned at each end. An axially stiff, pin-ended, link connects the leaning column to the lateral force-resisting system at each floor level. Thus, lateral displacements experienced by the leaning column induce lateral forces, $F_{i,P\text{-}\Delta}$, in order to maintain equilibrium of the leaning column. These lateral forces, act together with the lateral forces, $F_i$, applied in a static pushover analysis or induced by earthquake acceleration, to increase the internal forces and moments resisted by components of the lateral force-resisting system. They appear naturally in the structural analysis results. However, the externally applied forces, $F_i$, are lower in value. The sum of these forces is the base shear, $V_b$.

With additional modeling effort, the leaning column of Figure 14.10 could be used to represent the lateral strength and stiffness provided by the gravity framing.

## 14.7 DAMPING

As described in Chapter 5, damping is necessary to represent the decay in amplitude evinced by objects in free vibration. Sources of damping in reinforced concrete buildings include cracking of concrete and slip of reinforcing steel, friction between cracked surfaces, relative

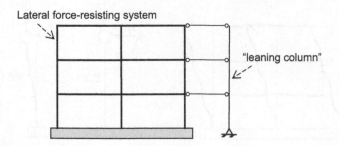

Figure 14.10 Use of a leaning column to represent $P$-$\Delta$ effects in two-dimensional (plane frame) analyses.

motion of interior partitions, exterior cladding, mechanical and electrical systems, deformation of soil supporting building foundations and basement walls, and radiation damping into supporting soils.

Various mathematical models are available for representing damping. Although it is not clear how to best model these sources of damping (e.g., as a function of velocity or displacement), the most common formulations, being Rayleigh and modal damping (Section 5.2.8), are functions of velocity. Because damping forces should be relatively minor, the form of the damping model has not been of great concern. However, unrealistically large damping forces can arise when conventional damping models are used in models of structures undergoing significant softening during nonlinear response; thus, the choice of damping model merits attention.

In Rayleigh damping, the damping matrix, $\mathbf{C}$, is determined as a linear combination of the mass, $\mathbf{M}$, and stiffness, $\mathbf{K}$, matrices. Thus, only two parameters may be selected, and these are usually selected to obtain specified levels of damping at two periods of vibration. As illustrated in Figure 5.6, mass-proportional damping leads to high damping at larger periods, while stiffness-proportional damping leads to high damping at the smaller periods (associated with higher modes).

As illustrated in Figure 14.11, mass-proportional damping acts on the velocities of the masses relative to the velocity of the base. Thus, rigid body motion (which may originate due to deformations of structural components at remote locations) is damped—as if the structure were moving through a viscous fluid. In contrast, stiffness proportional damping acts on the relative velocities of nodes that are interconnected by structural components. Both forms of damping impose forces at the nodes. These forces, along with those associated with mass acceleration, are not included in the evaluation of member forces, and so the member forces alone should not be expected to be in static equilibrium.

For structures responding inelastically, the standard Rayleigh damping model can result in unrealistically large damping forces as the tangent stiffnesses of some components reduce or may even become negative. This is apparent from a cursory inspection of Equation 5.10, but also can be seen in terms of critical damping—the reduction in tangent stiffnesses causes the instantaneous periods of vibration to lengthen, and the lengthening of periods shifts the percentage of critical damping that is present (Figure 5.6), resulting in much greater damping as the periods of vibration of the lowest mode(s) increase.

The PEER/ATC 72-1 (2010) report notes that the use of modal damping avoids large increases in damping due to period elongation and also limits damping in higher modes.

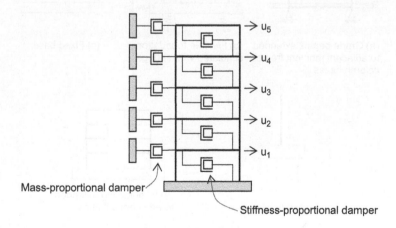

Figure 14.11 Mass- and stiffness-proportional damping. (Adapted from Charney, 2008.)

From a computational standpoint, there is no need to be restricted to the use of Rayleigh or modal damping because the equations of motions are integrated step-by-step in nonlinear response history analysis. Charney (2008) recommends the damping matrix be specified as a linear combination of the mass and tangent stiffness matrices (excluding the geometric stiffness), with coefficients determined to obtain a desired percentage of critical damping based on the current vibration periods (obtained from eigenvalue analysis at each time step). Ghannoum and Moehle (2012) report good correlation in simulating shake table responses using this approach, targeting 2% of critical damping in the first and second modes for a bare reinforced concrete frame. Values of 3%–5% are reasonable where partitions, cladding, and other nonstructural systems are present.

Finally, the user is cautioned to avoid so-called spurious damping, which may arise when very stiff elements reach a critical force level and open suddenly (such as those that occur when gap elements open or rigid-plastic hinges rotate), leading to large translational or rotational velocities in combination with the pre-event stiffness value. One strategy is to assign zero damping to such elements; another is to use explicit damping elements.

## 14.8 FOUNDATIONS AND SOIL–STRUCTURE INTERACTION

While the bases of columns and walls are often assumed to be fixed against translation and rotation (e.g., ASCE 7 Section 12.7.1), fixed-base models underestimate the period of vibration and yield displacement, and may alter the deformation demands computed in plastic hinges above the point of assumed fixity. A more realistic assessment of base fixity would consider the stiffness of interconnected grade beams and restraint provided by spread footings and pile caps (Figure 14.12).

Response of the soil–foundation–structural system is affected by the flexibility of the supporting soil and foundation. So-called soil–structure interaction (SSI) is considered in terms of three distinct effects:

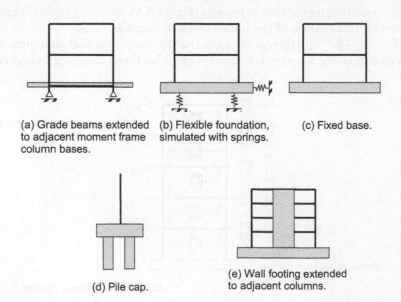

(a) Grade beams extended to adjacent moment frame column bases.

(b) Flexible foundation, simulated with springs.

(c) Fixed base.

(d) Pile cap.

(e) Wall footing extended to adjacent columns.

*Figure 14.12* (a)–(e) Illustrate options for modeling restraint at the bases of columns and walls.

- The flexibility of the foundations and supporting soil.
- Changes to input ground motions (filtering of ground motions due to kinematic effects), in which the stiffness of foundation elements causes their motion to deviate from free-field motion. Examples include base slab averaging, wave scattering, and embedment effects.
- Dissipation of energy through radiation damping and hysteretic losses within the supporting soil. This is termed inertial interaction, owing to the forces developed by the overlying structure, which induce movement of the foundation relative to the free field. The relative movement of the foundation and the supporting soil is associated with radiation of energy away from the foundation and hysteretic losses within the soil.

Generally, representation of the flexibility and damping of the soil is more important for stiffer structural systems on softer sites. This flexibility can cause an increase in the fundamental period and yield displacement of the system, and can alter the distribution of demands (e.g., interstory drifts and story shears), sequence of hinge formation, and yield mechanism, relative to a fixed based case. NIST GCR 11-917-14 (2011) recommends detailed evaluation of SSI for $h > 0.1 V_s \cdot T$, where $h = 2/3$ of the height of the structure, $T =$ fundamental period of vibration of the fixed base model, and $V_s =$ the shear wave velocity of the soil.

Input ground motions are reduced by kinematic interaction effects (base slab averaging and embedment). The reduction is greatest for short-period structures, and is negligible for periods greater than around 1 s. The reductions become significant for equivalent foundation widths, defined as $(\text{length} \times \text{width})^{0.5}$, exceeding about 40 m and for structures having two or more subterranean levels (NIST GCR 11-917-14, 2011).

SSI effects are generally modeled using either direct or substructure approaches. Direct approaches represent the soil and structure in the same model, using finite elements to represent the soil and transmitting boundaries where the soil elements terminate. Although the direct approaches have the potential of being the most accurate, they are more complex to implement.

In the substructure approach, the SSI problem is treated in three separate parts, involving an evaluation of the free-field motion, determination of foundation input motion, and characterization of the stiffness and damping characteristics at the soil–foundation interface. These parts are linked together through a response analysis of the combined structure–foundation system excited by the foundation input motions. The foundation input motions and characteristics of the soil–foundation interface are affected by nonlinearity in the soil and structure; in order to treat these separately, response is often approximated using equivalent-linear systems (in which the nonlinear system is replaced by a linear system having reduced stiffness and increased damping).

## 14.9 MODEL DEVELOPMENT AND VALIDATION

Modeling of structures for nonlinear analysis presents the analyst with a far richer set of features and options relative to those available in linear models. The analyst is advised to gradually build up his/her skill set and confidence in modeling when starting out with any software program. One should begin with small, simple models for which hand calculations or published results provide a means of validating the modeling and analysis, and then develop more complex models, exploring and validating the effects of different modeling options in separate investigations.

To begin, one can establish by hand calculation the expected period of vibration of a SDOF system and can use comparisons to known dynamic response results to validate the modeling and analysis results obtained with a new or unfamiliar program, first for a linear elastic oscillator and then for a yielding oscillator. This should be considered a necessary first step, as errors can exist even in mature, commercially available software.

In learning to use a nonlinear analysis program, one should pay particular attention to the use of unfamiliar options, solution convergence criteria, and error tracking (e.g., errors in energy balance). Computational experiments should be run using small models to understand the significance of these features and their influence on results. Experimentation with different damping models and validation of the use of leaning columns for handling $P$-$\Delta$ effects can be important. As well, discretizations of members that will respond inelastically, such as the segmentation of fiber models, may require validation against experimental results.

A number of checks should be done on any structural model to ferret out possible problems. A modal analysis should identify mode shapes, periods, modal participation factors, mass participation factors, total mass, and fraction of critical damping that are consistent with expectations. A basic pushover analysis should reveal a sensible pattern of plastic hinging, initial stiffness (period), yield displacement, and strength. For symmetric structures, deflected shapes under symmetric loading (e.g., gravity loads) should maintain the symmetry, and a pushover analysis in the negative direction should reveal a capacity curve that mirrors that obtained in the positive direction. A dynamic analysis with acceleration scaled down to assure linear elastic response should produce a peak roof displacement close to that estimated using an equivalent SDOF system.

## REFERENCES

Charney, F. (2008). Unintended consequences of modeling damping in structures, *ASCE Journal of Structural Engineering*, 134(4):581–592.

FEMA-440 (2005). Improvement of nonlinear static seismic analysis procedures, Project ATC-55, Applied Technology Council, Redwood City, California, p. 390.

Ghannoum, W., and Moehle, J. P. (2012). Dynamic collapse analysis of a concrete frame sustaining column axial failures, *ACI Structural Journal*, American Concrete Institute, 403–412.

Lepage, A. (1997). A method for drift control in earthquake-resistant design of reinforced concrete building structures, thesis submitted to the Graduate College of the University of Illinois, Urbana, IL, June 1997.

Nakashima, M., Ogawa, K., and Inoue, K. (2002). Generic frame model for simulation of earthquake responses of steel moment frames, *Earthquake Engineering and Structural Dynamics*, 31(3): 671–692.

NIST GCR 11-917-14 (2011). Soil-foundation-structure interaction for building structures, National Institute of Standards and Technology, Gaithersburg, Maryland, December, p. 270.

PEER/ATC 72-1 (2010). Modeling and acceptance criteria for seismic design and analysis of tall buildings, Pacific Earthquake Engineering Center, Richmond, California, and Applied Technology Council, Redwood City, California, October, p. 242.

# Section IV

# Reinforced Concrete Systems

# Chapter 15

# Component proportioning and design based on ACI 318

## 15.1 PURPOSE AND OBJECTIVES

This chapter summarizes key requirements and recommendations for design and detailing of reinforced concrete components in regions of high seismic hazard, as put forward by the American Concrete Institute and applied in American practice. These requirements may be used to detail structural components in conjunction with the system design approaches described elsewhere in this book (summarized in Chapter 18).

## 15.2 INTRODUCTION

Existing requirements and recommendations have developed with the benefit of many decades of experience and are both practicable and well-suited for obtaining ductile response. This body of knowledge is reflected in the seismic provisions in Chapter 18 of ACI 318-14 (2015). These design and detailing requirements are well-suited for use with performance objectives comparable to or more stringent than current code requirements. Only the most important requirements and recommendations, applicable to the most ductile building systems (special moment frames and special structural walls) are discussed in this chapter. Since design requirements and practice will continue to evolve, current information should always be sought.

Although ACI 318 Chapter 18 generally is intended for use with the results of linear analyses, some concepts of plastic mechanism analysis are embodied in the provisions. A capacity design approach is implemented in specific instances to establish required beam shear strengths and column flexural strengths, using the probable flexural strengths of the plastic hinge regions. The probable strengths are associated with the reinforcement provided at the critical section of each plastic hinge, and are established assuming that the reinforcement has yield strength equal to 1.25 times the specified minimum yield strength.

The capacity design approach is well-suited where there is little mass in between the plastic hinges. Applied to the simple analogy that a chain is only as strong as its weakest link (Section 11.2), the required strengths of the stronger, brittle links can be established based on the probable strengths of the weak, ductile links. An underappreciated shortcoming of this analogy is that where substantial mass is located between the links, the dynamic forces within the elastic links may significantly exceed the probable strength of the weak link. Thus, while the shears in a beam of a moment-resistant frame can be estimated with good accuracy based on the strength of the plastic hinges at the ends of the beam (e.g., Figure 15.4), the shears in a shear wall are driven higher than would be expected based on a simple mechanism analysis, due to higher mode excitations of the floor masses. For this reason, we prefer to use the results of dynamic analyses to establish the forces in force-protected members (Chapter 18), while nominal and reliable strengths can be assessed as described in this chapter.

## 15.3 STRENGTH REDUCTION FACTORS

In a deterministic analysis, uncertainties in strength are considered by use of fixed modification factors; in the case of ACI 318, the reliable strength is determined as $\phi$ times the nominal strength. The nominal strength is based on the specified material properties ($f_y$ and $f_c'$); $\phi$ is a strength reduction factor.

Strength reduction factors for flexure and axial load in columns and structural walls having rectilinear transverse reinforcement range between 0.65 and 0.90, based on the calculated net tensile strain in the cross section. The net tensile strain, $\varepsilon_t$, is the strain at the centroid of the extreme tension reinforcement when the compressive strain in the section reaches 0.003, in an analysis in which plane sections are assumed to remain plane and concrete is assumed to have no tensile capacity. Specifically, for $\varepsilon_t \geq 0.005$, $\phi = 0.90$, and for $\varepsilon_t \leq \varepsilon_y$ (taken equal to 0.002 for Grade 60 reinforcement), $\phi = 0.65$. For intermediate values of $\varepsilon_t$, $\phi$ is linearly interpolated between 0.65 and 0.90.

For shear in columns and in structural walls where sufficient shear strength is provided to develop the flexural strength, as well as for shear friction, $\phi = 0.75$. For shear in beam-column joints and in diagonally reinforced coupling beams, $\phi = 0.85$. For shear in diaphragms, $\phi$ cannot exceed the smallest value of $\phi$ used for shear in the design of the primary lateral force-resisting system (LFRS), which ordinarily will be 0.75.

In contrast to the use of fixed modification factors, uncertainties in strength can be considered explicitly. As discussed in Chapters 13 and 18, the probability of failure (demand exceeding performance limit) can be made acceptably low at a sufficiently high confidence level.

## 15.4 SPECIFIED MATERIALS

Primary longitudinal reinforcement in components of special moment frame and special structural wall systems is subject to restrictions on both minimum and maximum yield strength and minimum tensile (ultimate) strength. The limit on the maximum yield strength is intended to limit the shear and bond stresses that develop upon yielding of the longitudinal reinforcement; the minimum tensile strength ensures sufficient strain hardening is present to cause a spread of plasticity, which is needed to develop inelastic deformation capacity. ASTM A706 Grade 60 deformed reinforcement is acceptable. This grade of steel has yield strength between 60,000 and 78,000 psi (420 and 540 MPa) and ultimate strength at least 80,000 psi (550 MPa) and 1.25 times the actual yield strength. Grade 80 ASTM A706 reinforcement is not allowed, due to a lack of data substantiating its use with existing code provisions.

ASTM A615 reinforcement is generally more widely available than A706 Grade 60 reinforcement. ASTM A615 Grade 40 and 60 reinforcement is permitted as long as it satisfies additional requirements: (i) Grade 60 reinforcement must have actual yield strength not in excess of the specified yield strength plus 18,000 psi (12 MPa), (ii) the actual tensile strength must be at least 1.25 times the actual yield strength, and (iii) minimum elongation over an 8 in. (200 mm) length must be at least 14% for No. 3 through No. 6 bars, 12% for No. 7 through No. 11 bars, and 10% for Nos. 14 and 18 bars.

Less restrictive requirements pertain to all other reinforcement (e.g., transverse reinforcement and that used in gravity framing (not part of the LFRS)). Steel reinforcement satisfying ASTM A615 (carbon steel) Grade 40 or 60, A706 (low alloy steel), A955 (stainless steel), or A996 (axle steel and type R rail steel) is permitted. ASTM A1035 (low-carbon chromium steel) reinforcement is permitted for transverse reinforcement and can be useful for reducing congestion in columns of special moment frames, particularly those using higher strength concretes. The value of $f_{yt}$ used to determine the amount of transverse reinforcement used for confinement is capped at 100,000 psi (690 MPa).

In the design of transverse reinforcement for resisting shear, the value of $f_y$ or $f_{yt}$ used in design calculations is capped at 60,000 psi (414 MPa) in order to limit the width of shear cracks.

While deformed bars (in which deformations are rolled into the surface of the bar while it is hot) are generally required, plain bars and plain wire are permitted for spiral reinforcement used as lateral reinforcement for compression members, for resisting torsion, or for confining reinforcement at splices.

Concrete must have specified compressive strength of at least 3,000 psi (21 MPa). For specified compressive strengths above 10,000 psi (69 MPa), the value of $\sqrt{f_c'}$ used in determining shear strength and anchorage or development length is capped at 100 psi (0.69 MPa); this limit does not apply to beam-column joint shear strength or the development of bars at beam-column joints.

Special requirements constrain the use of lightweight concrete in ductile components of LFRSs. While lightweight concrete may be used throughout the structure to reduce weight, data to support its use within components of the LFRS is quite limited. Applicable requirements are not discussed herein.

Expected material strengths will generally exceed the nominal specified values. For reference, expected strengths are discussed in Section 17.4.

## 15.5 BEAMS OF SPECIAL MOMENT-RESISTANT FRAMES

### 15.5.1 Beam width and depth

The effective depth of the beam (distance from extreme compression fiber to centroid of tension reinforcement) must not exceed one-quarter of the clear span ($d < l_n/4$). Beam width, $b_w$, must be at least the larger of 0.3 times the height and 10 in. (250 mm). While beams typically are narrower than the columns that support them, wider beams are allowed, as long as the extension of the beam on either side of the column is at most the smaller of the width of the column and ¾ of the overall depth of the column. Beams deeper than 36 in. (900 mm) must be provided with skin reinforcement along the sides of the beam, to control crack widths.

### 15.5.2 Beam longitudinal reinforcement

Restrictions on longitudinal reinforcement over the beam span are illustrated in Figure 15.1. Reinforcement area anywhere along the span cannot exceed $0.025(b_wd)$. Ordinarily, the beam section will be governed by the negative moment at the face of support; to limit joint shear stresses the top reinforcement may need to be restricted to perhaps 1 to 1.5% of $b_wd$. The top reinforcement ordinarily is concentrated in the beam web and runs within the column cage at beam-column joints, although all reinforcement that can be developed within the effective width of the beam can be used to resist negative moment. Bottom reinforcement at the face of support must provide a nominal flexural strength equal to at least one half of the negative moment strength. Elsewhere along the span, reinforcement

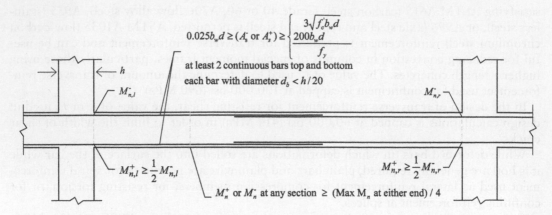

$$0.025b_w d \geq (A_s^- \text{ or } A_s^+) \geq \begin{cases} \dfrac{3\sqrt{f_c'}b_w d}{f_y} \\ 200b_w d \end{cases}$$

at least 2 continuous bars top and bottom
each bar with diameter $d_b < h/20$

$M_{n,l}^-$     $M_{n,r}^-$

$M_{n,l}^+ \geq \dfrac{1}{2} M_{n,l}^-$     $M_{n,r}^+ \geq \dfrac{1}{2} M_{n,r}^-$

$M_n^+$ or $M_n^-$ at any section $\geq$ (Max $M_n$ at either end) / 4

Figure 15.1 Longitudinal beam reinforcement in special moment frames.

must provide flexural strength at least equal to one-quarter of the largest nominal flexural strength along the span.

Lap splices are generally located near the beam midspan. Lap splices are not allowed within beam-column joints and within potential plastic hinge zones (within 2h of the face of support), or where analysis indicates flexural yielding is expected.

### 15.5.2.1 Member proportioning

The largest moments along the beam span are usually the negative moments at the support, resulting from the combination of gravity loads and lateral loads. We can use maximum reinforcement ratios to determine beam dimensions based on the negative moment at the support. For rapid initial proportioning, the plastic flexural strength can be estimated considering just the top reinforcement, as

$$M_p^- \approx \rho b d^2 f_{ye}\left(1 - 0.59\rho\frac{f_{ye}}{f_c'}\right) \tag{15.1}$$

where $\rho$ ($=A_{s,top}/(b_w d)$) is close to the maximum that can be accommodated in the beam. While $A_{s,top}$ is capped at $0.025b_w d$, a reasonable target is around $\rho = 0.01-0.015$. Equation 15.1 may be solved for $d$ assuming that $b = (1/2-2/3)d$.

Following ACI 318, positive moment reinforcement will normally be proportioned to provide just over half the negative moment strength. Because the floor slab creates a flange at the top of the beam, the depth of the compression zone for positive moment is very small, and a bottom steel area approximately equal to half the top steel area will normally suffice. A quick estimate of $M_p^+$ can be obtained considering only the bottom steel, as

$$M_p^+ \approx A_{s,bot} f_{ye} d\left(1 - 0.59\frac{A_{s,bot}f_{ye}}{b_{eff}df_c'}\right) \tag{15.2}$$

where $b_{eff}$ is the effective beam width (defined in Section 15.5.3 below). This equation does not consider the presence of compression reinforcement and assumes that the compression block does not extend below the flange into the web.

### 15.5.3  Beam probable flexural strength

The probable moment strength, as defined by ACI, is needed for determining the required beam shear strength, minimum column flexural strengths, and possibly column shear strength (depending on the approach taken). The probable moment strength, $M_{pr}$, is determined as the ACI nominal flexural strength considering the area of longitudinal steel provided and assuming yield strength equal to 1.25 times the specified minimum yield strength. Longitudinal steel that can be developed within the effective width should be included when determining required minimum column flexural strengths (Section 15.6.2). For Tee beams (having a flange on both sides of the web), the width of the flange overhanging on each side of the web is limited to the smaller of 1/8 of the beam span length, eight slab thicknesses, and half the distance to the next web. For edge beams, where a slab extends on only one side of the web, the width of the overhanging flange from the face of the web is limited to the smaller of 1/12 of the beam span length, six slab thicknesses, and half the distance to the next web.

### 15.5.4  Beam transverse reinforcement configuration

Transverse reinforcement consisting of hoops or stirrups is required over the entire length of a special moment frame beam. As shown in Figure 15.2, hoop reinforcement may be constructed of one or more closed hoops, and may be supplemented by crossties. Hoops

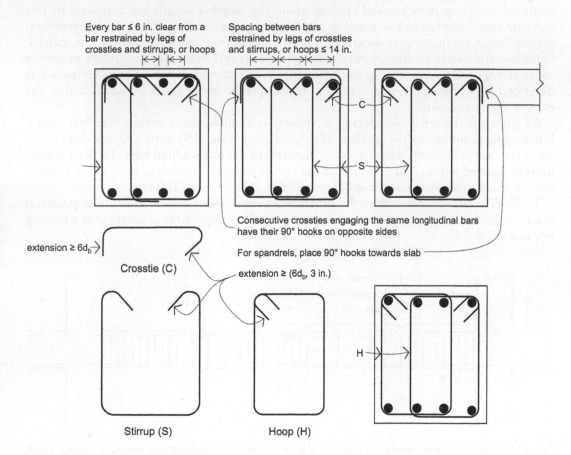

*Figure 15.2* Configuration requirements for transverse reinforcement in beams of special moment frames.

and crossties, where present, provide confinement, help to resist or delay buckling of longitudinal reinforcement, contribute to shear strength, and improve bond between reinforcing bars and concrete. For these reasons, hoops are required within potential plastic hinge zones and where longitudinal bars are spliced using lap splices. In contrast, stirrups are sufficient where only shear resistance is needed.

As an alternative to the use of closed hoops, an open stirrup supplemented by a crosstie having a 135° hook at one end and a 90° hook at the opposite end may be used; this may be preferable over a closed hoop to simplify the construction process. Where used, consecutive crossties engaging the same longitudinal bars must have their 90° hooks alternated end for end, except in a spandrel condition.

In locations where hoops are required, such as within potential plastic hinge regions, the hoops, possibly supplemented with crossties, must be configured so that the centers of longitudinal bars supported in the corner of a hoop or crosstie are no more than 14 in. (350 mm) apart; alternate longitudinal bars do not need to be supported in the corner of a hoop or crosstie, provided that the clear distance to a supported bar does not exceed 6 in. (150 mm). These requirements provide lateral support to help restrain longitudinal bars from buckling when acting in compression.

### 15.5.5 Beam transverse reinforcement spacing

The previous section addressed the configuration of transverse reinforcement; cross-sectional area requirements and spacing along the member length are described in this section. Hoop reinforcement must be provided at a maximum spacing within potential plastic hinge locations and locations where longitudinal reinforcement is lap spliced. Sufficient transverse reinforcement (as hoops or stirrups) must also be present to satisfy shear strength requirements. In the following, it is assumed that loading is restricted as described in Section 11.4, resulting in potential plastic hinge zones being located at the ends of the beam span.

As shown in Figure 15.3, spacing of transverse reinforcement within potential plastic hinge zones is limited to the smallest of $d/4$, $6d_b$, and 6 in. (150 mm) within a distance of 2h of the face of support, where $d_b$ is the diameter of the longitudinal bars. The first stirrup must be located within 2 in. (50 mm) of the face of support. Spacing of hoops over longitudinal bar lap slices must not exceed the smaller of $d/4$ and 4 in. (100 mm).

Transverse reinforcement must be sufficient to resist shear forces. Outside of potential plastic hinges and lap splices, transverse reinforcement is required to resist shear at a spacing not to exceed $d/2$.

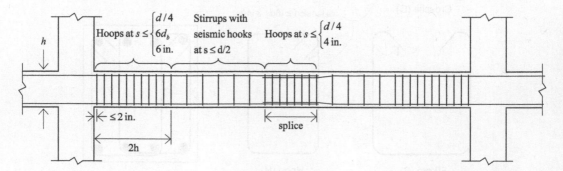

Figure 15.3 Spacing requirements for transverse beam reinforcement in special moment frames. Note: stirrups and crossties may be used in place of hoops.

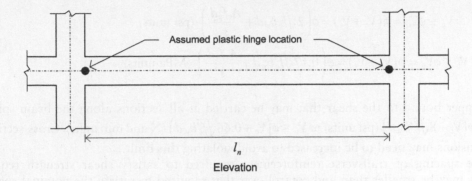

Elevation

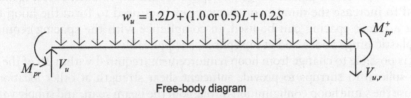

Free-body diagram

Figure 15.4 Evaluation of $V_u$.

The factored shear demand is determined assuming that the beam probable moments develop under seismic loading, in conjunction with the factored gravity load of $1.2D + (1.0$ or $0.5)L + 0.2S$[1] (Figure 15.4). For typical uniform gravity loads, summing moments about $V_{u,r}$ results in

$$V_u = \frac{M_{pr}^- + M_{pr}^+}{l_n} + \frac{w_u l_n}{2} \tag{15.3}$$

where $M_{pr}^-$ and $M_{pr}^+$ are the probable moment strengths defined earlier. While it is possible to evaluate the shear at a distance d away from the face of support, the gravity load contribution is typically small, resulting in a minor reduction in shear at this location. Thus, Equation 15.3 is usually used to evaluate $V_u$, for use over the entire span.

For beams having factored axial compressive force not exceeding $A_g f_c'/20$, the shear strength provided by hoops within the potential plastic hinge zone (within 2h from the face of support) is determined assuming $V_c = 0$. Thus

$$V_u \leq \phi V_n = \phi V_s = \phi \frac{A_v f_y d}{s} \tag{15.4}$$

is used in this case to determine the spacing of hoops for shear strength within potential plastic hinge zones. The strength reduction factor, $\phi$, for shear is 0.75.

Elsewhere along the span (away from potential plastic hinge zones), $V_c$ may be taken equal to $V_c = 2\sqrt{f_c'}b_w d$ (psi units[2]) or $V_c = 0.17\sqrt{f_c'}b_w d$ (N and mm units[3]). Thus, in these locations the transverse reinforcement is provided at a spacing that ensures:

---

[1] Based on the appropriate load combination in ASCE 7.
[2] Equations that indicate "pound and inch units" or "psi units" have embedded units, for which $f_c'$ and $f_y$ are specified in psi, dimensions in inches, and shear strength in pounds.
[3] Equations that indicate "N and mm units" have embedded units, for which $f_c'$ and $f_y$ are specified in N/mm² (=MPa), dimensions in mm, and shear strength in N.

$$V_u \leq \phi V_n = \phi\left(V_c + V_s\right) = \phi\left(2\sqrt{f_c'}b_w d + \frac{A_v f_y d}{s}\right) \text{ (psi units)}$$

$$V_u \leq \phi V_n = \phi\left(V_c + V_s\right) = \phi\left(0.17\sqrt{f_c'}b_w d + \frac{A_v f_y d}{s}\right) \text{ (MPa units)}$$

(15.5)

An upper bound to the shear that may be carried at all sections along the beam span is $V_u \leq \phi\left(V_c + 8\sqrt{f_c'}b_w d\right)$ (psi units) or $V_u \leq \phi\left(V_c + 0.66\sqrt{f_c'}b_w d\right)$ (N and mm units). Cross-sectional dimensions may need to be increased to avoid violating this limit.

The spacing of transverse reinforcement required to satisfy shear strength requirements may be smaller than and control over that required to satisfy the nominal spacing requirements at potential plastic hinge and splice locations. In such cases, it may be more economical to increase the number or diameter of bars used to form the hoop to increase $A_v$, so that a larger spacing can be used, in compliance with the spacing required within potential plastic hinge zones and at lap splices, as applicable.

While it is possible to change from hoop reinforcement (required within 2h of the beam ends and at lap splices) to stirrups to provide sufficient shear strength at other locations, it is far simpler to use the same hoop configuration throughout the beam span, and simply vary the spacing of hoops. This simplifies placement and inspection, and avoids potential placement errors.

## 15.6 COLUMNS OF SPECIAL MOMENT-RESISTANT FRAMES

Because columns typically support multiple floors, it is essential that their integrity be maintained. As discussed in Section 15.6.2, columns of special moment-resistant frames are proportioned anticipating the potential for some flexural yielding to occur at their ends. Because it can be difficult to integrate circular section columns with beam framing, rectangular section columns are more commonly used with moment-resistant frames and are emphasized herein. Confinement can be provided more effectively to circular section columns, however, by means of spiral reinforcement or circular hoops.

### 15.6.1 Section dimensions and reinforcement limits

To be considered a column, the aspect ratio of rectangular section columns must be within the limits $0.4 < h/b < 2.5$, where $h$ = depth of section and $b$ = width of section. While 12 in. (300 mm) is the minimum cross-sectional dimension of a special moment frame column, a more practical minimum is 16 in. (400 mm). In addition, the column depth must be at least 20 times the diameter of the beam longitudinal reinforcement, as discussed in Section 15.7.1.1.

The column longitudinal reinforcement ratio, $\rho_g$ (=$A_{s,total}/A_{gross}$) is limited to 1%–6%. However, to avoid heavily congested splice locations, $\rho_g$ generally is kept below 3%. Where higher reinforcement ratios are needed, the use of mechanical couplers should be considered to avoid congestion at splice locations.

#### 15.6.1.1 Column proportioning

Several considerations affect column proportioning:

1. Column flexural strength, curvature capacity, and curvature ductility are functions of axial force, $P$. Curvature ductility reduces substantially as the axial force approaches and exceeds the balanced axial load, $P_{bal}$ (Figure 15.5). While confinement provided by

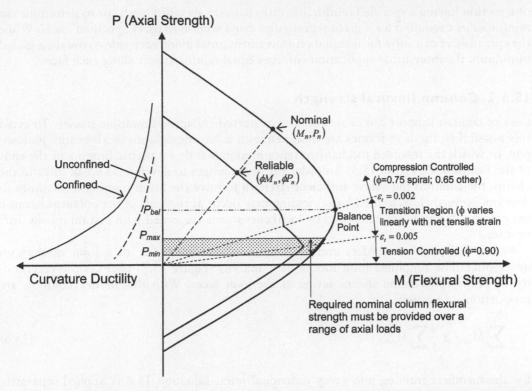

*Figure 15.5* Axial force–moment interaction—influence of axial force on flexural strength and curvature ductility capacity; application of $\phi$ to determine reliable strengths.

transverse reinforcement increases the curvature ductility of compression-controlled members ($P > P_{bal}$), the best practice is to keep axial force levels well below the balance point, because yielding of tension reinforcement generally will be more ductile than crushing of even well-confined concrete. The axial load ratio, $P/A_g f_c'$, provides a nondimensional index that can be used to ensure that $P < P_{bal}$, possibly requiring a change in cross-sectional area.

2. Requirements on minimum column flexural strength relative to the beam flexural strengths in ACI 318 Section 18.7.3, described in Section 15.6.2.

3. The beam-column joint requirements of ACI 318 Section 18.8 assert that the depth of the column be not less than 20 times the diameter of the largest beam longitudinal bar that frames into the column.

4. ACI Committee 352 (ACI 352R, 2002) recommends that the beam depth be at least 20 times the diameter of the column bars. Since the beam depth will usually be determined prior to sizing the columns, the effect of this recommendation typically is to cap the diameter of the column longitudinal reinforcement at $h/20$, where $h$ = depth of the beam.

A fast approach for preliminary proportioning of columns for prescribed values of $P$ and $M$ (in the case of uniaxial loading), or $P$, $M_x$, and $M_y$ (in the case of biaxial loading), is described by Aschheim et al. (2008). The spreadsheet tool[4] can be used to determine the dimensions

---

[4] The spreadsheet is available from http://users.ntua.gr/divamva/RCbook/aci.flexure.xls.

of a section having a specified reinforcing ratio to carry specified loads, or to determine the reinforcement required for a given rectangular cross section to carry specified loads. While the spreadsheet can solve for unequal reinforcement areas along each side to obtain a global minimum, the more usual application enforces equal reinforcement along each face.

### 15.6.2 Column flexural strength

Loss of column support can cause collapse or partial collapse of multiple stories. To avoid this possibility, moment frames are designed with a "strong-column weak-beam" philosophy, in which the intended mechanism relies predominantly on plastic hinging at the ends of the beams (e.g., Figure 15.6). In order for beam hinges to develop (as weak links in the chain), the columns must have sufficient strength to force the beam-hinging mechanism to develop. Nevertheless, limited local yielding may occur at the ends of the columns because the column demands are influenced by local curvatures associated with instantaneous drift profiles.

ACI 318 explicitly considers and aims to avoid the possibility of a joint mechanism developing. The simplified joint mechanism analysis (Figure 15.7) neglects the eccentricity of beam and column shears acting at the joint faces. With this model, columns are proportioned to satisfy

$$\sum M_{n,col} \geq \frac{6}{5} \sum M_{n,bm}$$
(15.6)

for the members framing into every individual joint. Equation 15.6 is applied separately in the vertical plane for each horizontal direction that contains special moment-resistant framing. Equation 15.6 must be satisfied in each horizontal direction for lateral sway in both "positive" and "negative" directions. Nominal beam flexural strengths should be determined considering reinforcement that is adequately anchored to develop $f_y$ within the portion of the slab that comprises the effective flange width.

Variations in axial load affect the nominal column flexural strength, $M_{n,col}$. The inequality of Equation 15.6 must be satisfied over the applicable range of axial loads for the load combinations that include earthquake effects. For columns with $P < P_b$, the critical condition (lowest flexural strength) may occur for the lowest possible axial load (Figure 15.5) that is present for the direction of loading that generates the beam and column moments.

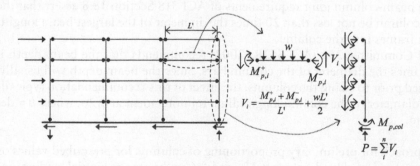

*Figure 15.6* Lateral forces induce axial forces in columns due to overturning effects. Free-body diagrams may be used to determine beam shears and maximum (or minimum) column axial forces, assuming the formation of beam plastic hinges (black circles).

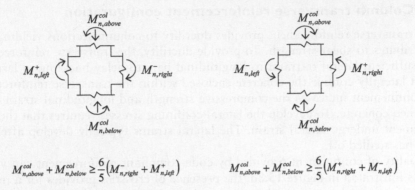

$$M_{n,above}^{col} + M_{n,below}^{col} \geq \frac{6}{5}\left(M_{n,right}^{-} + M_{n,left}^{+}\right) \qquad M_{n,above}^{col} + M_{n,below}^{col} \geq \frac{6}{5}\left(M_{n,right}^{+} + M_{n,left}^{-}\right)$$

*Figure 15.7* Required column flexural strength (Equation 15.6) must be satisfied over the range of column axial load, and for beam moments acting in either direction.

In typical practice, axial forces are established using the ASCE-7 load combinations, and thus are determined by superposition of elastic demands and design earthquake forces, divided by the code $R$ factor. Lateral loads typically have little effect on column axial forces at interior locations, but the end columns (at either end of the frame) are prone to develop significant overturning axial forces due to the lateral loads. Beam-hinging mechanisms limit the beam shears that can develop. Thus, the column axial force is limited by the shears developed in the beam-hinging mechanisms plus the column self-weight and any loads bearing directly on the column. It is possible that not all beams over the height of the building develop beam-hinging mechanisms at the same instant, and thus this approach may provide an upper bound to the column axial force. In this analysis, it is apparent that the column axial forces are a function of the actual reinforcement provided, which may result in a flexural strength that exceeds that required under a design load combination. The applicable ASCE 7 load combinations may be used to ensure code compliance, or simple mechanism analyses (including gravity loads) may be used to establish limits on the column axial force. Either approach should be adequate provided that column plastic rotations (computed in nonlinear dynamic analyses) are sufficiently small.

ACI 318 requires that if Equation 15.6 cannot be satisfied at a given location, the strength and stiffness of the noncompliant columns be excluded from the special moment frame. These columns must satisfy the requirements for columns not designated as part of the LFRS, also known also as "gravity columns" (Section 15.13).

As mentioned earlier, building drift profiles induce curvatures and moments in the columns that are not accounted for in Equation 15.6 (Bondy, 1996). Dynamic analyses have indicated that the 6/5 coefficient in Equation 15.6 is not sufficient to protect columns from yielding. Thus, columns must be detailed anticipating the possibility of some inelastic action near the joints, as well as at the base of the building where column plastic hinges are required to develop the mechanism of Figure 15.6. Suggestions have been made to design columns using a coefficient of 1.5 (rather than 6/5 = 1.2); nonlinear dynamic analyses may be used to ensure column plastic hinge rotations are within acceptable limits.

Axial forces in columns supporting the roof level are generally low. In the past, Equation 15.6 was not applied to column flexural strengths at this location; instead, plastic hinges were allowed to form in the columns just below the roof level, rather than requiring they be located in the roof beams. In such cases, the column plastic hinge region was detailed for ductile response over a length of twice the column depth.

### 15.6.3 Column transverse reinforcement configuration

Column transverse reinforcement provides ductility to column sections yielding in flexure and contributes to shear strength. To provide ductility, the transverse reinforcement must provide sufficient lateral restraint to longitudinal bars to delay buckling to larger strains, and must laterally confine the concrete enclosed within the transverse reinforcement. The lateral confinement increases the compressive strength and longitudinal strain capacity of the confined concrete. To develop the lateral confining stresses requires that the transverse reinforcement undergo lateral strain. The lateral strains typically develop after the cover concrete has spalled off.

The quality of confinement provided by code-compliant reinforcement can vary significantly. As illustrated in Figure 15.8d, the presence of crossties provides for a more ductile response than a single perimeter hoop even where the amount of transverse reinforcement is held approximately constant.

As shown in Figure 15.9, hoops, possibly in conjunction with crossties, must be provided so that all longitudinal bars are enclosed by transverse hoops and every corner longitudinal bar is laterally supported by the corner of a hoop. Beginning with the 2014 edition of ACI 318, the horizontal spacing, $x_i$, is measured between the centers of longitudinal bars that are laterally supported by the corner of a hoop or crosstie. While it is preferable to support every longitudinal bar in the corner of a hoop or crosstie, this is required in columns only where $P_u > 0.3A_g f_c'$ or $f_c' > 10,000$ psi (69 MPa). In this case, the largest of the $x_i$ values may not exceed 8 in. (200 mm). For other cases, this limit is relaxed to 14 in. (350 mm); every other longitudinal bar must have such lateral support, and the unsupported longitudinal bar must be within 6 in. (150 mm) clear of a supported longitudinal bar.

Where crossties are used, one or both ends must terminate with a 135° hook—to facilitate placement of the crosstie, a 90° hook may be used at one end of the crosstie, even though the leg of the 90° hook ceases to be anchored when the cover concrete spalls off. Where crossties having a 90° hook at one end are used, adjacent crossties must be alternated end for end to create a checkerboard pattern of 90° hooks.

### 15.6.4 Column transverse reinforcement spacing requirements

The spacing of transverse reinforcement along the length of the column must be adequate to separately satisfy requirements for confinement, shear strength, and lap splices—the most stringent of these controls at any location.

#### 15.6.4.1 Confinement in potential plastic hinge zones and at lap splices

In anticipation that some degree of flexural yielding may take place at the ends of columns of moment-resistant frames, a smaller spacing of transverse reinforcement is used within a distance $l_o$ of the column ends and at any other place where plastic hinging may be anticipated, in order to confine concrete and delay buckling of longitudinal reinforcement to larger deformations. The length $l_o$ must be at least the largest of the following:

$$l_o \geq \begin{array}{|l|} \text{depth of the member} \\ \text{clear span/6} \\ \text{18 in. (450 mm)} \end{array}$$  (15.7)

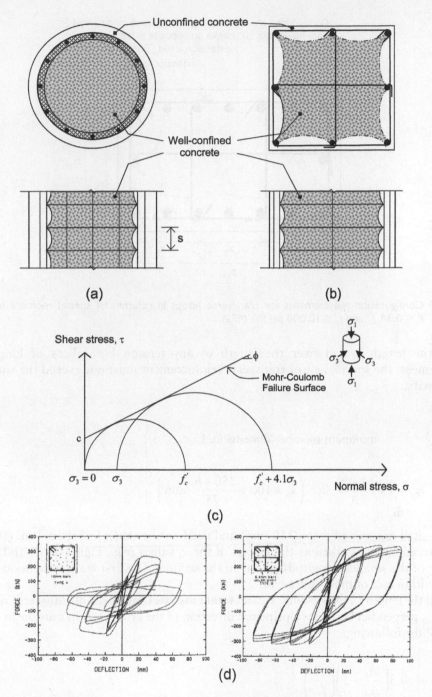

*Figure 15.8* Confinement on concrete columns: (a) restraint (confinement) provided by circular hoops and spirals; (b) restraint provided by rectangular hoops and crossties; (c) use of Mohr–Coulomb failure surface to illustrate effects of confining stress ($\sigma_3$) on compressive strength ($\sigma_1$); (d) effect of transverse reinforcement on lateral load–lateral displacement response—Specimens U3 and U6 had nearly the same spacing of transverse reinforcement and amount (as measured by $A_{st}f_{yt}/s$), but the presence of crossties in Specimen U6 allowed this column to maintain lateral strength to a much larger ductility level ($A_{st}$ is total transverse steel area within spacing s and $f_{yt}$ is the steel yield strength of transverse steel). (Adapted from Ozcebe and Saatcioglu, 1987.)

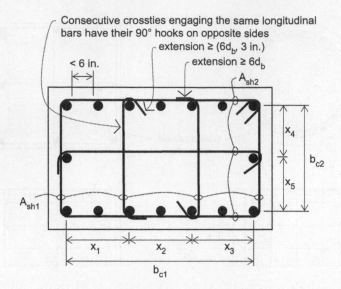

*Figure 15.9* Configuration requirements for transverse hoops in columns of special moment frames for $P_u \leq 0.3A_g f_c'$ and $f_c' \leq 10,000$ psi (69 MPa).

Within the length $l_o$ and over the length of any tension lap splices of longitudinal reinforcement, the spacing, $s_1$, of transverse reinforcement must not exceed the smallest of the following:

$$s_1 \leq \left|\begin{array}{c} \text{minimum member dimension}/4 \\ 6d_b \\ s_o = 4 + \dfrac{14-h_x}{3} \text{ in. } \left(s_o = 100 + \dfrac{350-h_x}{75}\text{ mm}\right) \end{array}\right| \tag{15.8}$$

where $s_o$ shall not exceed 6 in. (150 mm) and need not be taken less than 4 in. (100 mm). In the preceding, $h_x$ is taken as the largest of the $x_i$ values (e.g., Figure 15.9) and $d_b$ = the diameter of the smallest longitudinal bar in the section. The first transverse hoop must be located within $s_1/2$ of the column end.

Within the length $l_o$, the spacing, $s_1$, and total cross-sectional area of transverse reinforcement, $A_{sh}$, perpendicular to each principal direction of the cross section must be at least the largest of the following:

$$\frac{A_{sh}}{s_1 b_c} \geq \left|\begin{array}{c} 0.3\left(\dfrac{A_g}{A_{ch}} - 1\right)\left(\dfrac{f_c'}{f_{yt}}\right) \\ 0.09\left(\dfrac{f_c'}{f_{yt}}\right) \end{array}\right| \tag{15.9}$$

where $b_c$ = core dimension perpendicular to the tie legs that constitute $A_{sh}$ (see Figure 15.9), $A_{sh}$ = the total cross-sectional area of transverse reinforcement within spacing $s$ and perpendicular to $b_c$, $A_g$ = area of gross section, $A_{ch}$ = area of core, measured to outside edges of transverse reinforcement (neglecting cover), and $f_c'$ and $f_{yt}$ are the specified compressive strength of concrete and yield strength of transverse reinforcement, respectively.

Equation 15.9 must be satisfied in both principal directions of the cross section. Thus, with reference to Figure 15.9, $b_{c1}$ and $A_{sh1}$ would be used when the equation is applied in one direction, and $b_{c2}$ and $A_{sh2}$ would be used when applied in the orthogonal direction.

For columns where $P_u > 0.3 A_g f_c'$ or $f_c' > 10,000$ psi (69 MPa), a further restriction on the spacing, $s_1$, and total cross-sectional area of transverse reinforcement, $A_{sh}$, applies in addition to the requirements of Equation 15.9:

$$\frac{A_{sh}}{s b_c} \geq 0.2 k_f k_n \frac{P_u}{f_{yt} A_{ch}} \tag{15.10}$$

where $k_f = \dfrac{f_c'}{25,000} + 0.6 \geq 1.0$ (psi units) $\left( k_f = \dfrac{f_c'}{175} + 0.6 \geq 1.0 \text{ N and mm units} \right)$ and $k_n = \dfrac{n_1}{n_1 - 2}$, where $n_1$ = the number of longitudinal bars that are laterally supported by the corner of hoops or seismic hooks (hooks having a bend of at least 135° and an extension of at least $6d_b$ and not less than 3 in. (75 mm)).

No. 3, 4, and 5 bars (9.5, 12.7, and 15.9 mm diameter) are typically used for transverse reinforcement. No. 4 (12.7 mm) or larger must be used for hoops and crossties enclosing No. 11 (35.8 mm) or larger longitudinal bars.

Where column plastic hinges are required as part of the intended mechanism, such as at the base of the ground floor columns (e.g., Figure 15.6), it is recommended that transverse reinforcement be extended over a distance of $1.5l_o$ (rather than $l_o$) and the configuration of transverse reinforcement should be improved over what is required in ACI 318. Crossties with 90° hooks should be avoided altogether in this region; instead, crossties with 135° hooks at both ends or rectangular hoops are preferred.

Lap splices of longitudinal reinforcement are required to be located within the central half of the member and should not be located within potential plastic hinge regions. Lap splices are designed as tension lap splices.

A typical transverse bar set requires three layers of transverse reinforcement—the outer hoop sandwiched between crossties in one direction and in the other direction. An alternative to using separate hoops and crossties is to use equipment that can form relatively complex transverse reinforcement patterns out of a single bar. With either approach, the overall thickness of the hoop set can result in relatively small clear spacing within $l_o$. In some cases, reducing $h_x$ in Equation 15.8 by using additional crossties can allow for a larger clear spacing.

### 15.6.4.2 Transverse reinforcement outside of potential plastic hinge zones

Less stringent requirements apply to the spacing of transverse reinforcement outside potential plastic hinge regions defined by $l_o$ and away from lap splices. As illustrated in Figure 15.10, the spacing of transverse reinforcement in these locations, designated by the term $s_2$, may be greater than $s_1$. The spacing, $s_2$, may not exceed the smaller of 6 in. (150 mm) and $6d_b$. While the configuration requirements of Section 15.6.3 do not apply here, customary practice is to continue the configuration of transverse reinforcement used within $l_o$ throughout the length of the column, but at a greater spacing, $s_2$.

### 15.6.4.3 Transverse reinforcement for shear strength

The provisions of ACI 318 contain alternative approaches to evaluate column shear demands. The basic objective is to ensure a low probability of shear failure under seismic loading.

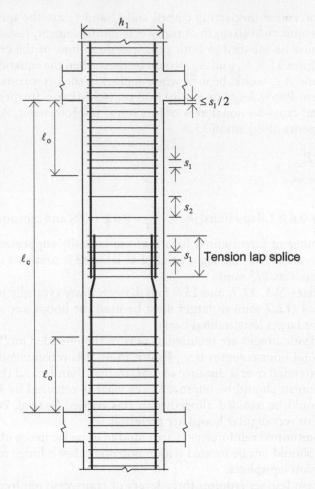

*Figure 15.10* Spacing of transverse hoops for confinement of columns of special moment-resistant frames.

In essence, this can be established by ensuring with suitably high confidence that the shear demand, $V_e$, obtained from nonlinear response history analyses is sufficiently less than the nominal shear strength, $V_n$. We prefer this approach, which is incorporated into Methods B and C of Chapter 18.

According to ACI 318, $V_e$ should not be taken lower than the shear obtained in a static analysis under the design forces (Figure 15.11a). NIST GCR 8-917-1 (2008) recommends amplifying these shears to account for overstrength, given as the average ratio of probable beam flexural strengths, $M_{pr}$, and beam design moments, $M_u$. We suggest a simpler alternative is to use the shears associated with the mechanism analysis for the design base shear at yield, $V_y$, amplified (perhaps by 10%) to account for the use of centerline dimensions in the mechanism analysis, and recognizing that expected material properties (rather than $f'_c$ and $1.25f_y$) are used in the analysis and member design. However, both of these approaches neglect contributions from higher modes to column curvature (and hence, bending moment).

An upper bound to $V_e$, recognized by ACI 318, is to assume that each column develops its probable flexural strength, $M_{pr}$, at each end. Then, by statics, $V_e = (M_{pr,i} + M_{pr,j})/l_n$, where $l_n$ = column clear length, and $M_{pr,i}$ and $M_{pr,j}$ are the probable flexural strengths at ends $i$ and $j$, respectively, taken at the largest values that can occur over the range of expected axial loads (Figure 15.11b).

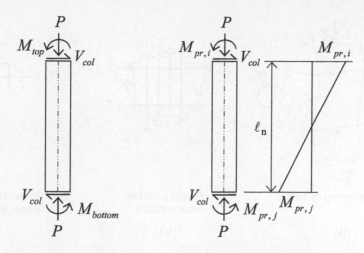

a) From analysis results    b) Plastic hinges at column ends

*Figure 15.11* Some approaches for evaluating column shear demand: (a) use of analysis results; (b) assumption of plastic hinging at the column ends.

The nominal shear strength is calculated as $V_n = V_c + V_s$. Within the length $l_o$, $V_c$ is taken equal to zero if the axial load is low ($P_u < A_g f_c'/20$) and the shear demand is primarily due to earthquake loading ($V_u > V_e/2$).

## 15.7 BEAM-COLUMN JOINTS IN SPECIAL MOMENT-RESISTANT FRAMES

Beam-column joints must satisfy dimensional requirements and shear strength requirements; attention to detailing is needed to ensure that reinforcement congestion is manageable during construction. Potential congestion problems arise because beam longitudinal reinforcement must pass through the joint within the column cage (the longitudinal column bars secured by transverse reinforcement).

The beams and column framing into a beam-column joint define both the dimensions of the joint and the moments and shears acting on the joint. It assumed that the beams will develop their probable moment strengths at the column faces, as illustrated in Figure 15.12. At a critical beam cross section (e.g., at the left side of the joint, or similarly, at the right side), horizontal equilibrium requires that the sum of the compression forces (in the concrete, $C_c$, and in the steel, $C_s$) equal the force carried by the reinforcement in tension ($T_s$), presuming the beam carries negligible axial force. Consequently, a beam longitudinal reinforcing bar extending through the joint is simultaneously pulled in tension on one side of the joint and pushed in compression in the same direction on the other side. This combined pulling and pushing requires that large bond stresses develop between the bar and the adjacent concrete to prevent the bar from pulling through. The force in the bar changes rapidly over its length. Strains associated with tension in the bar require that the bar elongate, leading to an anchorage slip contribution to beam deformation that is discussed in Section 17.8.1.2.

The column reinforcement in well-designed frames is expected to experience lower stresses than the beam reinforcement, due to (i) the application of capacity design principles that encourage the development of plastic hinges in the beams and (ii) the presence

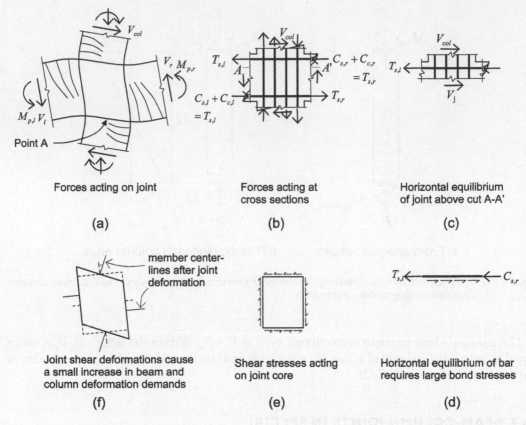

*Figure 15.12* Idealized behavior of beam-column joints: (a) forces acting on joint; (b) forces acting at cross sections; (c) horizontal equilibrium; (d) bond stress; (e) shear stress on joint core; (f) joint deformations increase deformation demands on beams and/or columns.

of compressive axial force in the columns, which increases the neutral axis depth and thus reduces tensile strains (or may induce compressive strains) in the longitudinal reinforcement. The column longitudinal reinforcement, together with the transverse reinforcement within the joint and the beams framing into the sides of the joint, help to confine the joint and therefore can contribute to the shear strength of the joint.

## 15.7.1 Joint proportioning

Joint shear strength is determined by the joint dimensions and the configuration of beam and column framing into the joint.

### 15.7.1.1 Joint dimensions

Bond stresses are limited in ACI 318 by ensuring the column dimension parallel to the beam reinforcement is at least 20 times the diameter of the largest beam longitudinal reinforcement.[5] In practice, this restriction usually has the effect of limiting the diameter of longitudinal bars rather than increasing the column depth.

---

[5] In addition, ACI Committee 352 recommends that beams have depth at least 20 times the largest diameter of column longitudinal reinforcement.

### 15.7.1.2 Joint shear strength

Shear stresses within the joint are limited by comparing the shear demand to a nominal shear capacity, $V_n$. With reference to Figure 15.12c, the horizontal shear force within the joint, $V_j$, can be calculated using the free-body diagram shown, as

$$V_j = T_{s,r} + T_{s,l} - V_{col} \tag{15.11}$$

where the tension forces in the beam reinforcement, $T_{s,r}$ and $T_{s,l}$, are based on probable strengths (equal to $1.25A_{s,r}f_y$ and $1.25A_{s,l}f_y$, respectively). ACI Committee 352 (ACI 352R, 2002) recommends that the longitudinal reinforcement within the effective width be used to establish the steel areas, while only the primary beam reinforcement must be considered according to ACI 318. For corner and exterior beam-column joints without transverse beams framing into the joint, the effective width is equal to the beam width plus a distance on each side of the beam equal to the length of the column cross section measured parallel to the beam generating the shear.

The shear in the column, $V_{col}$, is normally a relatively minor term in Equation 15.11. It can safely be neglected ($V_{col} = 0$) for simplicity. The conventional approach to evaluating $V_{col}$ assumes that column inflection points occur at midheight of the stories, where the story height is given by $h_i$. In this case, considering a free-body diagram of a column extending between the midheights of adjacent stories and including the moments and shears acting at the left and right faces of the beam-column joint (Figure 15.12a), we can sum moments about the midheight location of the lower story to determine $V_{col}$ as:

$$V_{col} = \frac{M_{p,l} + M_{p,r} + (V_l + V_r)(h/2)}{h_i} \tag{15.12}$$

where $h$ = column depth and $h_i$ = average of story heights above and below the joint. Of course, this equilibrium approach to evaluating $V_{col}$ can be adapted to other circumstances, such as where beams frame in from one side only or where the foundation stiffness would cause the inflection point locations to change.

It is required that the joint shear demand, $V_j$, be less than the reliable joint shear capacity, $\phi V_n$.

$$V_j \le \phi V_n = \phi\left(\gamma \sqrt{f_c'} A_j\right) \tag{15.13}$$

where $\phi = 0.85$ for joint shear, $\gamma$ is a strength coefficient that depends on the configuration of the joint (Figure 15.11), and $A_j$ is the effective joint area.

Note that the values of $\gamma$ shown in Figure 15.13 for joints at the roof level are recommended by ACI Committee 352, while ACI 318 allows those values established for typical floor levels to be used at the roof level.

The effective joint area, $A_j$ ordinarily is equal to the cross-sectional area of the column. Where a beam frames into a wider column, $A_j$ is defined to equal $w \cdot h$, where the effective joint width, $w$, is equal to the smaller of $b + h$ and $b + 2x$ and is bounded by the column width, as illustrated in Figure 15.14. Note that where joint shears are generated from beams running in both orthogonal directions, $A_j$ is determined separately for each direction of loading.

### 15.7.2 Transverse reinforcement

Transverse reinforcement must be provided within the joint to obtain the joint shear strengths given in Figure 15.13, and to improve the anchorage of beam and column reinforcement

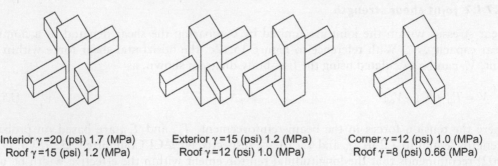

Interior γ =20 (psi) 1.7 (MPa)  Exterior γ =15 (psi) 1.2 (MPa)  Corner γ =12 (psi) 1.0 (MPa)
Roof γ =15 (psi) 1.2 (MPa)   Roof γ =12 (psi) 1.0 (MPa)   Roof γ =8 (psi) 0.66 (MPa)

*Figure 15.13* Interior, exterior, and corner joints, and associated values of the joint shear strength coefficient, γ. Note that (i) beams having width less than ¾ of the width of the column face at the joint are ignored when establishing γ, (ii) beam stubs terminating at least one beam depth away from the joint face may be included in establishing γ, and (iii) for roof configurations, the column need not extend beyond the top of the joint. Values provided are recommended by ACI 352R (2002).

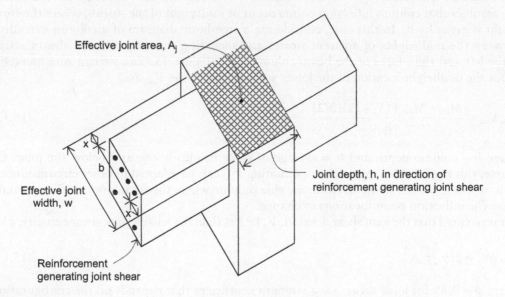

*Figure 15.14* Definition of effective joint area. (Adapted from NIST GCR 16-917-40, 2016.)

within the joint. In general, the transverse reinforcement provided in the adjacent column end regions is continued throughout the joint, but possibly using a modified configuration and spacing.

In the special case of an interior joint, in which beams frame into all four sides of the joint and each beam width is at least ¾ the column width, the transverse reinforcement within the depth of the shallowest beam framing into the joint may be reduced to one half of that required in the column end regions (Equations 15.9 and 15.10 as applicable); the spacing of the transverse reinforcement required by Equation 15.8 is permitted to be increased to a maximum of 6 in. (150 mm). For other cases, the spacing of transverse reinforcement within the column end regions is continued throughout the joint.

Figure 15.15 illustrates plan and elevation views of typical reinforcement passing through an interior joint. To avoid spatial interference of the beam reinforcement where orthogonal

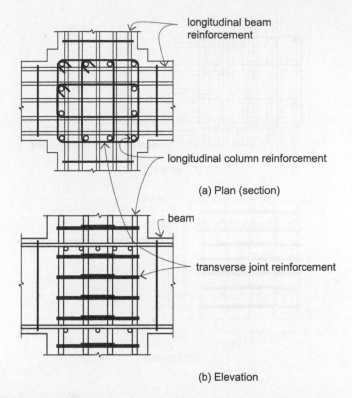

longitudinal beam
reinforcement

longitudinal column reinforcement

(a) Plan (section)

beam

transverse joint reinforcement

(b) Elevation

*Figure 15.15* (a) Plan and (b) elevation views illustrating transverse reinforcement at an interior beam-column joint. (Adapted from NIST GCR 16-917-40, 2016.)

beams intersect, beam longitudinal reinforcement can be defined to run in each direction within defined vertical intervals. Since the beam bars must be contained within the column cage, the column width will ordinarily be a few inches wider than the beams.

Figure 15.16 illustrates a typical exterior joint. Unlike the case of interior joints, the longitudinal beam reinforcement must be fully anchored within the joint core, and this ordinarily requires the use of hooked bars or headed reinforcement. The transverse reinforcement within the joint core provides confinement and improves the anchorage of the reinforcement within the joint. The visualization of internal compression struts reacting against the 90° hooks and concrete in compression at the face of the joint helps to explain why the bars are hooked as shown. Tails of hooked beam bars must be located within 2 in. (50 mm) of the edge of the confined core; the critical section for development of the hooks begins at the opposite edge of the confined core.

Figure 15.17 illustrates joints at the top of a building, or anywhere that a column terminates. In addition to transverse joint reinforcement, vertical joint reinforcement is recommended by ACI Committee 352 (ACI 352R, 2002) to help restrain vertical dilation of the joint. Column longitudinal reinforcement terminates with standard hooks anchored within the joint; horizontal reinforcement may terminate using standard hooks or using headed bars. Again, the visualization of internal compression struts helps to explain why the bars are hooked as shown.

To facilitate construction, the vertical transverse reinforcement can be composed of a combination of inverted U-bars and crossties, as shown in Figure 15.17c. The vertical bars should have development length less than the depth to the top beam reinforcement. The

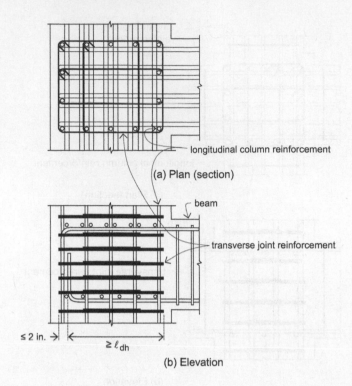

longitudinal column reinforcement

(a) Plan (section)

beam

transverse joint reinforcement

≤ 2 in.  ≥ $\ell_{dh}$

(b) Elevation

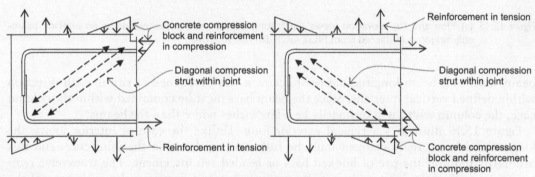

Reinforcement in tension

Concrete compression
block and reinforcement
in compression

Diagonal compression
strut within joint

Reinforcement in tension

Diagonal compression
strut within joint

Concrete compression
block and reinforcement
in compression

(c) Elevation showing development of internal diagonal compression struts

*Figure 15.16* (a) Plan and (b) elevation views illustrating typical reinforcement at an exterior beam-column; (c) development of internal diagonal compression struts within beam-column joint. (Portions adapted from NIST GCR 16-917-40, 2016.)

center-to-center spacing of the vertical transverse reinforcement may not exceed the smallest of the following:

$$s \leq \begin{vmatrix} \text{beam width/4} \\ 6d_b \\ 6 \text{ in. } (150 \text{ mm}) \end{vmatrix} \qquad (15.14)$$

where $d_b$ is the diameter of the beam longitudinal reinforcement. The configuration of vertical transverse reinforcement should provide restraint to each corner and alternate

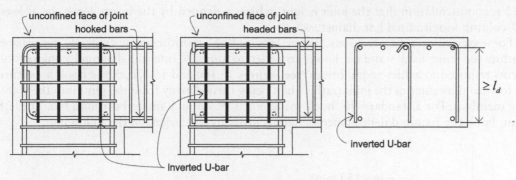

(a) Elevation (hooked beam bars)    (b) Elevation (headed beam bars)    (c) Vertical section through joint
(showing inverted U-bars)

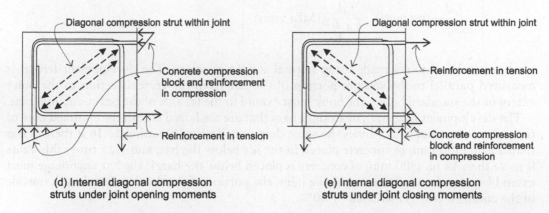

(d) Internal diagonal compression        (e) Internal diagonal compression
struts under joint opening moments        struts under joint closing moments

*Figure 15.17* (a) and (b) are elevation views and (c) is a cross-sectional view of a joint consisting of a discontinued column and beams framing into the joint from one or two orthogonal directions. Inverted U-bars are used to provide vertical confinement to the joint, in addition to the confinement provided by transverse reinforcement. Orthogonal beam bars (shown dashed) may be present. The development of internal diagonal compression struts under (d) opening moments and (e) closing moments. (Portions adapted from NIST GCR 16-917-40, 2016.)

longitudinal beam bar. The total cross-sectional area of vertical transverse reinforcement, $A_{sh}$, within a horizontal spacing s, should be at least

$$A_{sh} = 0.09\left(\frac{sh_e f_c'}{f_{yh}}\right) \tag{15.15}$$

where $h_e$ = core dimension of tied column, outside to outside edge of transverse reinforcement bars, perpendicular to the transverse reinforcement area $A_{sh}$ being designed.

### 15.7.3 Development of longitudinal reinforcement

Beam and column longitudinal reinforcement must be adequately anchored in order to be capable of developing the bending moments determined from analysis.

For interior joints, beam and column longitudinal reinforcement typically extends without interruption through the joint and is anchored beyond the joint. (This gives rise to the ACI 318 requirement that the joint depth, which is defined by the column dimension parallel to the beam reinforcement, be at least 20-beam longitudinal bar diameters, and the ACI

352 recommendation that the joint height, which is defined by the beam depth, be at least 20-column longitudinal bar diameters.)

For exterior and corner joints, beam longitudinal reinforcement typically terminates within the joint as a standard hook. In order to support internal diagonal compression struts required to achieve equilibrium (see Figures 15.16c and 15.17d,e), the hook must turn in toward the center of the joint (rather than being turned away from the joint into the framing member). For a standard 90° hook using No. 3 (9.5 mm diameter) through No. 11 (35.8 mm) bars, the required development length is given by the largest of the following:

$$l_{dh} = \left| \begin{array}{c} 6 \text{ in. (150 mm)} \\ 8d_b \\ \dfrac{f_y d_b}{65\sqrt{f_c'}} \text{ (psi units)} \quad \dfrac{f_y d_b}{5.4\sqrt{f_c'}} \text{ (MPa units)} \end{array} \right| \qquad (15.16)$$

for beam-column joints made using normal weight concrete. The development length is measured parallel to the straight portion of the bar from the critical section to the outer extent of the standard hook. The hook must extend to the far side of the beam-column joint.

The development length, $l_d$, of straight bars that are anchored within the confined core of the joint is set equal to 2.5 times the value determined using Equation 15.16 if there is less than 12 in. (300 mm) of concrete placed in the lift below the bar, and 3.25 times this value if more than 12 in. (300 mm) of concrete is placed below the bar. If the bar anchorage must extend beyond the confined core of the joint, the portion of the development length outside of the confined core is increased by 60%.

## 15.8 SPECIAL STRUCTURAL WALLS AND COUPLED WALLS

Reinforced concrete walls provide a relatively efficient and inexpensive means for providing lateral strength and stiffness (Figure 15.18). Sometimes referred to as "shear" walls and other times as "structural" walls, they typically cantilever up from a base. "Slender" walls are those having an aspect ratio (height/plan length) > 3 and are designed to yield in flexure just above the base. "Short" or "squat" walls have aspect ratio < 1.5 and are expected to fail in shear or slide at the base. Rocking at the base of slender walls is a viable means of resisting lateral load for some walls, particularly those with supplemental post-tensioning (see Sections 15.10 and 15.11).

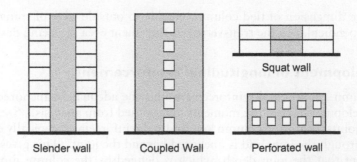

Figure 15.18 Types of structural walls. (Adapted from NIST GCR 11-917-11 REV1, 2012.)

Walls are considered solid if any openings that may be present do not influence wall strength or inelastic response. At the other extreme, perforated walls have a regular array of horizontal and vertical openings, which divide the wall into a series of vertical piers and deep beams.

Commonly used wall cross sections are rectangular, barbell-, I-, C-, and L-shaped (Figure 15.19). The more complex I-, C-, and L-shaped walls may be used to form party walls or to create internal cores (e.g., for elevators or stair towers). Since these more complex sections provide stiffness in each horizontal principal direction, their design should be based either on code-sanctioned 100–30 combinations of the results obtained from independent analyses in each principal direction, or preferably, the results of three-dimensional nonlinear analyses. Similarly, even where the simpler rectangular and barbell-shaped walls are used, if torsional irregularities are present, three-dimensional analysis may be necessary.

As illustrated in Figure 15.20, lateral displacement and interstory drifts at the top of a cantilever wall increase disproportionately as the height or number of stories increases. In contrast, coupled walls often are preferred for taller buildings, because the coupling beams create a more uniform pattern of interstory drifts. The coupling beams resist some of the overturning moment generated by the lateral forces.

Due to the relatively large cross-sectional area, wall axial load ratios generally will be well below the balance point. This results in flexural response involving yielding of tension reinforcement, which provides for greater curvature ductility than occurs when response is dominated by compression (e.g., under higher axial loads). In either case, relatively heavy transverse reinforcement is needed near the ends of the wall to confine concrete within the boundary elements and laterally support the longitudinal reinforcement. In addition, it is important to ensure that the wall section is not too narrow to avoid potential instabilities in compression.

Shear failures are relatively brittle, and should be avoided. Where walls having different lengths are used, the relative distribution of lateral forces to the walls varies with increasing deformation—the longer, stiffer walls will tend to soften and yield first, while the shorter walls continue to pick up load. Distributions determined in an elastic analysis (for one set of stiffness values) do not reflect the changes in relative stiffness that occur as differential

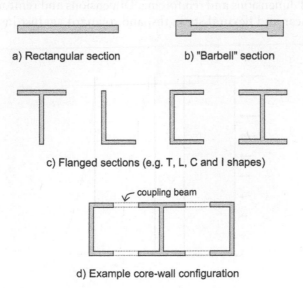

a) Rectangular section         b) "Barbell" section

c) Flanged sections (e.g. T, L, C and I shapes)

coupling beam

d) Example core-wall configuration

*Figure 15.19* Wall cross sections. (Adapted from NIST GCR 11-917-11 REV1, 2012.)

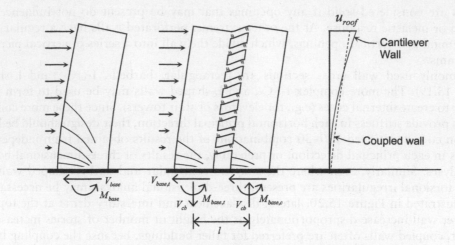

Figure 15.20 Static equilibrium and deformed shapes of cantilever walls and coupled walls.

softening and yielding occurs. A nonlinear static analysis, which can track changes in stiffness, is limited in not accounting for the presence of higher modes. (Dynamic analyses can produce wall shears that are several times those determined in nonlinear static (pushover) analyses.) Consideration should also be given to the large scatter in shear strength, degradation with cyclic loading, and the influence of cracking on shear demands. Depending on the fidelity of the model, some deference should be given to the time-honored approach of proportioning walls to have relatively low shear stresses and ample shear strength.

### 15.8.1 Proportioning of slender walls

Cantilever walls have shear and moment diagrams under lateral design forces as shown in Figure 15.21. Although peak shears and moments obtained during dynamic analysis will deviate from the static design values, the static values at the critical section typically are used for determining wall dimensions and reinforcing. Dimensions and reinforcing are selected to provide sufficient shear and flexural strengths, and to guard against instability of portions in compression.

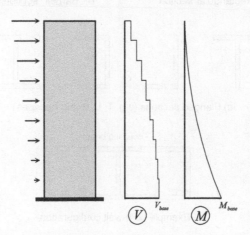

Figure 15.21 Shears and moments in a cantilever wall (based on static analysis).

Practical minimum widths for isolated cantilever walls are 12 in. (300 mm) for special boundary elements and 10 in. (250 mm) elsewhere. Two curtains of longitudinal reinforcement should be used to help resist instability of the compression zone, even though ACI 318 allows a single curtain for low levels of shear and for stocky (non-slender) walls. The reinforcement curtains should be located as close to the faces of the wall as cover requirements allow to reduce the loss of strength associated with cover spalling and to provide greater out-of-plane strength and stiffness.

Shear strength and average shear stress are based on the area of web, $A_{cv}$, defined as the product of the web thickness and length of section in the direction of shear force. While ACI 318 allows the factored shear, $V_u$, on individual wall segments to be as high as $10\phi\sqrt{f_c'} A_{cv}$ (lb and in. units) or $0.83\phi\sqrt{f_c'}A_{cv}$ (N and mm units), walls with lower shear stresses have greater ductility capacity. Thus, for initial proportioning, NIST GCR 11-917-11 REV1 (2012) recommends keeping the factored shear below approximately $(4-6)\phi\sqrt{f_c'}A_{cv}$ (lb and in. units) or $(0.33-0.50)\phi\sqrt{f_c'}A_{cv}$ (N and mm units).

For flexural strength, an approximate analysis can be used for initial proportioning of the walls, using trial values of wall dimensions and assumed locations of force resultants.[6] For simplicity, moments can be summed about the compression resultant (Figure 15.22); the longitudinal reinforcement is assumed to be at $f_{ye}$. One option is to distribute longitudinal reinforcement uniformly over the web, but more typically, longitudinal reinforcement is concentrated in boundary zones at the ends of the wall while distributed longitudinal reinforcement is used elsewhere (in the web). The minimum value for the distributed reinforcement is $\rho_l = 0.0025$; this reinforcement must be placed at a spacing not exceeding 18 in. (450 mm). Thus, if the minimum web reinforcement is provided, the amount of longitudinal boundary reinforcement needed to provide the required flexural strength can be determined. (Alternatively, if only distributed reinforcement will be used, the amount of distributed reinforcement can be solved directly, considering that the tension carried by the longitudinal boundary reinforcement, $T_{sb}$, is zero.) Iterations may be required to establish wall dimensions and reinforcement that provide sufficient flexural strength, using acceptable longitudinal web and boundary reinforcement ratios. Maximum longitudinal reinforcement within the boundary element is for columns of special moment frames (see Section 15.6.1). Web thickness can be adjusted to provide the required $A_{cv}$ for shear, which in turn may affect the amount of distributed web reinforcement. A thicker wall section generally will reduce reinforcement congestion, reduce shear stress levels, and improve earthquake performance, with little impact on cost or functionality.

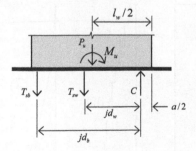

*Figure 15.22* Approximate analysis of wall flexural strength and required reinforcement. (Adapted from NIST GCR 11-917-11 REV1, 2012.)

---

[6] Alternatively, a spreadsheet built for this purpose, as described by Aschheim et al. (2008) is available at http://users.ntua.gr/divamva/RCbook/aci.flexure.xls.

## 15.8.2 Proportioning of coupled walls

Like slender cantilever walls, coupled walls are proportioned for shear and flexure, and to limit potential instability of portions in compression. Considerations for slender walls (Section 15.8.1) apply, but modified to account for the presence of coupling beams.

The simplest design idealization considers the coupling beams to individually resist identical shears, $V_{cb,i}$. Consequently, in addition to tributary gravity loads, each walls carries an axial force $V_{cb} = n \cdot V_{cb,i}$ at the base, where $n$ = the number of coupling beams. Figure 15.20 illustrates the $V_{cb}$ couple acting at the base of the walls. This couple resists a moment equal to $V_{cb} \cdot l$ at the base of the walls. Consequently, the moment that must be carried at the base of each cantilever wall is reduced to ½$(M_{base} - V_{cb} \cdot l)$. Repetitive application of this analysis at sections located higher up the wall reveals that the moments within the cantilever wall will typically reverse, as shown in Figure 15.23.

The coupling beams usually are proportioned to carry between 25% and 75% of the overturning moment resulting from the lateral forces, $M_{base}$. In order to avoid developing net tension in the plastic hinge zone of either wall, an upper bound on the coupling beam shears, $V_{cb}$, is the gravity load at the base of the wall. Because the walls generally will be proportioned with axial load below the balance point, they can be sized initially neglecting the effect of $V_{cb}$ on the axial force in the wall (i.e., considering only the tributary gravity load in the wall), since the reduction in flexural strength on the wall with reduced axial compression will be approximately compensated for by the increase in flexural strength of the wall with increased axial compression.

Assuming the walls are designed below the balance point, the wall experiencing increased compression (due to the effect of the coupling beam shears on the axial forces in the walls) will have greater flexural strength and stiffness, and therefore will experience larger shear forces. A reasonable approximation is to distribute the shear associated with hinging of the walls in proportion to the wall flexural strengths, considering the effects of coupling beam shears on the axial forces in the walls. (Because the lateral forces reverse direction, each wall is provided with the shear reinforcement required when it experiences increased compression.)

The minimum width of a coupling beam is approximately 14 in. (350 mm), although 16–18 in. (400–450 mm) is a more practical minimum where diagonally reinforced coupling beams are used. The walls typically have the same width as the coupling beams.

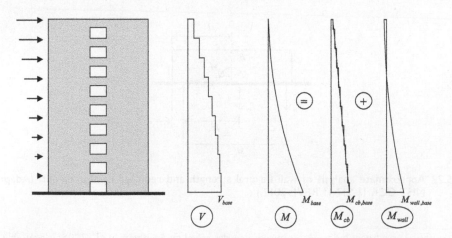

Figure 15.23 Distribution of overturning moment to coupling beams and walls.

## 15.8.3 Detailing of boundary zones

Boundary zones of unperforated walls are located at the ends of the wall cross section. For walls with openings, the boundary zones are located along the edges of the wall or opening. Boundary zones usually contain concentrated longitudinal reinforcement. In portions of the wall expected to undergo relatively large inelastic demands, relatively dense transverse reinforcement, similar to that required in columns of special moment frames, is provided in special boundary elements. This transverse reinforcement confines the concrete and helps to delay buckling of longitudinal reinforcement to later cycles. In barbell-shaped cross sections, the concentrated longitudinal reinforcement is located in the larger rectangular sections at the ends of the section. For barbell and flanged sections, the boundary element transverse reinforcement extends at least 12 in. (300 mm) into the web.

ACI 318 provides two alternative methods to determine whether a special boundary element is required. Of historical relevance is the evaluation of compression stresses in the extreme fiber of the section based on the superposition of normal stresses arising from axial load and moment, as determined using a linear elastic analysis. Of greater relevance today is a strain-based check applicable to wall segments that are effectively continuous over their height and designed to have a critical section at their base. In an indirect manner, the compressive strain developed in the extreme fiber of the critical section is evaluated based simply on displacement demand. First, the depth of the neutral axis, $c$, associated with the analytical evaluation of the nominal flexural strength, $M_n$, under the maximum (factored) axial compression force, $P_u$, is determined, according to the standard flexural analysis assumptions of ACI 318. Then, special boundary elements are required if

$$c \geq \frac{l_w}{600(1.5\delta_u/h_w)} \tag{15.17}$$

where $l_w$ = the plan length of the wall or wall segment, $h_w$ = the height to top of wall or wall segment from critical section, and $\delta_u$ = the design displacement at the top, determined conventionally as $C_d$ times the top displacement determined by analysis under the design forces (associated with the base shear $C_s W$, where $C_s = S_a/R$).

The width of the compression zone of a special boundary element is required to be at least $h_u/16$ to limit potential instability of compression zones, where $h_u$ is the laterally unsupported height of compression fiber of the wall or wall pier (i.e., usually the clear story height). It is recommended that a more stringent limit of $h_u/10$ be used in potential plastic hinge regions (NIST GCR 11-917-11 REV1, 2012). For walls subject to high axial load ratios, the calculated depth of the compression zone, $c$, may exceed $(3/8)(l_w)$, in which case the width of the compression zone must not be less than 12 in. (300 mm).

The transverse reinforcement within the special boundary element is required to extend horizontally from the wall edge a distance equal to the larger of $c - l_w/10$ and $c/2$, where $c$ is calculated at $M_n$ under the largest (factored) axial force $P_u$. For flanged sections, the special boundary element transverse reinforcement must include the effective flange width in compression and must extend horizontally at least 12 in. (300 mm) into the web.

Within the special boundary element, requirements for the configuration of transverse reinforcement follow those for special moment frame columns having $P_u < 0.3A_g f_c'$ and $f_c'$ < 10,000 psi (69 MPa). The required spacing, $s$, and total cross-sectional area of transverse reinforcement, $A_{sh}$, are as follows:

1. The spacing, $s$, of special boundary element transverse reinforcement may not exceed the smallest of

$$s \leq \begin{vmatrix} \text{minimum member dimension}/3 \\ 6d_b \\ s_o = 4 + \dfrac{14 - h_x}{3} \text{ in.} \quad \left( s_o = 100 + \dfrac{350 - h_x}{75} \text{ mm} \right) \end{vmatrix} \tag{15.18}$$

where $h_x$ is taken as the largest of the $x_i$ values (e.g., Figure 15.9) but is limited to the lesser of 14 in. (350 mm) and 2/3 of the boundary element thickness, and $d_b$ is the diameter of the smallest longitudinal bar within the boundary element.

2. The total cross-sectional area of transverse reinforcement, $A_{sh}$, perpendicular to each principal direction, within the special boundary elements must be at least the larger of:

$$A_{sh} \geq \begin{vmatrix} 0.09 \dfrac{sb_c f_c'}{f_{yt}} \\ 0.3 \left( \dfrac{A_g}{A_{ch}} - 1 \right) \dfrac{sb_c f_c'}{f_{yt}} \end{vmatrix} \tag{15.19}$$

where $s$ is the spacing of transverse reinforcement along the height of the special boundary element and other terms are defined as for Equation 15.9.

3. Horizontal reinforcement within the wall web has to extend to within 6 in. (150 mm) of the end of the wall and be anchored to develop $f_y$ within the confined core of the boundary element. Anchorage can be achieved using standard hooks or headed bars, or where sufficient length is available to develop the bars within the confined core, a straight anchorage may be used if $A_s f_y/s$ of the horizontal reinforcement does not exceed $A_s f_{yt}/s$ of the boundary element transverse reinforcement parallel to the horizontal web reinforcement. Figure 15.24a illustrates requirements for the use of standard hooks to anchor the horizontal reinforcement within the boundary element.

If special boundary elements are required, they must extend above and below the critical section at least $l_w$ and $M_{uCS}/4V_{uCS}$ (the larger value controls), where $M_{uCS}$ and $V_{uCS}$ are the factored moment and shear at the critical section, respectively. However, for critical sections near footings, mat foundations, and pile caps, the special boundary element transverse reinforcement need only extend at least 12 in. (300 mm) into the foundation (Figure 15.25). Where the edge of the boundary element is within one half the footing depth from an edge of the footing and for other support conditions identified in ACI 318, the boundary element transverse reinforcement must extend at least $l_d$ into the support, where $l_d$ is calculated for the longitudinal bars, using $1.25f_y$ in place of $f_y$ if the critical section occurs at the wall base.

Figure 15.25 also shows the use of ordinary boundary elements, where special boundary elements are not required and where the longitudinal reinforcing ratio within the boundary element exceeds $400/f_y$ (lb and in. units) or $2.8/f_y$ (N and mm units). An ordinary boundary element is illustrated in Figure 15.24b. The horizontal reinforcement typically terminates at the end of the wall with a standard hook; alternatively, U-shaped stirrups having the same size and spacing may be spliced to straight horizontal bars. Transverse reinforcement spacing within ordinary boundary elements cannot exceed the smaller of 8 in. (200 mm) and eight times the diameter of the longitudinal bars ($d_b$), except that within a distance equal to the larger of $l_w$ and $M_u/4V_u$ above and below the critical section where plastic hinging is anticipated, the spacing limit reduces to the smaller of 6 in. (150 mm) and $6d_b$.

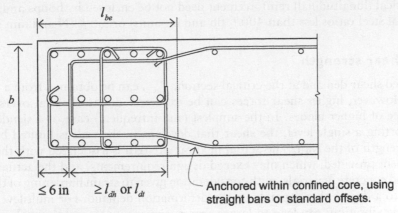

**(a) Special boundary element**

$\leq 6$ in      $\geq l_{dh}$ or $l_{dt}$      Anchored within confined core, using straight bars or standard offsets.

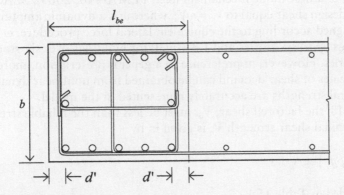

**(b) Ordinary boundary element where** $\rho_{be}\ (= A_{s,be}/(b \cdot l_{be})) > 400/f_y$

Figure 15.24 (a) Special and (b) ordinary boundary elements. (Adapted from NIST 11-917-11 REV1.)

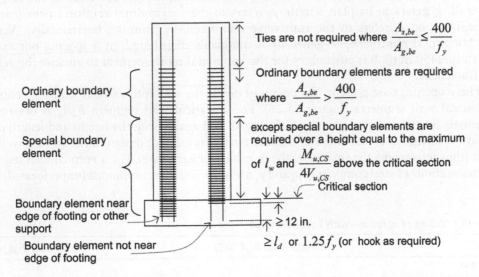

Ties are not required where $\dfrac{A_{s,be}}{A_{g,be}} \leq \dfrac{400}{f_y}$

Ordinary boundary elements are required

where $\dfrac{A_{s,be}}{A_{g,be}} > \dfrac{400}{f_y}$

except special boundary elements are required over a height equal to the maximum

of $l_w$ and $\dfrac{M_{u,CS}}{4V_{u,CS}}$ above the critical section

Critical section

Ordinary boundary element

Special boundary element

Boundary element near edge of footing or other support

Boundary element not near edge of footing

$\geq 12$ in.

$\geq l_d$ or $1.25 f_y$ (or hook as required)

Figure 15.25 Vertical extent of special and ordinary boundary elements for a continuous shear wall having critical section at the base of the wall. ($f_y$ in psi units. Adapted from ACI 318-14, 2015.)

The vertical (longitudinal) reinforcement need not be enclosed by hoops and crossties for longitudinal steel ratios less than $400/f_y$ (lb and in. units) or $2.8/f_y$ (N and mm units).

### 15.8.4 Shear strength

The factored shear demand at the critical section, $V_{uCS}$, can be obtained from a static elastic analysis. However, higher shear forces can be expected, due to flexural overstrength and the presence of higher modes. In the simplest (and infrequent) case of a slender cantilever wall supporting a single level, the shear that develops in the wall is limited by the actual flexural strength of the wall. The actual flexural strength is associated with the amount of reinforcement provided, which may exceed design requirements, and the actual strength of the steel and concrete materials, with consideration given to strain hardening of the steel reinforcement to a degree that depends on the deformation demand. For multilevel structures, higher mode vibrations can lead to larger shears, sometimes several times those associated with the development of a first-mode mechanism (e.g., FEMA-440, 2005). SEAOC (2008) recommends use of a design shear equal to $V'_u = \omega V_u$, where $\omega$ is a dynamic amplification factor. For buildings designed according to the equivalent lateral force procedure, $\omega$ is given by $0.9 + n/10$ for buildings up to six stories and $1.3 + n/30$ for buildings over six stories, where $n$ = the number of stories. However, in preference to a general prescription, more accurate, building-specific estimates of shear demand can be obtained from nonlinear dynamic analyses in which the flexural strengths are accurately represented in the model.

According to ACI 318, the factored shear, $V_u$, must be less than the reliable strength, $\phi V_n$, where $\phi$ = and the nominal shear strength $V_n$ is given in by

$$V_n = A_{cv}\left(\alpha_c \sqrt{f'_c} + \rho_t f_y\right) \tag{15.20}$$

Values of $\alpha_c$ are provided in Table 15.1.

Since $\rho_t$ = the ratio of cross-sectional area of distributed transverse reinforcement and gross concrete area perpendicular to that reinforcement, Equation 15.20 can be solved to determine the minimum value of $\rho_t$ to use in proportioning the transverse reinforcement.

For the design of a single, continuous wall without openings, $h_w$ refers to the height of the wall, $l_w$ refers to its plan length, $\rho_l$ refers to the longitudinal reinforcement (running vertically), and $\rho_t$ refers to the transverse reinforcement (running horizontally). Vertical and horizontal reinforcement generally is uniformly distributed, at a spacing not exceeding 18 in. (450 mm). It is customary for the horizontal reinforcement to enclose the vertical reinforcement.

Where openings are present, the openings divide the wall into vertical wall segments and horizontal wall segments (Figure 15.26). For a vertical wall segment $h_w/l_w$ is determined separately for the overall dimensions of the wall and again using the height and length of the vertical wall segment—the larger of these two ratios is used for design of the vertical wall segment (thus referencing the smaller of the $\alpha_c$ values for each case). For a vertical wall segment, the orientations of steel comprising $\rho_l$ and $\rho_t$ are as for a single continuous (unperforated) wall.

Table 15.1 Values of $\alpha_c$ for use with Equation 15.20[a]

| Units | Slender walls ($h_w/l_w > 2$) | Stocky walls ($h_w/l_w < 1.5$) |
|---|---|---|
| lb and in. | 2 | 3 |
| N and mm | 0.17 | 0.25 |

[a] $\alpha_c$ is to be determined by linear interpolation for intermediate values of $h_w/l_w$.

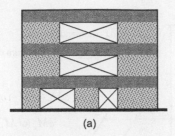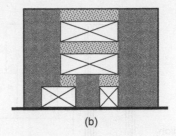

(a)                                    (b)

*Figure 15.26*  (a) Vertical and (b) horizontal wall segments (shaded). (Adapted from NIST 11-917-11 REV1.)

For all vertical wall segments resisting a common lateral force, the total nominal shear strength used in design is capped at $8A_{cv}\sqrt{f_c'}$ (lb and in. units) or $0.66A_{cv}\sqrt{f_c'}$ (N and mm units) where $A_{cv}$ is the combined gross area of all vertical wall segments. Any individual vertical wall segment is allowed to be designed to have a nominal strength as high as $10A_{cv}\sqrt{f_c'}$ (lb and in. units) or $0.83A_{cv}\sqrt{f_c'}$ (N and mm units). The common lateral force can refer to the shear resisted by a single wall, a line of walls in a single vertical plane, or the entire story shear.

Horizontal wall segments are, in effect, a vertical wall segment rotated by 90°. A coupling beam is one example of a horizontal wall segment. A horizontal wall segment is allowed to be designed to have a nominal shear strength as high as $10A_{cw}\sqrt{f_c'}$ (lb and in. units) or $0.83A_{cw}\sqrt{f_c'}$ (N and mm units), where $A_{cw}$ is the gross cross-sectional area (along a vertical cut) of the horizontally oriented segment. In the case of horizontal wall segments and coupling beams, $\rho_t$ refers to distributed vertical reinforcement and $\rho_l$ refers to distributed horizontal reinforcement.

## 15.8.5 Curtailment of reinforcement over the height

Nonlinear dynamic analyses reveal patterns of peak shears and peak moments that differ from the patterns generally used in design. Such differences may occur due to higher mode effects, differences in stiffness associated with the use of walls of different lengths, and torsional response of the building as a whole. Deviations of the peak shear and moment profiles from design values have implications for curtailment of longitudinal and transverse reinforcement over the height of the wall, as well as the extent of special boundary elements.

In many cases, the peak dynamic moments and shears will occur at the base of a continuous wall. Thus, the usual approach is to initially proportion the wall using values at a critical section at the base of the wall. Inelastic behavior would be expected within the plastic hinge region that forms proximate to the critical section at the base of the wall, presuming that the longitudinal reinforcement elsewhere provides sufficient flexural strength. In this circumstance, special boundary element details would be provided to support large inelastic demands in the vicinity of the critical section, while transverse reinforcement elsewhere would need to provide sufficient shear resistance.

Figure 15.27 illustrates this case, where longitudinal bars are provided to ensure $M_u < \phi M_n$ at the critical section. The longitudinal reinforcement is used to provide a region of constant flexural strength $\phi M_n$; the bars are gradually curtailed at various wall elevations. Curtailment of longitudinal reinforcement is based on two considerations: (i) each longitudinal bar must extend a distance not less than $l_d$ from the critical section or point where it is counted on for flexural strength[7] and (ii) a bar being cut must extend at least $0.8l_w$ from the point where the continuing bars provide sufficient flexural strength. Note that the

---

[7] The point where the reliable strength of the continuing bars, $\phi M_n$, equals the factored moment demand, $M_u$.

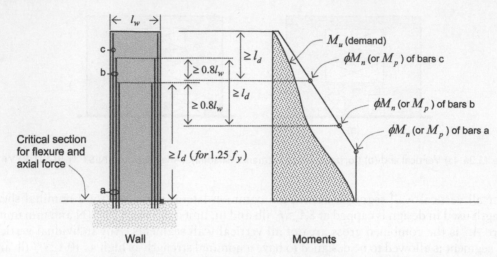

*Figure 15.27* Representative moment and shear demands over the height of a wall and criteria for curtailment of longitudinal reinforcement. Note: anchorage of reinforcement not shown. (Adapted from NIST 11-917-11 REV1.)

development length, $l_d$, is calculated for a stress of $1.25f_y$ at the critical section, and for $f_y$ at other cutoff points.

Since the moment diagram during seismic response cannot be determined with high precision, a simpler approach (not strictly in compliance with ACI 318) used by many engineers is to extend the bars a distance $l_d$ above the floor where the bars are no longer required for flexure (NIST GCR 11-917-11 REV1, 2012).

More generally, the gradual curtailment of longitudinal reinforcement over the height of the wall should avoid concentrating inelastic demands at any one location. Where nonlinear dynamic analysis is used, locations where flexural yielding may occur can be identified (other than at the critical section determined in a static analysis), and boundary element reinforcement can then be provided at these locations. For tall buildings in particular, walls may yield at locations above the base. In such cases, although not required by ACI 318, an intermediate level of boundary reinforcement may be provided in these regions, typically doubling the spacing of hoop sets that otherwise satisfy the requirements for special boundary elements (Section 15.8.3) (NIST GCR 11-917-11 REV1, 2012).

Shear demands over the height can be obtained using amplified shears ($V_u' = \omega V_u$ as described in Section 15.8.4). However, we consider it preferable to establish shear demands and horizontal shear reinforcement considering the potentially significant effects of higher modes as determined by nonlinear dynamic analysis.

## 15.8.6 Design of wall piers

A wall pier is a relatively narrow vertical wall segment, often formed by the placement of door or window openings in a shear wall. Wall piers are defined as having $h_w/l_w \geq 2.0$ and $l_w/b_w \leq 6.0$ and, where $b_w$, $l_w$, and $h_w$ are the width in plan, length in plan, and clear height of the wall segment, respectively. Wall piers are subject to the usual requirements for the design of vertical wall segments in addition to the specific requirements described in this section.

Wall piers with $l_w/b_w \leq 2.5$ must satisfy the requirements for columns of special moment frames contained in ACI 318 Sections 18.7.4, 18.7.5, and 18.7.6, with joint faces taken as the top and bottom of the clear height of the wall pier. These sections address longitudinal reinforcement limits and splices, transverse reinforcement, and shear strength.

Wall piers with $2.5 \leq l_w/b_w \leq 6.0$ may be designed in accordance with ACI Sections 18.7.4, 18.7.5, and 18.7.6 or according to the following:

- $V_u$ is determined as the shear required to develop $M_{pr}$ at both ends of the wall pier ($V_u = 2M_{pr}/h_w$). Shear strength, $V_n$, and distributed vertical reinforcement follow the normal requirements applicable to walls.
- Hoops are used for transverse reinforcement. The vertical spacing of transverse reinforcement is at most 6 in. (150 mm) and continues at least 12 in. (300 mm) above and below the clear height of the wall pier.
- Special boundary elements are provided if required by ACI 318 Section 18.10.6.3.

## 15.8.7 Anchorage and splices of reinforcement

### 15.8.7.1 Anchorage of longitudinal reinforcement

Vertical reinforcement at the interface between the wall and foundation (e.g., footing, pile cap, or mat foundation) must be fully developed into the foundation for tension. If yielding at this interface is expected, as would occur in the development of a plastic hinge at the base of the wall, development length is calculated with $1.25f_y$ substituted for $f_y$. Alternatively, if yielding is expected and standard 90° hooks are used, $l_{dh}$ is calculated using $1.25f_y$ in place of $f_y$, and the hook typically would extend to level of the horizontal reinforcement at the bottom of the foundation component.

### 15.8.7.2 Splices of longitudinal reinforcement

Longitudinal reinforcement may be spliced using lap splices for No. 11 and smaller bars or using mechanical and welded splices. Of these, only lap splices and mechanical splices may be used within or near sections where yielding of the reinforcement is anticipated. While lap splices are more common, because they result in a locally strengthened section, they can shift yielding above or below the lap splice location; thus, the splice is best located away from the plastic hinge region. If yielding of longitudinal reinforcement is expected, the splice length should be at least 1.25 times the tension development length determined for $f_y$. Because lap splices subjected to repeated inelastic load reversals may fail unless they are confined by closely spaced transverse reinforcement, closely spaced transverse reinforcement should be provided along the lap splice.

### 15.8.7.3 Anchorage of horizontal web reinforcement

Where non-special (i.e., ordinary) boundary elements are present, horizontal reinforcement terminating at the edges of a wall must be anchored to develop $f_y$ if $V_u \geq A_{cv}\sqrt{f_c'}$ (lb and in. units) or $0.083A_{cv}\sqrt{f_c'}$ (N and mm units). The horizontal reinforcement is required to either (i) hook around the end vertical bars, as shown in Figure 15.17 or (ii) be lap-spliced to U-shaped stirrups of the same bar size and spacing as the horizontal reinforcement.

Where special boundary elements are present, the horizontal reinforcement must be anchored within the confined core of the boundary element to develop $f_y$. The horizontal reinforcement must be extended to within 6 in. (150 mm) of the end of the section and is terminated using standard hooks or headed bars. As an alternative, the horizontal bars can be lapped to the special boundary element reinforcement provided that there is sufficient lap length and provided that the boundary element reinforcement provides strength $A_{sh}f_{yt}/s$ parallel to the web reinforcement at least equal to the strength of the web horizontal reinforcement $(A_v f_y/s)$. For this alternative, the special boundary reinforcement must be sufficient to satisfy the larger of (i) the requirements for special boundary elements and (ii) shear strength requirements.

### 15.8.8 Force transfer and detailing in regions of discontinuity

#### 15.8.8.1 Strut and tie models

Regions around openings are typically regions of discontinuity. Reinforcement needed in regions of discontinuity can be understood through the use of strut and tie models. For example, Figure 15.28 illustrates the modeling of shear and compression carried near the base of a punctured wall using diagonal compressive struts, and tension carried by reinforcement. Such regions of discontinuity should be reserved for regions away from plastic hinges. Stress limits applicable to compressive struts and tension ties are provided in ACI 318. Ordinary or special boundary element reinforcement may be needed to provide confinement in regions of compression. The need for such reinforcement can be evaluated based on elastic analysis (superposition of normal stresses arising from axial forces and bending moments, per ACI Section 18.10.6.3) or preferably, based on inelastic demands determined in nonlinear dynamic analyses.

Figure 15.28 also illustrates a column below a discontinued wall. Although ACI 318 allows discontinued walls to be supported by columns (and requires the transverse reinforcement normally within $l_o$ to be continued over the full height of the column), concentrated damage observed in past earthquakes suggests this configuration should be avoided.

#### 15.8.8.2 Detailing at boundaries of wall piers

Shear within wall piers at the edge of a wall must be transferred into adjacent wall segments above and below the wall pier. This transfer requires horizontal reinforcement, serving as collectors. As shown in Figure 15.29, the horizontal reinforcement must extend beyond the wall pier a sufficient distance to transfer the shear, $V_u$, in the wall pier into the adjacent segment. NIST GCR 11-917-11 REV1 (2012) recommends that the reinforcement extend a distance $V_u/\phi v_n$, where $v_n$ is the unit shear strength of the adjacent wall segment, and $\phi$ is the strength reduction factor for shear (=0.75 if the wall pier is designed for the shear corresponding to the development of the wall flexural strength).

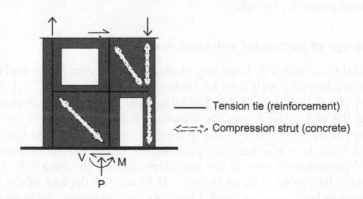

Figure 15.28 Strut and tie modeling in regions of discontinuity. (Adapted from NIST GCR 11-917-11 REV1, 2012.)

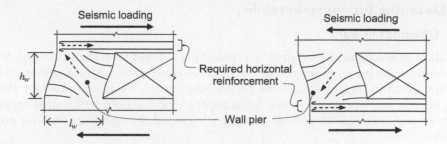

Figure 15.29 Horizontal reinforcement required to transfer $V_u$ into wall piers. (Adapted from NIST GCR 11-917-11 REV1, 2012.)

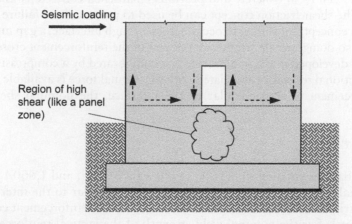

Figure 15.30 High shear and anchorage demand at the base of coupled walls. (Adapted from NIST GCR 11-917-11 REV1, 2012.)

### 15.8.8.3 Detailing at the base of coupled shear walls

At the base of coupled shear walls (Figure 15.30), and potentially at other locations where horizontal wall segments are present, transfer of tension and compression forces from longitudinal boundary elements into the horizontal wall segment can generate large shear and anchorage demands. Vertical longitudinal reinforcement requires good anchorage and preferably should run the full depth of the member. Portions of the horizontal wall segment subjected to high shears should be provided with transverse reinforcement for confinement and shear capacity. Strut and tie models may be useful for determining reinforcement to carry the loads to the foundation.

### 15.8.8.4 Detailing for transfer to collectors

Horizontal collectors are often required to transfer forces between walls and floor diaphragms. Longitudinal reinforcement within the collector is sized to carry the collector tensile force. If the longitudinal reinforcement is relatively light, lap splices can be used to transfer the tensile force to the horizontal wall reinforcement. For larger collector forces, the collector reinforcement should be embedded within the wall cross section to achieve adequate force transfer. An embedment length in excess of the development length is preferable.

### 15.8.9 Detailing for constructability

#### 15.8.9.1 Openings in walls

Coordination of mechanical, electrical, and plumbing systems with the structural design is needed to minimize penetrations through structural components. Minor penetrations can be accommodated in regions away from plastic hinge zones, avoiding boundary elements, coupling beams, and anchorage zones. Trim reinforcement should be detailed around the perimeter of any openings, to provide a load path around the opening and limit potential crack widths.

#### 15.8.9.2 Shear strength at construction joints (shear friction)

The interface between fresh concrete and previously hardened concrete is known as a construction joint. The shear friction concept can be used to prevent shear failure at such joints. According to this concept, for sliding to occur across a rough interface, a gap must form at the interface, and in so doing, tensile strain is developed in the reinforcement crossing this interface. The tension developed in the reinforcement is equilibrated by a compressive force across the interface. Frictional resistance associated with the normal force is available to resist shear.

Where reinforcement is perpendicular to the interface, the nominal shear strength is given by

$$V_n = \left( A_{vf} f_y + P \right) \mu_f \tag{15.21}$$

but not greater than the smallest of $0.2 f_c' A_{cv}$, $(480 + 0.08 f_c') A_{cv}$, and $1{,}600 A_{cv}$, where $A_{vf}$ is the cross-sectional area of reinforcement running perpendicular to the interface (typically consisting of distributed wall reinforcement and longitudinal reinforcement concentrated in boundary elements), $f_y$ is the nominal yield strength of the vertical reinforcement, and for normal weight concrete, $\mu_f = 1.0$ where the surface is intentionally roughened to full amplitude of ¼ in. (mm) or shear keys are provided, and 0.6 otherwise. Where permanent net compression force is present, $P$ may be taken equal to $N_u$ (positive in compression). Where transient net tension force acts perpendicular to the interface, $P = T_{u,net}$ (negative in tension).

## 15.9 COUPLING BEAMS

Damage to coupling beams can occur at relatively small building drifts; those located near the top of the building tend to be the most heavily damaged, as would be expected based on the deformation kinematics illustrated in Figure 15.20. Coupling beams may be reinforced with conventional longitudinal reinforcement or diagonal reinforcement. Diagonally reinforced coupling beams are preferred, as they provide stable hysteretic behavior to larger drifts compared with conventionally (orthogonally) reinforced coupling beams.

### 15.9.1 Proportioning of coupling beams

Following guidance from Paulay (2003), coupling beams over the height of a coupled wall may be designed for identical shears. The shear to be carried by the coupling beams can be determined as described in Section 15.8.2. Thus, the moment at the end of each coupling beam is $V_{cb,i} \cdot l_n / 2$.

Orthogonally reinforced beams are designed based on conventional moment and shear strength requirements and are indicated where the degree of coupling is relatively light.

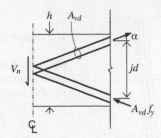

Figure 15.31 Truss model used for proportioning diagonal reinforcement in coupling beams.

Beam cross-sectional dimensions and reinforcement must comply with maximum longitudinal steel ratio limits. Orthogonal reinforcement must be used for relatively slender coupling beams, where $h < l_n/4$, where $h$ = height of coupling beam and $l_n$ = beam clear span.

Diagonally reinforced coupling beams are best suited where relatively strong coupling is desired. Diagonal reinforcement is required for relatively stocky coupling beams, where $h > l_n/2$ and $V_u \geq 4\sqrt{f_c'}A_{cw}$ (lb and in. units) or $0.33\sqrt{f_c'}A_{cw}$ (N and mm units). Diagonal reinforcement is proportioned based on the truss model of Figure 15.31, for which the nominal shear strength, $V_n$, is given by

$$V_n = 2A_{vd}f_y \sin\alpha \qquad (15.22)$$

For preliminary design, the distance $jd$ can be estimated as $h - 8$ in. ($h - 200$ mm). The nominal shear strength limit, $V_n \leq 10A_{cw}\sqrt{f_c'}$ (lb and in. units) or $0.83A_{cw}\sqrt{f_c'}$ (N and mm units) has the effect of providing an upper bound on the diagonal steel area, $A_{vd}$ (and a lower bound on the inclination of the reinforcement, $\alpha$).

Orthogonal or diagonal reinforcement may be used for beams of intermediate depth ($l_n/4 < h < l_n/2$). In addition to promoting better hysteretic behavior, diagonal reinforcement is more efficient for resisting higher shear levels, above approximately $(2-4)\sqrt{f_c'}A_{cw}$ (lb and in. units) or $(0.17$ to $0.33)\sqrt{f_c'}A_{cw}$ (N and mm units).

## 15.10 POST-TENSIONED CAST-IN-PLACE WALLS

Post-tensioned reinforced concrete walls provide an economical system for limiting interstory drifts while also providing a re-centering capability that results in small residual drifts. This system utilizes unbonded vertical post-tensioned reinforcement and gravity load in conjunction with conventional mild reinforcement to provide moment resistance (Figure 15.32). The post-tensioning contributes to moment resistance and provides an elastic restoring force that significantly reduces residual drifts. The reduction in mild reinforcement can improve constructability.

The unbonded post-tensioning steel strands extend vertically through the wall. In the case of a wall in which no continuous mild longitudinal steel is present, such as a wall formed from precast segments, moments developed at the base of the wall are resisted by movement of the vertical compressive reaction toward the toe of the wall. This fundamentally is a nonlinear contact problem (Figure 15.33a) for which both loading and unloading may be idealized as elastic, although there may be some dissipation of energy in rebound and damage within the confined compression toe of the wall. Thus, the moment–rotation response at the base of the wall may be idealized as bilinear elastic (Figure 15.33a), with

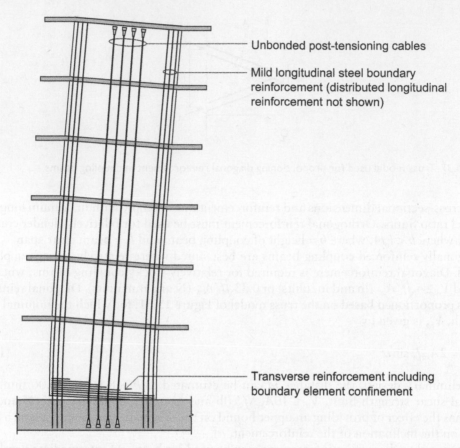

Unbonded post-tensioning cables

Mild longitudinal steel boundary reinforcement (distributed longitudinal reinforcement not shown)

Transverse reinforcement including boundary element confinement

*Figure 15.32* Schematic view of deformed post-tensioned cast-in-place wall. Elastic post-tensioning plus dead load provides an elastic restoring force, while mild longitudinal reinforcement provides hysteretic dissipation and limits inelastic behavior to the base of the wall.

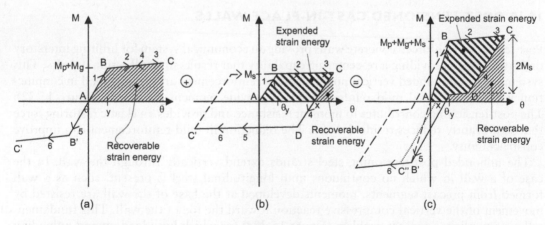

*Figure 15.33* Idealized moment–rotation responses of post-tensioned cast-in-place walls. Notice that (a) gravity load and post-tensioning are associated with nonlinear elastic response, represented here as a bilinear curve, while (b) yielding of mild reinforcement provides a more conventional hysteretic response. The combination of the two produces a flag-shaped hysteretic response (c).

the axial compression force equilibrating the applied post-tensioning force and gravity load. The post-tensioning strand is not grouted; elongation of the strand as the wall rotates is distributed over the length of the strand, resulting in relatively small increases in the strain and stress of the post-tensioning reinforcement. The lack of grout delays yield of the strand to relatively large rotations at the base of the wall. The small increase in post-tensioning force as the wall rotates results in a small increase in moment resistance.

The presence of continuous mild vertical reinforcement (e.g., a post-tensioned cast-in-place wall) causes additional moment resistance to develop associated with yielding and strain hardening of the mild reinforcement. The presence of bonded mild reinforcement causes cracking at the base of the wall to be distributed. If the depth of the compression zone is established based on equilibrium considerations, the moment resistance can be considered to be the sum of contributions due to (i) the post-tensioning and gravity load forces acting in concert with (ii) any mild reinforcement that is present.

If only mild reinforcement were present, a permanent or residual rotation would remain after unloading from a previous peak (Point 4 in Figure 15.33b). If the vertical force (due to post-tensioning and gravity load) is large enough to yield in compression the mild longitudinal reinforcement that had previously yielded in tension, then previous flexural cracks will close and the residual rotation will be close to zero. With reference to Figure 15.33, the closing of prior flexural cracks is ensured for $M_p + M_g > M_s$. The mild steel reinforcing index, $\beta_m$, was defined by Kurama (2005) as

$$\beta_m = \frac{M_s}{M_p + M_g} \tag{15.23}$$

Thus, to provide crack closure and re-centering, $\beta_m$ should be restricted to $0 < \beta_m < 1$. Panian et al. (2007) advise that $\beta_m$ values in the range of 0.75–1.0 provide a good combination of hysteretic energy dissipation (for limiting peak displacements) and a re-centering moment that will return the structure to a nearly plumb configuration. The relatively high stiffness provided by the wall system limits peak interstory drifts, while the small residual drifts simplify post-earthquake repairs.

The moment–rotation curve of Figure 15.33c is often referred to as a "flag" hysteresis, and the height of the flag is given by $2M_s$. The flag provides for hysteretic energy dissipation, which helps to reduce peak displacement amplitudes. This can be understood by considering the case of an SDOF oscillator represented by the hysteretic curves of Figure 15.33a–c, normalized to have the same flexural strength. First consider a case with no mild reinforcement ($\beta_m = 0$). Imagine at some instant during the response (Figure 15.33a), the rotation reaches point C. If the excitation were to cease at that instant, in the absence of damping, strain energy would propel the oscillator through point A with high velocity (the stored strain energy would be converted to kinetic energy $= \frac{1}{2}mv^2$) all the way to point C′, with oscillations gradually diminishing in amplitude (due to damping) and continuing until the oscillator comes to rest at point A. In contrast, an oscillator having the "full" hysteretic loops represented by Figure 15.33b, upon release from C, would oscillate about the abscissa ($M = 0$) with amplitude gradually diminishing until the oscillator ceases movement at the point marked "x." In the case of the flag hysteresis represented by Figure 15.33c, the stored (elastic) strain energy at point C propels the oscillator across the abscissa ($M = 0$) to reach a peak at point C″. Continued free-vibration with diminishing amplitude would result in a residual drift at the point marked "x." Of course, the ground motion likely continues, adding an element of random excitation in either direction, but the systematic benefit of adding the post-tensioning (nonlinear elastic behavior) to achieve reductions in residual drifts is clear.

Computational studies using SDOF oscillators having normalized hysteretic models that represent the load–displacement responses of Figure 15.33 show that peak displacements of the bilinear elastic model (a) tend to exceed those of the bilinear inelastic model (b), and that the hysteretic energy dissipation provided in (c) tends to mitigate the increase in displacement response associated with the bilinear elastic model (e.g., Christopoulos et al., 2002; Farrow and Kurama, 2003; Seo and Sause, 2005). In particular, Seo and Sause show very little increase in displacement response as $M_s$ approaches $(M_p + M_g + M_s)/2$. See Section 4.10.6 for additional information.

For preliminary design, one may simply consider the post-tensioning force in combination with the gravity load as a compressive force acting on a reinforced concrete section containing only mild longitudinal reinforcement, avoiding much of the complication of the model of Figure 15.33. Figure 15.34 presents a normalized form of an axial load–moment interaction surface for a wall having the parameters indicated, as described in more detail in Appendix 1. In theory, there exists an axial load–moment interaction curve for $\rho = 0$ (not shown), which necessarily passes through the point (0, 0). Such a curve would clearly indicate the moment resistance at a given axial load level that is due to compressive load (post-tensioning plus gravity loads) and that due to the steel reinforcement. We then would seek to ensure sufficient flexural strength under the constraint that $M_p + M_g > M_s$. An approximation for preliminary design, which ignores differences in neutral axis depth, $c$, associated with different axial load levels, is to evaluate the $M_s$ component at zero axial load, and find the required axial load level to ensure that $M_p + M_g > M_s$, as illustrated in the figure. In this figure, it is apparent that tensile yielding of longitudinal reinforcement (which requires an under-reinforced wall) with $\beta_m < 1$ is possible only for the low amounts of mild steel reinforcement ($\rho$ less than approximately 0.5% for the specific plot shown).

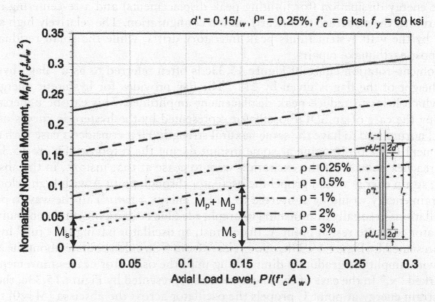

$d' = 0.15l_w$, $\rho'' = 0.25\%$, $f'_c = 6$ ksi, $f_y = 60$ ksi

Legend:
- - - - - $\rho = 0.25\%$
- - - - $\rho = 0.5\%$
- · - · - $\rho = 1\%$
- - - $\rho = 2\%$
——— $\rho = 3\%$

y-axis: Normalized Nominal Moment, $M_n/(f'_c t_w l_w^2)$

x-axis: Axial Load Level, $P/(f'_c A_w)$

*Figure 15.34* Example of normalized axial load–moment interaction curve. Moments are computed based on ACI 318 assumptions, using nominal material properties, as described in Appendix 1. For example, for $\rho = 0.5\%$, recentering can be obtained by providing a combination of prestress force and gravity load to obtain $P = 0.15f'_c A_w$, where $A_w$ = area of the rectangular section. To provide recentering capability, ensure the moment associated with mild reinforcement, $M_s$, is less than that associated with axial force $(M_p + M_g)$.

## 15.10.1  Guidelines for proportioning post-tensioned walls

Following Panian et al. (2007), the gravity load plus prestressing force should produce an average compressive stress over the cross section of $(0.1–0.25)f'_c$. Also, the tendons (consisting of groups of 270 ksi (1,860 MPa) seven-wire strands conforming to ASTM A416) should be grouped closely near the middle of the wall to minimize strains developed under large lateral drifts, with the resultant force aligned with the geometric centroid of the wall section.

Thus, the required base moment is carried through a combination of $M_s$ and $M_p + M_g$. Although the plots in Figure 15.34 and Appendix 1 were developed for the stated nominal material properties, they can be used to estimate flexural strength for preliminary design using expected properties. For example, Figure 15.34 would be used for $f'_c = f'_{ce}/1.25 = 6,000/1.25 = 4,800$ psi. Similarly, to maintain the correct value of $\rho \cdot f_{ye}$, reinforcement quantities would be the plotted percentages divided by 1.20 (for $f_{ye} = 1.20f_y$). (Keep in mind that the normalized quantities on the abscissa and ordinate utilize $f'_c = 6,000$ psi.)

Following Panian et al. (2007), the required prestressing force should be obtained using tendons stressed to approximately $0.6f_{pu}$, to delay yielding of the tendons to roof drifts of at least 2.5% of the building height. The increase in flexural strength associated with higher tendon stresses can be evaluated with plots such as those in Figure 15.34 and Appendix 1.

## 15.10.2  Modeling the load–displacement response of post-tensioned walls

A model for the lateral load–displacement response of unbonded post-tensioned (UPT) concrete walls was presented by Kurama et al. (1999). This design-oriented analytical model uses a set of equilibrium equations to estimate the nonlinear static (pushover) behavior of UPT walls. The model has been verified experimentally by other researchers (Pérez et al., 2007; PEER, 2011). Five points are identified on the capacity curve, representing different limit states (Figure 15.35). These points are as follows:

1. DEC: decompression at the base of the wall,
2. ELL: effective limit of the linear response of the wall,
3. SLP: spalling of concrete cover at the base of the wall,
4. LLP: yielding of the post-tensioning reinforcement, and
5. CCC: crushing of confined concrete at the base of the wall.

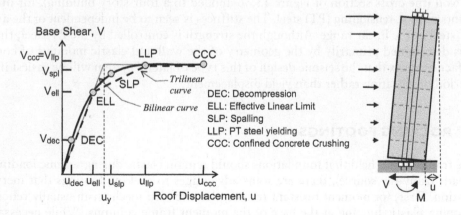

Figure 15.35  Generic capacity curve determined by nonlinear static (pushover) analysis of an UPT wall.

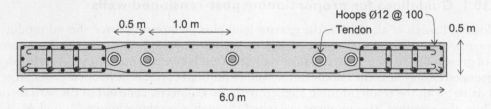

*Figure 15.36* Illustrative section of an UPT wall.

*Table 15.2* Points defining the bilinear curve

| Area of tendon (mm²) | $V_{ell}$ (kN) | $u_{ell}$ (mm) | $V_{llp}$ (kN) | $u_y$ (mm) |
|---|---|---|---|---|
| 1,400 | 1,765 | 7.0 | 3,253 | 13.1 |
| 2,800 | 2,463 | 9.8 | 4,548 | 18.1 |
| 4,200 | 2,990 | 11.9 | 5,716 | 22.8 |

A trilinear curve can be identified linking the origin with ELL, SLP, and CCC points. This trilinear curve constitutes the behavioral model proposed by Kurama et al. (1999).

In the case of UPT precast concrete walls, mild (deformed) steel reinforcement does not connect the wall to the foundation—only the post-tensioning reinforcement does. The initial stiffness of the wall is based on gross or transformed sections, until the point that cracking and/or gap opening occur. As a result, for the seismic design of buildings that utilize UPT walls as lateral force-resisting elements, the fundamental period of the building is determined largely by the UPT walls. Changes in prestress level and the quantity of post-tensioning reinforcement affect the strength of the wall but not the initial stiffness.

As an illustration, Figure 15.36 shows a UPT with five bundled tendons. Each tendon consists of some number of individually greased and sheathed seven-wire strands; the precise quantity to be determined based on the seismic performance objectives and site seismicity. The transverse hoops in the boundary zones at the ends of the wall section are 12-mm diameter at vertical spacing of 100 mm.

The five tendons are considered to be of the same cross-sectional area. Table 15.2 shows the values of $V_{ell}$, $u_{ell}$, $V_{llp}$, and $u_y$ for different values of the tendon cross section.

Figure 15.37 shows the bilinear curves determined using the Kurama et al. model for the UPT wall (the cross section of Figure 15.36 applied to a four-story building), for different amounts of post-tensioning (PT) steel. The stiffness is seen to be independent of the amount of PT steel in the linear range. Although the strength is controlled by the PT steel, the stiffness is determined primarily by the geometry of the wall and elastic modulus of concrete. This fact suggests that the seismic design of this type of lateral system will be easiest if based on period of vibration rather than yield displacement.

## 15.11 ROCKING FOOTINGS

While tradition has held that foundations should remain elastic during seismic loading (and for loads from any source), there are some advantages to rocking footings that merit consideration. First, for moment-resistant frames, the intended mechanism usually relies upon developing plastic hinging at the base of the moment frame columns. While necessary for mechanism development, the development of inelasticity in components that are crucial

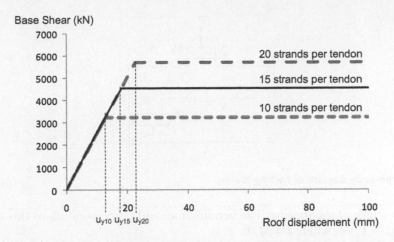

*Figure 15.37* Influence of area of post-tensioning reinforcement on the stiffness and strength of an UPT wall (using a simplified bilinear curve to characterize the response). In these cases, the PT is prestressed to $0.55f_{pu}$, where $f_{pu}$ is the nominal ultimate strength of the prestressing steel.

to the vertical stability of multiple supported floors is not an attractive proposition. Using capacity design principles to force the kinematically required rotations to occur at the soil-footing interface may be preferable to yielding the columns at their bases, where conditions allow their use. Rocking footings can also be used below slender walls, thereby avoiding plastic hinge development within the wall itself. Liu et al. (2013) find that allowing rocking reduces ductility demands in the overlying structure.

The main limitation in utilizing rocking footings is that the tuning of footing strength is limited by available gravity load and practical constraints on footing dimensions. In particular, the axial load (which affects the rocking moment) is given; in contrast, in the case of post-tensioned cast-in-place walls, the axial load is a design variable (Section 15.10). More generally, the lateral strength of conventional LFRSs can be determined independent of the gravity load. Additional considerations are (i) footing dimensions are limited to realistic, practicable values; (ii) possible restraint from grade beams and overlying floor slabs; (iii) to ensure rocking dominates over sliding, the shear span ($M/V$, acting at the footing–soil interface) has to exceed the length of the footing in the direction of the shear (preferably greater than $1.5L$); and (iv) to ensure vertical settlements are small, $L/L_c$ should be greater than 3 and preferably greater than 8. These constraints have the effect of limiting the applicability of rocking footings to buildings having a restricted range of stories for a given configuration and seismic hazard level.

## 15.11.1 Proportioning of rocking footings

Moment resistance is provided by the eccentricity of the axial load at the toe of the footing (Figure 15.38). This is effectively identical to the use of post-tensioning in precast walls (e.g., Figure 15.33a). In both cases, a nonlinear contact problem arises, and the response can be idealized as nonlinear elastic (Figure 15.33a), or as flag shaped (Figure 15.33c) if some inelastic deformation of the soil is considered.

Upon rocking, whether on cohesive or cohesionless granular soil, the soil-bearing capacity is mobilized at the toe of the footing (e.g., Deng and Kutter, 2012). The situation can be idealized much like the Whitney stress block for design of reinforced concrete members, in which the ultimate bearing capacity $q_u$ acts over an area of $L_c \cdot B$ to equilibrate the axial load,

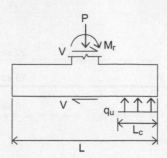

*Figure 15.38* Free-body diagram of rocking footing.

$P$, where $B$ = width of the footing. The bending moment corresponding to this condition is $M_r = P \cdot e = P(L - L_c)/2$, where $e = M_r/P$.

The required rocking capacity, $M_r$, for footings supporting moment-resistant frames can be determined as an extension of the discussion in Section 12.10.2 for avoiding weak-story behavior in non-rocking moment frames. For preliminary proportioning the axial gravity load in the end columns of the frame will be approximately half that in the intermediate columns (due to tributary areas), but the shears in these columns will also be about half (e.g., as assumed in the portal method of analysis), so the lengths of the footings below the end and intermediate columns will be about the same. Overturning axial forces will increase the rocking moment capacity at one end while reducing it by a similar amount at the other, and thus can be neglected in preliminary design (Figure 15.39).

Thus, the bearing length $L_c$ can be determined as

$$L_c = \frac{P}{q_u B} \tag{15.24}$$

and the required footing length can be established as

$$L = 2e + L_c \tag{15.25}$$

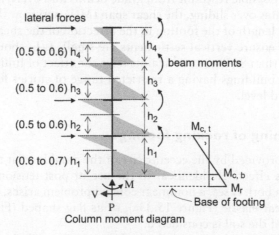

*Figure 15.39* Extension of column moment diagram to base of rocking footing, in plastic mechanism analysis for a strong column–weak beam (SCWB) moment frame, for determining the plastic moment strengths of the beams and rocking strength of the foundation. Design column moment strengths are increased above the values plotted to promote the SCWB mechanism.

## 15.12 FLOOR DIAPHRAGMS, CHORDS, AND COLLECTORS

Floor diaphragms typically consist of concrete floor slabs that interconnect vertical components of the LFRS. They must transfer inertial forces developed by the mass of the diaphragm itself along with the mass of attached materials to the vertical components of the LFRS. In addition, diaphragms may transfer forces between components of the LFRS, such as between moment frames and walls in dual systems, and between walls in systems for which wall strength and/or stiffness has a nonuniform distribution (Figure 15.40).

Away from discontinuities, diaphragms are commonly idealized as I-beams, with shears being approximated as uniformly distributed along the diaphragm itself (the web) and chords defined at the edges of the diaphragm (the flanges) to resist moment. Thus, consistent with this idealization, collectors are designed to "pick up" these shears and transmit them to the wall in cases where the wall does not extend the full width of the diaphragm. Chords are designed to carry the tension or compression needed for moment resistance within the diaphragm.

The conventional design intent is for diaphragms to remain essentially elastic. The forces used for diaphragm design, $F_{px}$, differ from those used for design of the vertical components of the LFRS. Damage to collectors in past earthquakes has led to the requirement that these components be designed for forces amplified by $\Omega_o$.

Where diaphragm spans between vertical components of the LFRS are large relative to their width (aspect ratios greater than perhaps 3), or where substantial openings are present in the diaphragm, the flexibility of the diaphragm should be modeled. Further details are available in NIST GCR 16-917-42 (2016).

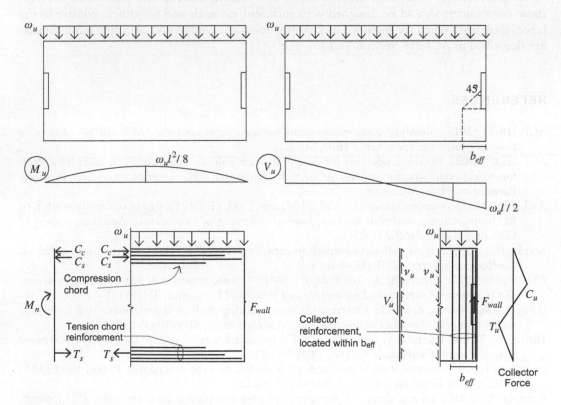

*Figure 15.40* Diaphragm components and idealized mechanisms of resistance.

## 15.13 GRAVITY FRAMING

Framing members that are not designated as part of the LFRS are often referred to as gravity framing (while recognizing that the LFRS also supports its tributary share of gravity load). Although gravity framing is not in itself required to be ductile, it should be recognized that the gravity framing must endure the same drifts that the LFRS experiences. As well, the gravity framing also must withstand forces developed as tributary dead loads are accelerated by the vertical component of the ground motion.

ACI Section 18.14 states requirements for such gravity framing, applicable to columns, beams, beam-column connections, slab-column connections, and wall piers of "gravity-only" framing. The governing idea is to ensure the gravity framing is capable of supporting the gravity loads at the design displacement. This is achieved by imposing modest requirements on longitudinal and transverse reinforcement in beams, enhancing the deformation capacity of columns that are expected to yield, providing reinforcement in slabs to reduce the likelihood of punching failures, and providing ductile details in wall piers.

## 15.14 FOUNDATIONS

Inelastic behavior within foundation components (e.g., footings, grade beams, pile caps, piles, and mats) is generally discouraged because of the difficulty of inspection and the cost of repair. Simply designing the foundation to resist the forces and moments obtained for the governing load combination is not sufficient to protect the foundation components from significant inelastic behavior. Rather, except where rocking footings are used (Section 15.11) these components should be designed with sufficient strength and toughness relative to the forces mobilized when the intended mechanism develops. Design and detailing requirements are described in ACI 318 Section 18.13.

## REFERENCES

ACI 318-14 (2015). Building code requirements for structural concrete (ACI 318-14), American Concrete Institute, Farmington Hills, MI.

ACI 352R (2002). Recommendations for design of beam-column connections in monolithic reinforced concrete structures, Joint ACI-ASCE Committee 352, American Concrete Institute, Farmington Hills, Michigan.

Aschheim, M., Hernández-Montes, E., and Gil-Martín, L. M. (2008). Design of optimally-reinforced RC beam, column, and wall sections, *Journal of Structural Engineering*, American Society of Civil Engineers, 134(2):231–239.

Bondy, D. (1996). A more rational approach to capacity design of seismic moment frame columns, *Earthquake Spectra*, 12(3):395–406.

Christopoulos, C., Filiatrault, A., and Folz, B. (2002). Seismic response of self-centering hysteretic SDOF systems, *Earthquake Engineering and Structural Dynamics*, 31(5):1131–1150.

Deng, L., and Kutter, B. (2012). Characterization of rocking shallow foundations using centrifuge model tests, *Earthquake Engineering Structural Dynamics*, 41(5):1043–1060.

Farrow, K. T., and Kurama, Y. C. (2003). SDOF demand index relationships for performance-based seismic design. *Earthquake Spectra*, 19(4):799–838.

FEMA 440 (2005). Improvement of nonlinear static seismic analysis procedures. Report No. FEMA-440, Applied Technology Council, Redwood City, CA.

Kurama, Y. (2005). Seismic design of partially post-tensioned precast concrete walls, *PCI Journal*, 50(4):100–125.

Kurama, Y. C., Pessiki, S., Sause, R., and Lu, L.-W. (1999). Seismic behavior and design of unbonded posttensioned precast concrete walls, *PCI Journal*, 44(3):72–89.

Liu, W., Hutchinson, T. C., Kutter, B. L., Hakhamaneshi, M., Aschheim, M., and Kunnath, S. K. (2013). Demonstration of compatible yielding between soil-foundation and superstructure components, *Journal of Structural Engineering*, American Society of Civil Engineers, 139(8)1408–1420.

NIST GCR 8-917-1 (2008). Seismic design of reinforced concrete special moment frames: A guide for practicing engineers, NEHRP Seismic Design Technical Brief No. 1, produced by the NEHRP Consultants Joint Venture, a partnership of the Applied Technology Council and the Consortium of Universities for Research in Earthquake Engineering, for the National Institute of Standards and Technology, Gaithersburg, MD.

NIST GCR 11-917-11 REV1 (2012). Seismic design of cast-in-place concrete special structural walls and coupling beams: A guide for practicing engineers, NEHRP Seismic Design Technical Brief No. 6, produced by the NEHRP Consultants Joint Venture, a partnership of the Applied Technology Council and the Consortium of Universities for Research in Earthquake Engineering, for the National Institute of Standards and Technology, Gaithersburg, MD.

NIST GCR 16-917-40 (2016). Seismic design of reinforced concrete special moment frames: A guide for practicing engineers, 2nd edition; NEHRP Seismic Design Technical Brief No. 1, produced by the NEHRP Consultants Joint Venture, a partnership of the Applied Technology Council and the Consortium of Universities for Research in Earthquake Engineering, for the National Institute of Standards and Technology, Gaithersburg, MD.

NIST GCR 16-917-42 (2016). Seismic design of cast-in-place concrete diaphragms, chords, and collectors: A guide for practicing engineers, NEHRP Seismic Design Technical Brief No. 3, produced by the NEHRP Consultants Joint Venture, a partnership of the Applied Technology Council and the Consortium of Universities for Research in Earthquake Engineering, for the National Institute of Standards and Technology, Gaithersburg, MD.

Ozcebe, G, and Saatcioglu, M. (1987). Confinement of concrete columns for seismic loading, *ACI Structural Journal, American Concrete Institute*, 84(4): 308–315.

Panian, L., Steyer, M., and Tipping, S. (2007). Post-tensioned concrete walls for seismic resistance, *PTI Journal, Post-Tensioning Institute, Phoenix, Arizona*, 5(1)7–16.

Paulay, T. (2003). Seismic displacement capacity of ductile reinforced concrete building systems, *Bulletin of the New Zealand Society for Earthquake Engineering*, 36(1):47–65.

PEER (2011). Design and instrumentation of the 2010 E-defense four-story reinforced concrete and post-tensioned concrete buildings, Report No. Report 2011/104, Pacific Earthquake Engineering Research Center, Richmond, CA.

Pérez, F. J., Sause, R., and Pessiki, S. (2007). Analytical and experimental lateral load behavior of unbonded posttensioned precast concrete walls, *Journal of Structural Engineering*, 133(11):1531–1540.

SEAOC (2008). Reinforced concrete structures, Article 9.01.010, SEAOC Blue Book—Seismic Design Recommendations, Seismology Committee, Structural Engineers Association of California.

Seo, C. Y., and Sause, R. (2005). Ductility demands on self-centering systems under earthquake loading, *ACI Structural Journal*, 102(2):275–285.

# Chapter 16

# Component proportioning and design requirements according to Eurocodes 2 and 8

## 16.1 PURPOSE AND OBJECTIVES

This chapter summarizes key requirements and recommendations for design and detailing of reinforced concrete components in regions of high seismic hazard as put forward by the European Committee for Standardization (CEN) and applied in European practice by means of Eurocodes 2 (2004) and 8 (2004). Eurocode 2 (henceforth EC2) is specialized to the design of concrete structures and Eurocode 8 (EC8) addresses design for earthquake resistance.

## 16.2 INTRODUCTION

Seismic design according to the Eurocodes (specifically EN1998) is based on two different performance objectives, aiming to provide life safety in a rare seismic event and limit damage in a more frequent one. These objectives, termed ultimate limit state (ULS) and serviceability limit state (SLS), respectively, are meant to correspond to ground motions having different probabilities of exceedance. More specifically:

- Life safety requirement (ULS). The structure must retain its structural integrity and a residual gravity load-carrying capacity. This seismic action associated with this performance level is recommended to have a probability of exceedance of 10% in 50 years (equivalent to a mean return period of 475 years[1]) for structures of ordinary importance.
- Damage limitation requirement (SLS). The structure may be damaged to a degree that is considered acceptable relative to the cost of the building itself. The seismic action associated with this performance level is recommended to have a probability of exceedance of 10% in 10 years (equivalent to a mean return period of 95 years) for structures of ordinary importance.

  Damage limits are usually measured in terms of interstory drift (relative displacement divided by the interstory height); limiting values range from 0.5% to 1% depending on the type of nonstructural component being evaluated.

Buildings are assigned to one of four importance classes, depending on the risk for human life, the importance for public safety and civil protection in the immediate post-earthquake

---

[1] Assuming a Poisson model for the occurrence of events (see Chapter 13), the following exponential model describes the relationship between the mean return period, $T_R$, the reference design lifetime, $T_L$, and the probability of exceedance $P$: $T_R = -T_L / \ln(1 - P)$

period, and the social and economic consequences of a collapse. The four importance classes and the associated importance factor ($\gamma_I$) in EC8 are summarized in Table 12.4.

Figure 16.1 shows a flowchart with the general procedure for seismic design of concrete buildings according to EC-8.

Compared with the design basis of North American codes (Chapter 15), two important differences may be perceived: (i) EN1998 (2004) employs a two-level design rather than the single level of, e.g., ASCE 7-16, and (ii) the interstory drift limit employed at the lower (SLS) level seems to be independent of the importance accorded to the building occupancy. In reality, though, these differences are more about the phrasing of the relevant provisions rather than their substance. Specifically, both the ULS and the SLS levels are keyed to the same 10% in 50-year design spectrum—the lower 10% in 10-year intensity of the SLS level is obtained as the product of the 10% in 50-year value and a *site-independent* factor (Section 16.4.2). Interestingly, this factor takes values that depend on the importance class: 0.5 for low and ordinary importance (Class I, II) and 0.4 for higher importance (Class III and IV). Thus, for buildings of ordinary importance, the SLS limit-state is numerically equivalent to a 10% in 50-year drift limit of 0.5%/0.5 = 1% and 1%/0.5 = 2% for buildings having brittle and ductile partitions, respectively. The latter case happens to coincide with the typical drift verification for North American codes. For important buildings, the corresponding 10% in 50-year equivalent drift limits are 0.5%/0.4 = 1.25% and 1%/0.4 = 2.5% for brittle and ductile partitions, respectively. Thus, the drift limit depends on important class and, intriguingly, is more permissive for important structures. Obviously, one can question whether the 0.4 and 0.5 multipliers are meant to modify the intensity hazard level or the drift limit, but in the end, in terms of the numerical limits used for design, the basis of codes on the two sides of the Atlantic are not very different. Note that this discussion applies to

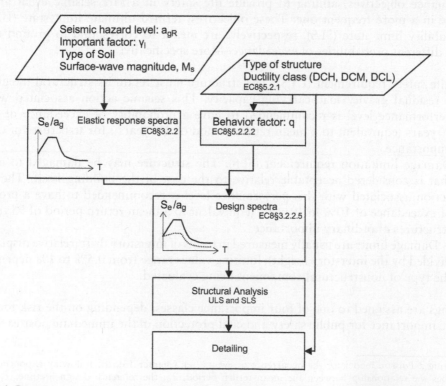

*Figure 16.1* Design process for reinforced concrete structures in accordance with EC8.

the first (2004) edition of EN1998; the newer drafts that are under preparation at the time of writing may differ.

## 16.3  THE SEISMIC ACTION IN EUROCODE-8

The seismic action to be considered for structural design is based on the assessment of the seismic hazard at particular location for a particular site soil conditions and for a given type of building.

In order to account for the influence of the different types of ground in the seismic action, EC8 distinguishes seven types of site soil conditions (EC8 Section 3.1.2): A, B, C, D, E, $S_1$, and $S_2$, ranging from rock (type A) to soils of lesser density (type B, consisting of very dense deposits with $N_{SPT} > 50$; type C, consisting of deep deposits with $N_{SPT}$ in the range 15–50; type D, consisting of loose-to-medium deposits with $N_{SPT}$ less than 15; type E, containing an alluvium layer); and including soft clays with high plasticity (type $S_1$) and deposits of liquefiable soils (type $S_2$). Three parameters are used in the classification of soil conditions provided in Table 3.1 of EC8: the value of the average shear wave velocity ($v_{s,30}$), the number of blows in the standard penetration test ($N_{SPT}$), and the undrained cohesive resistance ($c_u$). For example, ground type B has $360 < v_{s,30} < 800 \, \text{m/s}$, $N_{SPT} > 50$ blows/30 cm, and $c_u > 250 \, \text{kPa}$.

For the sake of simplicity (other ways may better describe the seismic action), the seismic hazard level is described in terms of a single parameter: the reference peak ground acceleration $a_{gR}$. This is the peak ground acceleration having a probability of exceedance of 10% in 50 years, for ground type A. Values of this parameter are plotted on maps; each is approved by each national authority. Figure 16.2 provides a description of the seismic hazard in Europe in terms of $a_{gR}$ from a consistent definition of seismic sources across Europe. Since the national authorities are free to choose their own maps, there is significant mismatch of the corresponding contours along the borders between European Union countries.

The design ground acceleration on ground type A ($a_g$) is equal to $a_{gR}$ times the importance factor ($\gamma_I$): $a_g = \gamma_I \cdot a_{gR}$. In cases of very low seismicity, in which the design ground acceleration on ground type A is smaller than 0.04 g, seismic action need not be considered (EC8 Section 3.2.1(5)).

The seismic action may be represented in terms of ground acceleration time-histories (EC8 Section 3.2.3.1). Nevertheless, with the intention of the application of modal analysis, the seismic action at ground level is also represented in EC-8 by an elastic ground acceleration response spectrum (the elastic response spectrum).

The seismic action can be considered in terms of its three orthogonal components: two horizontals and one vertical. The two horizontal components are assumed to be independent and can be represented by the same response spectrum.

The shape of the elastic response spectrum for the horizontal component given by EC8 (Section 3.2.2.2) is the piecewise function represented in Figure 16.3, where $S_e(T)$ is the elastic response spectrum, $T$ is the vibration period of a linear single-degree-of-freedom system, $a_g$ is the design ground acceleration on type A ground including the importance factor (i.e., $a_g = a_{gR}\gamma_I$), $S$ is the soil factor, $\eta$ is the damping correction factor with a reference value of 1 for 5% viscous damping, $T_B$ is the lower limit of the period of the constant spectral acceleration branch, $T_C$ is the upper limit of the period of the constant spectral acceleration branch and $T_D$ is the value defining the beginning of the constant displacement response range of the spectrum.

EC-8 (Section 3.2.2.2(2)) gives values of $S$, $T_B$, $T_C$, and $T_D$ for 5 ground types (A, B, C, D, and E) and for two types of elastic response spectrum to be used for the horizontal

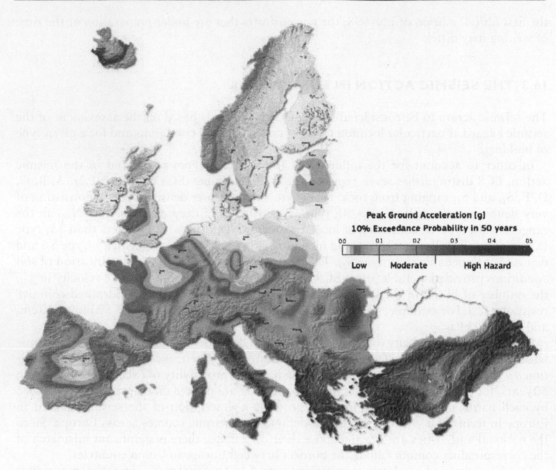

*Figure 16.2* European Seismic Hazard Map. (From SHARE, 2013.)

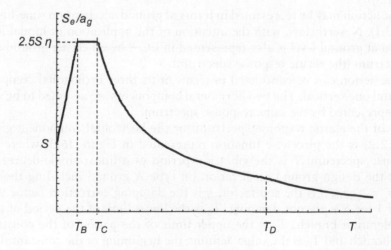

*Figure 16.3* Basic shape of the EC-8 elastic response spectrum.

components: Types 1–2. In case that deep geology is not accounted for, EC8 recommends the use of the two types of spectra. In case that the earthquakes that contribute most to the seismic hazard defined for the site for the purpose of probabilistic hazard assessment have a surface-wave magnitude $(M_s)$ greater than 5.5, the Type 1 spectrum is recommended. In case that $M_s$ is not greater than 5.5, Type 2 is recommended.

The shape of the response spectrum is the same for ULS and SLS.

For the three components of the seismic action, one or more alternative shapes of response spectra may be adopted. For different sources the use of more than one response spectrum is necessary to adequately represent the seismic action. For important structures $(\gamma_I > 1)$ topographic effects should also be considered (Annex A of EC-8 Part 5).

The vertical component of the seismic action is represented by an elastic response spectrum, $S_{ve}(T)$, in EC-8 (Section 3.2.2.3). Two types are provided, just as for the horizontal component. The basic shape of the vertical elastic response spectrum is similar to the horizontal one. It has four segments limited by periods $T_B$, $T_C$, and $T_D$. The vertical ground motion is not considered to be strongly influenced by the ground type, and is independent of the soil factor $(S)$.

## 16.3.1 Design spectrum

The capacity of structures to deform in a ductile manner allows for the design for forces smaller than those corresponding to the linear elastic response. To account for this reduction, EC-8 introduces the behavior factor $q$, similar in nature to the North American R-factor, defined as "an approximation of the ratio of the seismic forces that the structure would experience if its response was completely elastic with 5% viscous damping, to the seismic forces that may be used in the design, with a conventional elastic analysis model, still ensuring a satisfactory response of the structure" (EC8 Section 3.2.2.5(3)). In a practical sense, the behavior factor is a reduction of the ordinates of the elastic spectra; this reduction can be used if the structure did not collapse after the earthquake. For the horizontal and vertical components of the seismic action the design spectrum, $S_d(T)$, is defined as piecewise function given in EC8 Section 3.2.2.5(4). Figure 16.4 shows the horizontal design spectra Type 2 $(M_s < 5.5)$ for ground type C for different behavior factors (1.5, 2, 3, and 5).

The behavior factor of the structure $q$ may be chosen to be different in different horizontal directions, based on the selection of the ductility class in the two horizontal directions. The behavior factor for the vertical direction must be smaller than 1.5 unless justification is provided for a larger value.

## 16.3.2 Material safety factors and load combination in analysis

In Eurocode 0 (EC0, 2005) different load combinations are used for different design conditions. Seismic design in EC0 is considered to be an ULS design condition. The characteristic strength of materials (designated with a subscript k) is reduced by a material safety factor ($\gamma_c$ for concrete and $\gamma_s$ for steel) to obtain a reliable material strength for use in strength design calculations (designated with a subscript d). Thus:

$$f_{cd} = \frac{f_{ck}}{\gamma_c} \text{ and } f_{yd} = \frac{f_{yk}}{\gamma_s}$$

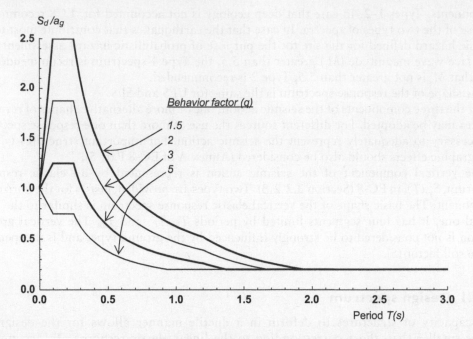

*Figure 16.4* Type 2 horizontal design spectra for different behavior factors for ground type C.

where $f_{ck}$ and $f_{cd}$ are the characteristic strength of concrete and the design strength of concrete and $f_{yk}$ and $f_{yd}$ are the characteristic strength of steel and the design strength of steel, respectively. Typical values of characteristic strengths are $f_{ck} = 25\,\text{MPa}$ or $30\,\text{MPa}$ for cast-in-place reinforced concrete and $f_{yk} = 400$ or $500\,\text{MPa}$.

Typical values of $\gamma_c$ and $\gamma_s$ are 1.5 and 1.15 for ULS for non-seismic loading, and 1.2 and 1.0 for ULS under seismic loading, respectively. Figure 16.5 illustrates how these material safety factors affect the cross sectional design capacity.

Every cross section of every member must have adequate strength. This typically focuses the engineer's attention on critical sections where the cross section size and reinforcement must be determined. Each critical section is designed for different combinations of loads, as mandated by EC0. The combinations of loads considered for seismic conditions, as determined by Eurocode 0 (Equations 6.12a,b of EC0) are defined by:

$$\sum_{j\geq 1} G_{kj} + P + A_{Ed} + \sum_{i\geq 1} \psi_{2i} Q_{ki} \tag{16.1}$$

where:
+ indicates to combine with
$\Sigma$ indicates the combined effect of
$G_{kj}$ is the characteristic value of permanent actions
$P$ is the characteristic value of prestressing action
$Q_{ki}$ is the characteristic value of variable action $i$
$A_{Ed}$ is the design value of the seismic action $= \gamma_I \cdot A_{Ek}$
$A_{Ek}$ is the characteristic value of the seismic action
$\gamma_I$ is the importance factor
$\Psi_{2i}$ is the quasi-permanent value of the combination coefficient of variable action $i$.
    Quasi-permanent loading is that considered to be present 50% of the time

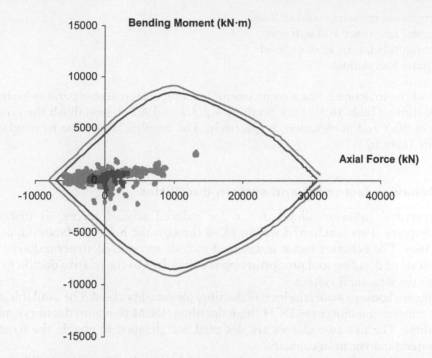

*Figure 16.5* Axial load–moment interaction diagram for a bridge deck. Load combinations are represented by points; design capacities by lines. Light grey represents seismic conditions and Dark grey represents non-accidental non-seismic conditions.

A smaller or coefficient ($\Psi_{Ei}$) can be used to calculate the inertial effects of the gravity loads. The following combination of actions is considered:

$$\sum_{j \geq 1} G_{kj} + \sum_{i \geq 1} \psi_{Ei} Q_{ki} \tag{16.2}$$

where $\Psi_{Ei} = \varphi \Psi_{2i}$. The value of $\varphi$, given in Table 4.2 of EC-8, ranges between 0.5 and 1.0. Coefficients $\Psi_{Ei}$ are also used for the reduced participation of masses in motion of the structure due to the nonrigid connection between them.

In general, the three components of the seismic action, two horizontals and one vertical, are taken as acting simultaneously. Nevertheless, some simplifications and combinations of loadings can be seen in EC8 Section 4.3.3.5. For example, 100% of one component can be combined with 30% of each of the other two components.

For the structural modeling, the use of cracked section stiffnesses is allowed. Alternatively, one half of the gross elastic flexural stiffness and the shear stiffness may be used.

## 16.4 PERFORMANCE OF THE STRUCTURAL SYSTEM

Basic principles are necessary in order to design in seismic regions. The structure must resist the seismic action retaining its integrity. Several principles should be recognized:

- Structural simplicity
- Uniformity, symmetry, and redundancy

- Bidirectional resistance and stiffness
- Torsional resistance and stiffness
- Diaphragm behavior at story level
- Adequate foundation

Regularity of the structure is not a requirement, but a regular structure permits higher levels of simplifications (Table 16.1). EC8 Sections 4.2.3.2 and 4.2.3.3 establish the criteria for regularity in plan and in elevation, respectively. The simplifications due to regularity are indicated in Table 16.1.

## 16.4.1 Behavior factor (q) and system ductilities

Ductile structural behavior allows design for reduced seismic forces, as discussed in previous chapters. This is achieved within EC-8 through the behavior factor, $q$, as shown in Figure 16.4. The behavior factor is assigned to code-recognized structural systems and relates the level of detailing and proportioning requirements to the relative ductility capacity accorded to the structural system.

The designer chooses *a priori* the level of ductility (or ductility class). The available ductility classes for concrete buildings are DCH (high ductility), DCM (medium ductility), and DCL (low ductility). The first two classes are designed and detailed to enable the structure to provide hysteretic dissipation capacity.

While lower design forces are used to design the higher ductility class buildings, better reinforcement details required for the higher ductility class buildings provide higher deformation capacities in critical regions. The lower design forces (and hence, lower strengths) would lead to an expectation of greater damage in critical regions during moderate earthquakes compared with the lower ductility class buildings that are designed for higher design forces. However, generous design minima often cause DCM buildings to experience similar damage as DCH buildings designed for the same site, while the DCH buildings have a higher margin of safety with respect to global or local collapse due to their inherently higher ductility capacity. Of course, this depends on the particular structural characteristics of each building, rendering any generalization practically untenable.

The $q$-factor (behavior factor) is formulated as (EC8 Section 5.2.2.2):

$$q = q_0 k_w \tag{16.3}$$

where, $q_0$ is the basic value of the behavior factor, which is a function of structural system and its regularity in elevation and $k_w$ is a factor reflecting the prevailing failure mode in

Table 16.1 Consequences of structural regularity on seismic analysis and design

| Regularity | | Allowed simplification | | Behavior factor (q) |
|---|---|---|---|---|
| Plan | Elevation | Model | Linear-elastic analysis | (for linear analysis) |
| Yes | Yes | Planar | Lateral force[a] | Reference value |
| Yes | No | Planar | Modal | Reduced value |
| No | Yes | Spatial[a] | Lateral force[a] | Reference value |
| No | No | Spatial | Modal | Reduced value |

[a] Additional conditions are needed.

structural systems with walls—for frame and frame-equivalent dual systems[2] $k_w = 1.00$; for wall, wall-equivalent[3] and systems that are torsionally flexible,[4] $k_w = (1 + \alpha_0)/3 \le 1$, but no less than 0.5, where $\alpha_0$ = the wall aspect ratio (height of the wall divided by length of the wall cross section).

For buildings that are regular[5] in elevation the value of $q_0$ is indicated in Table 16.2.

Table 16.2 requires knowledge of the overstrength coefficient $(\alpha_u/\alpha_y)$. The overstrength coefficient is defined in terms of a pushover analysis (Figure 16.6), although EC8 provides guidance for evaluating this coefficient in the absence of a pushover analysis—for example, for multistory, multi-bay buildings that are regular in plan, $\alpha_u/\alpha_y$ may be taken equal to 1.3 (EC8 Section 5.2.2.2(5)a). Note here, a clear difference relative to U.S. codes: EN1998 conservatively defines overstrength on the basis of the first yield point, while ASCE 7-16 uses the design base shear instead. Thus, the EN1998 definition disregards any overstrength from conservative sizing of cross sections, using centerline rather than clear dimensions, etc.,

Table 16.2 Basic value of the behavior factor, $q_0$, for systems regular in elevation

| Structural system | Ductility class DCM | Ductility class DCH |
|---|---|---|
| Frame system, dual system, coupled wall system | $3.0 \cdot \alpha_u/\alpha_y$ | $4.5 \cdot \alpha_u/\alpha_y$ |
| Uncoupled wall system | 3.0 | $4.0 \cdot \alpha_u/\alpha_y$ |

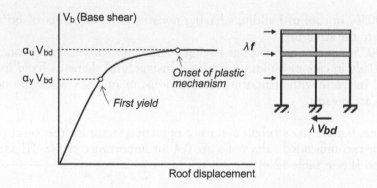

Figure 16.6 Capacity curve from pushover analysis, and overstrength coefficient.

[2] Dual systems are structural systems in which the vertical loads are mainly resisted by a framed structures and the lateral load is resisted in part by a wall.

[3] Wall-equivalent dual systems are those in which the shear resistance of the walls at the building base is higher than 50% of the total seismic resistance of the whole structural system.

[4] Dual or wall system not having a minimum torsional rigidity (EC8 Sections 5.2.2.1(4) and (6)), i.e.: $r_x \ge l_s$
$r_x$ is the square root of the ratio of the torsional stiffness to the lateral stiffness in the $y$ direction (torsional radius).
$l_s$ is the radius of gyration of the floor mass in plan (square root of the ratio of (i) the polar momentum of inertia of the floor mass in plan with respect to the center of mass of the floor to (ii) the floor mass).

[5] A building is considered to be regular in plan and in elevation if it fulfills a set of characteristics given by EC8 Sections 4.2.3.2 and 4.2.3.3, respectively, to assure that the building has a homogeneous behavior. Among others in terms of lateral stiffness: the building is regular in plan if the building structure is approximately symmetrical in plan with respect to the two orthogonal axes; the building is regular in elevation if lateral stiffness of individual stories remain constant or reduce gradually, without abrupt changes, from the base to the top.

resulting to $\alpha_u/\alpha_y$ factors that are clearly smaller than the corresponding $\Omega_0$ overstrength factors employed in North America.

Required details and proportioning requirements for structures designed for classes DCM and DCH have to be satisfied. EC8 Section 5.4 addresses requirements for DCM structures; EC8 Section 5.4 addresses requirements for DCH structures.

Low ductility structures are recommended only in regions of low seismicity. Low ductility structures must have Class B or C reinforcing steel (EC2, Table C1) and can have a behavior factor, $q$, up to 1.5 due to overstrength of the materials.

### 16.4.2 Story drift limits

Story drift limits pertain to the SLS-denominated "damage limitation requirement." The SLS seismic action has a larger probability of occurrence than that corresponding to the ULS—EC8 Section 2.1 recommends the SLS have a probability of exceedance of 10% in 10 years (equivalent to a mean return period of 95 years). Floor displacements induced under the seismic design actions ($d_s$) may be calculated as the behavior factor ($q$) times the displacements ($d_e$) obtained in a linear analysis using the design spectra for elastic analysis (EC8 Section 3.2.2.5): $d_s = q \cdot d_e$. This is in contrast to U.S. codes that stipulate the use of a system-type dependent $C_d$ factor, which is generally less than $R$, for estimating displacements. The corresponding interstory drift, $d_r$, is calculated as the difference of the average lateral displacement $d_s$ at the top and bottom of the story under consideration. The computed story drifts under the SLS are considered acceptable (EC8 Section 4.4.3.2) if the following are satisfied:

- $d_r v \leq 0.005h$, in case of buildings having nonstructural elements of brittle materials attached to the structure.
- $d_r v \leq 0.0075h$, in case of buildings having ductile nonstructural elements.
- $d_r v \leq 0.010h$, in case of buildings having nonstructural elements fixed in a way so as to not to interfere with structural deformations, or in cases where no nonstructural elements are present.

In the preceding, $h$ is the story height and $v$ is a reduction factor for the lower return period of the SLS. The recommended values of $v$ are 0.4 for importance classes III and IV and 0.5 for classes I and II (see Table 12.4).

## 16.5 DESIGN OF BEAMS AND COLUMNS IN DCM AND DCH STRUCTURES

Detailing practice for earthquake-resisting structures is similar in ACI and in EC8. In order to avoid repetition this chapter presents the design rules and detailing concisely in the form of tables supplemented with some discussion.

In earthquake design of structures, a distinction is made between primary and secondary structural members. Primary structural members are designated as part of the lateral force-resisting system; secondary members may deform under seismic action but are not part of the designated lateral force-resisting system.

Strong column–weak beam design is enforced in the design of DCM and DCH frames (including beams in wall systems and coupling beams) in order to avoid weak-story collapse mechanisms. To ensure the beam is the weak (or ductile, yielding) link in the chain

(Chapter 15), the following inequality must be satisfied at each beam-column joint of the lateral force-resisting system:

$$\sum M_{Rc} \geq 1.3 \sum M_{Rb} \tag{16.4}$$

where

$\Sigma M_{Rc}$ is the sum of the design values[6] of the flexural strengths of the columns framing into the joint. Because the column flexural strength is a function of the axial load, the smallest of the flexural strengths for the range of column axial forces produced in the seismic condition should be used. These may be calculated based on the relevant load combinations rather than being determined in nonlinear analysis of the structure.

$\Sigma M_{Rb}$ is the sum of design values of the flexural strengths of the beams framing the joint.

The inequality must be satisfied for lateral loading in either direction (e.g., left or right), which may induce different axial forces depending on whether the lateral loading induces overturning tension or overturning compression into the beam-column joint under consideration.

In the event that nonlinear dynamic analyses are used, the required column strengths can be determined to avoid a story mechanism. The presence or absence of a story mechanism can readily be determined by principal components analysis (Chapter 6) of the nonlinear dynamic floor displacement response data.

The main requirements for beams and columns (including beams attached to walls) focus on the ductile capacity of potential hinges, also named critical regions. The critical regions define locations where special details apply. As shown in Figure 16.7, critical regions are located at the ends of the beams and columns, extending a distance of $l_{cr} = 1.5h_w$ for DCH and $h_w$ for DCM and DCL from the column face where $h_w$ is the depth of the beam.

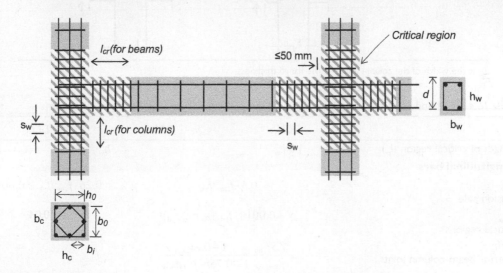

Figure 16.7 Critical regions in beam and column framing and transverse reinforcement spacing.

---

[6] The term "design values" indicates that material safety factors are applied to determine material properties used in determining calculated strengths; in contrast, "characteristic values" indicates that material safety factors are not applied in determining material properties.

Critical regions also exist at both sides of any other cross section with potential to yield under seismic loading. The length of critical regions for columns, $l_{cr}$, is detailed in Table 16.5.

### Design of beams

The top longitudinal reinforcement in the extremes of primary beams with T- or L-shaped cross sections must be placed within the effective flange width, $b_{eff}$, and may be concentrated mainly or entirely within the width of the web. Values of effective width, $b_{eff}$, are summarized in Table 16.3.

Table 16.4 summarizes the rules for detailing and dimensioning of primary beams according to EC2 and EC8, where:

**Table 16.3** Effective flange width, $b_{eff}$, for beams framing into columns

| Framing condition | Beam ending at an exterior column | Beam ending at an interior column |
|---|---|---|
| With transverse beam framing into the column | $b_c + 2h_f$ per side | $b_c + 4h_f$ per side |
| | | |
| Without transverse beam | $b_c$ | $b_c + 2h_f$ per side |
| | | |

$b_c$ is the width of the column and $h_f$ is the flange thickness.

**Table 16.4** EC-8 and EC-2 rules for detailing and dimensioning critical regions of primary beams[a]

| | DCH | DCM | DCL |
|---|---|---|---|
| Length of critical region ($l_{cr}$) | $1.5 \cdot h_w$ | $h_w$ | |
| **Longitudinal bars** | | | |
| $\rho_{min}$ tension side | $0.5 \cdot f_{ctm}/f_{yk}$ | | $0.26 \cdot f_{ctm}/f_{yk} \geq 0.0013$ |
| $\rho_{max}$ critical regions[b] | $\rho' + 0.0018 \cdot f_{cd}/(\mu_\varphi \varepsilon_{sy,d} f_{yd})^c$ | | 0.04 |
| $d_{bL}/h_c$ Interior beam-column joints | $\leq \dfrac{7.5 f_{ctm}}{\gamma_{Rd} f_{yd}} \dfrac{1 + 0.8\nu_d}{1 + 0.75 k_D \rho'/\rho_{max}}$ | | |
| | $\gamma_{Rd} = 1.2$ | $\gamma_{Rd} = 1.0$ | - |
| $d_{bL}/h_c$ Exterior beam-column joints | $\leq \dfrac{7.5 f_{ctm}}{\gamma_{Rd} f_{yd}}(1 + 0.8\nu_d)$ | | |
| | $\gamma_{Rd} = 1.2$ | $\gamma_{Rd} = 1.0$ | - |

(Continued)

Table 16.4 (Continued)  EC-8 and EC-2 rules for detailing and dimensioning critical regions of primary beams[a]

| | DCH | DCM | DCL |
|---|---|---|---|
| $A_{s,min}$, top and bottom | $2\Phi14$ | | - |
| $A_{s,min}$, top-span | $A_{s,top-supports}/4$ | | - |
| $A_{s,min}$, critical regions bottom | | $0.5 \cdot A_{s,top}$[c,d] | |
| $A_{s,min}$, supports bottom | | $A_{s,bottom-span}/4$ | - |
| **Transverse bars** | | | |
| *Outside critical regions* | | | |
| $s_w$ spacing | | $\leq 0.75d$ | |
| $\rho_w$ | | $\geq \left(0.08\sqrt{f_{ck}}\right)/f_{yk}$ in MPa | |
| *In critical regions* | | | |
| $d_{bw}$ | | $\geq 6\,mm$ | |
| $s_w$ spacing (Figure 16.7)$\leq$ | $6d_{bL}$, $h_w/4$, $24d_{bw}$, $175\,mm$ | $8d_{bL}$, $h_w/4$, $24d_{bw}$, $225\,mm$ | - |
| **Shear design** | | | |
| $V_{Ed}$, seismic | $1.2\dfrac{\sum M_{Rd}}{l_{cl}} \pm V_{o,g+\Psi_{2q}}$ | $\dfrac{\sum M_{Rd}}{l_{cl}} \pm V_{o,g+\Psi_{2q}}$ | - |
| | Capacity shear design of beams[e] | | |
| $V_{Rd,max}$, seismic[f] | | $= 0.3(1 - f_{ck}/250)b_w z f_{cd}\sin 2\theta$, $1 \leq \cot\theta \leq 2.5$ | |
| $V_{Rd,s}$, Outside critical regions | | $= b_w z \rho_w f_{ywd}\cot\theta$, $1 \leq \cot\theta \leq 2.5$ | |
| $V_{Rd,s}$, Critical regions | $\theta = 45°$ | $= b_w z \rho_w f_{ywd}\cot\theta$, $1 \leq \cot\theta \leq 2.5$ | |
| | If $\zeta = V_{Emin}/V_{Emax} < -0.5$ and $\left| V_E \right|_{max} > (2 + \zeta)$ $f_{ctd}b_w d$ inclined bars In two directions are needed. EC-8 §5.5.3.1.2(3). | - | |

[a] Adapted from Bisch et al. (2012). MPa indicates the use of mm, N, and MPa units.

[b] $\mu_\varphi$ is the curvature ductility factor that corresponds to the basic value of the behavior factor ($q_0$) used in design, which may be taken as $\mu_\varphi = 2q_0 - 1$ if $T \geq T_C$ and $\mu_\varphi = 1 + 2(q_0 - 1)T_C/T$ if $T < T_C$. These $q_0 - \mu_\varphi$ relations are proposed in EC8 Section 5.2.3.4(3), but EC8 allows the user to propose other relationships or modifications of these relationships.

[c] These conditions need not be checked if the curvature ductility factor, $\mu_\varphi$, is at least equal to the values specified in footnote (c), below. The curvature ductility factor is defined as the ratio of the post-ultimate strength curvature at 85% of the moment of resistance, to the curvature at yield, provided that the limiting strains of concrete and steel are not exceeded.

[d] The indicated minimum area is in addition to the compression steel relied upon for flexural strength.

[e] At a member end if the moment capacities around the joint satisfy $\Sigma M_{Rb} > \Sigma M_{Rc}$, $M_{Rb}$ is replaced in the calculation of the design shear force, $V_{Ed}$, by $M_{Rb}(\Sigma M_{Rc}/\Sigma M_{Rb})$. This is an approximate method to ensure that $V_{Ed}$ is calculated based on the plastic hinge formation at the end of the beam. $V_{o,g+\Psi_{2q}}$ represents the factored gravity load. This expression is similar to Equation 15.3 of Chapter 15.

[f] $\theta$ is the angle of the struts in the truss analogy for reinforced concrete member design. It can be freely chosen provided that $1 \leq \cot(\theta) \leq 2.5$, signifying inelastic behavior. If the orientation of the struts are chosen based on elasticity, then $\theta = 45°$ is a reasonable choice if no axial force is present.

$b_w$ = width of the web on T, I, or L beams
$h_w$ = cross-sectional depth of beam
$h_c$ = width of the column parallel to the bars
$d$ = effective depth of the cross section
$d_{bL}$ = diameter of the longitudinal bars
$d_{bw}$ = diameter of the hoop

*Table 16.5* EC-8 and EC-2 rules for detailing and dimensioning primary columns

| | DCH | DCM | DCL |
|---|---|---|---|
| Cross-sectional sides: $h_c, b_c \geq$ | 0.25 m $h_v/10$ if $P\delta/Vh > 0.1$ | | - |
| Length of critical region ($l_{cr}$) $\geq$ | $1.5b_c, 1.5h_c, 0.6\,m, l_{cl}/5$ | $b_c, h_c, 0.45\,m, l_{cl}/6$ | $b_c, h_c$ |
| **Longitudinal bars** | | | |
| $d_{bL}$ | | $\geq 8\,mm$ | |
| $\rho_{min} \geq$ | 1% | | $0.1 \cdot N_d/A_c f_{yd}, 0.2\%$ |
| $\rho_{max}$ | 4% (8% at laps) | | |
| Bars per side $\geq$ | 3 | | 2 |
| | In polygonal cross sections 1 per corner. 4 in circular cross sections | | |
| Spacing between longitudinal bars restrained in the corner of a hoop or cross tie bars$\leq$ | 150 mm | 200 mm | - |
| No bar within a compression zone should be further than 150 mm from a restrained bar | | | |
| **Transverse bars** | | | |
| *Outside critical regions* | | | |
| $d_{bw} \geq$ | | $6\,mm, d_{bL}/4$ | |
| Spacing $s_w \leq$ | | $20d_{bL}, h_c, b_c, 400\,mm$ | |
| Spacing at lap splices, if $d_{bL} \geq 14\,mm, s_w \leq$ | | $12d_{bL}, 0.6h_c, 0.6b_c, 240\,mm$ | |
| *In critical regions[a]* | | | |
| $d_{bw} \geq$ | $6\,mm, 0.4(f_{yd}/f_{ywd})^{1/2}d_{bL}$ | $6\,mm, d_{bL}/4$ | |
| Spacing $s_w \leq$ | $6d_{bL}, b_0/3, 125\,mm$ | $8d_{bL}, b_0/2, 175\,mm$ | - |
| $\omega_{wd} \geq$[b] | 0.08 | | - |
| $\alpha\omega_{wd} \geq$[b] | $30 \cdot \mu_\Phi \nu_d \varepsilon_{sy,d} b_c/b_0$ $- 0.035$ | | - |
| Critical regions at column base | | | |
| $\omega_{wd} \geq$[b] | 0.12 | 0.08 | - |
| $\alpha\omega_{wd} \geq$[b] | $30 \cdot \mu_\Phi \nu_d \varepsilon_{sy,d} b_c/b_0 - 0.035$ | | - |
| Verification for $M_x$–$M_y$–$N$ | Truly biaxial, or uniaxial with ($M_z/0.7, N$) and ($M_y/0.7, N$) | | |
| Axial load ratio $\nu_d = N_{Ed}/A_c f_{cd}$ | $\leq 0.55$ | $\leq 0.65$ | - |
| **Shear design** | | | |
| $V_{Ed}$, seismic | $1.3\dfrac{\sum M_{Rc}^{ends}}{l_{cl}}$ | $1.1\dfrac{\sum M_{Rc}^{ends}}{l_{cl}}$ | - |
| | Capacity shear design of columns[c] | | |
| $V_{Rd,max}$, seismic | $= 0.3(1 - f_{ck}/250)b_w z f_{cd}\sin(2\theta), \quad 1 \leq \cot(\theta) \leq 2.5$ | | |
| $V_{Rd,s}$, seismic[d] | $= b_w z \rho_w f_{ywd}\cot(\theta) + N_{Ed}(h - x)l_{cl}, \quad 1 \leq \cot(\theta) \leq 2.5$ | | |

[a] In case of DCM, for $q \leq 2$, the transverse reinforcement in critical regions of columns with axial load ratio ($\nu_d$) smaller than 0.2 may follow DCL design rules. In case of DCH, in the lower two storeys of the building, hoops designed for critical regions shall be provided beyond these regions for an additional length equal to half the length of these regions.

[b] In order to provide sufficient ductility adequate confinement is needed to compensate the lack of resistance due to the spalling of the cover. A study of the confinement is needed—see Sections 5.4.3.2.2 and 5.5.3.2.2.

[c] At a member end if the moment capacities around the joint satisfy $\Sigma M_{Rb} < \Sigma M_{Rc}$, $M_{Rc}$ is replaced in the calculation of the design shear force, $V_{Ed}$, by $M_{Rc}(\Sigma M_{Rb}/\Sigma M_{Rc})$. Likewise as in the previous table, this is an approximate method to ensure that $V_{Ed}$ is calculated based on the plastic hinge formation at the end of the beam.

[d] $x$ is the depth of the neutral axis at the end of the section in ULS of bending with axial load.

$f_{ck}$, $f_{cd}$, and $f_{ctm}$ are the characteristic strength of concrete, the design strength (=$f_{ck}/\gamma_c$ where $\gamma_c$ is the material safety coefficient of concrete) and the mean tensile strength of concrete, respectively.

$f_{yk}$, $f_{yd}$, and $f_{ywd}$ are the characteristic strength of steel, the design strength (=$f_{yk}/\gamma_s$ where $\gamma_s$ is the material safety coefficient of steel) and the design strength of the transverse reinforcement, respectively. In order to limit the strain to 0.002, the maximum value of the design strength of transverse reinforcement is limited to 400 MPa.

$k_D$ = factor reflecting the ductility class in the calculation of the required column depth for anchorage of beam bars in a joint, equal to 1 for DCH and 2/3 for DCM.

$l_{cl}$ = clear length of the column

$l_{cr}$ = length of the critical region

$V$ = shear force

$V_{Ed}$ = design value of the applied shear force

$V_{Rd}$ = shear resistance

$z$ = lever arm of internal forces

$\gamma_{Rd}$ = factor accounting for possible overstrength due to steel strain hardening.

$\rho$ = tension reinforcement ratio: reinforcement in the tension zone normalized by $b \cdot d$, where $b$ is the width of the compression flange of the beam.

$\rho'$ = compression reinforcement ratio: reinforcement in the compression zone normalized by $b \cdot d$, where $b$ is the width of the compression flange of the beam.

$\rho_l$ and $\rho_w$ = reinforcement ratio for longitudinal and shear reinforcement, respectively.

$\nu_d$ = axial force due in the seismic design condition ($N_{Ed}$), normalized by $A_c f_{cd}$

## Design of columns

Three criteria generally govern column design: (i) As noted in Chapter 15, the column section should be made sufficiently large that the design column axial force, $N_d$, is less than the balanced axial load, $N_{lim}$. This ensures that in the event of yielding of critical column sections, the curvature ductility capacity is based on yielding of the column longitudinal reinforcement. (ii) The column flexural strength under the range of design axial forces must satisfy Equation 16.4, to limit the potential for weak-story development. (iii) The column must be deep enough that joint shear stresses are not exceeded and that bond stresses on longitudinal beam reinforcement are not so large that excessive bar slip develops in the beam-column joint.

Similarly, Table 16.5 summarizes the rules for detailing and dimensioning of primary columns (secondary as in DCL) according to EC2 and EC8, where:

$h_v$ = the larger distance between the point of contraflexure and the ends of the columns, for bending within a plane parallel to the column dimension considered.

$b_c$ = breadth of the column.

$b_0$ = the width of confined core in a column or in the boundary element of a wall (to centerline of hoops)

$\omega_{wd}$ = the mechanical volumetric ratio of confining hoops within the critical regions. $\alpha$ in the expression $\alpha \cdot \omega_{wd}$ is the confinement effectiveness factor.

## 16.6 DESIGN OF WALLS IN DCM AND DCH STRUCTURES

As in other structural elements, the conditions imposed on detailing of boundary elements (edge regions of the wall cross section provided with confining reinforcement), are an indirect way to provide curvature ductility to the wall. Those conditions are specified in Tables 16.6 and 16.7, where (Figures 16.8–16.11):

*Table 16.6* EC-8 and EC-2 rules for detailing and dimensioning ductile walls

| | DCH | DCM | DCL |
|---|---|---|---|
| Web thickness, $b_{wo} \geq$ | 150 mm, $h_s/20$ | | - |
| Critical region length, $h_{cr}$ | $\geq \max(l_w, h_w/6), \leq 2l_w, \leq h_s$ for wall $\leq 6$ storys, $\leq 2h_s$ for wall $\geq 7$ storys | | - |
| **Boundary elements** (edge regions with confining reinforcement—see Figure 16.8) | | | |
| **a. Critical regions** | | | |
| length $l_c$ from edge $\geq$ | $0.15l_w$, $1.5b_w$, length over which $\varepsilon_c > 0.0035$ | | - |
| Thickness $b_w$ over $l_c$ | See Figure 16.9 | | |
| *Vertical reinforcement* | | | |
| $\rho_{min}$ over $A_c = l_c b_w$ | 0.5% | | 0.2% |
| $\rho_{max}$ over $A_c$ | 4% | | |
| *Confining hoops* | | | |
| Hoop bar diameter, $d_{bw} \geq$ | 6 mm, $0.4(f_{yd}/f_{ywd})^{1/2}d_{bL}$ | 6 mm | - |
| Spacing $s_w \leq$ | $6d_{bL}, b_0/3, 125$ mm | $8d_{bL}, b_0/2, 175$ mm | - |
| $\omega_{wd} \geq$[a] | 0.12 | 0.08 | |
| $\alpha\omega_{wd} \geq$[a] | $30 \cdot \mu_\Phi(\nu_d + \omega_v)\varepsilon_{sy,d}b_w/b_0 - 0.035$ | | - |
| **b. Over the rest of the wall height** | In parts of the section where $\varepsilon_c > 0.2\%$, then $\rho_{v,min} = 0.5\%$ elsewhere 0.2% | | |
| | In parts of the section where $\rho_L > 2\%$: distance of unrestrained bar in compression zone from nearest retrained bar $\leq 150$ mm | | |
| | Hoops with $d_{bw} \geq \text{Max}(6\,\text{mm}, d_{bL}/4)$ and spacing $s_w \leq \text{Min}(12d_{bL}, 0.6d_{wo}, 240\,\text{mm})$ up to a distance $4b_w$ above or below floor beams or slabs, or $s_w \leq \text{Min}(20d_{bL}, b_{wo}, 400\,\text{mm})$ beyond that distance | | |
| **Web** | | | |
| *Vertical bars (v)* | | | |
| $\rho_{v,min}$ | 0.5% if $\varepsilon_c > 0.002$, 0.2% elsewhere | | 0.2% |
| $\rho_{v,max}$ | 4% | | |
| $d_{bv} \geq$ | 8 mm | | |
| $d_{bv} \leq$ | $b_{wo}/8$ | | |
| Spacing $s_v \leq$ | $\text{Min}(25d_{bv}, 250\,\text{mm})$ | $\text{Min}(3b_{wo}, 400\,\text{mm})$ | |
| *Horizontal bars (h)* | | | |
| $\rho_{h,min}$ | 0.2% | $\text{Max}(0.1\%, 0.25\rho_v)$ | |
| $d_{bh} \geq$ | 8 mm | | - |
| $d_{bh} \leq$ | $b_{wo}/8$ | | - |
| Spacing $s_h \leq$ | $\text{Min}(25d_{bh}, 250\,\text{mm})$ | 400 mm | |
| Axial load ratio $\nu_d = N_{Ed}/A_c f_{cd}$ | $\leq 0.35$ | $\leq 0.4$ | - |
| Design moments $M_{Ed}$ | If $h_w/l_w \geq 2$, the shift rule of EC-2 Section 9.2.1.3(2) applies | | - |

[a] In order to provide sufficient ductility adequate confinement is needed. A study of the confinement is needed—see Sections 5.4.3.4.2 and 5.5.3.4.5.

$b_c$ width of the column
$h_s$ is the clear story height
$h_w$ is the height of the wall
$l_w$ is the length of the cross section of the wall

*Table 16.7* EC-8 and EC-2 rules for detailing and dimensioning ductile walls

**Shear design**

| | DCH | DCM | DCL |
|---|---|---|---|
| $V_{Rd,max}$ outside critical regions | EC2 applies, with lever arm $(z) = 0.8l_w$ and $\theta = 45°$ | - | |
| $V_{Rd,max}$ in critical regions | 40% of EC2 value | - | |
| $V_{Rd,s}$ if $\alpha_s = M_{Ed}/(V_{Ed}l_w) \geq 2$ | EC2 Section 6.2.3(1)–(7) apply with $z = 0.8l_w$ and $\theta = 45°$ | - | |
| $V_{Rd,s}$ if $M_{Ed}/(V_{Ed}l_w) < 2 \rightarrow \rho_h$ from | $V_{Ed} \leq V_{Rd,c} + 0.75\rho_h f_{yd,h}b_{w0}\alpha_s l_w$ According EC-2§6.2.3(8) | - | |
| $V_{Rd,s}$ if $M_{Ed}/(V_{Ed}l_w) < 2 \rightarrow \rho_v$ from | $\rho_h f_{yd,h}b_{w0}z \leq \rho_v f_{ydv}b_{w0}z + Min(N_{Ed})$ | - | |
| $V_{Ed}$, seismic[a] | $=\varepsilon V'_{Ed}$ where: $$\varepsilon = q\sqrt{\left(\frac{\gamma_{Rd}}{q}\frac{M_{Rd}}{M_{Ed}}\right)^2 + 0.1\left(\frac{S_e(T_C)}{S_e(T_1)}\right)^2}$$ and $1.5 \leq \varepsilon \leq q$ | $1.5V'_{Ed}$ | $V'_{Ed}$ |
| $V_{Ed}$ in dual systems[b] with $h_w > 2l_w$ | See Figure 16.10 | - | |
| Sliding shear failure | $V_{Ed} \leq V_{Rd,S}$ EC8 Section 5.5.3.4.4 | - | |
| $\rho_{min}$ at construction joints≥ | $0.0025, \dfrac{1.3f_{ctd} - N_{Ed}/A_c}{f_{yd} + 1.5\sqrt{f_{cd}/f_{yd}}}$ | - | |
| Design of coupling elements of coupled walls (see Figure 16.11) | Coupling beam can be designed as beams if either of the following is satisfied: $V_{Ed} \leq f_{ctd}b_w d$ or $l \geq 3h$ If a coupling beam cannot be designed as a beam, then $V_{Ed} \leq 2A_{si}f_{yd}Sin(\alpha)$ (See EC8 Section 5.5.3.5) | - | |

[a] $V'_{Ed}$ is the shear force obtained from the structural analysis. $q$ is the behavior factor used in the design. For the rest of the notation see EC8 Section 5.5.2.4.1(7).
[b] Dual systems are structural systems in which the vertical load is mainly resisted by a framed structure and the lateral load is partially resisted by a wall.

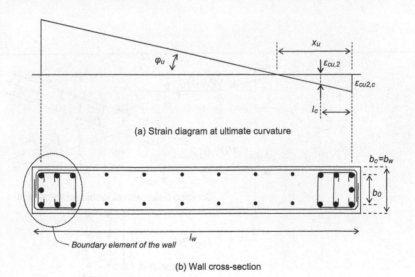

(a) Strain diagram at ultimate curvature

(b) Wall cross-section

*Figure 16.8* Confined boundary element of free-edge wall end. (a) Strain diagram and (b) boundary element locations. (Adapted from EC-8).

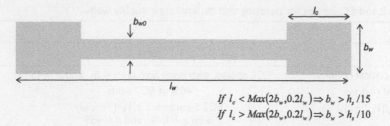

$$If\ l_c < Max(2b_w, 0.2l_w) \Rightarrow b_w > h_s/15$$
$$If\ l_c > Max(2b_w, 0.2l_w) \Rightarrow b_w > h_s/10$$

Figure 16.9 Minimum thickness for barbell cross-sectional walls. (Adapted from EC-8).

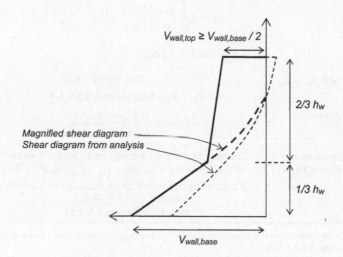

Figure 16.10 Design envelope of the shear force in slender walls of dual systems.

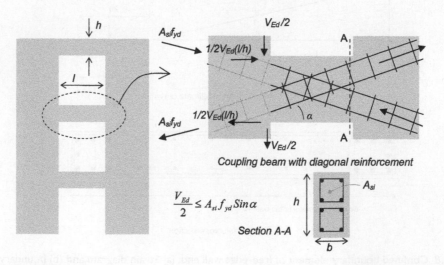

Coupling beam with diagonal reinforcement

$$\frac{V_{Ed}}{2} \le A_{si} f_{yd} Sin\alpha$$

Section A-A

Figure 16.11 Coupling beams.

$V_{\mathrm{Rd,S}}$ is the design value of shear resistance against sliding

$\varepsilon_{\mathrm{cu2}}$ ultimate strain capacity of unconfined concrete according to Table 3.1 of EC-2

$\varepsilon_{\mathrm{cu2,c}}$ ultimate strain capacity of confined concrete according to EC-2 Section 3.1.9

$\rho_{\mathrm{h}}$ is the reinforcement ratio of horizontal web bars

$\rho_{\mathrm{v}}$ is the reinforcement ratio of vertical web bars

## REFERENCES

Bisch, P., Carvalho, E., Degee, H., Fajfar, P., Fardis, M., Franchin, P., Kreslin, M., Pecker, A., Pinto, P., Plumier, A., Somja, H., and Tsiionis, G. (2012). Eurocode 8: Seismic design of buildings, Worked Examples, JRC Scientific and Technical Reports, Ispra, Italy.

EN 1990:2002+A1 (2005). Eurocode 0: Basis of structural design, European Committee for Standardization, Brussels.

EN 1992-1-1 (2004). Eurocode 2: Design of concrete structures-Part 1-1: General rules and rules for buildings, European Committee for Standardization, Brussels.

EN 1998-1 (2004). Eurocode 8: Design of structures for earthquake resistance – Part 1: General rules, seismic actions and rules for buildings, European Committee for Standardization, Brussels.

SHARE project (2013). Collaborative Project in the Cooperation programme of the Seventh Framework Program of the European Commission. www.share-eu.org.

# Chapter 17

# Component modeling and acceptance criteria

## 17.1 PURPOSE AND OBJECTIVES

Accurate modeling of the behavior of reinforced concrete components is necessary for obtaining reliable predictions of global (system) response and local (member) demands. The aim of this chapter is to provide context and depth to enable the engineer to exercise judgment in making reasonable modeling assumptions and in evaluating results of such models. Although limitations exist, modeling guidance and acceptance criteria are needed for use in nonlinear structural analysis. Thus, approaches for modeling reinforced concrete components are reviewed, along with available acceptance criteria (associated with component capacities at different performance levels).

## 17.2 INTRODUCTION

Analytical models for load–deformation response of reinforced concrete components typically have been developed by fitting analytical expressions to experimental data. A large number of experimental studies have been conducted, although each often is within a fairly narrowly prescribed parameter domain.

The response of reinforced concrete members to arbitrary combinations of axial load, shear, moment, and torsion in the inelastic regime is not well understood. Even for relatively simple loading in two dimensions, the inelastic response of reinforced concrete members is complex. Sources of inelastic deformation are manifold, including not only tensile and compressive strain of concrete and longitudinal reinforcement, but also slip of reinforcement associated with bond development and anchorage of reinforcing bars, shear deformation, and tensile cracking of concrete. Repeated cycles of inelastic deformation often lead to degradation of concrete and frictional slip along cracked surfaces and between the reinforcing steel and adjacent concrete. Sources of failure, which ultimately establish member deformation capacity, include low-cycle fatigue of longitudinal reinforcement (particularly after the reinforcement has buckled in prior compressive cycles), tensile fracture of transverse reinforcement, exhaustion of concrete in compression, failure of concrete in shear, and pullout of reinforcing bars. Observed failure modes can be sensitive to member geometry, reinforcement content or spacing, and loading history.

While flexural strength can be estimated with good accuracy, our ability to predict experimental behavior, or even the values of key parameters that characterize this behavior, is still quite limited. No model exists that can reliably predict behavior for members of arbitrary geometry and reinforcing, subjected to an arbitrary loading history. Even within particular domains, significant disparities exist among deformation capacity estimates obtained with different analytical models. In light of the poor precision available, simplifications may be introduced, but caution is needed, as unwarranted simplification can obscure underlying behavior and lead to incorrect conclusions about the effects of parameters on inelastic response.

This is a fertile area of study; modeling guidance and acceptance criteria can be expected to improve over time.

## 17.3 BACKGROUND

### 17.3.1 Moment–curvature response

The intended mechanisms used to resist earthquake actions in reinforced concrete lateral force-resisting systems all rely on the development of inelastic response in flexure, typically of members having no axial load, in the case of beams, or where axial load (expressed as $P/A_g f_c'$) is well below the balance point in walls and at the base of multistory columns. The simplest models determine the flexural load–deformation response at the member level by considering the response of individual cross sections along the member length.

Analytical approaches can be used to calculate the moment-curvature response of a beam, column, or wall cross section. The most generally applicable approach utilizes a so-called strain-compatibility analysis. In this approach, Bernoulli's assumption that plane sections remain plane is adopted, and iteration is used to establish static equilibrium ($\Sigma F_x = 0$) over the height of the cross section at a given curvature or reference strain value (Figure 17.1). The tensile capacity of the concrete is considered to obtain the cracking moment, $M_{cr}$, but is usually neglected at larger curvatures. The iterative process is concluded when horizontal equilibrium is achieved (within a specified tolerance), at which point the sum of moments is used to determine the external moment resisted by the cross section, for whatever externally imposed axial load, $P$, is acting on the section. This process is repeated successively at larger deformations, to obtain the moment–curvature response of the cross section.

Typically, some softening of concrete occurs prior to yielding of tension reinforcement. At larger curvatures, loss of concrete cover (the concrete outside of the transverse reinforcement) causes a loss of flexural strength, but strength increases subsequently as the longitudinal reinforcement strain hardens. The contribution of core concrete (the concrete contained within the transverse reinforcement) at large strains diminishes, causing the compressive force resultant to move lower, closer to the tension reinforcement. This reduction in lever arm causes the moment resistance to decrease. The ultimate curvature, $\phi_u$, is often defined at the curvature corresponding to a 20% reduction in peak moment, as shown in Figure 17.1.

Key features of moment–curvature response can be determined by simple calculation. For beams, the uncracked stiffness ($E_c I_{gt}$) and cracking moment, $M_{cr}$, can be determined by methods described in elementary reinforced concrete texts. For beams having a rectangular cross section, the neutral axis depth of the cracked beam section (Figure 17.1b) is $k{\cdot}d$, where $k$ can be determined for a tensile reinforcement ratio $\rho = A_s/b_w d$ and compressive reinforcement ratio $\rho' = A_s'/b_w d$ by application of first principles to the cracked, transformed area to obtain:

$$k = \sqrt{2\left(\rho n + \frac{\rho'(n-1)d'}{d}\right) + \left(\rho n + \rho'(n-1)\right)^2} - \left(\rho n + \rho'(n-1)\right) \tag{17.1}$$

where $n$, the modular ratio is $E_s/E_c$, and dimensions are as defined in Figure 17.1. Following methods described in elementary reinforced concrete texts, the moment of inertia of the cracked section can be computed. The flexural strength, $M_p$, can be established as that equal

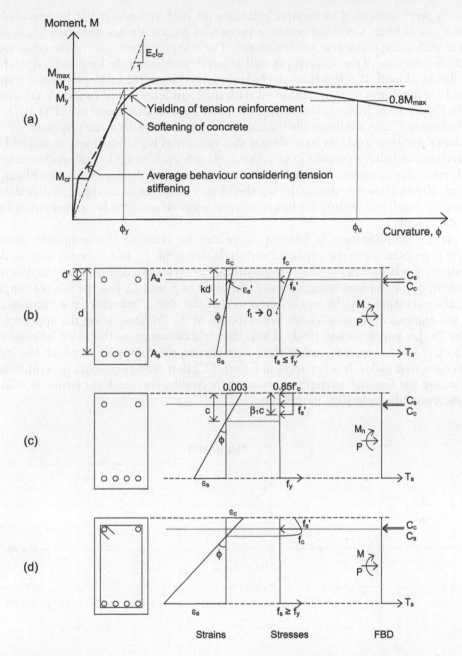

*Figure 17.1* (a) Moment–curvature response of beam cross section and bilinear approximation, (b) uncracked section analysis, (c) nominal flexural strength analysis, (d) post-nominal strength analysis.

to the value determined using the usual American Concrete Institute (ACI) approach to calculate the nominal moment, $M_n$, but using expected material properties $f'_{ce}$ and $f_{ye}$ instead of the nominal properties $f'_c$ and $f_y$ specified in the design. Thus, $\phi_y$ can be estimated as $\phi_y = M_p/E_c I_{cr}$, and the basic bilinear curve (Figure 17.1a) can be calculated by hand.

Typical software implementations of strain-compatibility analyses, such as BIAX-2 (Wallace, 1992), CUMBIA (Montejo and Kowalsky, 2007), or XTRACT (Chadwell et al., 2002), discretize the cross section into a large number of fibers, each representing a discrete concrete or steel element. Constitutive relations for each material must be specified and assigned to each fiber. Confined concrete properties frequently are specified for concrete contained within the transverse reinforcement. The computed behavior of the cross section may include softening of the concrete, as well as tensile yielding of the longitudinal reinforcement if the axial load, $P$, is less than the balanced axial load, $P_b$. Examples of the response computed for a column section reinforced with different amounts of longitudinal steel are shown in Figure 17.2. In the cases shown, the axial load was set equal to $0.10\,(A_g f_c')$. Also shown in Figure 17.2 is a bilinear idealization of the moment–curvature response.

Nonlinear dynamic analyses have shown that the initial higher stiffness associated with uncracked section behavior results in an appreciable reduction in peak displacement response for peak roof displacements less than about twice the yield displacement (Aschheim and Browning, 2008). However, due to drying shrinkage, temperature changes, settlement, and prior gravity, wind, and seismic loading, concrete components often have experienced some degree of cracking.

As a design simplification, a bilinear curve can be fitted to the computed moment–curvature responses. Since the expected materials strengths, $f_{ye}$ and $f_{ce}'$, were used to define the constitutive relations in the strain-compatibility analyses, we define $M_p$ as the ACI nominal moment, $M_n$, computed using $f_{ye}$ and $f_{ce}'$ in place of $f_y$ and $f_c'$. For the bilinear approximation, the curvature at yield was selected so that the elastic branch of the bilinear curve crosses the computed moment–curvature response at $0.75M_p$ (following the approach suggested by Paulay and Priestley, 1992). Thus, the yield curvature of the fitted bilinear curve is an effective yield curvature, rather than representing the curvature at which the tension reinforcement first yields. It is apparent in Figure 17.2 that even as changes in reinforcement content cause the flexural strength to more than double, the yield curvature is relatively stable, increasing by only 30% in this case.

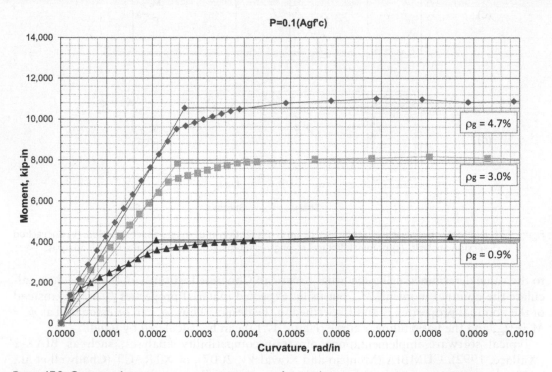

*Figure 17.2* Computed moment–curvature responses for a column cross section.

## 17.3.2 Plastic hinge models for load–deformation response of members

As indicated in Figures 17.1 and 17.2, moment–curvature response is nonlinear, with large curvatures developing for moments approaching and exceeding $M_p$. Experimental tests show that tensile elongation of reinforcement is accompanied by the development of cracks in the concrete at discrete locations. The cracks indicate that slip must be taking place between the reinforcement and the concrete (since the reinforcement is continuous). Slip represents a violation of the Bernoulli hypothesis, but nevertheless, moment–curvature analyses made with this hypothesis remain useful. For moments above the cracking moment, the results of a moment–curvature analysis apply strictly only to a section that has cracked in flexure. The concrete between cracked sections develops some tensile stress owing to bond between the concrete and reinforcement. This phenomenon, known as tension stiffening, causes a minor increase in stiffness relative to the cracked section behavior (see dashed curve in Figure 17.1). To determine the average moment–curvature response, the constitutive relationship for the reinforcement can be modified to represent tension-stiffening behavior. For example, based on Model Code 2010 (2012), the average tensile stress–strain response of concrete before the first yield in steel can be represented by

$$\sigma_{ct} = -\frac{\rho}{2} E_s \varepsilon + \sqrt{\left(\frac{\rho}{2} E_s \varepsilon\right)^2 + f_{ct}^{\ 2}(1 + n\rho)} \tag{17.2}$$

where $\rho$ is the reinforcement ratio and $n$ is $= E_s/E_c$ Hernández-Montes et al. (2017).

The simplest model for the response of a ductile reinforced concrete member is a plastic hinge model. The curvature distribution in a real cantilever beam may be irregular owing to the presence of cracks, the development of nonlinearity in the moment–curvature response, and slip of the reinforcement (Figure 17.3a). In the plastic hinge model, the curvatures are separated into an elastic component extending over the length of the beam and an inelastic component concentrated over a plastic hinge length, $l_p$.

Curvature over a differential length dx causes a differential rotation $d\theta = \phi \cdot dx$, where $\theta = du/dx$. With reference to the cantilever beam shown in Figure 17.3, if the moment diagram is linear and $E_c I_{cr}$ is constant, the curvature distribution is also linear.

$$\phi(x) = \frac{M(x)}{E_c I_{cr}} \tag{17.3}$$

In this case if the maximum curvature is smaller than the yield curvature ($\phi < \phi_y$) the displacement at the tip of the cantilever in Figure 17.3 is:

$$\phi(x) = \phi \frac{L - x}{L}$$

$$u = \int\left(\int \phi(x)dx\right)dx = \phi\left(\frac{x^2}{2} - \frac{x^3}{6L}\right)\Bigg|_0^L = \phi\frac{L^2}{3} \tag{17.4}$$

If the curvature is greater than the yield curvature ($\phi > \phi_y$) the peak displacement, or displacement capacity, is given by the sum of contributions from the elastic and inelastic curvature contributions:

$$u_u = u_y + u_p = \phi_y \frac{L^2}{3} + (\phi_u - \phi_y)(l_p)\left(L - \frac{l_p}{2}\right) \tag{17.5}$$

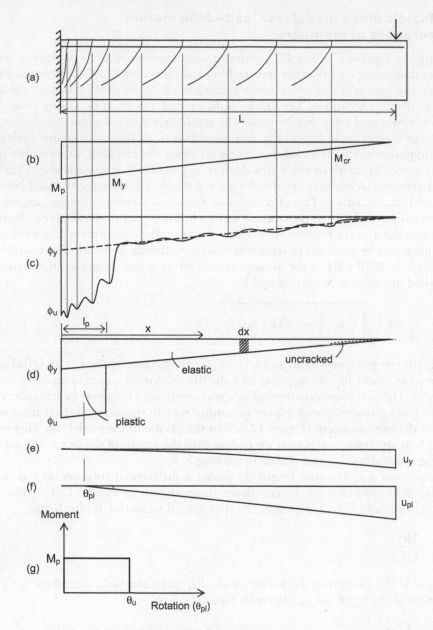

*Figure 17.3* Plastic hinge model for cantilever beams: (a) elevation view of cantilever beam, (b) moment diagram (neglecting self-weight), (c) schematic curvature diagram, (d) idealized curvature diagram, (e) displaced shape associated with the elastic portion of curvature diagram, (f) displaced shape assuming portion of curvature diagram is concentrated at a discrete location, and (g) relation between plastic moment and rotation.

One may wonder why the uncracked stiffness was ignored. The higher stiffness associated with uncracked behavior results in a reduction of curvature, represented by the small triangle close to the tip of the cantilever (see Figure 17.3d). Due to the short distance to the tip, this contribution is small and can be neglected.

More significant is the question of how to establish the values of $\phi_u$ and $l_p$. Research studies have generally calculated $l_p$ to fit experimental results, given $\phi_u$ defined as the analytically determined curvature at which the moment resistance drops 20% relative to its peak value. Thus, the plastic hinge length is defined numerically to match experimental data, rather than representing a specific physical hinge. Values of $l_p$ are often taken equal to $h/2$ in design, which is generally considered to be a conservative value for typical flexural members (having shear span/depth ratios of 2 or more).

The plastic hinge model is often implemented as a lumped plasticity model, with a moment–rotation spring defined at the end of the beam clear span (Figure 17.3f). Along the beam, the elastic flexural stiffness, $EI$, is given by $E_c I_{cr}$, while the plastic hinge is defined as rigid-plastic, with a rotation capacity $\theta_u$ equal to $(\phi_u - \phi_y)l_p$.

Although derived for a cantilever, the plastic hinge model is often applied in cases where the beam is a continuous member; that is, where bending moments are present at both ends of the beam. However, as indicated in Figure 17.4, changes in the end moments result in movement of the inflection point. Thus, the moment gradient ($dM/dx = V$) at the plastic hinge location may deviate from that occurring for the cantilever member. We expect that in the presence of larger moment gradients (higher shears), the physical extent of the plastic hinge will be reduced (along with its rotation capacity), while for smaller moment gradients (lower shears), the physical extent should be increased. At one extreme, the beam would be subjected to a uniform moment, and hence develop $\phi_u$ along its entire length, leading to a very large plastic rotation capacity, $\theta_p$. Thus, plastic deformation capacity may deviate from that determined in tests of cantilever beams.

## 17.3.3 Model fidelity

Variability in the empirically observed displacement capacity and the difficulty of accurately estimating displacement capacity are apparent in the following example. In this case, several models for the displacement capacity of columns under reversed cyclic loading were applied to a data set comprising test results obtained in various research studies (Inel et al., 2007).[1] The "simple" model is the plastic hinge model described previously, while the other models considered different contributions to displacement capacity (e.g., flexure, shear, and anchorage slip). Figure 17.5 illustrates that the drift capacity estimated by the different models varies significantly. Moreover, Figure 17.6 shows how different the plastic hinge contributions to peak displacement capacity can be—considering the sources of deformation accounted for in each model, the model values of plastic hinge rotation capacity can be as little as one half to several times the values derived from the experimental data, even for the relatively well-confined cases of interest in new design.

Also noteworthy is that the effect of displacement history on response is not well understood presently, and is not reliably represented in even the best currently available models that represent degradation. Even in the relatively simple and well-studied case of a cantilever column subjected to reversed cyclic lateral displacements in one horizontal direction under constant axial load, resistance at any displacement is a function of the prior displacement history (e.g., Figure 17.7).

Also poorly represented in so-called stick models is the effect of member elongation on local and system-level response. Member elongation arises because of the relatively small depth of the compression zone in typical flexural members. Consequently, tensile strain is

---

[1] The column database consisted of 29 rectangular cross section reinforced concrete cantilevered columns subjected to quasi-static reversed cyclic lateral loading, with axial load ratios held constant throughout the tests. Minimum section dimension was 300 mm, at least eight longitudinal bars were present, each bar was laterally supported by transverse reinforcement, aspect ratio $L_s/d$ ranged between 2.86 and 4.83, axial load ratio $P/A_c f_c'$ ranged between 0.1 and 0.77, $f_c'$ ranged between 22 and 47 MPa, and longitudinal reinforcement ratio quantity and strength ranged between $1.5\% \leq \rho \leq 3.3\%$ and $430 \leq f_y \leq 510$ MPa, respectively.

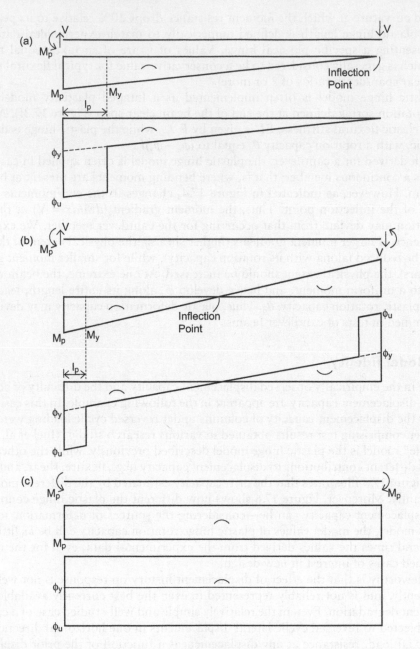

Figure 17.4 Effect of end moments on distribution of inelastic curvature. (a) Cantilever, (b) double curvature, and (c) uniform moment.

present along the member centerline. This causes longitudinal growth of beams, which may induce additional shears in first floor columns. Slender walls also elongate, thus picking up additional axial load as the beam framing is lifted up in the bays adjoining the walls.

Since most experiments are conducted quasi-statically, the question of strain rate effect is sometimes raised. Reinforced concrete members exhibit higher stiffness and strength when loaded rapidly. Reinforcement can sustain a stress greater than yield for a short period of time (known as "dwell time") before deformation associated with yielding develops

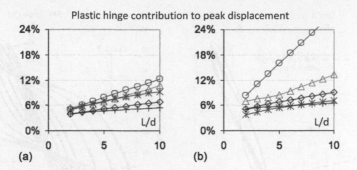

Figure 17.5 Drift capacity estimates for well-confined columns having axial load ratio $P/A_g f'_c$ equal to (a) 0.1 and (b) 0.5, as a function of aspect ratio (defined as shear span/member depth) (Inel et al., 2007).

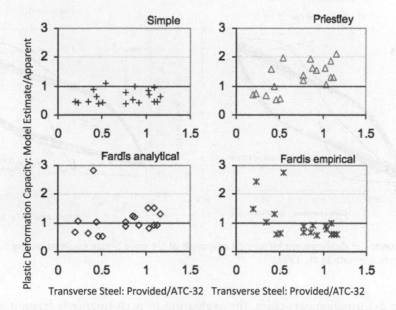

Figure 17.6 Contributions of plastic deformation to drift capacity: the ratio of the model estimate and the value apparent in the experimental result, after removing shear and anchorage slip estimates, is plotted (Inel et al., 2007).

(Massard and Collins, 1958). These effects are appreciable at the velocities associated with blast loading, but are usually small and disregarded at the velocities associated with earthquake loading. This is reasonable for members whose loading is associated with the relatively large deformation demands associated with the relatively slow first-mode response. It is possible that members whose force demands have significant higher mode components (such as the shears in multistory shear walls) may exhibit higher strength and stiffness due to strain rate effects.

## 17.3.4 Robust design in the context of modeling uncertainty

Performance evaluation asks the question, "Are demands less than capacities associated with the limit state of interest, at the desired level of confidence?" Given uncertainty in the accuracy of computed demands (due to modeling inaccuracy), and the relatively large

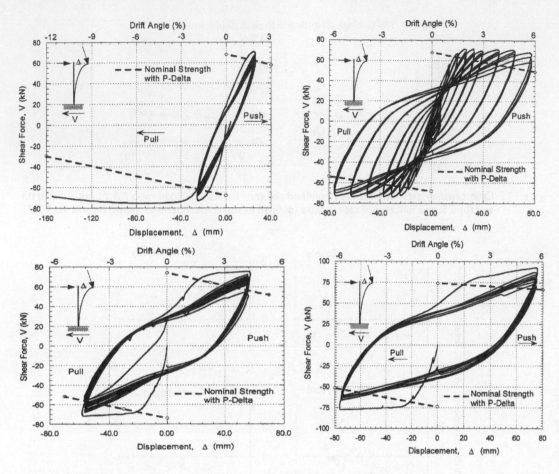

*Figure 17.7* Influence of displacement history on response of 1/4 scale bridge columns having ductile details. (From Kunnath et al., 1997.)

variability in deformation capacities, the evaluation of performance is fraught with uncertainty, particularly at the more extreme limit-states (approaching collapse). Nevertheless, nonlinear models provide results of much higher fidelity than can be obtained with the linear elastic models. Improvements in modeling recommendations will improve the fidelity of demand estimates; limit-state capacities may one day be described by their statistical distributions, rather than simply by point values.

To cope with the present uncertainties, it is worth noting that variability in flexural strength is much lower than variability in displacement capacity ($\beta$ values of around 0.15 for flexural strength versus perhaps 0.4 for displacement capacity). This is because well-designed reinforced concrete earthquake-resisting structures have axial loads well below the balance point, resulting in tension-controlled sections whose flexural strength primarily is a function of the longitudinal reinforcement.[2] Since the coefficient of variation of

---

[2] For members below the balance point, changes in concrete compressive strength have a relatively small effect on the lever arm, $jd$, used to compute the flexural strength. This is apparent in the formula for the flexural strength of a singly reinforced rectangular section: $M_n = A_s f_y d(j) = A_s f_y d\left(1 - 0.59\dfrac{A_s f_y}{bd f_c'}\right)$ where the concrete term has a relatively small effect.

reinforcement yield strength is relatively small (around 0.10–0.12), the variability of section flexural strength also is relatively small.

Thus, the relatively low variability in flexural strength makes design for plastic mechanisms feasible while allowing force-controlled members to be proportioned with sufficient strength to remain elastic. In effect, uncertainty in loading (represented by the seismic hazard curve and record-to-record variability in demands) is converted by the ductile mechanism(s) into uncertainty in inelastic deformation capacities relative to demands in the primary lateral force-resisting system. Because a degree of redundancy is present in the lateral force-resisting system, premature loss of lateral load resistance in a small proportion of elements does not cause failure (but instead leads to redistribution of load). Thus, the relatively large uncertainty in component deformation capacity is transformed to a smaller uncertainty in the response of redundant systems. Hence, there is value in providing a redundant lateral force-resisting system, capable of maintaining gravity load support even as some components reach their deformation capacities, complemented by a capable gravity load-resisting system.

## 17.4 EXPECTED MATERIAL PROPERTIES

Ultimate strength design is now commonly used for the design of reinforced concrete members. In ACI 318 for conventional, non-seismic design, nominal strengths are determined on the basis of the minimum materials strengths specified in design, and a strength reduction ($\phi$) factor is applied to obtain reliable strengths. In Eurocode 2, the strengths of the individual materials are reduced. The concrete design strength $f_{cd} = f_{ck}/\gamma_c$, where $f_{ck}$ = characteristic strength and $\gamma_c$ = concrete material safety factor ($\gamma_c = 1.5$ is non-seismic or accidental situations). Similarly, for steel $f_{yd} = f_{yk}/\gamma_s$, where $\gamma_s = 1.15$ in the same situations. In order to come closest to approximating the behavior of the actual structure, and to avoid distorting performance expectations by introducing biases in strengths, median strengths should be modeled, determined using median material properties and without application of a strength reduction factor. In the case of Eurocode 2, $f_{cm}$ is used ($f_{cm} = f_{ck} + 8$ MPa). Expected (mean) strengths are typically close enough to median values so that mean values may be used where median values are not available.

For both steel and concrete, actual material strengths generally exceed the nominal (minimum) values specified in design. This occurs because material suppliers are contractually bound to deliver material having strength greater than the minimum specified strength. For seismic applications, ACI 318 requires use of reinforcement for which lower and upper limits on yield strength are satisfied, along with a lower limit on ultimate (tensile) strength.

For reinforcing steel, PEER/ATC 72-1 (2010) recommends simply that the expected yield strength, $f_{ye}$, be taken equal to $1.2f_y$, where $f_y$ is the nominal, specified minimum yield strength. Thus, $f_{ye} = 1.2(60) = 72$ ksi for Grade 60 reinforcement. More detailed and slightly different recommendations are given by Caltrans (2006). With reference to Figure 17.8, Caltrans identifies expected properties for ASTM A706 Grade 60 reinforcement (having specified yield strength equal to 414 MPa) as given in Table 17.1.

Continued hydration of cement causes concrete strengths to increase beyond the usual 28- or 56-day time specified for cylinder tests. For concrete, PEER/ATC 72-1 (2010) recommends assuming the expected compressive strength $f'_{ce} = 1.25\ f'_c$, where $f'_c$ = the specified compressive strength. Caltrans (2006) recommends $f'_{ce} = 1.3\ f'_c$. For normal weight concrete, the elastic modulus can be estimated as $E_c = 57{,}000\ (f'_{ce})^{0.5}$ for $f'_{ce}$ and $E_c$ expressed in lb and in. units, and as $4{,}700\ (f'_{ce})^{0.5}$ for $f'_{ce}$ and $E_c$ expressed in MPa units.

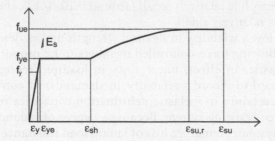

Figure 17.8 Expected stress–strain curve for steel reinforcement.

Table 17.1 Expected properties for mild steel reinforcement (Caltrans, 2006)

| Property | Value | |
|---|---|---|
| | Imperial units | SI units |
| Modulus of elasticity, $E_s$ | 29,000 ksi | 200 GPa |
| Expected yield strength, $f_{ye}$ | 68 ksi | 475 MPa |
| Expected yield strain, $\varepsilon_{ye}$ | 0.0023 | 0.0023 |
| Expected tensile strength, $f_{ue}$ | 95 ksi | 655 MPa |
| Onset of strain hardening, $\varepsilon_{sh}$ | | |
| #8 (#25m) | 0.0150 | 0.0150 |
| #9 (#25m) | 0.0125 | 0.0125 |
| #10, #11 (#32m, #36m) | 0.0115 | 0.0115 |
| #14 (#43m) | 0.0075 | 0.0075 |
| #18 (#57m) | 0.0050 | 0.0050 |
| Reduced ultimate strain, $\varepsilon_{su,r}$ | | |
| #10 (#32m) and smaller | 0.090 | 0.090 |
| #11 (#36m) and larger | 0.060 | 0.060 |
| Ultimate strain, $\varepsilon_{su}$ | | |
| #10 (#32m) and smaller | 0.120 | 0.120 |
| #11 (#36m) and larger | 0.090 | 0.090 |

## 17.5 PROPERTIES OF CONFINED CONCRETE

While building codes establish component strengths on the basis of unconfined concrete strengths, detailing requirements ensure that transverse steel present in potential plastic hinge zones will restrain the lateral dilation of the core concrete. This restraint of lateral expansion is known as confinement. Much as sand confined within a barrel displays greater bearing capacity, confined concrete is stronger and sustains compressive stress through substantially larger strains than unconfined concrete (e.g., Figure 17.9a). Confinement effects are particularly important for members undergoing substantial compressive strain, such as frame columns subjected to axial loads above the balance point and the cores of columns below discontinued shear walls. Since transverse confining stress also increases shear strength (Figure 17.9b), portions of members sustaining relatively high shears under inelastic reversals (such as plastic hinge locations) also benefit from transverse reinforcement. The transverse reinforcement is understood to contribute in two ways: (i) by providing

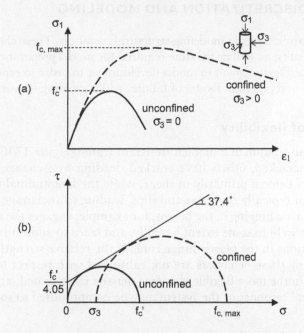

*Figure 17.9* Effects of confinement on concrete: (a) stress–strain response, and (b) representation of strength using Mohr–Coulomb failure surface (Aschheim, 2000).

confinement, it improves the shear resistance of the concrete even as the concrete degrades under reversed cyclic loading and (ii) the transverse reinforcement contributes directly to carrying shear (e.g., Aschheim, 2000; Pujol, 2002).

Various models for confined concrete are available. The motivated analyst may wish to explicitly account for confinement in establishing component backbone curves. Several confined concrete models are available in existing moment–curvature software programs. Note, however, that confinement effects on strength are relatively minor where axial loads are kept below the balance point, while effects on deformation capacity are already accounted for in the somewhat rudimentary modeling and acceptance criteria contained within this chapter.

## 17.6 NOMINAL, RELIABLE, AND EXPECTED STRENGTHS

ACI considers the "nominal strength" to be one calculated using code-prescribed formulas or assumptions, while the "reliable strength" is reduced considering a degree of uncertainty represented in the value of the strength reduction factor, $\phi$, which varies with the type of action and even the strains associated with that action. Thus, in flexure, the reliable flexural strength is given as the product of $\phi$ and the nominal flexural strength, $M_n$, where $\phi$ varies between 0.65 and 0.9 based on the tensile strain established in the sectional analysis. In many cases, it will be convenient to estimate the expected strength using the same provisions applicable in determination of the nominal strength, but with expected material strengths used in place of the nominal values and material safety or strength reduction factors set equal to 1.0.

## 17.7 ELEMENT DISCRETIZATION AND MODELING

The analyst faces many choices in modeling structural members. These choices affect the accuracy of computed results, as well as the time required for model preparation and computation. The analyst must exercise judgment in model development in order to represent the important sources of flexibility and potential modes of failure, while avoiding unwarranted complexity.

### 17.7.1 Sources of flexibility

Consider a beam and column of a moment-resistant frame (Figure 17.10). While some cross sections may be uncracked, others have cracked, leading to a change in elastic stiffness. Beam-column joints deform primarily in shear, while the longitudinal reinforcement passing through the joint typically elongates and slips, leading to anchorage slip rotations at the faces of the joint. Plastic hinging in the beams, for example, engages the slab (cast monolithically with the beam web) to some extent laterally, and leads to additional inelastic rotations and shear deformations in the plastic hinge zone. If the relative strengths and stiffnesses of elements representing these behaviors are not calibrated with respect to each other, deformations computed in the more flexible elements may be exaggerated, and the fidelity of the analysis of the global response of the system may be compromised to some degree.

### 17.7.2 Hysteretic behavior

Most tests of reinforced concrete components impose cyclic displacement histories. Several cycles of loading are usually imposed before advancing to a larger displacement amplitude. The envelope of response obtained under reversed cyclic loading is known as a skeleton curve. The skeleton curve is formed by connecting the peak points obtained for the first loading cycle at each increment in displacement amplitude (Figure 17.11).

In comparison to a full, bilinear hysteretic curve, reinforced concrete components exhibit degradation in stiffness and strength during reversed cyclic loading (Figure 17.11). As a result, the load resisted at a given displacement depends greatly on the stiffness represented in the model. In many cases, even where stiffness degradation occurs, the strength of the member can be attained if displacements are further increased (e.g., for displacement ductilities less than $\mu_8$ in Figure 17.11), but in other cases, strength degrades as well. If the strength obtained in subsequent cycles is reduced, this is termed cyclic degradation (e.g., for displacement ductilities greater than $\mu_8$ in Figure 17.11). Concrete cracking, frictional slip, and bond slip can contribute to cyclic degradation.

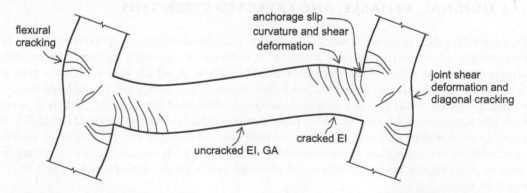

*Figure 17.10* Sources of flexibility in reinforced concrete framing (deformations exaggerated for clarity).

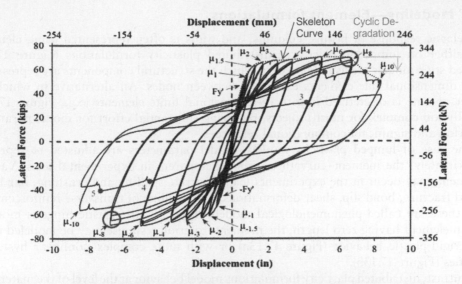

*Figure 17.11* Lateral load–displacement results from testing of a cantilever column, showing (i) definition of skeleton curve as that obtained from connecting peak responses at each new displacement amplitude, (ii) loss of stiffness apparent in consecutive cycles, and (iii) cyclic strength degradation associated with buckling of longitudinal reinforcement, for displacement ductility greater than $\mu_8$. (Adapted from Goodnight et al., 2012.)

Strength may also reduce as displacements increase within any given cycle, termed in-cycle degradation (see Figure 17.12). In-cycle degradation may result from concrete crushing, shear failure, buckling or fracture of reinforcement, and splice failure. In-cycle strength degradation is of particular concern if it allows a global mechanism to develop having negative post-yield tangent stiffness, since the negative stiffness would exacerbate displacement response, just as it occurs in the presence of $P$-$\Delta$ effects. For new, well-detailed reinforced concrete buildings, sources of in-cycle degradation are associated with fairly large drifts and would be particularly important to model for collapse assessment, but are of less significance for routine design.

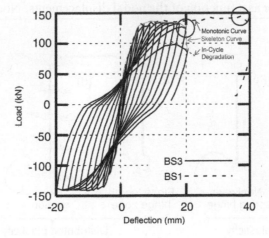

*Figure 17.12* Load–displacement response of a reinforced concrete column: dashed curve shows response under monotonic loading; solid curve shows response under reversed cyclic biaxial loading. (Adapted from Umemura and Ichinose, 2004.)

### 17.7.3 Modeling—Element formulations

The inelastic response of beams, columns, and walls is often represented using elements based either on lumped plasticity or distributed plasticity formulations (Figure 17.13). So-called stick models are usually used, in which the structural components are represented by one-dimensional line elements extending between nodes. An alternative in which the member volume is discretized into three-dimensional finite elements (e.g., Figure 17.13e) is usually too complex for most projects, requiring substantial effort for model parameter calibration and significant computational resources.

In the case of lumped plasticity models, model parameters are adjusted to represent (approximately) the moment–curvature response apparent in experimental tests. Various phenomena may occur in the experiments (e.g., concrete spalling and crushing, bar buckling and fracture, bond slip, shear deformations, and yielding of transverse reinforcement); hence, these are called phenomenological models. Typically, the plastic hinge is modeled using an element having zero length; the moment–rotation response may be modeled using simple rigid-plastic behavior (Figure 17.13a) or with more complex nonlinear hysteretic properties (Figure 17.13b).

In contrast, distributed plasticity formulations model behavior at the level of the material—in the case of a fiber model (Figure 17.13d), axial and flexural behavior is represented by discretizing the member into fibers, the response of each fiber is based on the appropriate material constitutive relation, and internal integration of the fiber responses results in the member response. Thus, while distributed plasticity formulations manifest effects at the member level that derive from the behavior of the constituent fibers, the models are more computationally intensive. Consensus recommendations on modeling phenomena such as buckling and fracture of reinforcing bars and the effects of shear on fiber stresses and strains have not been established; thus, the analyst typically must calibrate such models to experimental results.

#### 17.7.3.1 Distributed plasticity elements

Distributed plasticity models allow nonlinearities to develop anywhere along the member length. Displacement- and force-based formulations are available. The classic displacement-based formulation uses cubic Hermitian shape functions to express displacement fields along the length of the member as a function of the nodal displacements. Nonlinearity in response

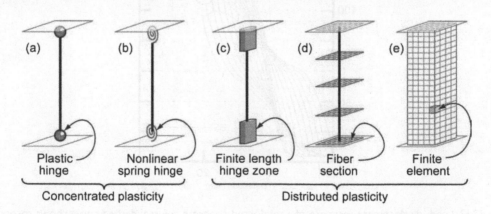

Figure 17.13 Beam-column element models: (a) and (b) illustrated concentrated plasticity models; (c) – (e) illustrate distributed plasticity models. (Courtesy of NIST GCR 10-917-5, 2010.)

is characterized by disproportionately large strains and curvatures as the moments approach the flexural strength. Since the cubic Hermitian shape functions can only represent linear variations of curvature, representing the displacement field accurately requires dividing regions of inelasticity into multiple elements. In contrast, in the more recently developed force-based formulation, internal forces (e.g., moments and shears) over the member length are related to the nodal forces. Because the internal forces are related by static equilibrium, a single element can be used to represent the complex, internal, distribution of deformation over the length of the structural member (Spacone et al., 1996a,b).

Fiber elements are especially useful for representing the flexural response of reinforced concrete members. In particular, reinforced concrete beams may have asymmetric behavior due to the presence of a floor slab and differences in positive and negative moment strengths (and stiffnesses) at critical sections; flexural stiffness can vary as the bending moment changes sign over the beam span, and thus can vary as the location of the inflection point changes under lateral loading. Also, fiber elements naturally represent the influence of axial load on flexural strength and stiffness. Because the concrete fibers are able to develop tensile strain, elongation of the centerline due to flexural cracking is also represented. Note, for example, that elongation along the centerline of a wall can generate additional axial compression in the wall due to restraint provided by other framing (Figure 17.14). Also shown in Figure 17.14 are very stiff or rigid links that run transverse to the fiber element to the face of the member.

Results computed using fiber elements display sensitivity to the number of sections (or integration points) at which fiber stresses and strains are monitored. Regularization techniques applicable to force-based fiber elements having perfectly plastic or strain-softening moment–curvature responses and post-processing of computed curvatures are described by Coleman and Spacone (2001). Post-processing is needed to establish performance relative to acceptance criteria, which are usually specified in terms of plastic rotations at discrete plastic hinges.

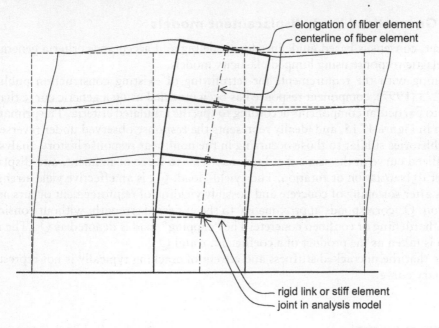

*Figure 17.14* Elongation of a fiber element along the wall centerline, and use of rigid links or stiff elements to connect to beam elements.

### 17.7.3.2 Lumped plasticity elements

In the lumped plasticity approach, discrete nonlinear moment–rotation hinges are placed at the ends of otherwise linear elastic elements. The hinge elements may have zero length, and connect nodes located at identical points in space. Discrete plastic hinges provide a computationally efficient and convenient approach for representing plastic hinging, and are often used, particularly for beam hinges. Some element formulations allow the axial load to influence the flexural strength of the plastic hinge; in others, the flexural strength is held constant throughout the analysis.

Hysteretic behavior is modeled using rules that describe the current (tangent) stiffness with respect to loading history. The most common models for representing hysteretic behavior are discussed in Chapter 4.

The lumped plasticity models usually represent moment and rotation at the end of a beam-column member. Hysteretic model properties are defined using a backbone curve, which is a boundary that defines the region that contains the load–deformation response. Initially, prior to the onset of degradation, the backbone curve resembles that obtained under monotonic loading (see Figure 17.12). Degradation that sets in during reversed cyclic loading causes the branches of the backbone curve to move toward the origin; thus, the so-called cyclic backbone curve generally varies with loading history. (In contrast, the skeleton curves trace the peaks obtained during a particular loading history, and also can be expected to vary with loading history.) Various approaches exist for modeling degradation (e.g., Ibarra et al., 2005; Sivaselvan and Reinhorn, 2000), and modeling parameters generally must be tuned to accurately represent relevant experimental results.

While more accurate simulations can be obtained with properly calibrated degrading models, simpler models may be adequate, in which the backbone curve is adjusted to approximate the skeleton curve and the details of cyclic degradation are not represented. These simplifications would be most applicable for the design of new structures where degradation is limited by good detailing practice and accurate assessment of collapse is not an objective.

## 17.7.4 Generalized load–displacement models

At present, consensus-based guidelines for modeling and acceptance criteria generally consider inelastic response using lumped plasticity models.

Beginning with the requirements for retrofitting of existing construction published as FEMA-273 (1997), component response has been modeled using a generic curve that is specialized to particular components according to specific tabulated criteria. The primary curve is shown in Figure 17.15, and ideally represents the response observed under reversed cyclic loading histories similar to those occurring in the nonlinear response history analysis.[3] It is a generalized curve in the sense that load may refer to force or moment, and displacement may refer to translation or rotation. The "yield" load, $Q_y$, is an effective yield strength that develops after softening of concrete and possibly yielding of reinforcement occurs at a critical section. $Q_y$ corresponds approximately to the expected strength, without consideration of strain hardening or confined concrete. The "capping" load is denoted as $Q_c$. The residual strength is taken as the product of a coefficient, $c$ and $Q_y$.

Notice that the uncracked stiffness and advent of cracking typically is not represented in the primary curve.

---

[3] Alternatively, a more complex approach models the monotonic backbone curve and degradation in strength and stiffness so as to represent results obtained under various reversed cyclic displacement histories.

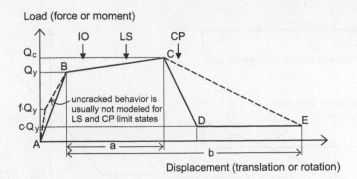

*Figure 17.15* Generalized load–displacement (backbone) curve for representing beam, column, joint, and wall component response, based on ASCE/SEI 41 (2013). Acceptability limits are shown schematically for primary components of the lateral force-resisting system for the IO, LS, and CP Levels.

Prior loading (due to settlement, gravity, wind loads, or a prior seismic event) may have caused moments in excess of the cracking moment to develop prior to the seismic loading under consideration. However, cracks in members such as columns and walls may close in the presence of compressive axial loads, resulting in stiffness comparable to the uncracked stiffness at small drifts. It appears that modeling of the uncracked stiffness is only important for peak roof drifts less than approximately twice the roof yield displacement (observed in a first-mode pushover analysis) (Aschheim and Browning, 2008). Thus, for relatively small drifts, conservative estimates of displacement response will be obtained if the uncracked stiffness is not modeled.

The relatively sharp drop in strength from Point C to Point D can create difficulties with convergence in the numerical solution of the dynamic response history. In these instances, some have recommended the primary curve be modified to run from Point C directly to Point E.

Parameters *a*, *b*, and *c* can be established based on the modeling and acceptance criteria reported in Section 17.8. Attempts to simulate collapse necessarily would require modeling degradation in strength and stiffness. Acceptance criteria values must be regarded with a degree of uncertainty, and can be expected to improve as nonlinear assessment becomes more common.

## 17.8 COMPONENT MODELING

### 17.8.1 Beams and Tee beams

Reinforced concrete beams typically are cast integrally with the floor slab and hence function as Tee beams.

#### 17.8.1.1 Effective stiffness

Because gravity loads must be limited to prevent the formation of plastic hinges along the span of the beam (see Section 15.5), and also will be small due to the relatively small tributary area for moment frames located at the building perimeter, moment diagrams will range between pure gravity load at one extreme and plastic hinging under lateral loads at the other extreme (Figure 17.16). For the plastic hinging case, the location of the inflection point is a

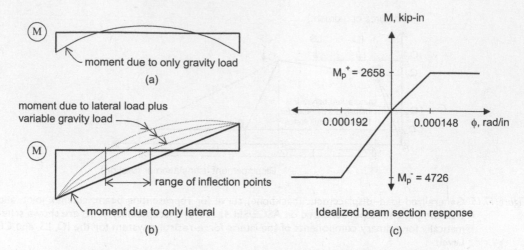

*Figure 17.16* Moments along the beam span under (a) gravity and (b) combined gravity and lateral load (upon formation of plastic hinges at the ends of the beam) illustrating the range of inflection point locations for one direction of lateral loading. The idealized moment–curvature response for an example section in (c) cannot be represented using typical discrete plastic hinge models, but may be represented using fiber element models.

function of the gravity load. Both the negative moment strength $M_p^-$ and yield curvature $\phi_y^-$ typically are greater than the corresponding positive moment values. The negative moment stiffness ($E_c I^- = M_p^-/\phi_y^-$) ordinarily will be a little larger than the positive moment stiffness. Because the sign of the moments varies along the span, the flexural stiffness will vary along the span (as a function of the location(s) of the inflection point(s)). Since the inflection point(s) will move under lateral load reversals, the overall stiffness and spatial variation of stiffness vary over time.

Further complicating this situation is that the lateral extent of the slab that contributes to the strength and stiffness of the Tee beam is not constant. Under negative moment, tensile cracks develop in the slab, running perpendicular to the longitudinal beam reinforcement. These cracks widen and lengthen with further deformation, developing tension in slab reinforcement over a greater horizontal extent, through shear lag. Shear lag effects, therefore, increase the effective width of the beam to include part of the slab as a flange; however, the extreme steel is not coupled to the beam web with as much stiffness as a plane sections hypothesis would imply. For simplicity and for compatibility with code provisions, an invariant effective flange width is usually assumed for defining the lateral extent of the Tee beam (the effective flange width). Specific recommendations for the effective flange vary. PEER/ATC 72-1 (2010) recommends setting $b_{eff}$ equal to the width of the web plus 1/8 of the beam span on each side of the web. ASCE/SEI 41 (2013) limits $b_{eff}$ to the width of the web plus 1/5 of the beam span on each side of the web, and not exceeding a distance of eight slab thicknesses on each side of the web. ACI 318 Section 18.7.3 cites the definition of effective width given in ACI 318 Section 6.3.2.1, which specifies the overhanging flange for beams with slab extending to one side of the web be limited to the smaller of 1/12 of the span and six slab thicknesses; for beams with slab extending on both sides of the web, the overhanging flange is limited to the smaller of eight slab thicknesses and $l_n/8$ on each side of the web. In all cases, $b_{eff}$ should not extend beyond the actual flange width and should not extend more than halfway toward the next beam (to avoid double-counting the slab).

The most accurate way to represent the beam stiffness is to use fiber elements to represent the effective Tee beam, using expected material properties to define the steel and concrete fiber material properties, because the appropriate flexural stiffness is represented at each section monitored by the element.

However, if a simple elastic beam-column element is to be used, a single value of EI is needed. Some guidelines recommend taking an average of the EI values determined under positive and negative moment. However, where gravity loads are relatively light and $M_p^-$ is larger than $M_p^+$, and given the significance of stiffness near the support on deflections and the possibility that a greater portion of the beam is under negative moment during lateral response, more weight should be given to the negative moment stiffness. One possibility is to use a weighted average, where the positive and negative moment strengths are used as the weights:

$$EI = \frac{EI^- M_p^- + EI^+ M_p^+}{M_p^- + M_p^+}$$
(17.6)

Beams of special moment-resisting frames will have $M_p^+ \approx \frac{1}{2} M_p^-$ to satisfy ACI code requirements, those designed to EC2 are subject to a similar requirement (Section 5.4.3.1.2(4)). This allows the weighted EI to be approximated on the basis of the negative moment alone. Using the following estimates of yield curvatures from Paulay (2003):

$$\phi_y^- = \frac{1.8\varepsilon_y}{h_b} \text{ (flange in tension)}$$
(17.7)

$$\phi_y^+ = \frac{1.4\varepsilon_y}{h_b} \text{ (flange in compression)}$$
(17.8)

the estimated weighted $EI$ value to be used in a prismatic representation of the beam span (in the absence of diagonal cracking) is

$$EI = \frac{M_p}{\phi_y} \approx \frac{M_p^- h_b}{2\varepsilon_y}$$
(17.9)

Inclined or diagonal cracking results when flexural cracks occur in the presence of shear. As shown in Figure 17.17, an internal truss can be considered, composed of diagonal compression struts and longitudinal and transverse reinforcement. Equilibrium requires that the tension force at any perpendicular cross section be elevated relative to the force in the compression chord. The elevated tension force results in larger tensile strains and hence greater curvature than would be calculated ignoring this so-called "tension shift." The additional strain in the tensile steel is associated with additional deformation or, in effect, a reduction in the flexural stiffness of the member. This effect can be accounted for by introducing the term $(1 + d/a)$ to reduce the flexural stiffness, resulting in the following estimate of the flexural stiffness for use in a prismatic representation of a Tee beam:

$$EI = \frac{M_p^- h_b}{2\varepsilon_y \left(1 + \dfrac{d}{a}\right)} \approx \frac{M_p^- h_b}{2\varepsilon_y \left(1 + \dfrac{h_b}{L/2}\right)}$$
(17.10)

where $d$ = depth to the tension reinforcement, $h_b$ = the overall height of the Tee beam, and $a$ = shear span, which may be taken as half the beam clear span. Expected material

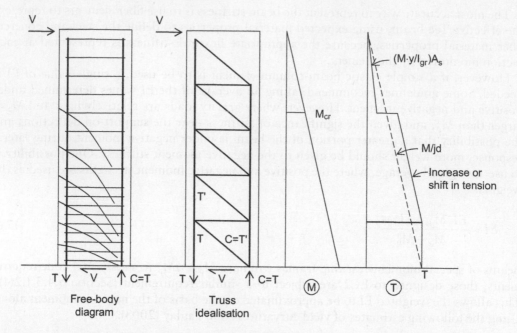

*Figure 17.17* Tension shift in reinforced concrete members, arising where inclined cracks due to combined flexure and shear are present. (Note: this model requires a sudden change in tension force at $M_{cr}$, which would require infinite bond strength.)

properties are used to determine $M_p^-$ and $\varepsilon_y$ ($=1.2f_y/E_s$). The approximation in which $h_b$ and $L$ = column spacing are used is intended to simplify preliminary design.

The choice to simplify modeling by selecting the stiffness associated with negative moment (Equation 17.9) means that $b_{eff}$ is significant only in that it determines the amount of longitudinal reinforcement within the slab to add to the beam top reinforcement in calculating $M_p^-$. Limiting the overhanging flange width to 1/12–1/8 of the span and six to eight slab thicknesses on each side of the web is considered reasonable for interstory drift limits on the order of 2% of the story height; smaller widths should be considered where smaller drift limits are used.

Shear stiffness is customarily represented considering elastic properties of the gross concrete section and is given by $G_c A_v$, where $G_c = E_c/(2(1 + \nu)) \approx 0.4E_c$ and $A_v$ is taken as $(5/6)b_w h_b$ for rectangular or Tee beam sections.

### 17.8.1.2 Beam plastic hinge (and anchorage slip)

Plastic hinge rotations and anchorage slip at the end of the beam can be represented using a discrete plastic hinge located at each beam end. If a zero-length element is used, the hinge element is typically located at the face of the column. Alternatively, a fiber hinge may extend from the face of the column some distance into the beam. Because the elastic deformations of the beam are already accounted for within the beam element, the beam plastic hinge element can have rigid-plastic behavior (Figure 17.18a). Because the plastic moment strengths typically differ (e.g., $M_p^- > M_p^+$), the rigid-plastic hinge element should be capable of representing unequal strengths. If such an element is not available, a fiber hinge element can be tuned to have the required moment–rotation behavior.

Reinforcement anchored within the adjoining member (e.g., beam-column joint or wall) will slip out from the face of the adjoining member, due to strains associated with anchorage

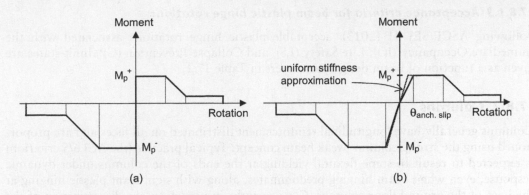

*Figure 17.18* Idealized beam plastic hinge backbone curves: (a) rigid-plastic, representing only the plastic hinge, and (b) including slip of reinforcement at the anchorage.

of the reinforcing bars within the adjoining member. Some deformation of the compression face also occurs. These anchorage effects introduce a concentrated rotation at the beam end. This component of rotation can be incorporated into the moment–rotation behavior of the plastic hinge, as shown in Figure 17.18b, as described in the following.

The displacement at the tip of a cantilever due to anchorage slip can be estimated, following Pantazopoulou and Syntzirma (2010) as

$$\Delta_{y,slip} = \left(\phi_y \frac{L_b}{2}\right) L_s \qquad (17.11)$$

where $L_s$ is the length of the cantilever and $L_b$ is the idealized bond length. $L_b$ is given by

$$L_b = \frac{d_b f_{ye}}{4 f_b} \qquad (17.12)$$

The average bond stress when the bars (having diameter $d_b$) yield is given by

$$f_b = \eta_1 f_t' \qquad (17.13)$$

where $\eta_1 = 2.25$ for ribbed bars and the tensile strength of the concrete can be taken as

$$f_t' = \begin{array}{l} 6\sqrt{f_c'} \text{ lb and in. units} \\ 0.5\sqrt{f_{ck}} \text{ MPa units} \end{array} \qquad (17.14)$$

Bringing the preceding together, the rotation of the beam due to anchorage slip, at plastic hinging, can be estimated as

$$\theta_{anchslip} = \begin{array}{l} \dfrac{\phi_y d_b f_y}{96\sqrt{f_c'}} \text{ lb and in. units} \\[2mm] \dfrac{\phi_y d_b f_y}{8\sqrt{f_{ck}}} \text{ N and mm units} \end{array} \qquad (17.15)$$

### 17.8.1.3 Acceptance criteria for beam plastic hinge rotations

Following ASCE/SEI 41 (2013), acceptable plastic hinge rotations associated with the Immediate Occupancy (IO), Life Safety (LS), and Collapse Prevention (CP) limit-states are given as a function of beam design parameters in Table 17.2.

## 17.8.2 Columns

Columns generally have longitudinal reinforcement distributed on all faces and are proportioned using the strong column–weak beam concept. Typical practice (the ACI 6/5 criterion) is expected to result in some flexural yielding at the ends of the columns under dynamic response, even when beam hinging predominates, along with significant plastic hinging at the base of the ground floor columns. Consequently, in modeling, the discretization of the structure should provide for the possibility of forming plastic hinges at the column ends.

### 17.8.2.1 Column stiffness

Column flexural stiffness is influenced by longitudinal reinforcement content and axial load. Axial force in a column may vary as the structure deforms laterally, particularly for the end columns of a perimeter moment-resisting frame. Column axial force also varies due to the vertical component of ground motion, although this tends to be at a relatively high frequency, affecting force levels but having less effect on deformations, which require time to develop. Where long beam spans are present, the beams may be excited by vertical components of the ground motion, and will induce variations in column axial forces.

The use of a fiber model to represent elastic behavior of the beam-column element is recommended, as the fiber model will directly represent the effects of longitudinal reinforcement and axial force on the flexural and axial stiffness of the member. Alternatively, one may use beam-column elements that allow the flexural strength to be specified as a function of axial force.

Where beam-column elements are used to represent the column, an elastic flexural stiffness is needed. In the United States, recent recommendations for modeling column flexural stiffness (PEER/ATC 72-1, 2010; ASCE/SEI 41, 2013) emphasize the dependence of $EI_{eff}$ on axial load ratio $(P/A_g f_c')$ and ignore possible dependence on longitudinal steel ratio.

*Table 17.2* Modeling and acceptance criteria for beam plastic hinges

| Condition | | Modeling parameters[a] | | | Acceptance criteria[a] | | |
|---|---|---|---|---|---|---|---|
| | | Plastic rotation angle, radians | | Residual strength ratio | Acceptable plastic rotation angle, radians | | |
| $\dfrac{\rho - \rho'}{\rho_{bal}}$ | $\dfrac{V}{b_w d \sqrt{f_c'}}$[b,c] | | | | Performance level | | |
| | psi (MPa) units | a | b | c | IO | LS | CP |
| ≤0.0 | ≤3 (0.25) | 0.025 | 0.05 | 0.2 | 0.010 | 0.025 | 0.050 |
| ≤0.0 | ≥6 (0.50) | 0.02 | 0.04 | 0.2 | 0.005 | 0.020 | 0.040 |
| ≥0.5 | ≤3 (0.25) | 0.02 | 0.03 | 0.2 | 0.005 | 0.020 | 0.030 |
| ≥0.5 | ≥6 (0.50) | 0.015 | 0.02 | 0.2 | 0.005 | 0.015 | 0.020 |

Source:   Excerpted from ASCE/SEI 41, 2013; Table 10-7 (2013).

[a]  Values between those listed in the table should be determined by linear interpolation.
[b]  The strength provided by the hoops ($\phi V_s$) must be at least 3/4 of the design shear, $V$.
[c]  $V$ is the design shear unless determined by a nonlinear analysis.

This approach causes the yield curvature to increase (unrealistically) in proportion to the flexural strength, as a function of the longitudinal steel content (Figure 17.19).

Even simple moment–curvature analyses (e.g. Figure 17.20) demonstrate that the flexural stiffness changes with changes in longitudinal steel ratio. Bilinear fits to the computed moment–curvature response are also shown in Figure 17.21. The bilinear curves have flexural strength given by the ACI nominal flexural strength (computed using expected material properties), and having initial stiffness adjusted to intersect the computed moment–curvature response at 75% of the flexural strength (Figure 17.21). It is apparent that the flexural stiffness, $EI$, given by the slope of the initial branch of the bilinear curve ($EI = M/\phi$), varies with reinforcement ratio for axial loads less than approximately $0.4A_g f'_c$. It is evident that yield curvatures, defined by the corner point of the bilinear curves, vary to some extent as well.

Consequently, one may consider two alternative approaches to estimate the flexural stiffness, derived considering only flexural deformations:

1. Perhaps the simplest involves interpolation based on the results derived for the initial branch of the bilinear curve of the 20 in. square column (Figure 17.22).

   For design, the member flexural stiffness is determined as the value of $E_c I_{eff}$ estimated by interpolation of Figure 17.22, reduced to account for tension shift as before, by dividing by $(1 + d/a)$.
2. The yield curvatures (as defined in Figure 17.21) may be estimated by

$$\phi_y = \left(1.8 - 1.3\frac{P}{A_g f'_c} + 9\left(\rho_g - 0.025\right)\right)\frac{\varepsilon_y}{d} \tag{17.16}$$

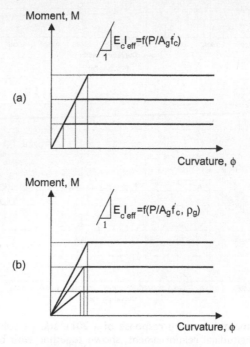

Figure 17.19 Implications of flexural stiffness assumptions on yield curvatures (a) using $EI = f(P/A_g f'_c)$ and (b) idealized based on moment–curvature analysis for fixed axial load ratio ($P/A_g f'_c$).

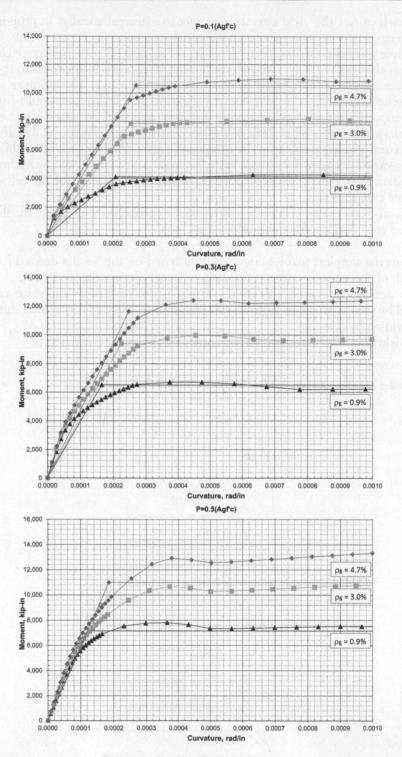

*Figure 17.20* Computed moment–curvature response of a 20 in. square column section having different amounts of longitudinal reinforcement, shown together with bilinear fits to the moment–curvature response. Column section has $f_c' = 6$ ksi, $f_y = 60$ ksi, and was modeled using $f_{ce}' = 1.25$ $f_c' = 7.5$ ksi, $f_{ye} = 1.2f_y = 72$ ksi.

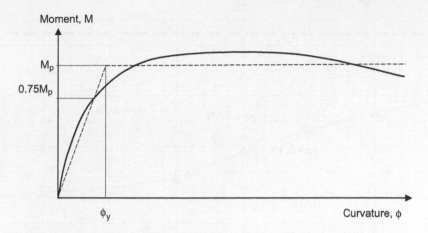

*Figure 17.21* Criteria for fitting bilinear curve from computed moment–curvature response, after Paulay and Priestley (1992).

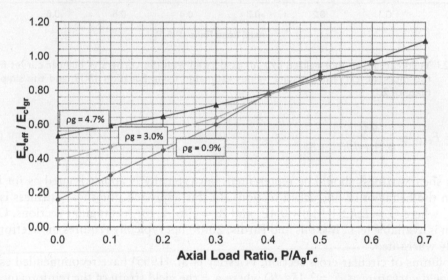

*Figure 17.22* Column section effective stiffness as a fraction of $E_c I_{gr}$, as a function of axial load ratio (considering only flexural deformation).

where $P/A_g f_c'$ is based on nominal material properties, $\rho_g = A_s/A_g$, where $A_s$ = total area of longitudinal reinforcement and $A_g$ = gross area of column section, the yield strain $\varepsilon_y$ is given by $f_{ce}'/E_s$, and $d$ = depth to the centroid of the extreme layer of reinforcement. As before, $M_p$ can be estimated as the nominal moment computed using ACI 318 assumptions, but using expected concrete and steel properties in place of nominal values.

Equation 17.16 provides a reasonable fit to the moment–curvature data, as shown in Figure 17.23. The fit is improved over the simpler estimate $\phi_y = 2.1\,\varepsilon_y/d$, provided by Priestley et al. (2007).

For design, the member flexural stiffness (associated with the initial branch of the bilinear curve) can be estimated as $E_c I_{eff} = M_p/\phi_y$, modified to account for tension shift:

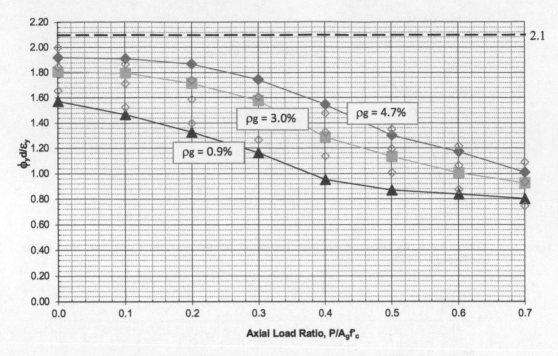

*Figure 17.23* Yield curvatures (normalized by $\varepsilon_y/d$) associated with the corners of bilinear curves fitted to the moment–curvature results: values estimated using Equation 15.17, and the simple estimate of 2.1.

$$EI = E_c I_{eff} = \frac{M_p}{\phi_y(1 + d/a)} \tag{17.17}$$

Column shear stiffness should be represented, but typically is approximated as for beams based on the behavior of uncracked sections. In this approach, the shear stiffness is given by $G_c A_v$, where $G_c = E_c/(2(1 + \nu)) \approx 0.4E_c$ and $A_v = (5/6)A_g$ for rectangular sections. Greater precision is available (e.g., Biskinis and Fardis, 2008) but typically requires more effort than is usually warranted.

For columns of circular cross section, Priestley et al. (1995) have recommended estimating the yield curvature as $\phi_y = 2.45\varepsilon_y/D$ where $\varepsilon_y$ = the yield strain of the reinforcement and $D$ = the diameter of the circular section concrete pier. This result is nearly identical to the recommendation $\phi_y = 2.3\varepsilon_y/d$, made by Hernández-Montes and Aschheim (2003) for columns of circular section with axial load ratio, $P/A_g f_c'$, between 0 and 0.40, where $d$ = the depth to the extreme tension reinforcement.

### 17.8.2.2 Column plastic hinge (and anchorage slip)

Just as for the beam plastic hinge, it is convenient to include anchorage slip in the moment–rotation relationship used for the column plastic hinge. With reference to Figure 17.18, the rotation associated with anchorage slip can be estimated according to Equation 17.15, using the yield curvature estimated according to Equation 17.16.

The column plastic hinge strength is a function of axial force, which may vary during the dynamic response. This can be represented directly with a properly calibrated fiber hinge model. Alternatively, some zero-length plastic hinge models allow P–M interaction surface to be specified. In the case of regular, symmetric moment frames, overall behavior can usually be represented using an invariant plastic hinge strength, associated with the gravity load acting at the time of the dynamic response; as effects of overturning axial forces on flexural strength are not represented, local behavior may deviate from that obtained in the model.

### 17.8.2.3 Acceptance criteria for column plastic hinge rotations

Following ASCE/SEI 41 (2013), acceptable plastic hinge rotations associated with the IO, LS, and CP limit-states are given as a function of column design parameters in Table 17.3.

## 17.8.3 Beam-column joints

### 17.8.3.1 Joint stiffness

Tensile strain in the longitudinal beam and column reinforcement, in conjunction with slip of the reinforcement associated with the development of bond within the joint, causes the longitudinal reinforcement to elongate and "poke" out of the joint. This so-called "anchorage slip" has already been represented in the beam and column plastic hinges. Thus, the

Table 17.3 Modeling and acceptance criteria for plastic hinges of rectangular cross-section columns

| Condition | | | Modeling parameters[a] | | | Acceptance criteria[a] | | |
|---|---|---|---|---|---|---|---|---|
| | | $\dfrac{V}{b_w d \sqrt{f'_c}}$ [d] | Plastic rotation angle, radians | | Residual strength ratio | Acceptable plastic rotation angle, radians | | |
| | | | | | | Performance level | | |
| $P/(A_g f'_c)$ [c] | $\rho_v = A_v/(b_w s)$ | psi (MPa) units | a | b | c | IO | LS | CP |
| **Condition i. $V_e/\phi V_n \leq 0.8$**[b] | | | | | | | | |
| ≤0.1 | ≥0.006 | NA | 0.035 | 0.060 | 0.2 | 0.005 | 0.045 | 0.060 |
| ≥0.6 | ≥0.006 | NA | 0.010 | 0.010 | 0.0 | 0.003 | 0.009 | 0.010 |
| ≤0.1 | =0.0036[e] | NA | 0.030[e] | 0.044[e] | 0.2 | 0.005 | 0.034 | 0.044 |
| ≥0.6 | =0.0036[e] | NA | 0.007[e] | 0.007[e] | 0.0 | 0.002 | 0.006 | 0.007 |
| **Condition ii. $0.8 \leq V_e/\phi V_n \leq 1.0$**[b] | | | | | | | | |
| ≤0.1 | ≥0.006 | ≤3 (0.25) | 0.032 | 0.060 | 0.2 | 0.005 | 0.045 | 0.060 |
| ≤0.1 | ≥0.006 | ≥6 (0.50) | 0.025 | 0.060 | 0.2 | 0.005 | 0.045 | 0.060 |
| ≥0.6 | ≥0.006 | ≤3 (0.25) | 0.010 | 0.010 | 0.0 | 0.003 | 0.009 | 0.010 |
| ≥0.6 | ≥0.006 | ≥6 (0.50) | 0.008 | 0.008 | 0.0 | 0.003 | 0.007 | 0.008 |
| ≤0.1 | =0.0036[e] | ≤3 (0.25) | 0.023[e] | 0.039[e] | 0.2 | 0.005 | 0.030 | 0.039 |
| ≤0.1 | =0.0036[e] | ≥6 (0.50) | 0.017[e] | 0.036[e] | 0.2 | 0.005 | 0.028 | 0.036 |
| ≥0.6 | =0.0036[e] | ≤3 (0.25) | 0.007[e] | 0.007[e] | 0.0 | 0.003 | 0.006 | 0.007 |
| ≥0.6 | =0.0036[e] | ≥6 (0.50) | 0.005[e] | 0.005[e] | 0.0 | 0.002 | 0.004 | 0.005 |

Source:   Adapted in part from ACI 374 (2016).

NA = not applicable.
[a]  Values between those listed in the table should be determined by linear interpolation.
[b]  The strength provided by the hoops or spirals ($\phi V_s$) must be at least 3/4 of the design shear force, $V_e$.
[c]  Design axial force, $P$, should be based on the maximum expected axial load due to gravity and earthquake loads.
[d]  $V$ is the design shear force, $V_e$, per ACI 318 Section 18.7.6.1.1 unless determined by a nonlinear analysis.
[e]  The transverse reinforcement ratio values in ASCE/SEI 41 were modified per the minimum requirement in ACI 318 and the modeling parameter values were modified by linear interpolation.

primary behavior to represent in the joint model is deformation within the joint, which is considered to be due primarily to shear. Joints designed meeting modern code requirements are unlikely to fail, and thus the joint shear behavior is often approximated as linear elastic.

As described by Bonacci and Pantazopoulou (1993), joint shear deformations can be modeled as contributing an amount $\Delta_j$ to story drift:

$$\Delta_j = \gamma H \left( 1 - \frac{h_b}{H} - \frac{h_c}{L} \right) \tag{17.18}$$

where $\gamma$ = shear strain within joint defined by height equal to the beam height, $h_b$, width equal to the column depth, $h_c$, and $H$ and $L$ are the story height and beam spans, respectively.

The shear strain, $\gamma$, is given simply as

$$\gamma = \frac{\tau}{G} = \frac{V}{A \cdot G} = \frac{2M_b/h_c}{h_b b_c G} = \frac{2M_b}{h_b h_c b_c G} \tag{17.19}$$

where the shear, $V$, can be obtained by the change in moment over the width of the joint, given by $2M_b/h_c$, where $M_b$ is the beam moment acting at the face of the joint, and the shear modulus, $G$, is given by $E_c/[2(1+\nu)] \approx 0.4E_c$.

Several models are available for representing joint shear deformations. The simplest to apply, where available, is a panel zone joint model (e.g., PERFORM (CSI, 2006), illustrated in Figure 17.25a. The panel zone is defined by the actual joint dimensions, $h_b$ and $h_c$, while the panel zone stiffness is given by $G \cdot b_c$.

Where a panel zone element is not available, similar behavior can be represented with a rigid link corner spring model (Figure 17.25b), consisting of rigid links connected at their ends by moment–rotation springs. This model requires definition of the nodes at the ends of each link and at the locations of the beam and column plastic hinges, the specification of rigid (or stiff) link elements, and definition of the moment–rotation springs that connect each link.

Using this model, to obtain the same contribution to story drift, $\Delta_j$, as provided in Equation 17.18, the stiffness of each moment–rotation spring is given by

$$\frac{M}{\theta} = \frac{h_b h_c b_c G}{4} \tag{17.20}$$

A third option provides joint flexibility by introducing a moment–rotation spring between rigid links defined within the joint along the beam and column centerlines (Figure 17.25c). Using this model, to obtain the same contribution to story drift, $\Delta_j$, as provided in Equation 17.18, the stiffness of the moment–rotation spring is given by

$$\frac{M}{\theta} = \frac{h_b h_c b_c G}{\left(1 - \dfrac{h_c}{L}\right)\left(1 - \dfrac{h_b}{H} - \dfrac{h_c}{L}\right)} \tag{17.21}$$

A fourth option, recommended by ASCE/SEI 41 (2013), models joint flexibility using a flexible link within the joint along the beam centerline and a rigid link along the column centerline (Figure 17.25d). If the beam link is simply assigned the beam $EI$, then the joint flexibility increases as the column depth, $h_c$, increases. A more rigorous approach establishes the flexural stiffness, $EI$, of the beam link to result in the joint contribution to interstory drift equal to that given by Equation 17.18, resulting in:

$$EI = \frac{h_b h_c{}^2 b_c G \left( 6 - \dfrac{\left( \dfrac{h_c}{L} \right)^2}{1 - \dfrac{h_c}{L}} \right)}{24 \left( 1 - \dfrac{h_b}{H} - \dfrac{h_c}{L} \right)} \approx \frac{h_b h_c{}^2 b_c G}{4 \left( 1 - \dfrac{h_b}{H} - \dfrac{h_c}{L} \right)} \tag{17.22}$$

assuming that the link has infinite (or very large) shear stiffness, GA.

Joint shear strength can be set equal to $V_n$ as given in Section 15.7.1.2, with residual strength $cV_n$ using $c = 0.2$ as set by ASCE/SEI 41 (2013).

### 17.8.3.2 Acceptance criteria for beam-column joint deformations

Design criteria for beam-column joints in new construction should result in little or no inelastic deformation occurring in the analytical structural model. As suggested in Figure 17.24, the model of Figure 17.25a imposes slightly larger deformations in the beam and/or column framing than would be expected with the models of Figure 17.25 (b) and (c). If this difference is neglected, the plastic hinge rotations in the moment-rotation springs of (b) and (c) provide deformation equivalent to the plastic shear deformations of (a). Therefore, the acceptance criteria (Table 17.4) for use with the ASCE/SEI 41(2013) model containing flexible beam links (Figure 17.25d) may be used with the models of Figure 17.25a–c.

## 17.8.4 Walls and coupled walls

### 17.8.4.1 Stiffness of elastic wall elements

Fiber elements can be used to accurately represent flexural stiffness for walls of arbitrary cross section, since the fibers naturally represent the amount of longitudinal reinforcement and the influence of axial force on stiffness. Fiber elements also represent the elongation that occurs during flexural response of the wall, and resulting effects on loads carried by the wall as the adjacent framing is deformed (Figure 17.14). While the gravity load should be based on area tributary to the wall, only the concrete and reinforcement within the effective flange

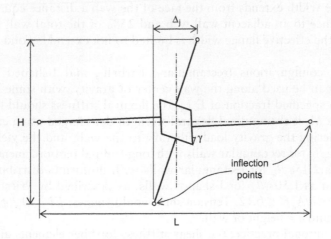

*Figure 17.24* Influence of shear deformation of beam-column joints on drift.

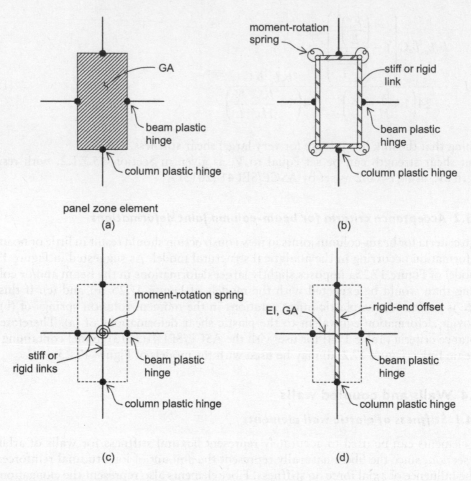

*Figure 17.25* Beam-column joint models (a) panel zone and (b) rigid link corner spring model, (c) rigid link centerline spring model and (d) flexible beam link model.

width should be included in the fiber element model. In NIST GCR 11-917-11 REV-1 (2012), the effective flange width extends from the face of the web a distance equal to the lesser of one half the distance to an adjacent wall web and 25% of the total wall height above the level in question; the effective flange width is limited to not extend beyond the actual flange width.

For symmetric configurations (rectangular-, barbell-, and I-shaped sections) beam-column elements can be used along the wall center of gravity, with some loss of accuracy. Rather than use a specified fraction of $E_c I_{gr}$, the flexural stiffness should be determined as $EI = M_p/\phi_y$, where $M_p$ is the nominal flexural strength determined using expected material properties, considering the gravity load tributary to the wall, and the yield curvature, $\phi_y$, estimated as $1.85\varepsilon_y/l_w$ for rectangular walls with longitudinal reinforcement concentrated at the boundaries, as $2.15\varepsilon_y/l_w$ for rectangular walls with uniformly distributed longitudinal reinforcement, and as $1.50\varepsilon_y/l_w$ for I-shaped walls, as described by Priestley et al. (2007) for walls having $0 \le P/A_g f_c' \le 0.12$. Tension shift would suggest $EI = M_p/(\phi_y(1 + d/a))$ where $d$ = section depth and $a$ = height of wall.

Following conventional practice, the shear stiffness for fiber elements and beam-column line elements can be set equal to $G_c A_{cv}$, where $G_c = 0.4E_c$ and $A_{cv}$ = area of the web extending

*Table 17.4* Modeling and acceptance criteria for beam-column joints of special moment frames

| Condition | | Modeling parameters[a] | | | Acceptance criteria[a] | | |
| | | | | | Acceptable plastic rotation angle, radians | | |
| | | Plastic rotation angle, radians | Residual strength ratio | | | | |
| | | | | | Performance level | | |
| $P/(A_g f'_c)$[b] | $V/V_n$[c] | a | b | c | IO | LS | CP |
|---|---|---|---|---|---|---|---|
| **Condition i. Interior joints (Note: for classification of joints, refer to Figure 15.13)** | | | | | | | |
| ≤0.1 | ≤1.2 | 0.015 | 0.030 | 0.2 | 0.0 | 0.020 | 0.030 |
| ≥0.4 | ≤1.2 | 0.015 | 0.025 | 0.2 | 0.0 | 0.015 | 0.025 |
| **Condition ii. Other joints (Note: for classification of joints, refer to Figure 15.13)** | | | | | | | |
| ≤0.1 | ≤1.2 | 0.010 | 0.020 | 0.2 | 0.0 | 0.015 | 0.020 |
| ≥0.4 | ≤1.2 | 0.010 | 0.025 | 0.2 | 0.0 | 0.020[d] | 0.025[d] |

Source:   Adapted in part from ACI 374 (2016).

[a]  Values between those listed in the table should be determined by linear interpolation.
[b]  $P$ is the design axial force on the column above the joint, and $A_g$ is the gross cross-sectional area of the joint.
[c]  The modeling parameters only can be used if the joint transverse reinforcement is spaced at ≤$h_c/3$ within the joint, where $h_c$ = average height of the beams framing into the joint in the direction of applied shear.
[d]  Acceptance criteria values inferred.

over the full length of the wall, given by $l_w b_w$. Modeling of inelastic response in shear may be necessary to improve the fidelity of wall shear demands.

### 17.8.4.2 Wall plastic hinges

Since different failure modes can develop within a plastic hinge (e.g., low-cycle fatigue of reinforcing bars, exhaustion of the concrete in compression, and diagonal tension or diagonal compression failures in the web), we generally use simpler models that represent overall expectations without representing each potential failure mode.

Discrete plastic hinges should be located in the model at the base of the wall and between the elastic wall elements wherever flexural yielding is anticipated. If fiber hinges are used to represent inelastic flexural response, they can be included within the fiber element used to model elastic wall response.

Whether a fiber hinge or a zero-length plastic hinge element is selected to represent flexural yielding of the wall, provision for anchorage slip should be included. The approach used for beam plastic hinges also is applicable here (Figure 17.18). Yield curvature estimates for symmetric walls were provided in Section 15.8.8; for other cases, moment–curvature analyses can be run to determine the secant stiffness using the criteria illustrated in Figure 17.16.

Axial load in a coupled wall varies with the direction of application of the lateral load, and hence the flexural strength (and stiffness) varies due to changes in axial load associated with the coupling beams. Variation of flexural strength and stiffness with axial load is inherently represented with fiber hinge models. Some concentrated hinge elements allow axial load–moment yield surfaces to be specified.

### 17.8.4.3 Acceptance criteria for wall plastic hinges

Table 17.5 provides modeling and acceptance criteria for plastic hinges in special structural walls.

*Table 17.5* Modeling and acceptance criteria for plastic hinges of special structural walls

| $\dfrac{(A_s - A_s')f_y + P}{A_{cv}f_c'}$ | $\dfrac{V}{A_{cv}f_c'}$ psi (MPa) [b] | Confined boundary[c] | Plastic rotation angle, radians $a$ | $b$ | Residual strength ratio $c$ | Acceptable plastic rotation angle, radians Performance level IO | LS | CP |
|---|---|---|---|---|---|---|---|---|
| ≤0.1 | ≤4 (0.33) | Yes | 0.015 | 0.020 | 0.75 | 0.0050 | 0.017[d] | 0.020 |
| ≤0.1 | ≥6 (0.50) | Yes | 0.010 | 0.015 | 0.40 | 0.0040 | 0.012[d] | 0.015 |
| ≥0.25 | ≤4 (0.33) | Yes | 0.009 | 0.012 | 0.60 | 0.0030 | 0.010[d] | 0.012 |
| ≥0.25 | ≥6 (0.50) | Yes | 0.005 | 0.010 | 0.30 | 0.0015 | 0.007[d] | 0.010 |
| ≤0.1 | ≤4 (0.33) | No | 0.008 | 0.015 | 0.60 | 0.0020 | 0.010[d] | 0.015 |
| ≤0.1 | ≥6 (0.50) | No | 0.006 | 0.010 | 0.30 | 0.0020 | 0.008[d] | 0.010 |
| ≥0.25 | ≤4 (0.33) | No | 0.003 | 0.005 | 0.25 | 0.0010 | 0.004[d] | 0.005 |
| ≥0.25 | ≥6 | No | 0.002 | 0.004 | 0.20 | 0.0010 | 0.002 | 0.004 |

Source:   Adapted in part from ACI 374 (2016).

Assumes shear strength is adequate to ensure components are dominated by flexural response. (Transverse reinforcement should have a strength of closed stirrups Vs ≥ 3/4 of required shear strength of the coupling beam to use modeling parameters defined in this table.)

[a]   Values between those listed in the table should be determined by linear interpolation.
[b]   $V$ is the design shear force, $V_e$, per ACI 318, 18.7.6.1.1, unless determined by a nonlinear analysis. The design axial force, $P$, should be based on the maximum expected axial demand due to gravity and earthquake loads.
[c]   Walls should be considered to have confined boundaries when special boundary elements are provided in accordance with ACI 318 Section 18.10.6.
[d]   Acceptance criteria values inferred.

## 17.8.5  Coupling beams

Damage to coupling beams can occur at relatively small building drifts; those located near the top of the building tend to be more heavily damaged, as would be expected based on the deformation kinematics illustrated in Figure 15.20. Coupling beams may be reinforced with conventional longitudinal reinforcement or diagonal reinforcement. Diagonally reinforced coupling beams are generally preferred, as they have stable hysteretic behavior to larger ductilities in comparison to conventional, orthogonally reinforced, coupling beams.

### 17.8.5.1  Proportioning of coupling beams

Following guidance from Paulay (e.g., Paulay, 2003), the coupling beams over the height of a coupled wall may be designed for identical shears. The shear to be carried by the coupling beams can be determined as described in Section 15.8.2. Thus, the moment at the end of each coupling beam is $V_{cb,i} \cdot l_n / 2$.

Orthogonally reinforced beams are designed based on conventional moment and shear strength requirements and are indicated where the degree of coupling is relatively light. Beam cross-sectional dimensions and reinforcement must comply with maximum longitudinal steel ratio limits. Orthogonal reinforcement must be used for beams having height, $h$, less than $l_n/4$, where $l_n$ = coupling beam clear span.

Diagonally reinforced coupling beams are best suited where relatively heavy coupling is desired; diagonal reinforcement is required for $h > l_n/2$. Diagonal reinforcement is proportioned based on the truss model of Figure 15.31 for which the nominal shear strength, $V_n$, is given by

$$V_n = 2A_{vd}f_y \sin \alpha \tag{17.23}$$

The distance $jd$ can be estimated as $h-8$ inches for preliminary design. The nominal shear strength limit, $V_n \leq 10\sqrt{f_c'}\, A_{cw}$ (lb and in. units) and in effect provides an upper bound on the diagonal steel area, $A_{vd}$ (and a lower bound on the inclination of the reinforcement, $\alpha$.

Either orthogonal or diagonal reinforcement can be used for intermediate beam depths, $l_n/4 < h < l_n/2$, with diagonal reinforcement being more efficient for higher shear levels, above approximately $(2-4)\sqrt{f_c'}A_{cw}$ (lb and in. units) or.

### 17.8.5.2 Elastic stiffness

As noted by Wallace (2012), beam-column elements are generally preferred over fiber models for modeling coupling beams, especially where the coupling beams are diagonally reinforced.

In typical conditions, the coupling beam is integral with the floor slab. The elastic flexural stiffness can be represented as defined for beams in Section 17.8.1. PEER/ATC 72-1 (2010) recommends that shear stiffness GA be based on $G_c = 0.4E_c$ for $l_n/h \geq 2$ and $G_c = 0.1E_c$ for $l_n/h \leq 1.4$, with linear interpolation for intermediate values of $l_n/h$.

### 17.8.5.3 Coupling beam plastic hinge

For a conventionally reinforced coupling beam, the plastic hinges at the face of the wall should represent both plastic hinging and anchorage slip, as was recommended for beams (Figure 17.18).

### 17.8.5.4 Acceptance criteria for coupling beam plastic rotations

Modeling and acceptance criteria for plastic hinge rotations for coupling beams are provided in Table 17.6 and in part reflect recommendations by Naish (2010) (Figure 17.26).

## 17.8.6 Post-tensioned reinforced concrete walls

### 17.8.6.1 Modeling of post-tensioned walls

The wall component may be modeled using beam-column elements or fiber elements. Beams framing into the wall should have rigid end offsets or rigid links over the depth of the wall.

Table 17.6 Modeling and acceptance criteria for coupling beams

| Condition | | Modeling parameters[a] | | | Acceptance Criteria[a] | | |
|---|---|---|---|---|---|---|---|
| | | Chord rotation, radians | | Residual strength ratio | Acceptable plastic rotation angle, radians | | |
| | $\dfrac{V}{A_{cv}f_c'}$ | | | | Performance level | | |
| Reinforcement configuration | psi (MPa)[b] | d | e | c | IO | LS | CP |
| Conventionally reinforced | ≤3 (0.25) | 0.025 | 0.050 | 0.75 | 0.010 | 0.025 | 0.050 |
| | ≥6 (0.50) | 0.020 | 0.040 | 0.50 | 0.005 | 0.020 | 0.040 |
| Diagonally reinforced | NA | 0.030 | 0.050 | 0.80 | 0.006 | 0.030 | 0.050 |

Source: Adapted in part from ACI 374 (2016) and Naish (2010).

Assumes shear strength is adequate to ensure components are dominated by flexural response. (Transverse reinforcement should have a strength of closed stirrups Vs ≥ 3/4 of required shear strength of the coupling beam to use modeling parameters defined in this table.)

[a] Values between those listed in the table should be determined by linear interpolation.
[b] V is the design shear force, $V_e$, per ACI 318, 18.7.6.1.1, unless determined by a nonlinear analysis. The design axial force, P, should be based on the maximum expected axial demand due to gravity and earthquake loads.

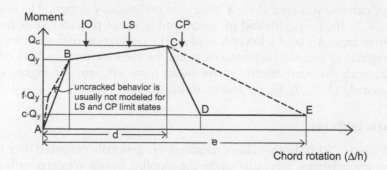

Figure 17.26 Generalized load–displacement (backbone) curve for representing coupling beam response, based on ASCE/SEI 41 (2013) and ACI 374 (2016). Acceptability limits are shown schematically for primary components of the lateral force-resisting system for the IO, LS, and CP Levels.

Plastic hinge modeling can utilize fiber elements with post-tensioning simulated via loads applied at the post-tensioning anchorage, or using prestressed elements. For dynamic response, use of discrete elements to represent the pre-tensioned tendons and the change in the tendon force with elongation is preferred.

### 17.8.6.2 Acceptance criteria

Acceptance criteria may need to be formulated on a project-specific basis; the extremes are bounded by criteria for rocking footings (lenient) and criteria for conventionally reinforced slender walls.

## 17.8.7 Collectors, floor diaphragms, and chords

Collectors and chords are normally horizontal elements that are cast integrally with the floor slab. These components are intended to remain elastic and thus can be modeled fairly simply where needed. For two-dimensional models, interaction of the vertical elements of the lateral force-resisting system with each other and with the floor diaphragm normally is not modeled. (Note that where these vertical elements along a horizontal line of resistance vary in strength and/or stiffness, differences in their relative flexural and shear stiffnesses can lead to redistribution between these elements via collectors and/or floor diaphragms (e.g., Beyer et al., 2014).) Forces in floor diaphragms and diaphragm chords can be determined in three-dimensional models. Such models are needed where plan irregularities lead to the potential for plan torsion, and where flexible diaphragms or irregular diaphragm geometry leads to the need to consider diaphragm deformations and forces.

## 17.8.8 Rocking footings as plastic hinges

### 17.8.8.1 Modeling and acceptance criteria for rocking footings

ASCE/SEI 41 (2013) provides for modeling the moment–rotation response at the soil–footing interface of various footing shapes, using the generic curve of Figure 17.27. Recommended values and acceptance criteria for rectangular footings are summarized in Table 17.7.

$$M_c = \frac{L_f P}{2}\left(1 - \frac{q}{q_c}\right)$$

(17.24)

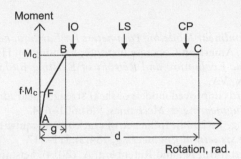

*Figure 17.27* ASCE/SEI 41 (2013) recommendations for modeling rocking footings. Note the stiffness AF is four times that of AB, and initial stiffness AF is expected to be approximately $300M_c$.

*Table 17.7* Modeling and acceptance criteria for rocking footings, rectangular in plan

| Condition | | Modeling parameters[a] | | | Acceptance Criteria[a] | | |
|---|---|---|---|---|---|---|---|
| | | Footing rotation angle, radians | | Elastic strength ratio | Total footing rotation angle, radians[b] | | |
| | | | | | Performance Level | | |
| $b/l_c$ | $A_c/A_f$ | $g$ | $d$ | $f$ | IO | LS | CP |
| ≥10 | 0.02 | 0.009 | 0.1 | 0.5 | 0.020 | 0.080 | 0.100 |
| | 0.13 | 0.013 | 0.1 | 0.5 | 0.015 | 0.080 | 0.100 |
| | 0.50 | 0.015 | 0.1 | 0.5 | 0.002 | 0.003 | 0.004 |
| | 1.00 | 0.015 | 0.1 | 0.5 | 0 | 0 | 0 |
| 3 | 0.02 | 0.009 | 0.1 | 0.5 | 0.020 | 0.068 | 0.085 |
| | 0.13 | 0.013 | 0.1 | 0.5 | 0.011 | 0.060 | 0.075 |
| | 0.50 | 0.015 | 0.1 | 0.5 | 0.002 | 0.003 | 0.004 |
| | 1.00 | 0.015 | 0.1 | 0.5 | 0 | 0 | 0 |
| 1 | 0.02 | 0.009 | 0.1 | 0.5 | 0.020 | 0.056 | 0.070 |
| | 0.13 | 0.013 | 0.1 | 0.5 | 0.007 | 0.040 | 0.050 |
| | 0.50 | 0.015 | 0.1 | 0.5 | 0.002 | 0.003 | 0.004 |
| | 1.00 | 0.015 | 0.1 | 0.5 | 0 | 0 | 0 |
| 0.3 | 0.02 | 0.009 | 0.1 | 0.5 | 0.010 | 0.040 | 0.050 |
| | 0.13 | 0.013 | 0.1 | 0.5 | 0.007 | 0.024 | 0.030 |
| | 0.50 | 0.015 | 0.1 | 0.5 | 0.001 | 0.003 | 0.004 |
| | 1.00 | 0.015 | 0.1 | 0.5 | 0 | 0 | 0 |

Source:  Adapted from ASCE/SEI 41 (2013); Table 8-4).

Assumes rigid foundations modeled using uncoupled springs. Rocking assumed to dominate over sliding, for $M/V > L_f$

[a]  Values between those listed in the table should be determined by linear interpolation.
[b]  Allowable story drift > 1%.

where $M_c$ is the rocking capacity of the footing, $P$ is the vertical load on the footing, $L_f$ is the length of the footing in the direction of the lateral force, $q$ is the vertical bearing pressure, and $q_c$ is the expected bearing capacity of the soil.

# REFERENCES

ACI 374 (2016). *Guide to Nonlinear Modeling Parameters for Earthquake-Resistant Structures*, ACI Report No. 374.3R-16, American Concrete Institute, Farmington Hills, MI.

ASCE/SEI 41 (2013). *Seismic Evaluation and Retrofit of Existing Buildings*, American Society of Civil Engineers, Reston, VA.

Aschheim, M. (2000). Towards improved models of shear strength degradation in reinforced concrete members, *Structural Engineering & Mechanics*, 9(6):601–614.

Aschheim, M., and Browning, J. (2008). Influence of cracking on equivalent-SDOF estimates of RC frame drift, *Journal of Structural Engineering*, 134(3):511–517.

Beyer, K., Simonini, S., Constantin, R., and Rutenberg, A. (2014). Seismic shear distribution among interconnected cantilever walls of different lengths, *Earthquake Engineering & Structural Dynamics*, 43(10):1423–1441.

Biskinis, D. E., and Fardis, M. N. (2008). Cyclic deformation capacity, resistance and effective stiffness of RC members with or without retrofitting, *14th World Conference on Earthquake Engineering*, October 12–17, 2008, Beijing, China.

Bonacci, J. F., and Pantazopoulou, S. J. (1993). Parametric investigation of joint mechanics, *ACI Structural Journal*, 90(1):61–71.

Caltrans (2006). *Seismic Design Criteria, Version 1.6*, California Department of Transportation, Sacramento, CA.

Chadwell, C.B., and Imbsen & Associates (2002). XTRACT: Cross section analysis software for structural and earthquake engineering, www.imbsen.com/xtract.htm.

Coleman, J., and Spacone, E. (2001). Localization issues in force-based frame elements, *Journal of Structural Engineering*, 127(11):1257–1265.

FEMA-273 (1997). NEHRP guidelines for the seismic rehabilitation of buildings, *Prepared by the Applied Technology Council for the Federal Emergency Management Agency*, Washington, DC.

FIB, Model Code (2010). *Model Code 2010 – Final draft*, vol. 1, Fib Bulletin No. 65, International Federation for Structural Concrete, Lausanne.

Goodnight, J. C., Kowalsky, M. J., and Nau, J. M. (2012). Effect of load history and design variables on the behavior of circular bridge columns, *Presentation to ACI Committee 341*, October 22, 2012.

Hernández-Montes, E., and Aschheim, M. (2003). Estimates of the yield curvature for design of reinforced concrete columns, *Magazine of Concrete Research*, 55(4):373–383.

Hernández-Montes, E., Fernández-Ruíz, M. A., Carbonell-Márquez, J. F., and Gil-Martín, L. M. (2017). Theoretical experimental in-service long-term deflection response of symmetrically and non-symmetrically reinforced concrete piles, *Archives of Civil and Mechanical Engineering*, 17:433–445.

Ibarra, L. F., Medina, R. A., and Krawinkler, H. (2005). Hysteretic models that incorporate strength and stiffness deterioration, *Earthquake Engineering and Structural Dynamics*, 34(12):1489–1511.

Inel, M., Aschheim, M., and Pantazopoulou, S. (2007). Seismic deformation capacity indices for concrete columns: Model estimates and experimental results, *Magazine of Concrete Research*, 59(4):297–310.

Kunnath, S., El-Bahy, A., Taylor, A. W., and Stone, W. C. (1997). *Cumulative Seismic Damage of Reinforced Concrete Bridge Piers*, Report NISTOR 6075, National Institute of Standards and Technology, Gaithersburg, MD.

Massard, J. M., and Collins, R. A. (1958). *The Engineering Behavior of Structural Metals Under Slow and Rapid Loading*, University of Illinois Engineering Experiment Station, College of Engineering, University of Illinois at Urbana-Champaign, Champaign, IL.

Montejo, L., and Kowalsky, M. (2007). *CUMBIA, Set of Codes for the Analysis of Reinforced Concrete Members*, Technical Report IS-07-01, North Carolina State University, Raleigh, NC.

Naish, D. (2010). Testing and modeling of reinforced concrete coupling beams, *Doctoral Thesis*, Department of Civil & Environmental Engineering, University of California, Los Angeles, CA.

NIST GCR 10-917-5 (2010). Nonlinear structural analysis for seismic design: A guide for practicing engineers, by Deierlein, G. G., Reinhorn, A. M., Willford, M. R., Report No. GCR 10-917-5, National Institute of Standards and Technology, Gaithersburg, MD, October 2010, 32 pages.

NIST GCR 11-917-11 Rev-1 (2012). Seismic design of cast-in-place concrete special structural walls and coupling beams: A guide for practicing engineers, by Moehle, Ghodsi, Hooper, Fields, and Gedhada, National Institute of Standards and Technology, Gaithersburg, MD, March 2012, 41 pages.

Pantazopoulou, S. J., and Syntzirma, D. V. (2010). Deformation capacity of lightly reinforced concrete members: Comparative evaluation. In Fardis, M. N. (ed.), *Advances in Performance-Based Earthquake Engineering* (pp. 359–371), Springer, Dordrecht.

Paulay, T., and Priestley, M. J. N. (1992). *Seismic Design of Reinforced Concrete and Masonry Buildings*, John Wiley & Sons, New York.

Paulay, T. (2003). Seismic displacement capacity of ductile reinforced concrete building systems, *Bulletin of the New Zealand Society for Earthquake Engineering*, 36(1):47–65.

Pujol, S. (2002). Drift capacity of reinforced concrete columns subjected to displacement reversals, *Doctoral Thesis*, Purdue University, Hammond, IN.

PEER/ATC 72-1 (2010). *Modeling and Acceptance Criteria for Seismic Design and Analysis of Tall Buildings*, Report No. 72-1, Applied Technology Council, Redwood City, CA, 242 pages.

PERFORM CSI (2006). *3D: Nonlinear Analysis and Performance Assessment for 3D Structures, Version 4*. Computers and Structures, Inc., Berkeley, CA.

Priestley, M. J. N., Seible, F., and Calvi, M. (1995). *Seismic Design and Retrofit of Bridges*, John Wiley & Sons, New York.

Priestley, M. J. N., Calvi, G. M., and Kowalsky, M. J. (2007). *Displacement-Based Seismic Design of Structures*, IUSS Press, Pavia.

Sivaselvan, M., and Reinhorn, A. M., 2000, Hysteretic models for deteriorating inelastic structures, *Journal of Engineering Mechanics, American Society of Civil Engineers*, 126(6):633–640.

Spacone, E., Filippou, F. C., and Taucer, F. F. (1996a). Fibre beam column model for nonlinear analysis of R/C frames. I: Formulation, *Earthquake Engineering and Structural Dynamics*, 25(7):711–725.

Spacone, E., Filippou, F. C., and Taucer, F. F. (1996b). Fibre beam column model for nonlinear analysis of R/C frames. II: Applications, *Earthquake Engineering and Structural Dynamics*, 25(7):727–742.

Umemura, H., and Ichinose, T. (2004). Experimental study on the effects of loading history on the ductility capacity of reinforced concrete members, *13th World Conference on Earthquake Engineering*, Vancouver, BC, Canada, August 1–6, 2004.

Wallace, J. (1992). BIAX-2: Analysis of reinforced concrete and reinforced masonry sections, http://nisee.berkeley.edu/elibrary/getdoc?idBIAX2ZIP (accessed October 3, 2017).

Wallace, J. (2012). Behavior, design, and modeling of structural walls and coupling beams—Lessons from recent laboratory tests and earthquakes, *International Journal of Concrete Structures and Materials*, 6(1):3–18.

# Section V

# Design methods and examples

# Chapter 18

# Design methods

## 18.1 PURPOSE AND OBJECTIVES

This chapter integrates material presented in earlier chapters to create three distinct approaches for design and assessment: a straightforward code comparable approach and two more sophisticated approaches that require nonlinear response history analyses (NLRHAs) for design validation and determination of forces in force-controlled members. Examples illustrating the application of the three design approaches are presented in Chapter 19.

## 18.2 INTRODUCTION

Of the many ways to integrate the concepts presented in this book for design, we provide three distinct straw-man proposals for design. We begin with an "entry-level" approach that can be considered a substitute for current code equivalent lateral-force methods. For those cases where additional analysis is justified, two additional approaches are articulated, having increased power (or sophistication) but requiring more effort. These arc summarized in Table 18.1.

The choice of design approach will depend on the degree to which conformance with existing code provisions is required, the user's comfort with more advanced tools and concepts (described in this book), the desired precision and the degree to which compliance with seismic performance objectives (POs) needs to be demonstrated, and the available time and budget.

Readers unfamiliar with the methods described in this book may wish to begin with Method A (termed "Quasi-code"), which can be implemented in the format of an equivalent lateral force procedure, a mainstay of current codes. Method A introduces (i) the concept of design based on the yield displacement, (ii) the use of yield point spectra to determine the required strength, and (iii) the use of a plastic mechanism analysis for the design of the yielding components of the lateral force-resisting system. This approach addresses limits on interstory drift and system ductility. Constraints on system ductility provide explicit consideration of short-period displacement amplification (which is neglected in the North American practice of using $R$ factors that are independent of period).

Methods B and C rely on a probabilistic description of inelastic response of the equivalent single-degree-of-freedom (SDOF) system, termed yield frequency spectra (YFS) (Section 13.5), to determine the base shear coefficient required to satisfy one or more POs with specified confidence. The $\beta$ distribution of lateral forces is an option for design (as described by Goel and Chao (2008), and summarized in Section 12.8), although users of conventional design software may prefer a code lateral force distribution. Once a preliminary design has been completed, NLRHAs are used to assess performance. In Method B, the

*Table 18.1* Summary of design approaches A, B, and C

| Design approach | Method A: Quasi-code | Method B: Simplified dynamic | Method C: Dynamic |
|---|---|---|---|
| Hazard information | Code spectrum[a] | Code spectrum[a] with prescriptive hazard slopes | Site-specific hazard surface |
| Design tool | YPS | YFS or YFS-T | YFS or YFS-T |
| Design POs | Based on limits on drift and ductility related to building code provisions (Serviceability performance criteria can be added) | Based on limits on drift and ductility related to building code provisions or as adopted for design; multiple POs may be considered | Based on limits on drift and ductility related to building code provisions or as adopted for design; multiple POs may be considered |
| Lateral force distribution | Code | $\beta$ or code | $\beta$ or code |
| Mechanism strength basis | Use material safety factors[a] per code, but with expected materials strengths. Design and analysis for expected load conditions | Use material safety factors[a] set to 1.0 and expected material properties | Use material safety factors[a] set to 1.0 and expected material properties |
| Response history analysis | None. A linear static analysis is sufficient to validate design; a nonlinear static (pushover) analysis is optional | NLRHA using ground motions scaled to one intensity level per MAF involved in defining the POs | NLRHA using ground motions scaled to two intensity levels per MAF involved in defining the POs |
| Verification of compliance with design and/or additional POs | Code compliance as usual, safety and overstrength factors with nominal material strengths Confidence as implied by code (i.e., unknown) No verification of collapse capacity | Single-stripe analysis ($b = 1$) with custom safety factors to (i) Check ductile members (ii) Design force-protected members Stated confidence of achieving POs Verification of collapse possible | Double-stripe analysis ($b$ determined from analysis) with custom safety factors to (i) Check ductile members (ii) Design force-protected members Stated confidence of achieving POs Verification of collapse possible |

[a] $\phi$ for ACI, $\gamma_M$ for Eurocodes.
[b] For U.S. practice, the code design spectrum is a risk-targeted spectrum. This spectrum is 2/3 of the spectrum corresponding to a 2% chance of exceedance in 50 years (return period of 2,475 years), roughly equivalent to the 10% in 50-year uniform hazard spectrum of older U.S. codes, capped by any applicable deterministic limits associated with the physical characteristics of causative faults, and further adjusted for risk. For the Eurocodes, the 10% in 50-year uniform hazard spectrum (or essentially its smoothed equivalent) is employed instead.

NLRHAs are conducted with ground motions scaled to a single intensity level; in Method C, the ground motions are scaled to two intensity levels to improve accuracy.

All three approaches use expected material properties for design of the ductile components of the lateral force-resisting system. Method A employs conventional strength reduction or material safety factors. These factors are taken equal to 1.0 in Methods B and C because safety against failure in these approaches is achieved at the system level through the use of probabilistic approaches (with demands determined by NLRHA).

Also common to all three approaches is the use of elastic design spectra (whether developed for a building code, extrapolated using an assumed hazard slope ($k$), or determined by site-specific hazard analysis), an assumed yield displacement,[1] assumed values of modal

---

[1] In the case of unbonded post-tensioned walls, an assumed period is used rather than an assumed yield displacement.

parameters (which may be refined after completing the preliminary design), and the use of a plastic mechanism analysis to establish the required strengths of yielding components of the lateral force-resisting system.

All three approaches easily accommodate design for multiple POs. In the case of Method A, all such objectives are keyed to the code spectrum hazard level (typically close to 10% in 50 years). Methods B and C can also employ POs at different levels of intensity, with Method C using a more accurate representation of the site hazard. The largest of the yield strength coefficients required to satisfy each PO is selected (Sections 10.6 and 13.5) for the estimated yield displacement.[2] This is equivalent to designing for the PO that requires the greatest stiffness.

Note that wide disparities exist in modeling of the stiffness of reinforced concrete members at present. The overestimation of stiffness in conventional practice (e.g., relative to the recommendations in Chapter 17) results in lower periods and drift expectations when following code design approaches, compared with what can be expected. Stiffness assumptions will directly influence the slope of a first-mode capacity curve and hence the estimated yield displacement of the model. While the yield displacement can be expected to be a stable parameter in design iterations, the "goodness" of the initial yield displacement estimate relative to that determined in a pushover analysis will be influenced by the stiffness assumptions introduced in modeling.

As a practical matter, we mention some considerations related to code enforcement, construction practices, and inspection. We suggest that system ductility be used as a yoke (or constraint) on design as a simple and robust measure to protect the public from an overly weak system (much as $R$ and $q$ factors achieve today), along with constraints on drift or interstory drift to protect against overly flexible systems. As well, while we recognize that the most general form of performance-based seismic design would allow member detailing to be tailored to computed demands, we consider this to be too bold of a step—structural details that have been developed over the years are well understood in practice, and are easily recognized and enforced by building officials. Thus, we utilize current code requirements for structural details, along with the acceptance criteria of Chapter 17, to make use of established practice and to avoid the possibility of lesser details being used inadvertently in some projects.

## 18.3 DESIGN METHOD A (QUASI-CODE)

*Description*: Figure 18.1 outlines the design process according to Method A. Code design response spectra are used as a basis for establishing yield point spectra (YPS). An estimate of the yield displacement (that would be obtained in a first-mode pushover analysis) is used along with estimated modal parameters to estimate the yield displacement, $D_y^*$, of the equivalent SDOF (ESDOF) system. Limits on system ductility are determined based on (i) a system ductility limit and (ii) a peak interstory drift limit; both derived to be approximately consistent with current code requirements (Chapter 12). YPS are used to establish the minimum required yield strength coefficient for the ESDOF system, $C_y^*$. If multiple POs are being considered, yield strength coefficients to satisfy each PO are determined and the largest of these is used for design. The period associated with the largest $C_y^*$ is calculated.

The required base shear strength at yield, $V_y = (\alpha_1 C_y^*)W$, is used in a plastic mechanism analysis to determine the required strengths of the plastic hinges. The code distribution

---

[2] For unbonded post-tensioned walls, designed on the basis of constant period, a variation of YFS called YFS-T may be used for the estimated period of vibration.

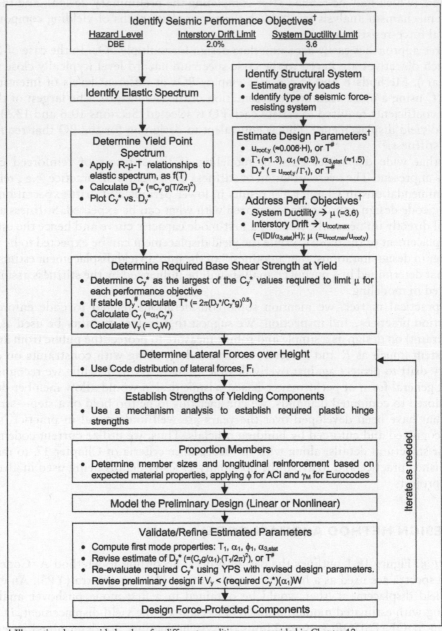

† Illustrative data provided; values for different conditions are provided in Chapter 12.

\# Most structural systems are designed assuming $D_y$ is relatively stable during design iterations. Exceptions such as unbonded post-tensioned wall systems are designed assuming T is relatively stable during design iterations.

*Figure 18.1* Flowchart for design Method A (quasi-code).

of lateral force over the height of the structure is used in the mechanism analysis. Plastic hinges are sized on the basis of their expected flexural strengths. Force-protected members are sized considering overstrength, using the provisions of the NEHRP or EN1998 codes. Because physical plastic hinges are offset from member centerlines, the use of centerline dimensions in the plastic mechanism analysis will introduce a slight conservatism. Expected material strengths are used together with strength reduction or material safety factors for the design of yielding components of the lateral force-resisting system, while nominal material strengths are used for the design of force-protected members.

Having established a preliminary design, a linear elastic model of the structure is prepared and its first-mode properties $(T_1, \alpha_1, \Gamma_1, \phi_1, \alpha_{3,\text{stat}})$ are determined in a modal analysis. If the values of parameters assumed in developing the preliminary design are accurate, the first-mode period will match the period of the ESDOF system. The computed first-mode period allows the estimated yield displacement to be refined, considering the design base shear strength at yield and computed first-mode period, as $D_y^* = \left(\dfrac{C_y g}{\alpha_1}\right)\left(\dfrac{T_1}{2\pi}\right)^2$. Using this yield displacement and the computed modal parameters, refinements to the required base shear strength, $V_y$, can be determined. Any significant increase in $V_y$ would require redesign of plastic hinge locations, while small decreases in $V_y$ may be neglected.

While not required, engineers wishing to better understand the characteristics of the structure (as well as this design approach) may choose to do a nonlinear static (pushover) analysis. A traditional first-mode or energy-based pushover analysis will provide sufficient information to assess yield strength and yield displacement, and to observe the stability of the yield displacement throughout the design iterations. With a good initial estimate of the yield displacement, there may not be a need for any iteration.

_Advantages of Method A_: This introductory approach provides individuals familiar with equivalent lateral force methods exposure to a design approach in which the yield displacement assumes a primary role, and period is seen as the outcome of the strength required to satisfy performance requirements. In particular, for flexible structures such as moment frames, whose design often is controlled by drift, Method A provides a direct means to establish the strength (and stiffness) required to satisfy interstory drift limits, avoiding the iteration typical of current (period-based) approaches to design. This approach provides accurate estimates of peak displacements over the height of the structure, explicitly accounting for short-period displacement amplification (through the use of $R$–$\mu$–$T$ relationships), unlike current code approaches. This approach also introduces the explicit use of plastic mechanism analyses in design, and provides the freedom to assign the distribution of plastic hinge strengths associated with a desired mechanism. In short, Method A relies on familiar code approaches while being faster (less iteration, due to the stability of the yield displacement), more accurate (with respect to peak displacements), and more explicit in the development of an intended mechanism.

_Limitations of Method A_: As with the current code approaches, expected inelastic behavior is not explicitly evaluated. Thus, there is no validation (by nonlinear dynamic analysis) that the intended pattern of plastic hinges forms, or that other POs are satisfied. The lack of performance assessment is of particular concern for some irregular structures, including those vulnerable to plan torsion. Also, potential vulnerabilities associated with prescriptive code criteria may be overlooked—for example, shears in structural (shear) walls and interstory drifts in moment frames.

## 18.4 DESIGN METHOD B (SIMPLIFIED DYNAMIC)

_Description_: Method B allows for the explicit treatment of specified uncertainties in establishing the required base shear strength, the strengths of yielding components of the lateral force-resisting system, and the strength of strength-protected components to meet stated POs with specified levels of confidence. Figures 18.2 and 18.3 outline the design

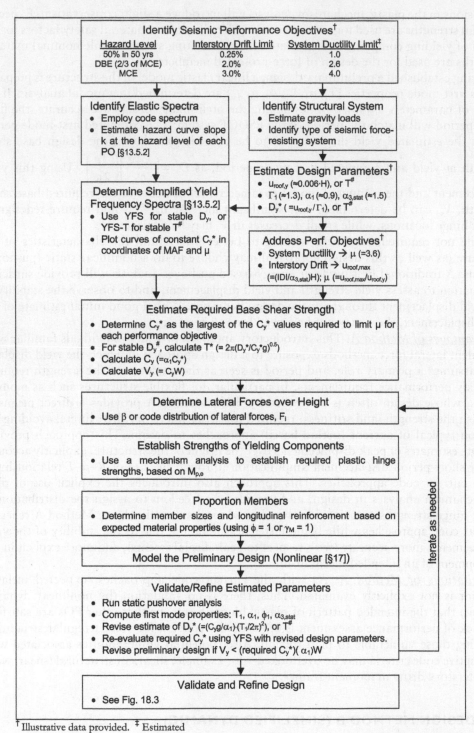

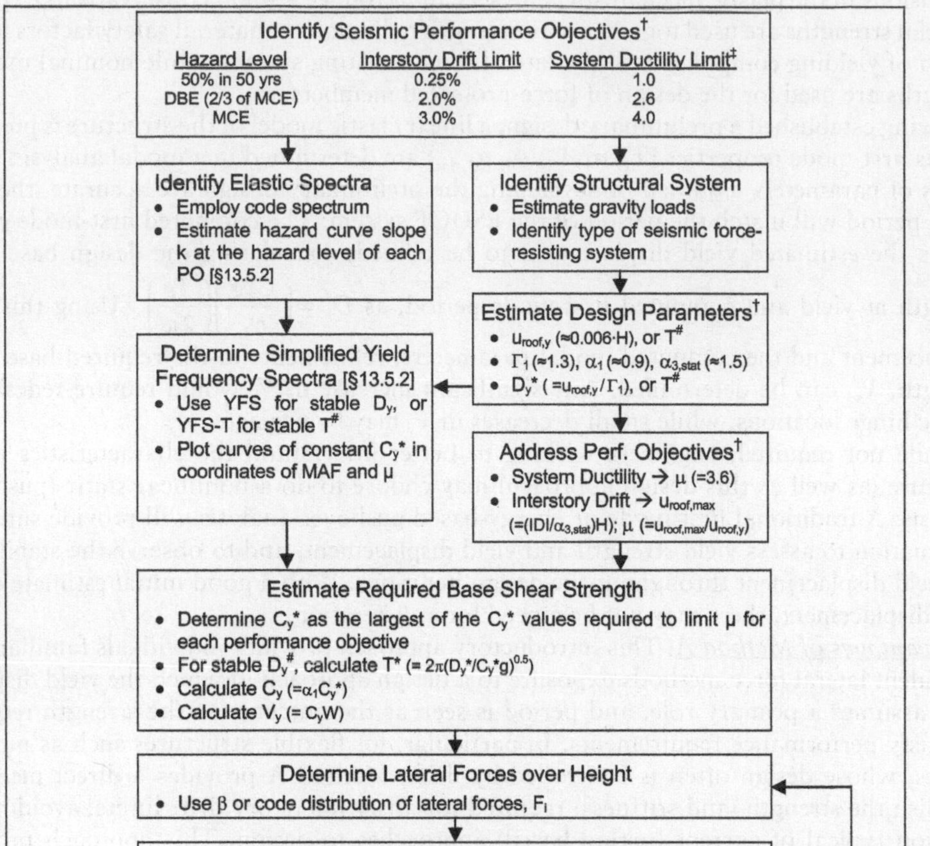
Identify Seismic Performance Objectives†
Hazard Level | Interstory Drift Limit | System Ductility Limit‡
50% in 50 yrs | 0.25% | 1.0
DBE (2/3 of MCE) | 2.0% | 2.6
MCE | 3.0% | 4.0
Identify Elastic Spectra
• Employ code spectrum
• Estimate hazard curve slope k at the hazard level of each PO [§13.5.2]
Identify Structural System
• Estimate gravity loads
• Identify type of seismic force-resisting system
Estimate Design Parameters†
• $u_{roof,y}$ (≈0.006·H), or $T^{\#}$
• $\Gamma_1$ (≈1.3), $\alpha_1$ (≈0.9), $\alpha_{3,stat}$ (≈1.5)
• $D_y^*$ ( $=u_{roof,y}/\Gamma_1$), or $T^{\#}$
Determine Simplified Yield Frequency Spectra [§13.5.2]
• Use YFS for stable $D_y$, or YFS-T for stable $T^{\#}$
• Plot curves of constant $C_y^*$ in coordinates of MAF and μ
Address Perf. Objectives†
• System Ductility → μ (=3.6)
• Interstory Drift → $u_{roof,max}$ ($=(|D|/\alpha_{3,stat})H$); μ ($=u_{roof,max}/u_{roof,y}$)
Estimate Required Base Shear Strength
• Determine $C_y^*$ as the largest of the $C_y^*$ values required to limit μ for each performance objective
• For stable $D_y^{\#}$, calculate $T^*$ ($= 2\pi(D_y^*/C_y^*g)^{0.5}$)
• Calculate $C_y$ ($=\alpha_1 C_y^*$)
• Calculate $V_y$ ($= C_y W$)
Determine Lateral Forces over Height
• Use β or code distribution of lateral forces, $F_i$
Establish Strengths of Yielding Components
• Use a mechanism analysis to establish required plastic hinge strengths, based on $M_{p,e}$
Proportion Members
• Determine member sizes and longitudinal reinforcement based on expected material properties (using $\phi = 1$ or $\gamma_M = 1$)
Model the Preliminary Design (Nonlinear [§17])
Validate/Refine Estimated Parameters
• Run static pushover analysis
• Compute first mode properties: $T_1$, $\alpha_1$, $\phi_1$, $\alpha_{3,stat}$
• Revise estimate of $D_y^*$ ($=(C_y g/\alpha_1)\cdot(T_1/2\pi)^2$), or $T^{\#}$
• Re-evaluate required $C_y^*$ using YFS with revised design parameters.
• Revise preliminary design if $V_y$ < (required $C_y^*$)( $\alpha_1$)W
Validate and Refine Design
• See Fig. 18.3
Iterate as needed

† Illustrative data provided. ‡ Estimated

\# Most structural systems are designed assuming $D_y$ is relatively stable during design iterations. Exceptions such as unbonded post-tensioned wall systems are designed assuming T is relatively stable during design iterations.

*Figure 18.2* Flowchart for design Method B (simplified dynamic).

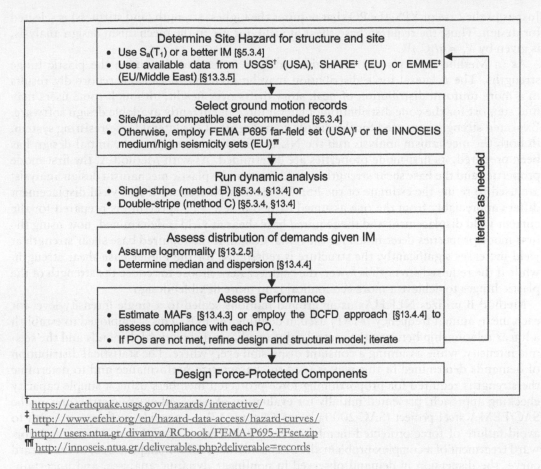

Determine Site Hazard for structure and site
- Use $S_a(T_1)$ or a better IM [§5.3.4]
- Use available data from USGS[†] (USA), SHARE[‡] (EU) or EMME[‡] (EU/Middle East) [§13.3.5]

Select ground motion records
- Site/hazard compatible set recommended [§5.3.4]
- Otherwise, employ FEMA P695 far-field set (USA)[¶] or the INNOSEIS medium/high seismicity sets (EU)[¶¶]

Run dynamic analysis
- Single-stripe (method B) [§5.3.4, §13.4] or
- Double-stripe (method C) [§5.3.4, §13.4]

Assess distribution of demands given IM
- Assume lognormality [§13.2.5]
- Determine median and dispersion [§13.4.4]

Assess Performance
- Estimate MAFs [§13.4.3] or employ the DCFD approach [§13.4.4] to assess compliance with each PO.
- If POs are not met, refine design and structural model; iterate

Design Force-Protected Components

Iterate as needed

[†] https://earthquake.usgs.gov/hazards/interactive/
[‡] http://www.efehr.org/en/hazard-data-access/hazard-curves/
[¶] http://users.ntua.gr/divamva/RCbook/FEMA-P695-FFset.zip
[¶¶] http://innoseis.ntua.gr/deliverables.php?deliverable=records

*Figure 18.3* Flowchart for design validation and refinement for Methods B (simplified dynamic) and C (dynamic).

process according to Method B, and Sections 18.6 and 18.7 discuss uncertainty and confidence levels, respectively.

Uniform hazard spectra (UHS) are used to define discrete hazard levels used for one or more performance limits. The UHS may already be available (e.g., provided by the U.S. Geological Survey) or may be extrapolated from a code uniform hazard spectrum using an approximate value for the slope of the hazard curve, $k$, as described in Section 13.3.

The yield displacement of the roof and corresponding displacement of the ESDOF system are estimated as for Method A (initially using suggestions in Section 12.5 or relying on past experience with similar systems). YFS are prepared using the approach described in Section 13.4, which relies on linearly interpolating (and extrapolating where needed) in natural log space the UHS at the 50%, 10%, and 2% in 50-year probabilities of exceedance to determine additional hazard estimates within the range of interest. Note that where extrapolation is required, i.e., for the very frequent or truly rare ground motions, the linear fit in log-space introduces some conservatism to the process, due to the concave shape of the hazard curve in log-space (see Figure 13.14).

Each YFS shows the strength–ductility relationship across the entire hazard for a structure having the specified yield displacement, $D_y^*$, at a specified level of confidence. Thus, the yield strength coefficient, $C_y^*$, required to satisfy each PO can be determined by interpolation between the curves of constant $C_y^*$ using a YFS determined at the desired confidence level.

Just as in the case of YPS, the PO that requires the highest strength (and stiffness) is selected for design. Thus, the required base shear at yield, for use in the mechanism design analysis, is given by $V_y = (\alpha_1 C_y^*)W$.

As in Method A, a plastic mechanism analysis is used to determine the plastic hinge strengths. The $\beta$ lateral force distribution may be preferred because it reportedly results in a more uniform distribution of peak story drifts over height, although some users may find support for the code distribution of lateral force in currently available design software. Expected strengths are used for the ductile components of the lateral force-resisting system, in both the mechanism analysis and the NLRHAs. Once a model of the initial design has been prepared, its first-mode properties are determined. As with Method A, the first-mode properties and the base shear strength at yield (used in the plastic mechanism design analysis) are used to refine the estimate of the ESDOF yield displacement. If this yield displacement differs appreciably from the one assumed earlier, a new YFS spectrum is prepared for the current yield displacement and the required base shear at yield is determined, now using the first-mode properties determined for the initial design. If the required base shear strength at yield increases significantly, the structure is redesigned for the higher base shear strength, while if the required strength is lower, the engineer may choose to reduce the strength of the plastic hinges to achieve a more economical (and more flexible) design.

Method B utilizes NLRHAs using ground motions scaled to a single intensity level for each mean annual frequency (MAF) used in defining the POs. This is employed to establish a linear relationship between the central value (median) of structural demands and the seismic intensity, while assuming a constant dispersion everywhere. The statistical distribution of demands determined in this way is used to characterize performance and to determine the strengths required for proportioning force-protected members using a simple capacity checking approach, presented initially for evaluation of new and existing structures in the SAC/FEMA steel project (SAC 2000a,b, see Chapter 13), to provide sufficient strength to avoid failure of force-protected members at a stated level of confidence. This straightforward treatment of a complex problem allows consideration of the shape (slope) of the hazard curve, the dispersion in demand observed in nonlinear dynamic analyses, and uncertainties associated with system modeling and member strengths. This approach is known to be conservative due to the linear fit (in log-space) of the hazard curve (indicated by the slope, $k$), even as it tends to smear out the uncertainty over the entire structure, ignoring possible dependencies (or correlations) between, for example, the distribution of plastic hinge strengths and the associated force demands in force-protected members.

While nonlinear static (pushover) analysis is not required in this approach, it is useful for characterizing the lateral behavior of the structure, providing a direct confirmation of the yield displacement, and as an aid to validate the computational model prior to running the nonlinear dynamic analyses. Although not explicitly required, analysis of the displacement history obtained in the NLRHAs using principal components analysis (Chapter 6) is recommended for identifying the development of any undesirable collapse mechanisms.

*Advantages of Method B*: The "simplified dynamic" approach provides a quick (albeit conservative) way to approach design at stated confidence levels. Starting with a reasonable estimate of the yield displacement allows for rapid convergence of the design of the ductile portions of the lateral system, with nonlinear dynamic analysis then being used to both characterize response and to establish the strengths required for design of the force-protected components. Thus, Method B provides a relatively fast way to arrive at a somewhat conservative design that complies with the performance requirements including confidence levels.

*Limitations of Method B*: This approach only employs limited hazard data, essentially extrapolating the design spectrum to both higher and lower return periods via fixed adjustment factors that are period independent. Thus, slopes required for estimating the MAF of exceedance are assumed rather than computed, leading to some inaccuracy. In addition, Approach B assumes a linear relationship between the seismic intensity and the central value of any structural demand of interest. This is a fairly accurate assumption for deformation and displacement response quantities of moderate and long-period structures far from global collapse, akin to the equal displacement rule (Veletsos and Newmark, 1960). It will generally become conservative for short-period structures and for limit-states approaching collapse. In such cases, the double-stripe assessment of Method C is considered superior.

## 18.5 DESIGN METHOD C (DYNAMIC)

_Description_: Similar to Method B, Method C allows for the explicit treatment of specified uncertainties while demands from NLRHAs are used to establish the required strengths of strength-protected components to achieve a user-specified confidence that the MAF of demands exceeding capacities is sufficiently low. Figures 18.3 and 18.4 outline the design process. Sections 18.6 and 18.7 discuss uncertainty and confidence levels, respectively.

Method C differs from Method B in that NLRHAs are performed at two intensity levels for each hazard level used in the performance criteria, and the change in median engineering demand parameter values is used to more accurately capture the linear or nonlinear evolution of engineering demand parameters with shaking intensity, for use in calculating the MAF of exceedance. Greater accuracy is also achieved by using actual site hazard data at multiple periods and annual frequencies of exceedance, rather than extrapolating from the design spectrum. Finally, the two stripes of Method C (with the potential addition of a third one for greater accuracy), allow assessment of the global collapse risk.

_Advantages of Method C_: Method C provides greater accuracy in performance assessment and design because it uses site-specific hazard data and employs more comprehensive structural analysis response data from double-stripe NLRHAs, in lieu of the slopes assumed in Method B for hazard and response at different levels of intensity.

_Limitations of Method C_: Method C is the most rigorous of the approaches identified in this chapter; limitations exist primarily in assumptions made to allow for tractable solutions, such as the assumption of log-normal distributions, the assumptions that the distributions of response quantities are independent of one another, and the reliability of structural models to represent expected structural response. Method C still employs the conservative demand-capacity factored design approach of SAC/FEMA (SAC 2000a,b) for simplified assessment, with the same limitations as Method B, but with an improved power law model of the median demand versus intensity relationship, allowing for higher accuracy for short-period structures and limit-states closer to collapse. It also allows for assessment of the global collapse risk, although in many cases an additional stripe at higher intensities may be required.

## 18.6 TREATMENT OF UNCERTAINTY

A seismic PO is a triplet of (i) a threshold limit of a response parameter of interest, (ii) a specified MAF of exceedance, and (iii) a confidence level with which we want to be assured that the threshold limit is not exceeded. In typical code applications, the MAF does not apply to the exceedance of the threshold limit per se but rather to the exceedance of a ground motion intensity level employed for design (e.g., through specified uniform hazard spectrum), essentially substituting the integration of demand over the entire seismic hazard

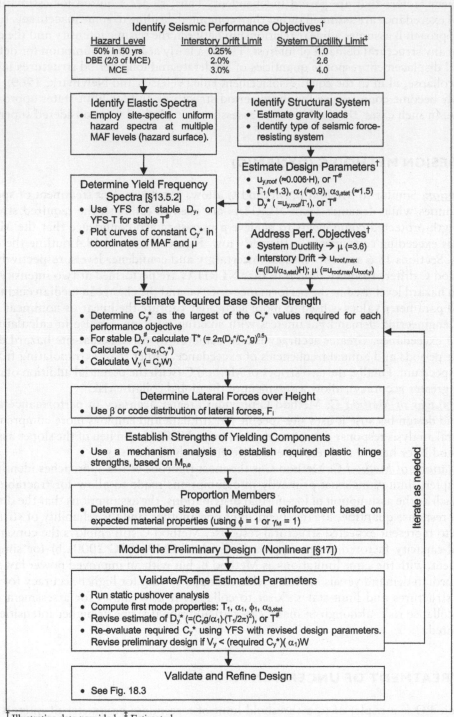

Figure 18.4 Flowchart for design Method C (dynamic).

The flowchart contains the following blocks:

**Identify Seismic Performance Objectives[†]**

| Hazard Level | Interstory Drift Limit | System Ductility Limit[‡] |
|---|---|---|
| 50% in 50 yrs | 0.25% | 1.0 |
| DBE (2/3 of MCE) | 2.0% | 2.6 |
| MCE | 3.0% | 4.0 |

**Identify Elastic Spectra**
- Employ site-specific uniform hazard spectra at multiple MAF levels (hazard surface).

**Identify Structural System**
- Estimate gravity loads
- Identify type of seismic force-resisting system

**Estimate Design Parameters[†]**
- $u_{y,roof}$ ($\approx 0.006 \cdot H$), or $T^{\#}$
- $\Gamma_1$ ($\approx 1.3$), $\alpha_1$ ($\approx 0.9$), $\alpha_{3,stat}$ ($\approx 1.5$)
- $D_y^*$ ($= u_{y,roof}/\Gamma_1$), or $T^{\#}$

**Determine Yield Frequency Spectra [§13.5.2]**
- Use YFS for stable $D_y$, or YFS-T for stable $T^{\#}$
- Plot curves of constant $C_y^*$ in coordinates of MAF and $\mu$

**Address Perf. Objectives[†]**
- System Ductility → $\mu$ ($=3.6$)
- Interstory Drift → $u_{roof,max}$ ($=(IDI/\alpha_{3,stat})H$); $\mu$ ($=u_{roof,max}/u_{roof,y}$)

**Estimate Required Base Shear Strength**
- Determine $C_y^*$ as the largest of the $C_y^*$ values required for each performance objective
- For stable $D_y^{\#}$, calculate $T^*$ ($= 2\pi(D_y^*/C_y^* g)^{0.5}$)
- Calculate $C_y$ ($=\alpha_1 C_y^*$)
- Calculate $V_y$ ($= C_y W$)

**Determine Lateral Forces over Height**
- Use $\beta$ or code distribution of lateral forces, $F_i$

**Establish Strengths of Yielding Components**
- Use a mechanism analysis to establish required plastic hinge strengths, based on $M_{p,e}$

**Proportion Members**
- Determine member sizes and longitudinal reinforcement based on expected material properties (using $\phi = 1$ or $\gamma_M = 1$)

**Model the Preliminary Design (Nonlinear [§17])**

**Validate/Refine Estimated Parameters**
- Run static pushover analysis
- Compute first mode properties: $T_1$, $\alpha_1$, $\phi_1$, $\alpha_{3,stat}$
- Revise estimate of $D_y^*$ ($=(C_y g/\alpha_1) \cdot (T_1/2\pi)^2$), or $T^{\#}$
- Re-evaluate required $C_y^*$ using YFS with revised design parameters. Revise preliminary design if $V_y < $ (required $C_y^*$)($\alpha_1$)W

**Validate and Refine Design**
- See Fig. 18.3

*Iterate as needed*

[†] Illustrative data provided.　[‡] Estimated

[#] Most structural systems are designed assuming $D_y$ is relatively stable during design iterations. Exceptions such as unbonded post-tensioned wall systems are designed assuming T is relatively stable during design iterations.

with a point estimate, while confidence is generally not incorporated explicitly. The threshold limit may refer to an absolute numerical value (e.g., an interstory drift or a plastic hinge rotation) or a condition such as demand exceeding a capacity (e.g., the force computed in a force-protected member exceeding the strength of the member), the latter recognizing that both capacities and demands may be characterized by statistical distributions. Generally, without this being a requirement or a constraint, a PO (e.g., an interstory drift limit or failure of a force-protected member) will be specified at a single MAF of exceedance, meaning that all component checks associated with this PO are based on the same MAF of exceedance.

Threshold limits represent capacities that are grounded in physical reality. Some, like plastic hinge rotation capacities or shear strengths, may be directly related to physically observable phenomena. Others, such as interstory drift limits, may be based more on accepted good practice rather than being associated with a discrete, physically manifested event. Uncertainty is present, both in establishing the threshold limits and in evaluating the demands. Uncertainty in demand evaluation results from uncertainty in the evaluation of seismic hazard, variability in ground motions (and structural response) associated with a given hazard level, approximations in the modeling of structural components and assumptions about material properties.

The three design methods articulated in Table 18.1 represent a gradation from a relatively simple approach anchored in current building codes to relatively complex approaches that are more reliant on analysis results and statistical inference.

Method A does not explicitly consider uncertainty. Rather, we infer limits on interstory drift and system ductility that are consistent with the code (ASCE-7 and Eurocode 8) and apply code-specified strength reduction or material safety factors in the design. We explicitly design to achieve a desired mechanism strength using expected material strengths rather than the nominal strengths usually used in code-based design. However, conventional code strength-reduction or material safety factors are used to maintain the general hierarchy of failure modes intended by code developers. Method A provides explicit treatment of drift limits in design and provides for short-period displacement amplification (through the use of YPS). An estimate of the yield drift is used initially. As indicated in Chapter 17, stiffness estimates should account for the amount of reinforcement present, rather than assuming the member stiffness is a fixed proportion of the gross stiffness.

Uncertainty is explicitly considered in Methods B and C. This uncertainty arises in estimation of the seismic hazard, material properties, modeling, and motion-to-motion variability in response. The uncertainties are reflected in the use of statistical distributions that reflect dispersion inherent in the values of estimated parameters, in contrast to the use of point estimates in Method A. An explicit accounting of uncertainty can allow the structure to be designed such that the MAF of exceeding a target value of a parameter of interest (e.g., a plastic hinge rotation demand) can be estimated, and a desired MAF may be targeted in the design process. While Method A may utilize ground motions that have a stated MAF of exceedance (input), the focus in Methods B and C is on the MAF of failing to meet a PO (output).

The main difference between Methods B and C is the quality of the data employed to power this estimation. In Method B, simple approximations of both the seismic hazard and the intensity-to-response relationship are employed. Method C uses more accurate data for hazard (assuming they are available) and an approximation of the intensity–response relationship that is twice as expensive computationally, due to the use of two stripes rather than one stripe, requiring twice as many NLRHA runs.

Of a more subjective nature is the degree of confidence with which we wish to achieve a PO, which is addressed below.

## 18.7 CONFIDENCE LEVELS IN DESIGN AND CAPACITY ASSESSMENT

One could view the design objective as targeting a MAF of failure (to meet a desired PO) without regard to the type of failure or its consequences. However, we prefer failures to be biased toward gradual loss of lateral strength and away from potential collapse owing to loss of gravity load support, particularly given limitations in precision owing to uncertainties. Further, the presence of uncertainties in demands and capacities means that simply satisfying a mere inequality does not mean the stated POs are achieved with complete (i.e., 100%) confidence. Greater confidence in achieving a PO generally requires more material and therefore comes at greater cost. We suggest that the degree of confidence used in design should be justified on the basis of the consequence of failure. The potential consequences and thus evaluation of appropriate confidence level requires engineering judgment, preferably founded on a wealth of experience in field reconnaissance and experimental testing.

The overall concept is to proportion structures in such a way that failure modes are segregated by the consequences for the system. A hierarchy of failure mode probabilities is intended, with a reduction of lateral load resistance being more likely than the catastrophic loss of gravity load support. At one extreme, the failure of a small proportion of components that do not carry superimposed gravity load within a highly redundant lateral force-resisting system can be readily tolerated. At the other extreme, failures that jeopardize the support of gravity loads and risk the development of a progressive (or catastrophic) collapse should be assigned the highest confidence levels in design (along with a suitably low MAF). One may view this approach as a refinement of the conventional focus on promoting inelastic deformation in the lateral force-resisting system while providing ample strength to force-protected members.

The probabilistic approach for considering confidence levels amplifies inherent demand (or equivalently reduces capacity) by a factor associated with the desired confidence level. Because the use of a 50% level of confidence has no numerical influence on design values, we generally recommend the use of confidence levels above approximately 60% and generally not exceeding perhaps 95%. Specifically, we suggest the following framework for establishing confidence levels:

- Advisory performance target, for which consequences of exceedance are minimal. Examples include interstory drift limits and system ductility limits, which are intended to be satisfied in a general way but need not be satisfied precisely. For such cases we recommend a confidence level of 50%.
- Failure of limited consequence. This includes components prone to isolated failure, which do not have a disproportionate consequence on the response of the structure to lateral loads, and which pose minor risk to the support of gravity loads. Examples include: (i) the exhaustion of plastic rotation capacity of a beam plastic hinge in a highly redundant system and (ii) the shear failure of an isolated beam or slab, where redistribution or catenary action can be relied upon to support the floor gravity loads without risk of a progressive (catastrophic) collapse. For such cases, we recommend a confidence level of 60% for design or application of acceptance criteria.
- Failure jeopardizes response of the structure to lateral loads (with minimal risk to the support of gravity loads). An example is the exhaustion of plastic hinge rotation capacity at the base of a shear wall for a lateral system that lacks redundancy. The reduction in base shear resistance and development of torsional eccentricity associated

with the loss of flexural strength at the base of the wall would compromise the response of the structure to lateral loads. For such cases, we recommend a confidence level of approximately 75% for design or application of acceptance criteria.

• Failure jeopardizes the support of gravity loads. The component may be part of the gravity or lateral system. Examples include: (i) the shear failure of a column or wall supporting multiple (overlying) stories and (ii) the punching failure of a slab-column connection that lacks continuous bottom bars. For such cases, we recommend a confidence level of approximately 90%–95% for design or application of acceptance criteria.

In some cases, the failure to meet a PO is based on a comparison of deformation (e.g., plastic rotations) demands and capacities; in others, it is based on a comparison of strength demands and capacities.

The failure of some components may jeopardize both lateral resistance and gravity load support, and thus should be assigned the higher applicable confidence level.

Ideally, a distinction should be made between the (nearly) simultaneous failure of multiple components (associated with uniform demand-to-capacity ratios on multiple components for which there is little uncertainty or high correlation in component capacities) and a progressive failure associated with redistribution to components that are unable to withstand the modified demands (also referred to as an "unzipping" or "domino" effect). Both issues can be viewed as impacting the "robustness" of the structure, viewed as its capability to retain load-carrying capacity in the face of uncertainties (Vamvatsikos, 2015). Properly capturing simultaneous collapses in general requires a detailed probabilistic model of the structure, where localized (i.e., component-level) sources of uncertainty and associated correlation among different components play a major role and need to be directly incorporated. The formulation described in Chapter 13 does not explicitly account for such correlations. Still, this is an issue of relatively secondary importance for modern capacity-designed structures (see, e.g., Kazantzi et al., 2014), as even the (nearly) simultaneous failure of multiple components does not necessarily imply a catastrophic collapse. Nevertheless, one should be aware of this issue in highly optimized structures and consider increasing the level of confidence employed whenever multiple elements are close to failure at any limit-state verification. Cases where a progression of failures may occur due to the inability to redistribute forces would also benefit from a detailed probabilistic model, yet this can be represented reasonably well with a good (deterministic) numerical model of the structure that can simulate the progression of said failures. Still, this may be difficult where shear and/or axial failures need to be accounted for. At the very least, some sensitivity analysis can help one understand whether a collapse mechanism is overly sensitive to small changes in the model parameters and lead to the adoption of a higher level of confidence for the overly impacted limit-state verifications.

## REFERENCES

Goel, S. C., and Chao, S. (2008). *Performance-Based Plastic Design Earthquake-Resistance Steel Structures*, International Code Council, Washington, D.C.

Kazantzi, A. K., Vamvatsikos, D., and Lignos, D. G. (2014). Seismic performance of a steel moment-resisting frame subject to strength and ductility uncertainty. *Engineering Structures*, 78:69–77.

SAC Joint Venture (2000a). Recommended seismic design criteria for new steel moment-frame buildings, FEMA-350, prepared for the Federal Emergency Management Agency, Washington, DC.

SAC Joint Venture (2000b). Recommended seismic evaluation and upgrade criteria for existing welded steel moment-frame buildings, FEMA-351, prepared for the Federal Emergency Management Agency, Washington, DC.

Vamvatsikos, D. (2015). A view of seismic robustness based on uncertainty. Proceedings of the 12th International Conference on Applications of Statistics and Probability in Civil Engineering, ICASP12, Vancouver, Canada.

Veletsos, A. S., and Newmark, N. M. (1960). Effect of inelastic behavior on the response of simple systems to earthquake motions, *Proceedings of the 2nd World Conference on Earthquake Engineering*, Japan, vol. 2, 895–912.

# Chapter 19

# Design examples

## 19.1 PURPOSE AND OBJECTIVES

This chapter develops six examples to illustrate a variety of ways to use the design approaches of Chapter 18. ASCE-7 (ASCE, 2010) and ACI-318 (ACI, 2014) are followed for the first three examples, and the Eurocodes (CEN, 2004, 2005) are followed for the second set of three examples.

## 19.2 INTRODUCTION

The first set of examples consists of three designs of a four-story moment-resistant frame, each having different performance objectives (POs), and designed following Methods A, B, or C. The second set of three examples illustrates the design of different wall systems: a coupled wall using Method A, a cantilever wall using Method B, and a post-tensioned wall using Method C. The same site and nominal hazard are used for all examples. For the first set of examples, the representation of hazard follows ASCE-7 (ASCE, 2010) and members are designed following ACI 318 (ACI, 2014). The second set of three examples utilizes an EC8 (CEN, 2005) representation of the hazard and EC2 (CEN, 2004) requirements for member design.

Table 19.1 provides a summary of the design examples, identifying POs and the resulting design parameters, as well as characteristics of the preliminary design and its success or not in meeting the performance criteria. Note that the code-like approach of Method A does not require nonlinear analysis (although it was performed herein for reasons of comparison), while Methods B and C use some form of nonlinear response history analysis (NLRHA), for verification. Therefore, in case of failing to meet a criterion in a structure designed via Methods B and C, one would expect that this will be detected by the user and a corrective will be taken. For Method A this is not necessarily the case. Still, as Table 19.1 plainly shows, none of the examples designed by Methods B and C required any revision.

## 19.3 APPLICATION OF YIELD FREQUENCY SPECTRA AND PERFORMANCE ASSESSMENT METHODOLOGIES

In the application of the performance assessment approaches outlined in Section 13.4.2, some assumptions are introduced. First, the limit-state (LS) threshold values (plastic rotation, drift, etc.) are interpreted/employed probabilistically by taking them to be the median value of a lognormally distributed random variable. Furthermore, verification is always done at the output level of risk (i.e., the mean annual frequency (MAF), or equivalently via

*Table 19.1* Overview of design examples

| Design method | Moment frame examples (American codes) | Wall examples (Eurocodes) |
|---|---|---|

**A**

**Example 1**
4-story moment frame

POs
- System ductility limit = 3.6 for 2/50 hazard level (*governs design*)
- Interstory drift limit = 2% of story height (under static lateral forces)

Target properties
$C_y = 0.211$
$T_1 = 1.08$ s

Computed properties
$C_y = 0.211$
$T_1 = 1.08$ s

Computed performance

**Example 4**
12-story coupled wall

POs
- System ductility limit = 3.3 (*governs design*)
- Interstory drift limit = $0.010 \cdot h/v$

Target properties
$C_y = 0.095$
$T_1 = 1.54$ s

Computed properties
$C_y = 0.178$
$T_1 = 1.51$ s

Computed performance

| | | NLRHA method | | | | NLRHA method | |
|---|---|---|---|---|---|---|---|
| **PO** | **IDA** | **Single stripe** | **Double stripe** | **PO** | **IDA** | **Single stripe** | **Double stripe** |
| System ductility | ✗ | ✗ | ✗ | System ductility | ✓ | ✓ | ✓ |
| Interstory drift ratio | ✓ | ✓ | ✓ | Interstory drift ratio | ✓ | ✓ | ✓ |
| Column plastic hinge rotation | ✓ | ✓ | ✓ | Wall plastic hinge rotation | ✓ | ✓ | ✓ |
| Beam plastic hinge rotation | ✗ | ✓ | ✓ | Beam plastic hinge rotation | ✓ | ✓ | ✓ |

**B**

**Example 2**
4-story moment frame

POs
- MAF of interstory drift exceeding 2% of the story height = $2.11 \times 10^{-3}$ (during dynamic response) at 68% confidence

Target properties
$C_y = 0.275$
$T_1 = 0.91$ s

Computed properties
$C_y = 0.247$
$T_1 = 1.01$ s

Computed performance

**Example 5**
7-story cantilever wall

POs
- MAF of interstory drift exceeding 2% of story height = $2.11 \times 10^{-3}$ during dynamic response at 70% confidence (*governs design*)
- MAF of plastic hinge rotation demand exceeding capacity at the base of the wall = $2.11 \times 10^{-3}$ at 85% confidence

Target properties
$C_y = 0.181$
$T_1 = 1.47$ s

Computed properties
$C_y = 0.253$
$T_1 = 1.12$ s

Computed performance

| | | NLRHA method | | | | NLRHA method | |
|---|---|---|---|---|---|---|---|
| **PO** | **IDA** | **Single stripe** | **Double stripe** | **PO** | **IDA** | **Single stripe** | **Double stripe** |
| Interstory drift ratio | ✓ | ✓ | ✓ | Interstory drift ratio | ✓ | ✓ | ✓ |
| Column plastic hinge rotation | ✓ | ✓ | ✓ | Wall plastic hinge rotation | ✓ | ✓ | ✓ |
| Beam plastic hinge rotation | ✓ | ✓ | ✓ | | | | |

(*Continued*)

*Table 19.1 (Continued)* Overview of design examples

| Design method | Moment frame examples (American codes) | | | Wall examples (Eurocodes) | | | |
|---|---|---|---|---|---|---|---|
| C | **Example 3** 4-story moment frame | | | **Example 6** 4-story post-tensioned wall | | | |

POs

**Moment frame examples (American codes):**
- MAF of system ductility demand exceeding $1.5 = 1.39 \times 10^{-2}$ at 70% confidence
- MAF of interstory drift exceeding 2% of story height $= 2.11 \times 10^{-3}$ at 70% confidence
- MAF of collapse $= 2.01 \times 10^{-4}$ at 90% confidence *(governs design)*

**Wall examples (Eurocodes):**
- MAF of gap opening at wall base exceeding 2 cm $= 1.39 \times 10^{-2}$ at 50% confidence
- MAF of interstory drift exceeding 0.7% of the height of the structure $= 1.39 \times 10^{-2}$ at 50% confidence
- MAF of CCC $= 2.01 \times 10^{-4}$ at 90% confidence *(governs design)*

Target properties

$C_y = 0.364$
$T_1 = 0.79$ s

Target properties

$C_y = 0.422$
$T_1 = 0.28$ s

Computed properties

$C_y = 0.360$
$T_1 = 0.79$ s

Computed properties

$C_y = 0.448$
$T_1 = 0.28$ s

Computed performance

Computed performance

| PO | **NLRHA method** | | | PO | **NLRHA method** | | |
| | IDA | Single stripe | Double stripe | | IDA | Single stripe | Double stripe |
|---|---|---|---|---|---|---|---|
| System ductility | ✓ | ✓ | ✓ | System ductility | ✓ | ✗ | ✓ |
| Interstory drift ratio | ✓ | ✓ | ✓ | Interstory drift ratio | ✓ | ✓ | ✓ |
| Collapse | ✓[a] | NA | NA | Crushing of confined concrete | ✓[a] | NA | NA |
| Column plastic hinge rotation | ✓ | ✓ | ✓ | | | | |
| Beam plastic hinge rotation | ✓ | ✓ | ✓ | | | | |

[a] After correcting for the spectral shape of ground motion records.

the demand-capacity factored design (DCFD) approach of Section 13.4.2.2), rather than the input level of intensity, taking full account of the probabilistic nature of demand and capacity as per Section 13.4. This is maintained even when LS values are employed within the context of the code, i.e., via Design Method A, where design is based on checking, subjecting all approaches to the same high standard of performance assessment. Thus, a 0.7% drift limitation at 10/10 for EN1998 is interpreted as a requirement for the MAF of the drift response exceeding 0.7% to be lower than 10/10, rather than the MAF of the intensity that causes drift response of 0.7% to be lower than 10/10. This is a subtle but important difference, and constitutes a stricter test for performance that is fully compatible with performance-based earthquake engineering (Section 13.4).

The logarithmic dispersion (standard deviation of the log of the data) applied to LS threshold (or capacity) values is $\beta_{CR} = 0.2$. Overall additional uncertainty on demand and capacity is taken to be $\beta_{TU} = 0.2$ in engineering demand parameter (EDP) terms, to be employed for the DCFD format of Section 13.4.2.1. An approximately equivalent dispersion in intensity measure (IM) terms of $\beta_{IM,U} = 0.25$ is instead adopted for the MAF format (Section 13.4.2.2) to incorporate both of the aforementioned dispersion values, as explained in Section 13.4.3.1. Uncertainty for the generation of yield frequency spectra (YFS) in Design Methods B and C, was handled via a higher resolution approach with ductility dependence, rather than the response-independent approach adopted in assessment. Specifically, $\beta_{\text{IM,U}}$ is taken equal to zero at the zero point, 0.10 at the yield point, and 0.30 at the ultimate ductility. Between the zero and the yield points a linear interpolation is employed, while between the yield and the ultimate point a linear interpolation on the logarithm of ductility is used to estimate the uncertainty dispersion. Due to the nature of the interpolation employed, most values of ductility that are greater than 1.5 receive a $\beta_{\text{IM,U}}$ that is closer to the upper value of 0.30. This increased dispersion vis-à-vis the 0.25 employed in assessment via the MAF format is taken to account for the fact that a less accurate equivalent single-degree-of-freedom (ESDOF) approximation is employed in YFS, compared to the more reliable multi-degree-of-freedom (MDOF) model used for verification.

Overall, the importance of $\beta_{\text{TU}}$ and $\beta_{\text{IM,U}}$ in using YFS and, more generally, in performance-based design and assessment, is determined by the confidence level adopted for each LS. As discussed in Section 13.2.8, higher confidence levels go closer to the tails of the capacity and demand distributions, thus putting more importance on the magnitude of the uncertainty dispersion and demanding more safety out of the design.

## 19.4 SITE SEISMIC HAZARD AND GROUND MOTIONS

A site characterized by relatively high seismic design values was selected for use with all the examples. It is located in San Jose, California (latitude = 37.33659° and longitude = −121.89056°) on Site Class D soils.

The seismic hazard (as established in the 2008 National Seismic Hazard Mapping Project[1]) is illustrated in Figure 19.1. The corresponding uniform hazard spectra are shown in Figure 19.2. The ASCE-7 (2010) hazard data for this site, representing the direction of maximum horizontal spectral response accelerations modified for Site Class D soils, has 2/50-year values of $S_{\text{MS}} = 1.50$ g and $S_{\text{M1}} = 0.90$ g. These values match those used by Haselton (Chapter 6, 2006), this being the main reason for selecting the site. The smoothed design spectra according to ASCE 7 are shown in Figure 19.3, and are based on

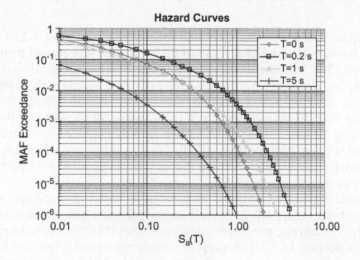

*Figure 19.1*    Seismic hazard for the site in San Jose, California.

---

[1] Available at https://earthquake.usgs.gov/hazards/interactive/.

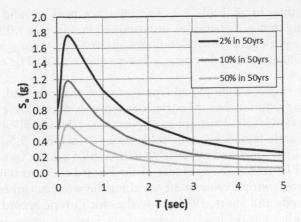

*Figure 19.2* Uniform hazard spectra for the site in San Jose, California.

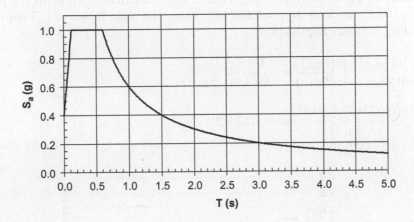

*Figure 19.3* ASCE-7 design spectrum for the site in San Jose, California (reduced for short periods, as would be done for a modal analysis).

$S_{DS} = 2/3(S_{MS}) = (2/3)(1.500\,g) = 1.00\,g$ and $S_{D1} = (2/3)(S_{M1}) = (2/3)(0.90\,g) = 0.60\,g$. ASCE-7 hazard data will in general differ from the 2008 National Seismic Hazard Mapping Project data (at the same nominal MAF or return period level), as (i) the latter has not been modified for risk-targeting (see Section 13.3.7) and (ii) it represents the arbitrary component of horizontal spectral acceleration (rather than the maximum). However, this hazard data is a better representation of the actual site hazard, and thus is better suited for performance-based assessment or design via YFS, as employed in Design Method C. For all examples, we used the mean hazard curve to account for uncertainty in hazard at a predefined level, consistent with the mean.

Note here that for the verification of Design Method A, we employ code design criteria associated with a 10/50 level of risk (see also Section 19.3 on applying this MAF at the response level for design checking). The reason is that this was the original intention behind adopting these values, at least for the West Coast of the United States, when the 10/50 intensity was replaced by 2/3 of 2/50 intensity.

A compatible elastic spectrum was selected among those established in EC-8 (CEN, 2005). The horizontal seismic action is represented by a Type 1 elastic response spectrum ($M_s > 5.5$, EC-8 §3.2.2.2). For soil type C (EC-8 Table 3.1): $T_B = 0.2\,s$, $T_C = 0.6\,s$, $T_D = 2.0\,s$

and $S = 1.15$, according EC-8 Table 3.2. The reference peak ground acceleration, $a_{gR}$, is 0.35 g. For a building that is classified as importance class II, $\gamma_I = 1.0$ (EC-8 Table 4.3 and §4.2.5(5). Thus, the peak ground acceleration is given by $a_g = \gamma_I \cdot a_{gR} = 0.35$ g. For viscous damping equal to 5% of critical damping, $\eta$ is set equal to 1.0. The resulting spectrum is shown in Figure 19.4.

Design Method B employs a simplified representation of the hazard, based on the design UHS together with appropriate hazard curve log-slopes, $k$, that allow extrapolating to both lower and higher hazard levels. Thus, for its application, 2/50 and 50/50 spectra were established based on the scale factors given in Table 19.2—a factor of 0.5 is used for the 50/50 spectrum along the lines suggested by EN1998 (CEN, 2005), and a factor of 1.5 for the 2/50 spectrum as implied by ASCE 7. The slopes of the hazard curve at these MAFs of exceedance are also tabulated, with a value of 3 typical for the western United States at the 10/50 level (see, e.g., Kennedy and Short, 1994), as well as for Europe according to EN1998. The value of $k$ for the other hazard levels is selected via a secant approximation of the slope. For the serviceability level earthquake of 50/50 (or actually the similar 10/10 as EN1998 typically mandates), the corresponding MAF is $\lambda_{SLE} = 0.010536$, while for the design level earthquake (DLE), we have $\lambda_{DLE} = 0.002107$. Then, the scale factor of 0.5 connecting the corresponding $S_a$ values implies that

$$k = -\frac{\ln(\lambda_{DLE}) - \ln(\lambda_{SLE})}{\ln(Sa_{DLE}) - \ln(Sa_{SLE})} = -\frac{\ln(\lambda_{DLE}/\lambda_{SLE})}{\ln(Sa_{DLE}/Sa_{SLE})}$$

$$= -\frac{\ln(0.002107/0.010536)}{\ln(2)} = 2.32 \approx 2$$

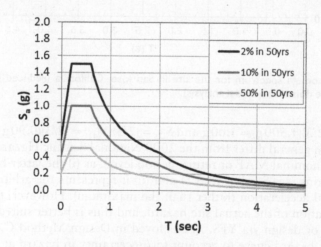

Figure 19.4   Uniform hazard spectra generated using the assumptions of Table 19.2.

Table 19.2 Assumptions used to estimate hazard on the basis of a 10/50 reference code spectrum

| Probability of exceedance in 50 years (%) | MAF of exceedance | Mean return period (year) | Scale factor | Hazard slope k |
|---|---|---|---|---|
| 50 | $1.39 \times 10^{-2}$ | 72 | 0.5 | 2 |
| 10 | $2.11 \times 10^{-3}$ | 475 | 1.0 | 3 |
| 2 | $4.04 \times 10^{-4}$ | 2,475 | 1.5 | 4 |

For the maximum considered earthquake (MCE) of 2/50, the MAF is $\lambda_{MCE} = 0.000404$; we can estimate the slope between DLE and MCE as

$$k = -\frac{\ln(\lambda_{MCE}/\lambda_{DLE})}{\ln(Sa_{MCE}/Sa_{DLE})} = -\frac{\ln(0.000404/0.002107)}{\ln(1.5)} = 4.07 \approx 4$$

The resulting uniform hazard spectra are plotted in Figure 19.4 and they are a reasonable, although imperfect, match of the actual ones appearing in Figure 19.2. Obviously, the use of such simplifications is only meant to cover cases where detailed hazard information is not available, thus such a comparison is of little significance. Having better information will always convey better results.

Note that when assessing the performance of any of the structures designed for the design spectra of ASCE 7 or EN1998, there is a minor issue of hazard incompatibility. Specifically, accurate assessment will require using the actual hazard data for the site, resulting from probabilistic seismic hazard analysis (Section 13.3), regardless of the design spectrum that may have been employed to approximate the uniform hazard spectrum (UHS) at the 10/50 probability of exceedance. In general, design spectra shapes that allow for better fits, such as those provided in ASCE 7 (2010) that employ two anchor periods and the corresponding spectral acceleration values (Figure 19.3), will better match the true UHS (Figure 19.2). The EN1998 spectrum (Figure 19.5), being fitted only by one parameter (the PGA, i.e., $a_g$) and having a rigid shape, is more disadvantaged. Thus, when employing Design Methods A and B, where design is based on code spectra rather than the true UHS, discrepancies will exist between the design spectrum and the true UHS, either in favor or against the structural performance as determined in the assessment. To provide a fair basis for judging the capability of a design approach to fulfill the targeted POs regardless of such discrepancies, we introduce an additional step of scaling the hazard to match the $10/50 \cdot S_a(T_1)$ value implied by the design spectrum employed each time. Such scaling is not necessary in practical applications. It is also not required for Design Method C, as no approximation of the hazard takes place during design.

For all examples, assessment is conducted via multiple nonlinear response history analyses using the far-field ground motion set of FEMA P695 (FEMA, 2009a). This set comprises 22 ground motions (each having two horizontal components) resulting in a total of 44 accelerograms, applied individually to a two-dimensional (planar) model of the lateral

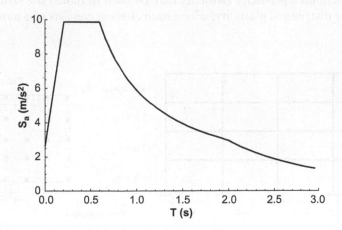

*Figure 19.5* EC8 design spectrum comparable to that shown in Figure 19.3 for the site in San Jose, California.

force-resisting system. They were mainly selected from large magnitude, relatively short distance events to be able to represent high-intensity near-collapse level ground motions as much as possible. Still, care was exercised to avoid having waveforms with long-duration characteristics or near-source directivity pulses, hence they can be considered as "ordinary" far-field accelerograms that can challenge even a modern structure. For assessment via incremental dynamic analysis (IDA), all 44 records were employed for maximum accuracy, while for the more frugal single-/double-stripe assessment a reduced set of 17 records, each from a different recording, was selected. The full set of accelerograms is available from the book website http://users.ntua.gr/divamva/RCbook/FEMA-P695-FFset.zip (as FEMA, 2009b).

## 19.5 MATERIAL PROPERTIES

The moment frame examples (Examples 1–3) were designed using concrete having $f'_c = 5\,ksi$ and expected strength $f'_{ce} = 6.5\,ksi$, and steel reinforcement having yield strength $f_y = 60\,ksi$ and expected strength $f_{ye} = 69\,ksi$. Thus, the expected yield strain, $\varepsilon_{ye}$, is 0.00238.

The wall examples (Examples 4–6) were designed using B500 longitudinal steel reinforcement (having characteristic strength $f_{yk} = 500\,MPa$) and C-30 concrete (having characteristic strength $f_{ck} = 30\,MPa$). Expected strengths are $f_{ye} = 575\,MPa$ and $f_{ce} = 38\,MPa$. Thus, the expected yield strain, $\varepsilon_{ye}$, is 0.00288.

## 19.6 MOMENT FRAME PLAN, ELEVATION, AND MODELING (EXAMPLES 1–3)

The moment frame is part of the perimeter of a four-story reinforced concrete (RC) frame structure (Figure 19.6). The height of the first story is 15 ft and each of the remaining stories is 13 ft high, resulting in a total height of 54 ft. The frame has four bays, each with 30 ft spans on center.

This example was originally developed by Haselton (2006), who determined dead loads totaling 175 psf smeared over the floor plate. Therefore, the weight per story is (180 ft)·(120 ft)·(175 psf) = 3,780 kips (16,800 kN). The total weight of the building is 15,120 kips, with half of this (7,560 kips or 33,600 kN) tributary to each frame. We assume design live loads of 50 psf.

Lumped or distributed plasticity elements may be used to model the structure. Fiber elements provide for distributed plasticity, where each element consists of a number of sections,

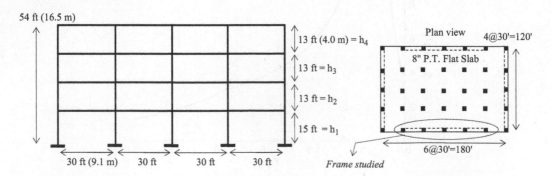

Figure 19.6  Plan and elevation of four-story building used in Examples 1–3.

and each section is composed of a number of fibers. Each fiber consists of a simple spring that represents the uniaxial stress–strain behavior of reinforcing steel or concrete (either confined or unconfined). At the expense of higher computational complexity, fiber elements allow representation of phenomena such as the cracking of concrete, the spread of plasticity and axial–moment interaction at the level of the section as an inherent consequence of the modeling of the individual fibers and sections. At the same time, fiber models typically cannot reproduce well the behavior beyond the maximum strength (or "capping" point), and generally do not accurately represent some phenomena such as rebar buckling. Shear is not represented within the fiber response, but is typically modeled elastically. In contrast, lumped plasticity models represent observed phenomena at a high level, providing concentrated plasticity at predefined plastic hinge locations. Typically a rotational spring is used with moment–rotation characteristics defined using a multi-linear backbone (see Chapter 4). This approach can capture yielding, capping, and the post-capping region well enough, but often fails to reproduce the cracking of concrete and the gradual plastification of sections. Therefore, the lack of a transition from the initial uncracked stiffness to the cracked effective stiffness (and associated changes in instantaneous period) may affect the accuracy of performance assessment. Still, lumped plasticity models are far less complex, numerically more robust, and generally better suited to assessing collapse or near-collapse performance (see also Haselton, 2006).

A two-dimensional lumped plasticity model of the structure was used for assessment of the moment frame examples. To overcome the initial versus effective stiffness issue, a distributed plasticity model was also created and subjected to static pushover analysis. The elastic stiffness of concentrated plasticity elements was then tuned (i.e., typically increased vis-à-vis a "cracked section" assumption) so that the pre-yield behavior better matched the distributed plasticity model. In both models, a leaning column was added to simulate the effect of the columns that carry the gravity loads. The leaning column was pinned at the foundation and modeled using linear elastic elements. These have area and moment of inertia that match at each story the corresponding aggregated cross-sectional properties of one half of the columns that do not participate in the moment frames explicitly incorporated in the model (i.e., gravity columns and columns that belong to moment frames acting in the other direction). Only half of these columns are incorporated in the leaning column as only one out of two moment frames was modeled in the direction of interest.

## 19.6.1 Distributed plasticity model

To simulate a rigid diaphragm, the frame nodes at any given floor (including the leaning column) were rigidly connected by stiff truss elements. Such a rigid constraint can impose a condition of zero axial strain on the beams, which is a restraint against the natural elongation of beams that are modeled using fiber elements (since tensile strain would naturally develop along the beam centerline). Imposing this constraint can produce a fictitious axial compression that increases the moment rotation capacity of the beam plastic hinges, due to the moment–axial interaction displayed by fiber sections. Therefore, one end of each beam element was provided with a low stiffness axial (i.e., horizontal) spring at the connection to the column to avoid generating significant compressive force in the beams. The structural model is presented in Figure 19.7 along with some diaphragm modeling details.

Beams and columns were modeled using force-based distributed plasticity—fiber elements, discretized into longitudinal steel and concrete fibers. Typical fiber section discretization is shown in Figure 19.8. A uniaxial nonlinear model, proposed by Kent, Scott, and Park and realized in OpenSees (McKenna et al., 2000), was employed for unconfined cover concrete (dark gray color in Figure 19.8), while confinement-related parameters were specified

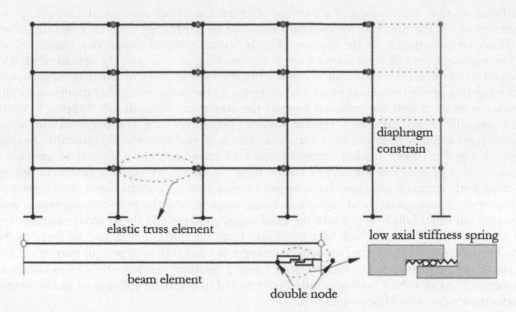

*Figure 19.7* Distributed plasticity model of the moment-resisting frame.

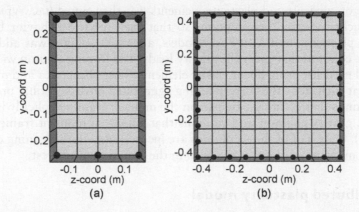

*Figure 19.8* Typical section fiber discretization used to model (a) a 14 × 22 in. beam and (b) a 36 × 36 in. column, showing reinforcing steel (black), unconfined concrete fibers (dark gray), and confined concrete fibers (light gray). (Dimensions in m).

for confined core concrete (light gray color in Figure 19.8), based on the Mander model (Mander et al., 1988). Steel reinforcing bars (black color in Figure 19.8) were modeled using a bilinear constitutive law accounting for pinching and stiffness degradation. The strength of steel and concrete materials was set at their expected value of $f_{ye}$ = 69 ksi and $f_{ce}$ = 6.5 ksi, respectively, rather than at the nominal characteristic strengths.

## 19.6.2 Lumped plasticity model

In lumped plasticity models, rigid diaphragm behavior can be imposed at each floor using rigid kinematic constraints on all nodes of a floor, thus enforcing the same lateral

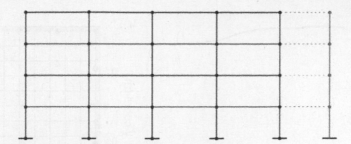

*Figure 19.9*   Lumped plasticity model of the moment-resisting frame.

displacement. In the structural model (Figure 19.9), the horizontal displacement of all nodes at a given floor level were constrained to that of the leftmost beam-column joint.

In the lumped plasticity approach, beams and columns were modeled using a single force-based beam-column element per member, with a concentrated plastic hinge located at each end. Moment–rotation relationships for the plastic hinges were defined based on the generalized backbone curve presented in Chapter 17. As shown in Figure 17.15 plastic hinge behavior can be defined by five points (A, B, C, D, and E in the figure). The stiffness properties of the interior elastic part of the member were determined during the calibration of the model. Specifically, an increased (vis-à-vis the "cracked section") moment of inertia was employed for both beams and columns by averaging the initial "uncracked" stiffness and the nominal "cracked" stiffness at yield, as derived from moment–rotation analyses of the fiber sections (rather than using prescriptive formulas). For columns with different reinforcement ratios at the two ends, an average value of the moment of inertia of the two end sections was used. This calibration allowed for better matching of the period and stiffness of the lumped and distributed plasticity models, reducing differences in model response at low to moderate levels of deformation.

### 19.6.2.1 Columns

Modeling parameters *a*, *b*, and *c* that establish the rotation at the capping point (point C), ultimate rotation (point E), and residual strength, respectively, were determined according to ASCE SEI 41 (Table 17.3). Hardening and capping stiffnesses were considered to be equal to 1% and 90% of the initial flexural stiffness, respectively. The (nominal) yield rotation, $\theta_y$, was estimated as the product of yield curvature, calculated according to Equation 17.17, multiplied by the plastic hinge length, which was assumed to be equal to 1.5 times the member depth.

In lieu of an approximation, the plastic hinge flexural moment at point B was calculated by moment–curvature analysis for each section. The material properties and section fiber discretization are described in Section 19.6.1 on distributed plasticity. Column axial loads were based on expected service gravity loads. Figure 19.10 presents the moment–curvature diagram for a first-story end column (above floor section) designed according to Design Method A, along with section fiber discretization. The plastic hinge flexural moment, $M_p$, was set equal to the maximum moment from the moment–curvature analysis (3,282 kN·m).

### 19.6.2.2 Beams

Typically, negative and positive plastic moment strengths differ ($M_p^+ < M_p^-$), thus the plastic hinge moment–curvature relationship used for beams is asymmetrical. Modeling parameters

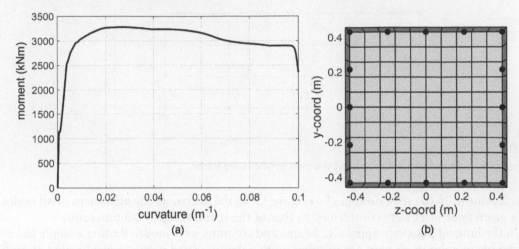

*Figure 19.10* End column of the first story of Example 1: (a) moment–curvature diagram and (b) section fiber discretization.

$a$, $b$, and $c$ that establish capping rotation (point C), ultimate rotation (point E), and residual strength, respectively, were determined according to Table 17.2 separately for positive and negative moment. Hardening and capping stiffnesses in both cases were considered to be equal to 1% and 90% of the initial flexural stiffness associated with the positive or negative moment, respectively. Yield rotations $\theta_y^+$ and $\theta_y^-$ were estimated as the product of yield curvature, calculated according to Equations 17.7 and 17.8, respectively, multiplied by the plastic hinge length (assumed to be equal to twice the member depth). The effect of reinforcement slip at the anchorage was also considered; the yield rotation was increased by $\theta_{anchslip}$ calculated by Equation 17.15.

Moment–curvature analyses of beam sections were computed for both positive and negative curvatures. Material properties and fiber section discretization used in the analyses were described in Section 19.6.1 on distributed plasticity. Axial load was equal to zero. Figure 19.11 shows the moment–curvature diagrams for positive and negative moments, along with section fiber discretization, for a first- and a fourth-story beam designed according to Design Method A (Example 1). The latter is discussed in detail in the following section.

### 19.6.2.3 Example of calculating beam modeling parameters and assessment criteria

To better illustrate the application of Chapter 17 in determining modeling parameters and acceptance criteria, the example of a single fourth-story beam is presented, using dimensions and reinforcement calculated for the final Example 1 design (Section 19.7).

- Modeling parameters and acceptance criteria

$$\rho' = \frac{A_{compression}}{b \cdot d} = \frac{2.37 \text{ in.}^2}{14 \text{ in.} \cdot 0.9 \cdot 22 \text{ in.}} = 0.0085$$

$$\rho = \frac{A_{tension}}{b \cdot d} = \frac{6.0 \text{ in.}^2}{14 \text{ in.} \cdot 0.9 \cdot 22 \text{ in.}} = 0.0216$$

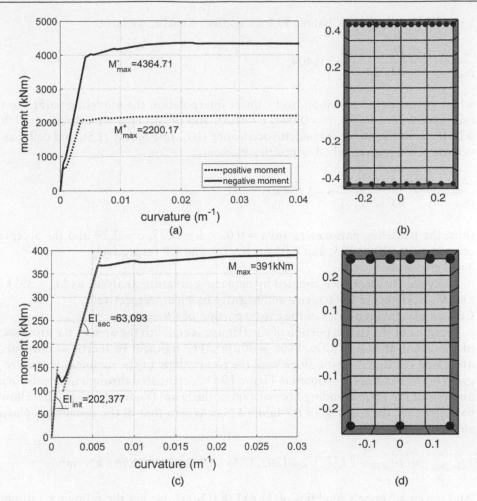

*Figure 19.11* Beam modeling for Example 1: (a) moment–curvature diagram and (b) section fiber discretization for the first-story beams, (c) moment–curvature diagram, and (d) section fiber discretization for the fourth-story beams.

$$\rho_{bal} = \frac{0.85 \cdot \beta_1 \cdot f_c'}{f_y} \cdot \frac{87,0000}{87,0000 + f_y} = \frac{0.85 \cdot 0.725 \cdot 6,500 \text{ psi}}{69,000 \text{ psi}} \cdot \frac{87,0000}{87,0000 + 69,000 \text{ psi}}$$

$$= 0.0324,$$

where

$$\beta_1 = 0.85 - 0.05(f_c' - 4) = 0.85 - 0.05(6.5 \text{ ksi} - 4) = 0.725 > 0.65 \text{ as } f_c' > 4 \text{ ksi.}$$

The shear force that acts on the beam ends is estimated using the portal frame method for the lateral forces of Table 19.11, plus the shear force due to the gravity loads, thus $V = 58.73$ kips.

$$\frac{V}{b_w \cdot d \cdot \sqrt{f_c}} = \frac{58.73 \text{ kip}}{14 \text{ in.} \cdot 0.9 \cdot 22 \text{ in.} \sqrt{\dfrac{6.5}{1,000}}} = 2.63 < 3$$

Using Table 17.2 from Chapter 13 for negative moment, we have:

$$\frac{\rho - \rho'}{\rho_{bal}} = \frac{0.0216 - 0.0085}{0.0324} = 0.404,$$

which is within 0.0 and 0.50, so by linear interpolation the modeling parameters are estimated as $a = 0.021$, $b = 0.034$, $c = 0.20$, and the acceptance criteria are 0.0060, 0.0210, and 0.0338 for immediate occupancy (IO), Life Safety (LS), and collapse prevention (CP), respectively. For positive moment:

$$\frac{\rho - \rho'}{\rho_{bal}} = \frac{0.0085 - 0.0216}{0.0324} = -0.404 < 0$$

thus, the modeling parameters are $a = 0.025$, $b = 0.05$, $c = 0.20$ and the acceptance criteria are 0.010, 0.025, and 0.05 for IO, LS, and CP respectively.

- Yield curvature

  The yield curvature is estimated by moment–curvature analysis, as $M_y^+ = 391\,kN{\cdot}m$ and $M_y^- = 959\,kN{\cdot}m$ for positive and negative moment, respectively.

- Cross-sectional properties of the elastic portion of the element

  The area of the elastic portion of the element is equal to the area of the gross section of the beam, thus $A = 22\,in.{\cdot}14\,in. = 308\,in^2$. The moment of inertia is estimated by averaging the initial elastic slope with the secant slope of the moment–curvature diagram of the beam (as illustrated in Figure 19.11c), estimated through moment–rotation analysis of the corresponding fiber section of the beam (Figure 19.11d). This allows us to better tune the stiffness of the lumped plasticity to that of the distributed plasticity element. Thus, the

$$EI_{\text{elastic beam}} = \left(EI_{\text{initial}} + EI_{\text{secant}}\right)/2 = (202{,}377 + 63{,}093)/2 = 132{,}735\,kN \cdot m^2$$

And, given a Young's modulus of $31{,}681{,}000\,N/m^2$, we get the following estimate of the moment of inertia for the elastic segment of each beam.

$$I_{\text{elastic beam}} = 132{,}735/31{,}681{,}000 = 0.0042\,m^4$$

- Yield rotation

  The yield rotation is estimated using Equations 17.7 and 17.8 for positive and negative moment as:

$$\phi_y^- = \frac{1.8 \cdot \varepsilon_y}{h_b} = \frac{1.8 \cdot 0.00238}{22\,in.} = 1.95 \cdot 10^{-4}\,in.^{-1}$$

$$\phi_y^+ = \frac{1.4 \cdot \varepsilon_y}{h_b} = \frac{1.4 \cdot 0.00238}{22\,in.} = 1.51 \cdot 10^{-4}\,in.^{-1}$$

The anchorage slip at the end of the beam is estimated using Equation 17.15 as:

$$t_{\text{slip}}^- = \frac{\phi_y^- \cdot d_b \cdot f_y}{96 \cdot \sqrt{f_c'}} = \frac{1.95 \cdot 10^{-4}\,in.^{-1} \cdot 1.128\,in. \cdot 69{,}000}{96 \cdot \sqrt{6{,}500}} = 0.0020$$

$$t_{slip}^+ = \frac{\phi_y^+ \cdot d_b \cdot f_y}{96 \cdot \sqrt{f_c'}} = \frac{1.51 \cdot 10^{-4} \text{ in.}^{-1} \cdot 1 \text{ in.} \cdot 69{,}000}{96 \cdot \sqrt{6{,}500}} = 0.0014$$

The yield rotation is estimated as:

$$t_y^- = 2 \cdot d \cdot \phi_y^- + t_{slip} = 2 \cdot 19.8 \text{ in.} \cdot 1.95 \cdot 10^{-4} \text{ in.}^{-1} + 0.0020 = 0.0097$$

$$t_y^+ = 2 \cdot d \cdot \phi_y^+ + t_{slip} = 2 \cdot 19.8 \text{ in.} \cdot 1.51 \cdot 10^{-4} \text{ in.}^{-1} + 0.0014 = 0.0073$$

## 19.7 EXAMPLE I: MOMENT-RESISTANT FRAME DESIGNED USING METHOD A

### 19.7.1 POs

Method A seeks performance comparable to that obtained in current codes. For this example, performance requirements comparable to those in ASCE-7 are used as follows:

- Table 12.4 indicates a system ductility limit at the 2/50-year level of $3.6/I_e = 3.6$.
- Table 12.9 indicates an interstory drift ratio limit of 0.02.

### 19.7.2 Use of nonlinear response analysis in this example

Although nonlinear analysis is not required for Method A, the performance of the initial design realization is evaluated using both nonlinear static (pushover) analysis and nonlinear response history analysis (consisting of IDA and the simpler single- and double-stripe approaches), to provide rigorous data for use in comparing the results obtained using Methods A, B, and C, and to provide a blueprint for applying these available assessment options.

### 19.7.3 Required base shear strength

Design Step 1. Estimated roof displacement at yield: From Section 9.5.1, we estimate the roof drift at yield, $D_y$, to be about 0.55% of the height:

$$D_y = \frac{0.55}{100} h = 0.297 \text{ ft} = 3.56 \text{ in.}$$

Design Step 2. Based on the system ductility limit of 3.6, the peak roof displacement limit is

$$D_{u,ductility} = \mu D_y = 3.6(3.56 \text{ in.}) = 12.8 \text{ in.}$$

Based on the interstory drift limit of 0.02, following Equation 12.14, the peak roof displacement limit at the 2/50-year level is estimated to be:

$$D_{u,drift} = \frac{3}{2} C_1 \left( \frac{\Delta_a}{h_{sx}} \right) \left( \frac{R_{code}/I_e}{C_d \rho} \right) \frac{\alpha_1}{\alpha_{3,stat}} h_n = \frac{3}{2} 1(0.020) \left( \frac{8/1}{5.5} \right) \frac{0.88}{1.3} (54 \text{ ft} \cdot 12 \text{ in./ft})$$

$$= \frac{3}{2} 12.7 \text{ in.} = 19.1 \text{ in.}$$

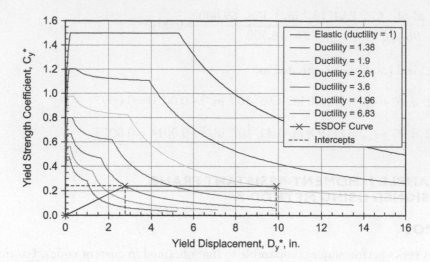

*Figure 19.12*    YPS for initial proportioning of Example 1.

The more restrictive of the drift limits control (12.8 in.), resulting in $\mu = D_u/D_y = 12.8/3.56 =$ 3.6 at the 2/50-year level.

Design Step 3. From Table 12.1, we estimate $\Gamma_1 = 1.30$ and $\alpha_1 = 0.88$, so

$$D_y^* = \frac{D_y}{\Gamma_1} = 2.75 \text{ in.}$$

Design Step 4. A yield point spectra representation of the 2/50-year hazard was developed using the $R$–$\mu$–$T$ relations of Section 4.10.3. As shown in Figure 19.12, for $D_y^* = 2.75$ in. and $\mu = 3.6$ the corresponding $C_y^* = 0.24$.

The corresponding period is:

$$T^* = 2\pi \sqrt{\frac{D_y^*}{C_y^* g}} = 1.08 \text{ s}$$

Design Step 5. Since $C_y = \alpha_1 C_y^*$, the yield strength coefficient is estimated as $C_y = 0.88 \cdot 0.24 = 0.211$. The design base shear at yield for a single frame is given by $V_y = C_y W/2 = 0.211 \cdot 7,560 = 1,597$ kips (7,100 kN).

## 19.7.4 Design lateral forces and required member strengths

Design Step 6. The equivalent lateral force method of ASCE-7 is used in design, wherein lateral forces at each floor are applied in proportional to $w_x \cdot h_x^k$, where $k$ varies linearly between 1 and 2 as $T$ varies between 0.5 and 2.5 s. The effective height of the resultant of the ASCE-7 lateral forces, for a period of 1.08 s, is $0.779 h_n$. The height of the resultant of the first-mode forces, $h_1^*$, is estimated to be $0.72 h_n$ according to Table 12.1. Thus, the base shear to be used with the ASCE-7 lateral force distribution, to obtain the same resistance to overturning moments as is needed for response in the first mode, is estimated to be:

$$V_y \frac{h_1^*}{h_{\text{eff,ASCE7}}} = 1,476 \text{ kips}$$

The lateral forces associated with this base shear at yield according to the ASCE 7 distribution (for a period of 1.08 s) are given in Table 19.3.

Design Step 7. To determine the design moments in the columns and beams, two assumptions are made. The first is the applicability of the portal frame method to distribute shears to the end and intermediate columns (in proportion to $V$ and $2V$, respectively). The second is to specify the locations of the inflection points in the moment diagrams of the columns. Statics is used to determine the column moments. The beam moments are established based on equilibrium at each beam-column joint. Although other values may be selected, we have chosen to locate the inflection points at 0.7 times the height of the first story and at 0.6 times the height of the other stories.

With respect to Figure 19.13, a general expression may be deduced for the shear, where $n_b$ is the number of bays of moment-resistant framing:

$$V = \frac{1,476}{2n_b} = 184.5 \text{ kips}$$

Locating the inflection point at 0.7 of the story height in the first story, the plastic moment at the base of the column, $M_c$, is calculated:

$$M_c = V \cdot 0.7 \cdot 15 = 1,937 \text{ k-ft}$$

Enforcing zero moment at the inflection points and working from either the top-down or the bottom-up, a series of uncoupled equilibrium relationships are obtained for the column and beam moments. In Table 19.4 and Figure 19.14, the horizontal forces resisted by the exterior column are indicated—these are the forces of Table 19.3 divided by $2n_b$.

Table 19.3 Lateral design forces at each floor level, Example 1

| Level | $F_i$ (kips) |
|---|---|
| 4 | 636 |
| 3 | 446 |
| 2 | 272 |
| 1 | 122 |
| Total | 1,476 |

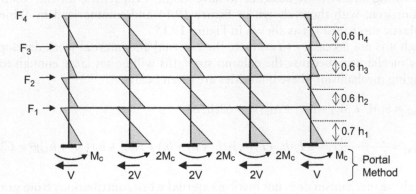

Figure 19.13   Distribution of lateral forces and application of the portal frame method for determining column shears for Example 1; and assumed locations of inflection points.

Table 19.4 Lateral forces tributary to an end column, Example I

| Level | Lateral force tributary to end column (kips) |
| --- | --- |
| 4 | 79.5 |
| 3 | 55.7 |
| 2 | 34.1 |
| 1 | 15.2 |

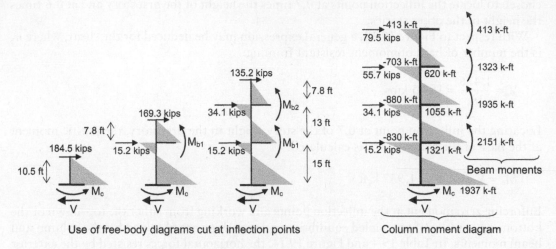

Use of free-body diagrams cut at inflection points          Column moment diagram

Figure 19.14   Bending moment diagram for an end column, Example I.

Table 19.5 Equilibrium moments, Example I

|  | Moment (kip-ft) |
| --- | --- |
| $M_c$ | 1,937 |
| $M_{b1}$ | 2,151 |
| $M_{b2}$ | 1,935 |
| $M_{b3}$ | 1,323 |
| $M_{b4}$ | 413 |

If the framing members are designed to have moment capacities, for the column and the beams, consistent with those shown in Figure 19.14 and summarized in Table 19.5, the implicit plastic mechanism is as shown in Figure 19.15.

Although it is not necessary to evaluate the external and internal work developed during the plastic mechanism (because the column strengths will be set large enough to enforce a beam-hinging mechanism), these quantities are given below:

$$W_{internal} = 8(M_c + M_{b2} + M_{b3} + M_{b4})\theta = 62,076\theta$$

$$W_{external} = \frac{h_1^*}{h_{eff,ASCE7}}(F_1 h_1 + F_2(h_1 + h_2) + F_3(h_1 + h_2 + h_3) + F_4(h_1 + h_2 + h_3 + h_4))\theta = 62,076\theta$$

Since the above mechanism does not involve external work contributions from gravity loads, the beam moments are a function only of the lateral loads (which have a positive contribution to external work).

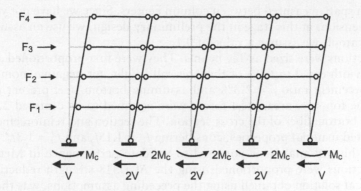

Figure 19.15    Mechanism compatible with the moment diagram shown in Figure 19.13.

The mechanism shown in Figure 19.15 (strong column–weak beam) can be achieved if the columns are designed for moments greater than those associated with beam hinging (e.g., Figure 19.14) at locations above the base—the only intended column plastic hinges are at the base, where they cannot be avoided for a kinematically admissible mechanism. Clearly, if the beam flexural strengths are increased, a different plastic mechanism may result, involving the formation of plastic hinges in the columns (e.g., a weak-story mechanism).

We assume some live load is present during the seismic response. The design of the beams recognizes the presence of gravity loads, by adjusting the beam plastic moment strengths as described below. The amount of live load to consider may be adjusted based on occupancy; we assume gravity loads of $D + 0.25L$ in lieu of a more refined estimate of the expected floor live load, while only $D$ is assumed at the roof. These gravity loads are considered present when designing the columns, and when conducting the nonlinear static (pushover) analysis.

### 19.7.5 Sizing of RC members

Design Step 8. The bending moments of Figure 19.14 were determined based on member centerline dimensions, while beam flexural strengths only need to be adequate to resist the bending moments at the column faces. An additional adjustment of bending moments is made to account for the negative bending moment induced by gravity loads in the beams near the column faces—to assure good behavior under service conditions, we increase the negative moment strengths and reduce the positive moment strengths at the column faces. Because the moments at both ends of a beam go through the same rotation ($\theta$) in the mechanism analysis, the end moments can be adjusted such that the sum of the negative and positive moments remain equal to $2M_p$ (thereby ensuring the same amount of virtual work, $2M_p\theta$, is done by the hinges at both ends of a beam). Thus, for a negative moment strength equal to twice the positive moment strength, the required plastic moment strengths at the face of the column are

$$M_p^+ = \frac{2}{3} M_{pb}\left(1 - \frac{h_{col}}{L}\right)$$

$$M_p^- = \frac{4}{3} M_{pb}\left(1 - \frac{h_{col}}{L}\right)$$

where $L$ = beam span, measured between column centers. Since we have not yet determined the column dimensions at this stage in the preliminary design, we use an assumed value for $h_{col}/L$; in this example we assume a value of 0.1.

The beam sections were sized as Tee beams. They were first proportioned as rectangular beams having width equal to 60% of their overall height, for negative moment (top steel) having a reinforcement ratio $\rho$ = 2.3% and assuming bottom steel present in an amount equal to half the top steel area. Steel forces were assumed to be centered 2.5 in. (64 mm) from the top or bottom fiber of the cross section. The section and reinforcement were sized based on expected material properties, considering $f_{ye} = 1.15 f_y$ and $f'_{ce} = 1.3 f'_c$. Because non-linear response history analysis is not used to validate performance in Method A, beam and column sections were proportioned using the ACI 318 strength reduction ($\phi$) factor. The mathematical solution obtained using the preceding assumptions, was then adjusted to obtain practicable dimensions and reinforcing. The top steel could be rounded up or down, while the bottom steel was sized a little generously to comply with the ACI 318 requirement that $M_n^+ \geq \frac{1}{2} M_n^-$. The mathematical and preliminary design solutions for the beam sections are given in Tables 19.6 and 19.7, respectively.

Design Step 9. For the preliminary design of the columns, the overall column depth, $h_{col}$ was set greater than 20 times the diameter of the beam reinforcement, to satisfy requirements for beam-column joints of moment-resistant frames. Column sections were first sized for the $2M_c$ moments at the base of the intermediate columns (Figure 19.13) under the axial load corresponding to $D + 0.25L$, to determine a cross section having practicable dimensions and $\rho_g$ close to the practical upper limit of 4%–5%. At the ground level, the tributary dead load is $D$ = (175 psf)(30 ft)(15 ft)(4 stories) = 315,000 pounds = 315 kips = 1,401 kN, and the live load is 0.25(50 psf)(30 ft)(15 ft)(3 stories) = 16,875 pounds = 16.9 kips = 75.1 kN. Expected material properties were used along with the strength reduction factors ($\phi$) given by ACI. Above this level, the columns were sized to satisfy the ACI 318 requirement that

$$\sum M_{n,col} \geq \frac{6}{5} \sum M_{n,bm}$$

except that expected material properties were used for the beams and columns. The sum of column moments was distributed unequally, with about 53%–58% of this sum assigned to

Table 19.6  Mathematical solution for beam dimensions and reinforcement, Example 1

| | Required strengths | | Mathematical solution | | | |
|---|---|---|---|---|---|---|
| Level | $M_p^-$ (kip-in.) | $M_p^+$ (kip-in.) | $b_w$ (in.) | $h$ (in.) | $A_{s,top}$ (in.$^2$) | $A_{s,bot}$ (in.$^2$) |
| 4 | 5,947 | 2,974 | 12.92 | 21.54 | 5.66 | 2.00 |
| 3 | 19,051 | 9,526 | 18.43 | 30.72 | 11.96 | 5.07 |
| 2 | 27,864 | 13,932 | 20.77 | 34.62 | 15.34 | 6.67 |
| 1 | 30,974 | 15,487 | 21.43 | 35.72 | 16.38 | 7.48 |

Table 19.7  Preliminary design (beam dimensions and reinforcement), Example 1

| Level | $b_w$ (in.) | $h$ (in.) | Top bars | Bottom bars |
|---|---|---|---|---|
| 4 | 14 | 22 | (6) No. 9 | (3) No. 8 |
| 3 | 18 | 32 | (12) No. 9 | (7) No. 8 |
| 2 | 24 | 36 | (16) No. 9 | (9) No. 8 |
| 1 | 24 | 36 | (16) No. 9 | (10) No. 8 |

the column cross section just below the level under consideration, and the remainder to the column moment above, in recognition of the greater axial load in the section below the level; since these columns have relatively light axial loads, the higher axial load level results in a larger flexural strength. (For the column supporting the roof, 100% of the sum was assigned to the column section below the roof.) In sizing the gross column section and reinforcement, it was assumed that reinforcement splices were at mid-height of each story; therefore, gross section dimensions are constant between floor levels, while reinforcement is constant at the floor level but may change at the midheight of any story.

Note that beams frame into both sides of an intermediate column, with negative moments (equal to 4/3 of the beam moments shown in Figure 19.14) acting on one side and positive moments (equal to 2/3 of the moments in Figure 19.14) acting on the other side. Thus, critical sections of intermediate columns framing into a beam-column joint were designed for 6/5 of twice the moments shown in Figure 19.14. For the end columns framing into a beam-column joint, it is assumed that 4/3 of the beam moments shown in Figure 19.14 act, corresponding to negative beam moments—thus the critical sections of end columns were designed for 6/5 of 4/3 of the beam moments shown in Figure 19.14.

With the mathematical solution in hand, a preliminary bar configuration was selected that contained steel area slightly greater than that required mathematically. The gross dimensions of the columns at the ends of the frame were selected to match those of the intermediate columns. Equation 15.6 was applied, but just using the (larger) negative moment for the beam. Axial loads were taken as $D + 0.25L$, that is, neglecting any overturning forces, since we assume the columns are below the balance point—thus, the column subject to overturning tension at one end of the frame and that subject to overturning compression at the other end have nearly compensating changes in moment strength—the result is considered to be a negligible change on plastic mechanism strength. Column designs are given in Tables 19.8–19.11.

### 19.7.6 Preliminary evaluation of the initial design

To quickly assess the basic characteristics of the preliminary design, a fiber element model was developed in SeismoStruct (Seismosoft, 2014). If the yield displacement and values of

Table 19.8 Preliminary design of intermediate columns just above floor levels, Example 1

| Floor level | $M_c$ or $M_{pc}$ (k-in.) | $D + 0.25L$ (k) | H (in.) | $A_s$ (in.²) | Bars | $A_{s,provided}$ (in.²) | $P/A_g f'_{ce}$ | $\rho_g$ % |
|---|---|---|---|---|---|---|---|---|
| 4 | 0 | NA | NA | NA | NA | NA | NA | NA |
| 3 | 15,431 | 78.8 | 26 | 23.79 | (20) No. 10 | 25.40 | 1.79% | 3.76% |
| 2 | 23,573 | 163.1 | 30 | 29.84 | (24) No. 10 | 30.48 | 2.79% | 3.39% |
| 1 | 25,089 | 247.5 | 32 | 27.70 | (24) No. 10 | 30.48 | 3.72% | 2.98% |
| Footing | 46,488 | 331.9 | 36 | 47.73 | (36) No. 10 | 45.72 | 3.94% | 3.53% |

Table 19.9 Preliminary design of intermediate columns just below floor levels, Example 1

| Floor level | $M_c$ (k-in.) | $D + 0.25L$ (k) | H (in.) | $A_s$ (in.²) | Bars | $A_{s,provided}$ (in.²) | $P/A_g f'_{ce}$ | $\rho_g$ % |
|---|---|---|---|---|---|---|---|---|
| 4 | 10,705 | 78.8 | 26 | 15.25 | (20) No. 10 | 25.40 | 1.79% | 3.76% |
| 3 | 18,861 | 163.1 | 30 | 22.61 | (20) No. 10 | 25.40 | 2.79% | 2.82% |
| 2 | 26,582 | 247.5 | 32 | 29.83 | (24) No. 10 | 30.48 | 3.72% | 2.98% |
| 1 | 30,665 | 331.9 | 36 | 28.01 | (24) No. 10 | 30.48 | 3.94% | 2.35% |
| Footing | NA | NA | NA | NA | NA | NA | NA | NA |

*Table 19.10* Preliminary design of end columns just above floor levels, Example 1

| Floor level | $M_c$ (k-in.) | $D + 0.25L$ (k) | $H$ (in.) | $A_s$ (in.²) | Bars | $A_{s,provided}$ (in.²) | $P/A_g f_{ce}$ | $\rho_g$ % |
|---|---|---|---|---|---|---|---|---|
| 4 | 0 | NA | NA | NA | NA | NA | NA | NA |
| 3 | 9,602 | 78.8 | 26 | 13.36 | (12) No. 10 | 15.24 | 1.79% | 2.25% |
| 2 | 15,047 | 163.1 | 30 | 16.97 | (16) No. 10 | 20.32 | 2.79% | 2.26% |
| 1 | 16,726 | 247.5 | 32 | 16.17 | (16) No. 10 | 20.32 | 3.72% | 1.98% |
| Footing | 23,244 | 331.9 | 36 | 19.23 | (16) No. 10 | 20.32 | 3.94% | 1.57% |

*Table 19.11* Preliminary design of end columns just below floor levels, Example 1

| Floor level | $M_c$ (k-in.) | $D + 0.25L$ (k) | $H$ (in.) | $A_s$ (in.²) | Bars | $A_{s,provided}$ (in.²) | $P/A_g f_{ce}$ | $\rho_g$ % |
|---|---|---|---|---|---|---|---|---|
| 4 | 7,137 | 78.8 | 26 | 9.23 | (12) No. 10 | 15.24 | 1.79% | 2.25% |
| 3 | 13,260 | 163.1 | 30 | 14.41 | (12) No. 10 | 15.24 | 2.79% | 1.69% |
| 2 | 18,390 | 247.5 | 32 | 18.40 | (16) No. 10 | 20.32 | 3.72% | 1.98% |
| 1 | 20,443 | 331.9 | 36 | 16.01 | (16) No. 10 | 20.32 | 3.94% | 1.57% |
| Footing | NA | NA | NA | NA | NA | NA | NA | NA |

*Table 19.12* First-mode force vector, Example 1

| Level | $F_{1st}$ (kip) |
|---|---|
| 4 | 639 |
| 3 | 443 |
| 2 | 276 |
| 1 | 118 |

first-mode properties determined with this model suggest acceptable performance, a lumped plasticity model may be developed for use in detailed evaluation, in accordance with the design method. In the SeismoStruct model, the fiber element sections are composed of individual concrete and steel fibers that can respond inelastically, allowing the section as a whole to respond inelastically to both flexural and axial loads. Shear is resisted elastically.

A nonlinear static (pushover) analysis was conducted using a lateral force vector proportional to the first-mode distribution of lateral forces as shown in Table 19.12 (i.e. $\mathbf{F} \propto \mathbf{M}\boldsymbol{\phi}_1$, where $\mathbf{F}$ is the vector of lateral forces, $\mathbf{M}$ is a diagonal matrix containing the masses at each floor level, and $\boldsymbol{\phi}_1$ is the first-mode shape).

The analysis was conducted in the presence of gravity loads associated with the $D + 0.25L$ load combination. The resulting capacity curve, representing a plot of base shear as a function of roof displacement, is shown in Figure 19.16. Two analyses are shown; one in which the concrete is assigned tensile strength and the other in which the concrete is assigned zero tensile strength. The gravity loads cause pre-compression of the columns prior to any application of lateral loads. As normal stresses associated with flexure relieve this pre-compression, some cracking occurs and the capacity curve softens.

The bending moments of Figures 19.13 and 19.14 are based on a mechanism analysis in which plastic hinges form at the ends of each beam; the positive and negative plastic hinge strengths are assumed equal within each floor level. Beams were proportioned to have larger negative and smaller positive moment strengths, in recognition of service-level gravity loads. The largest bending moments in the end columns occur when lateral forces are applied in the direction that mobilizes the adjacent beam plastic hinges in negative bending;

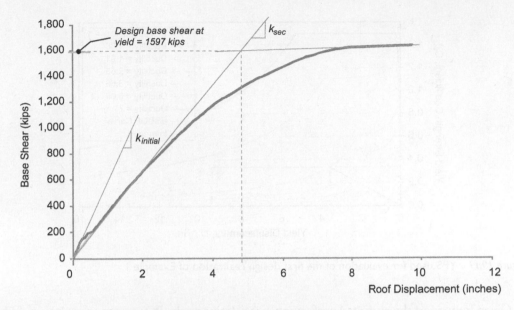

*Figure 19.16*   Capacity curve of the structure observed in a first-mode pushover analysis of the first design realization, Example 1.

this loading was assumed when proportioning the end columns. (In contrast, the sum of the beam moments acting on the beam-column joint of an intermediate column is unchanged.) The pushover analysis results can be used to establish whether the intended mechanism actually developed. The moments in the end columns did not exceed those associated with the loading that mobilized the adjacent beam negative moments, and the moments in the intermediate columns did not exceed those of Figure 19.14, above the base of the first floor. Similarly, the reinforcement yield strain was not exceeded in the column longitudinal reinforcement at these locations.

From an eigenvalue analysis prior to the nonlinear static analysis, $T_1 = 0.787$ s, $\Gamma_1 = 1.36$, and $\alpha_1 = 0.63$. With reference to Figure 19.16, the period associated with the secant stiffness representative of cracked section behavior can be estimated as

$$T_{sec} = T_{initial}\sqrt{\frac{K_{initial}}{K_{sec}}} = 0.787 \text{ s}\sqrt{\frac{590 \text{ k/in.}}{330 \text{ k/in.}}} = 1.05 \text{ s}$$

which compares well to $T^* = 1.08$ s. The design base shear at yield, 1,597 kips (associated with first-mode response) compares quite well to the base shear at yield inferred from the first-mode pushover analysis, as indicated in Figure 19.16. The estimated yield displacement of the roof was 3.56 in., while the value inferred from the capacity curve is 4.8 in.

Given the increase in yield displacement and reduction in $\alpha_1$ (from the assumed value of 0.88), we may wish to check the performance of this initial design realization. We estimate $D_y^* = D_y/\Gamma_1 = 4.8/1.36 = 3.53$ in. Since $V_y$ is approximately 1,600 kips, $C_y = V_y/W = 1,600/7,560 = 0.211$. Therefore, $C_y^* = C_y/\alpha_1 = (0.211/0.63) = 0.336$. This point is plotted on the YPS below (using new constant ductility curves), in Figure 19.17. The yield point lies on the $\mu = 2.65$ contour; the ductility demand of 2.65 is less than the design limit of 3.6. The peak roof displacement is estimated to be 4.8(2.65) = 12.7 in. Thus, the resulting design complies with the constraints on system ductility (3.6) and interstory drift (19.1 in.).

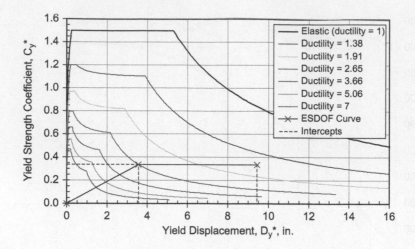

*Figure 19.17*   YPS used for evaluation of the first design realization of Example 1.

Given how good the expected performance is relative to the POs, some economy could be achieved in a refinement of the initial design that makes use of updated estimates of design parameters ($\Gamma_1 = 1.36$, $\alpha_1 = 0.63$, and $D_y = 4.8$ in.), to achieve a design having a lower base shear strength with performance more closely matching the POs.

Note that a pushover analysis is not necessary to evaluate the initial design. Rather, if members are modeled using cracked sections (stiffness estimated are provided in Section 17.8), an eigenvalue analysis of the linear elastic model may be used to determine $\Gamma_1$, $\alpha_1$, and $T_1$. This information, together with the first-mode mechanism design strength, $V_y$, is sufficient to determine the parameters of the ESDOF system ($C_y = V_y/W$ and $C_y^* = C_y \cdot \alpha_1$). Then,

$$D_y^* = \left(\frac{T^*}{2\pi}\right)^2 C_y^* g$$

These values of $C_y^*$ and $D_y^*$ may be plotted on a yield point spectra to estimate the resulting ductility ($\mu$) and drift demands ($D_u^* = \mu D_y^*$); the peak roof displacement is estimated as $D_u = \mu \cdot D_y = \mu(D_y^*)(\Gamma_1)$.

### 19.7.7 Nonlinear modeling and acceptance criteria

Although not required by the methodological prescription of Method A (Table 18.1), we conducted a nonlinear response history analysis to evaluate the performance of the moment frame design and provide a basis for comparison with the other two design approaches. In Tables 19.13 and 19.14, the modeling parameters for columns and beams, respectively, are summarized. In Tables 19.15 and 19.16 the acceptance criteria for columns and beams are presented.

As noted in Section 19.6, distributed plasticity and lumped plasticity models are available. The fundamental period of the system using lumped plasticity models, after the calibration step of Section 19.6.2, is $T_1 = 1.05$ s, which is somewhat larger than the initial period of the distributed plasticity model, $T_1 = 0.85$. Our aim is not to perfectly represent the initial period, as it only pertains at small deformations, but instead to represent the effective stiffness (and secant period) that characterizes the majority of the pre-yield segment. The nonlinear static

*Table 19.13* Plastic hinge modeling parameters for columns, Example 1

| | End columns | | | | Intermediate columns | | | |
| Story | $M_p$ (kN·m) | a | b | c | $M_p$ (kN·m) | a | b | c |
| --- | --- | --- | --- | --- | --- | --- | --- | --- |
| 4 | 1,563 | | | | 2,378 | | | |
| 3 | 1,955 (top) | | | | 2,917 (top) | | | |
| | 2,433 (bot) | 0.030 | 0.044 | 0.20 | 3,416 (bot) | 0.030 | 0.044 | 0.20 |
| 2 | 2,732 | | | | 3,785 | | | |
| 1 | 3,282 | | | | 4,469 (top) | | | |
| | | | | | 6,262 (bot) | | | |

*Table 19.14* Plastic hinge modeling parameters for beams, Example 1

| | Negative moment | | | | Positive moment | | | |
| Story | $M_y^-$ (kN·m) | a | b | c | $M_y^+$ (kN·m) | a | b | c |
| --- | --- | --- | --- | --- | --- | --- | --- | --- |
| 4 | 959 | 0.0210 | 0.0338 | | 391 | | | |
| 3 | 2,864 | 0.0211 | 0.0346 | | 1,359 | | | |
| 2 | 4,338 | 0.0215 | 0.0359 | 0.20 | 1,981 | 0.025 | 0.050 | 0.20 |
| 1 | 4,365 | 0.0218 | 0.0371 | | 2,200 | | | |

*Table 19.15* Column plastic hinge rotation limits for IO, LS, and CP POs, Example 1

| | Positive rotation (rad) | | |
| Story | IO | LS | CP |
| --- | --- | --- | --- |
| $\theta_{p,max}$ (rad) | 0.005 | 0.034 | 0.044 |

*Table 19.16* Beam plastic hinge rotation limits for IO, LS, and CP POs, Example 1

| | Positive rotation (rad) | | | Negative rotation (rad) | | |
| Story | IO | LS | CP | IO | LS | CP |
| --- | --- | --- | --- | --- | --- | --- |
| 4 | | | | 0.0060 | 0.0210 | 0.0338 |
| 3 | | | | 0.0061 | 0.0211 | 0.0346 |
| 2 | 0.010 | 0.025 | 0.050 | 0.0065 | 0.0215 | 0.0359 |
| 1 | | | | 0.0068 | 0.0218 | 0.0371 |

(pushover) capacity curve resulting from a first-mode-proportional lateral load pattern is presented in Figure 19.18 for both lumped and distributed plasticity models. The effect of the calibration of member elastic stiffness is obvious in the matching of the pre-yield segments of the two curves, leading to the determination of $D_y$ shown in Figure 19.19. The curves of Figure 19.18 compare reasonably well to the capacity curve obtained with fiber models in Seismostruct (Figure 19.16). Further discussion on the calibration of elastic stiffnesses occurs in the context of shear wall models in Sections 19.10 and 19.11, where only fiber elements are used and the $T_1$ corresponds to the initial stiffness (and is applicable only for very small deformations)—in this case, to achieve better fidelity in the performance assessment, an appropriate increased period, which is more akin to the $T_1$ obtained with the lumped plasticity model herein, is recommended.

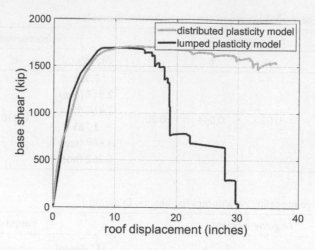

*Figure 19.18*  Static pushover curves of the distributed (black color) and lumped plasticity (gray color) models of Example 1.

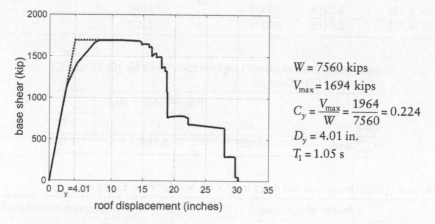

$W = 7560$ kips

$V_{max} = 1694$ kips

$C_y = \dfrac{V_{max}}{W} = \dfrac{1964}{7560} = 0.224$

$D_y = 4.01$ in.

$T_1 = 1.05$ s

*Figure 19.19*  Determination of $D_y$, Example 1.

## 19.7.8 Performance evaluation of the initial design by nonlinear dynamic analysis

For this example, performance was assessed using all three methods introduced in Section 13.4: (i) by convolving the seismic hazard curve with the fragility curve (IDA), and by (ii) single- and (iii) double-stripe analyses. The stripe analyses were done at the 2/50 level for the ductility limit and at the 10/50 level for the other performance criteria. A summary of the acceptability of peak dynamic interstory drifts, plastic hinge rotations, and ductility demands is provided in Tables 19.17 and 19.18. For IDA, the first two values of each PO correspond to estimated and allowable MAFs (see MAF format, Section 13.4.2.1), while for the stripe analyses (DCFD format, Section 13.4.2.2) they correspond to factored demand (FD) and factored capacity (FC) (expressed as demand/capacity ratios (DCRs) for plastic hinge rotations and as roof drift ratios for system ductility), respectively.

IDA was employed to estimate the MAF of exceeding the maximum interstory drift ratio limit of $\theta_{max} = 2\%$; the fragility curve was calculated based on the $S_a(T_1, 5\%)$ values that correspond to 2% drift limit (Figure 19.20). The median value of the lognormal distribution

Table 19.17 Verification of the maximum interstory drift and beam plastic hinge rotations (in terms of DCR) for Example I

| | IDRª | | | Beam plastic hinge rotation | | |
|---|---|---|---|---|---|---|
| | IDA (year⁻¹) | Single (m/m) | Double (m/m) | IDA (year⁻¹) | Single (rad/rad) | Double (rad/rad) |
| Demand | 0.0018 | 0.0177 | 0.0173 | 0.0024 | 0.9490 | 0.9304 |
| Capacity | 0.0021 | 0.0190 | 0.0193 | 0.0021 | 0.9524 | 0.9594 |
| Check | ✓ | ✓ | ✓ | ✗ | ✓ | ✓ |
| Confidence | | 50% | | | 60% | |

ª IDR = maximum interstory drift ratio over the height of the building.

Table 19.18 Verification of column plastic hinge rotations (in terms of DCR) and system ductility (expressed in terms of roof drift) for Example I

| | Column plastic hinge rotation | | | System ductility | | |
|---|---|---|---|---|---|---|
| | IDA (year⁻¹) | Single (rad/rad) | Double (rad/rad) | IDA (year⁻¹) | Single (m/m) | Double (m/m) |
| Demand | 0.0006 | 0.3426 | 0.2727 | 0.0008 | 0.0263 | 0.0259 |
| Capacity | 0.0021 | 0.9540 | 0.9872 | 0.0004 | 0.0208 | 0.0210 |
| Check | ✓ | ✓ | ✓ | ✗ | ✗ | ✗ |
| Confidence | | 60% | | | 50% | |

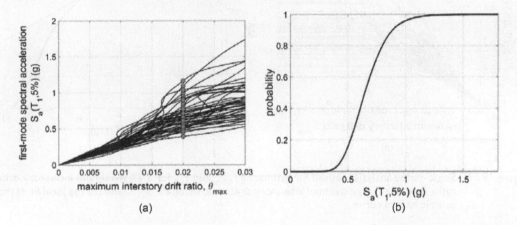

Figure 19.20 IDA approach for estimating MAF of exceeding a 2% maximum interstory drift ratio, Example 1: (a) the $S_a(T_1, 5\%)$ values calculated for a 2% maximum interstory drift ratio limit, $\theta_{max}$ and (b) corresponding fitted lognormal fragility curve.

is equal to 0.64 g and the log standard deviation is 25%. The hazard curve is presented in Figure 19.21, and was scaled to the design spectrum to offer a fair assessment of Design Method A, as discussed in Section 19.4. Applying Equation 13.58 (Section 13.4.2.1), the estimated MAF of exceeding the maximum interstory drift limit of 2%, at 50% confidence, is 0.0018, which is lower than the corresponding code value of 0.0021 (10/10 probability of exceedance). This means that the PO is satisfied.

For the stripe analysis, FD was calculated and compared to FC. Interstory drift ratio values were calculated for 17 records for $S_a^{design} = 0.57$ g, as shown in Figure 19.22a. The points

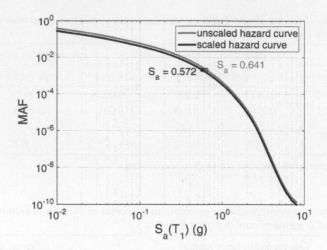

Figure 19.21   Seismic site hazard curve scaled to match the 10/50-year value of $S_a(T_1)$ as provided by the design spectrum, Example 1.

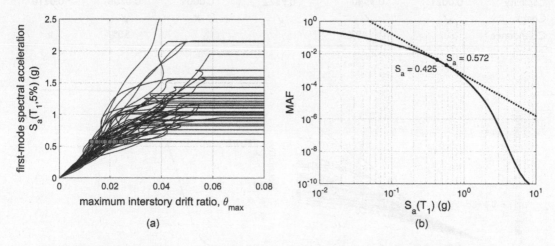

Figure 19.22   Single-stripe analysis method for estimating $F_{DRPo}$ and $F_{CR}$ for a 2% maximum interstory drift ratio, Example 1: (a) maximum interstory drifts calculated for $S_a^{design}$ and (b) the local fit of the seismic hazard curve.

of the hazard curve needed to estimate the slope $k$ (=2.52) are presented in Figure 19.22b. FD and FC at 50% confidence are estimated as:

$$FD_{RPo} = \theta_{max,50} \exp\left(0.5 \cdot k \cdot \beta_{\theta max|Sa}{}^2/b\right) = 0.017 \exp\left(0.5 \cdot 2.52 \cdot 0.22^2/1\right) = 0.0177$$

$$FC_R = \theta_{max,C} \exp\left(-0.5 \cdot k \cdot \beta_{CR}{}^2/b\right) = 0.02 \exp\left(-0.5 \cdot 2.52 \cdot 0.20^2/1\right) = 0.0190$$

Since the FC of 0.0190 is greater than the FD of 0.0177 the result is satisfactory, at a 50% confidence level.

For the double-stripe analysis, the interstory drift ratios were calculated for 17 records at $S_a^{design} = 0.57\,g$ (dark gray points) and at $1.1S_a^{design} = 0.63\,g$ (light gray points), as shown in

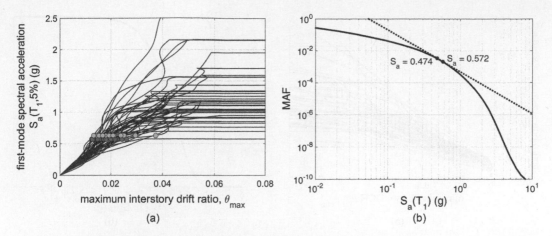

*Figure 19.23*   Double-stripe analysis method for estimating $F_{DRPo}$ and $F_{CR}$ for a 2% maximum interstory drift ratio, Example 1: (a) maximum interstory drifts calculated for $S_a^{design}$ (dark gray points) and for $1.1S_a^{design}$ (light gray points) and (b) the local fit of the seismic hazard curve.

Figure 19.23a. The points of the hazard curve needed to estimate the slope $k$ are presented in Figure 19.23b. FD and FC at 50% confidence are estimated as:

$$FD_{RPo} = \theta_{max,50} \exp\left(0.5 \cdot k \cdot \beta_{\theta max|Sa}{}^2/b\right) = 0.017 \exp\left(0.5 \cdot 2.63 \cdot 0.22^2/1.58\right) = 0.0173$$

$$FC_R = \theta_{max,C} \exp\left(-0.5 \cdot k \cdot \beta_{CR}{}^2/b\right) = 0.02 \exp\left(-0.5 \cdot 2.63 \cdot 0.20^2/1.58\right) = 0.0193$$

Since the FC of 0.0193 is greater than the FD of 0.0173, the result is satisfactory, at a 50% confidence level.

Due to differences in beam plastic rotation capacities, the maximum $DCR = \left(\theta_{demand}^{pl}/\theta_{capacity}^{pl}\right)_{max}$ determined in each NLRHA was used to facilitate comparison. The MAF of exceeding DCR = 1.0 was estimated using a lognormal fragility curve that was fitted to the $S_a(T_1, 5\%)$ values that correspond to the DCR limit, according to the IDA results (Figure 19.24a. The median value of the lognormal distribution equals 0.62 g and the log standard deviation is 28% (Figure 19.24b). Following the approach of Section 13.4.2.1, the estimated MAF of exceeding DCR = 1.0 at 60% confidence equals 0.0024, which is higher than the corresponding code value of 0.0021 (10% probability of exceeding PO in 50 years); thus, the PO is marginally not met.

For the stripe analysis, FD was calculated and compared to the FC. The maximum DCR for beam plastic hinge rotations was calculated for 17 records at $S_a^{design} = 0.57$ g, as shown in Figure 19.25a. Note therein that interpolating the IDA curves for near-zero or zero values of the plastic rotation causes a minor overshoot and the appearance of seemingly negative rotations, which has negligible consequence due to our interest in much larger demand values. The points of the hazard curve that define the slope $k$ (=2.44) are presented in Figure 19.25b. FD and FC are estimated as:

$$FD_{RPo} = DCR_{max,50} \exp\left(0.5 \cdot k \cdot \beta_{\theta max|Sa}{}^2/b\right) = 0.091 \exp\left(0.5 \cdot 2.44 \cdot 0.33^2/1.0\right) = 0.9021$$

$$FC_R = DCR_{max,C} \exp\left(-0.5 \cdot k \cdot \beta_{CR}{}^2/b\right) = 1.0 \exp\left(-0.5 \cdot 2.44 \cdot 0.20^2/1.0\right) = 0.9524$$

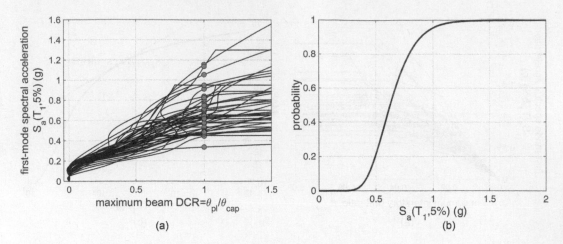

*Figure 19.24* Analytical approach for estimating MAF of exceeding demand capacity ratio limit of 1.0 for beam plastic hinge rotations, Example 1: (a) the $S_a(T_1, 5\%)$ values calculated for DCR = 1 and (b) corresponding fitted lognormal fragility curve.

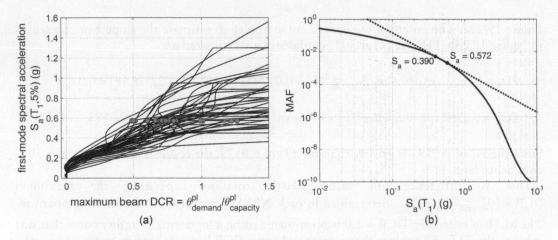

*Figure 19.25* Single-stripe analysis method for estimating $F_{DRPo}$ and $F_{CR}$ for a DCR = 1 of beam plastic hinge rotations, Example 1: (a) maximum beam DCR calculated for $S_a^{design}$ and (b) the local fit of the seismic hazard curve.

For a confidence level of 60%, the lognormal standard variate is $K_x = 0.253$ and the evaluation inequality becomes:

$$FC_R > FD_{RPo} \exp(K_x \cdot \beta_{TU}) = 0.9021 \exp(0.253 \cdot 0.20) = 0.9490$$

Since the FC of 0.9524 is lower than the FD of 0.9490 the result is marginally satisfactory at the 60% confidence level.

For the double-stripe analysis, the DCR values were calculated for 17 records for $S_a^{design} = 0.57\,g$ (dark gray points) and for $1.1 S_a^{design} = 0.63\,g$ (light gray points). The points of the hazard curve needed to estimate the slope $k$ are presented in Figure 19.26a,b, respectively. FD and FC are estimated as:

$$FD_{RPo} = DCR_{max,50} \exp(0.5 \cdot k \cdot \beta_{\theta max|Sa}^2 / b) = 0.791 \exp(0.5 \cdot 2.47 \cdot 0.33^2 / 1.20) = 0.8844$$

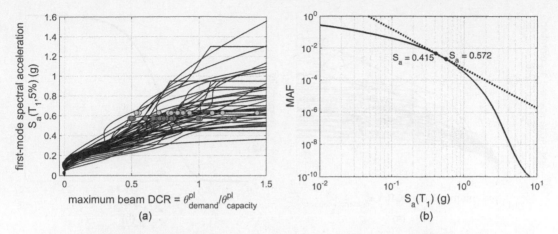

*Figure 19.26* Double-stripe analysis method for estimating $F_{DRPo}$ and $F_{CR}$ for a DCR = 1 of beam plastic hinge rotations, Example 1: (a) maximum beam DCR calculated for $S_a^{design}$ (dark gray points) and for $1.1 S_a^{design}$ (light gray points) and (b) the local fit of the seismic hazard curve.

$$FC_R = DCR_{max,C} \exp\left(-0.5 \cdot k \cdot \beta_{CR}^2/b\right) = 1.0 \exp\left(-0.5 \cdot 2.47 \cdot 0.20^2/1.20\right) = 0.9594$$

For a confidence level of 60%, the lognormal standard variate is $K_x = 0.253$ and the evaluation inequality becomes:

$$FC_R > FD_{RPo} \exp\left(K_x \cdot \beta_{TU}\right) = 0.8844 \exp\left(0.253 \cdot 0.20\right) = 0.9304$$

Since the FC of 0.9594 is greater than the FD of 0.9304 the result is marginally satisfactory at a 60% confidence level. Comparing the results from all three methods, namely IDA, single and double stripe, it becomes obvious that the beam plastic hinge rotation demand is very close to the available capacity. In this case, IDA is the most accurate approach, as it employs numerical integration and a larger number of records to find that the PO is not satisfied. At a lower accuracy level, both single- and double-stripe approaches find the design to be adequate. Clearly, one should be careful with such close comparisons, as they are subject to considerable uncertainty. This is where the introduction of the appropriate confidence level works in the engineer's favor, safeguarding the validity of assessment. In the present case, if the exceedance of beam plastic hinge rotation was an important objective, it should have been tested at a 90% confidence level, where all three approaches would find the design to be inadequate.

IDA curves in terms of the first-mode spectral acceleration $S_a(T_1, 5\%)$ and maximum $DCR = \left(\theta_{demand}^{pl}/\theta_{capacity}^{pl}\right)_{max}$ for column plastic hinge rotations are presented in Figure 19.27a. The MAF of exceeding DCR = 1.0 was estimated using the lognormal fragility curve that was fitted to the $S_a(T_1, 5\%)$ values that correspond to the DCR limit, according to IDA results, and is shown in Figure 19.27b. The median value of the lognormal distribution equals 0.99 g and the log standard deviation is 27%. Following the approach of Section 13.4.2.1, the estimated MAF of exceeding DCR = 1.0 at 60% confidence is 0.0006, which is lower than the corresponding code value of 0.0021 (10% probability of exceeding PO in 50 years). This means that the PO was met.

For the stripe analysis, FD was calculated and compared to the FC. The maximum DCR for the column plastic hinges was calculated for 17 records for $S_a^{design} = 0.57$ g, as shown

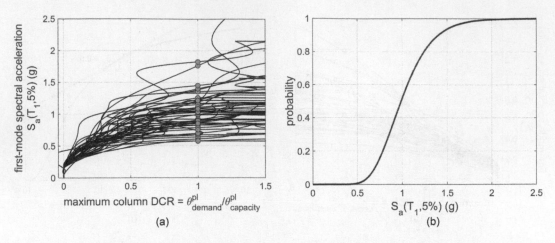

*Figure 19.27*   Analytical approach for estimating MAF of exceeding demand capacity ratio limit of 1.0 for column plastic hinge rotations, Example 1: (a) the $S_a(T_1, 5\%)$ values calculated for DCR = 1 and (b) corresponding fitted lognormal fragility curve.

in Figure 19.28a. The points of the hazard curve that define the slope $k$ are presented in Figure 19.28b. FD and FC are estimated as:

$$FD_{RPo} = DCR_{max,50} \exp\left(0.5 \cdot k \cdot \beta_{\theta max|Sa}{}^2/b\right) = 0.238 \exp\left(0.5 \cdot 2.36 \cdot 0.52^2/1.0\right) = 0.3257$$

$$FC_R = DCR_{max,C} \exp\left(-0.5 \cdot k \cdot \beta_{CR}{}^2/b\right) = 1.0 \exp\left(-0.5 \cdot 2.36 \cdot 0.20^2/1.0\right) = 0.9540$$

For a confidence level of 60%, the lognormal standard variate is $K_x = 0.253$ and the evaluation inequality becomes:

$$FC_R > FD_{RPo} \exp\left(K_x \cdot \beta_{TU}\right) = 0.3257 \exp\left(0.253 \cdot 0.20\right) = 0.3426$$

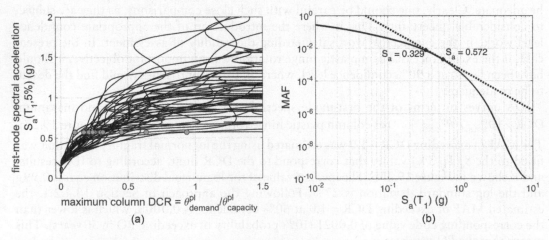

*Figure 19.28*   Single-stripe analysis method for estimating $F_{DRPo}$ and $F_{CR}$ for a DCR = 1 of column plastic hinge rotations, Example 1: (a) maximum column DCR calculated for $S_a^{design}$ and (b) the local fit of the seismic hazard curve.

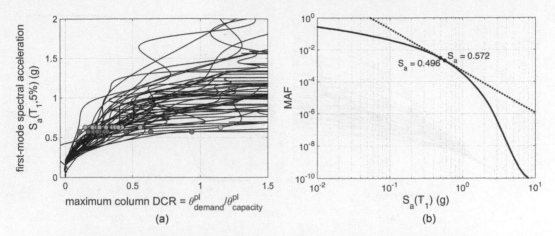

*Figure 19.29* Double-stripe analysis method for estimating $F_{DRPo}$ and $F_{CR}$ for a DCR = 1 of column plastic hinge rotations, Example 1: (a) maximum column DCR calculated for $S_a^{design}$ (dark gray points) and for $1.1 S_a^{design}$ (light gray points) and (b) the local fit of the seismic hazard curve.

Since the FC of 0.9540 is greater than the FD of 0.3426 the result is satisfactory at a 60% confidence level.

For the double-stripe analysis, DCR values were calculated for the column plastic hinge rotations for 17 records for $S_a^{design} = 0.57\,g$ (dark gray points) and for $1.1 S_a^{design} = 0.63\,g$ (light gray points) (Figure 19.29a). The points of the hazard curve needed to estimate the slope $k$ are presented in Figure 19.29b. FD and FC are estimated as:

$$FD_{RPo} = DCR_{max,50} \exp\left(0.5 \cdot k \cdot \beta_{\theta max|Sa}{}^2/b\right) = 0.238 \exp\left(0.5 \cdot 2.49 \cdot 0.52^2/3.87\right) = 0.2593$$

$$FC_R = DCR_{max,C} \exp\left(-0.5 \cdot k \cdot \beta_{CR}{}^2/b\right) = 1.0 \exp\left(-0.5 \cdot 2.49 \cdot 0.20^2/3.87\right) = 0.9872$$

Note that rather large value of $b = 3.87$ is estimated from the two stripes and employed above. This is in general an indication that this verification is performed relatively close to the onset of global collapse, as also indicated by the proximity of the second stripe to the flatlines in Figure 19.29a. In such cases, some care should be exercised to make sure that no more than 20% of the runs are non-convergent (i.e., indicative of global collapse), otherwise the approximation underlying the DCFD approach (see Section 13.4.2.2) will fail. In such cases IDA is the recommended approach. Still, there is considerable margin of safety between the FD and the FC; therefore, this potential inaccuracy will be of little consequence.

For a confidence level of 60%, the lognormal standard variate is $K_x = 0.253$ and the evaluation inequality becomes:

$$FC_R > FD_{RPo} \exp\left(K_x \cdot \beta_{TU}\right) = 0.2593 \exp(0.253 \cdot 0.20) = 0.2727$$

Since the FC of 0.9872 is greater than the FD of 0.2727 the result is satisfactory at 60% confidence level.

The acceptability of the ductility limit of 3.6 was assessed at the 2/50 level. This ductility limit is equivalent to a roof drift ratio limit of $\theta_{roof} = 3.6 \cdot D_y/H_{tot} = 3.6 \cdot 4.01$ in./648 in. = 0.022. IDA curves in terms of the first-mode spectral acceleration $S_a(T_1, 5\%)$ and roof drift ratio, $\theta_{roof}$ are presented in Figure 19.30a. The MAF of exceeding $\theta_{roof} = 0.022$

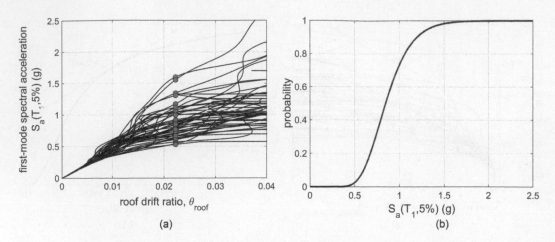

Figure 19.30 Analytical approach for estimating MAF of exceeding a ductility limit of 3.6, Example 1: (a) the $S_a(T_1, 5\%)$ values calculated for $\theta_{roof} = 0.022$ and (b) corresponding fitted lognormal fragility curve.

was estimated using the lognormal fragility curve that was fitted to the $S_a(T_1, 5\%)$ values that correspond to the roof drift limit (dark gray points in Figure 19.30a), and is presented in Figure 19.29b. The median value of the lognormal distribution equals $0.85\,g$ and the log standard deviation is 27%. Following the approach of Section 13.4.1, the estimated MAF of exceeding $\theta_{roof} = 0.022$ at 50% confidence is 0.0008, which is higher than the limit of 0.000404 (2/50 years). This means that the PO was not met. Note that this relatively large difference in MAFs corresponds to a much smaller difference in terms of $S_a(T_1)$ due to their exponential relationship, meaning that a safe design is achievable with only a small increase in member strengths. This will become apparent in the DCFD checks that follow.

For the stripe analysis, the maximum roof drift ratio was calculated for 17 records at an acceleration $S_a^{2\%/50} = 0.95\,g$, as shown in Figure 19.31a. This spectral acceleration value corresponds to MAF = 0.000404 and is calculated via the hazard curve. The points of the hazard curve defining slope $k$ are presented in Figure 19.31b. FD and FC are estimated as:

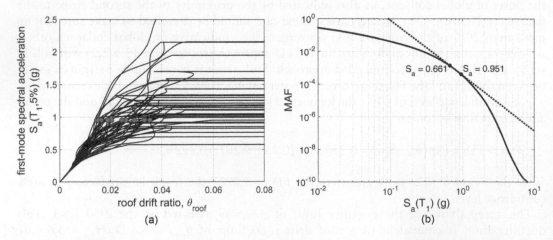

Figure 19.31 Single-stripe analysis method for estimating $F_{DRPo}$ and $F_{CR}$ for a ductility limit of 3.6, Example 1: (a) roof drift ratio calculated for $S_a^{2\%/50}$ and (b) the local fit of the seismic hazard curve.

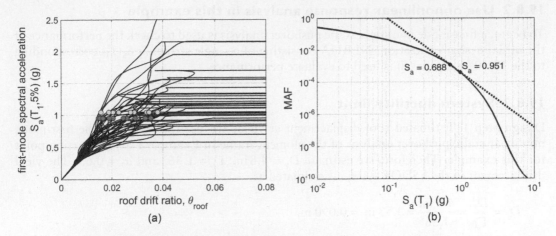

*Figure 19.32* Double-stripe analysis method for estimating $F_{DRPo}$ and $F_{CR}$ for a ductility limit of 3.6, Example 1: (a) roof drift ratio calculated for $S_a^{2\%/50}$ (dark gray points) and for $1.1S_a^{2\%/50}$ (light gray points) and (b) the local fit of the seismic hazard curve.

$$FD_{RPo} = \theta_{roof,50} \exp\left(0.5 \cdot k \cdot \beta_{\theta max|Sa}^2/b\right) = 0.023 \exp\left(0.5 \cdot 3.38 \cdot 0.30^2/1.0\right) = 0.0263$$

$$FC_R = \theta_{roof,C} \exp\left(-0.5 \cdot k \cdot \beta_{CR}^2/b\right) = 0.022 \exp\left(-0.5 \cdot 3.38 \cdot 0.20^2/1.0\right) = 0.0208$$

Since the FC of 0.0208 is lower than the FD of 0.0263 the result is not satisfactory at even a 50% confidence level.

For the double-stripe analysis, the $\theta_{roof}$ values were calculated for 17 records for $S_a^{2\%/50} = 0.95\,g$ (dark gray points) and for $1.1S_a^{2\%/50} = 1.05\,g$ (light gray points) (Figure 19.32a). The points of the hazard curve that define the slope $k$ are presented in Figure 19.32b. FD and FC at 50% confidence ($K_x = 0$) are estimated as:

$$FD_{RPo} = \theta_{roof,50} \exp\left(0.5 \cdot k \cdot \beta_{\theta max|Sa}^2/b\right) = 0.023 \exp\left(0.5 \cdot 3.38 \cdot 0.30^2/1.12\right) = 0.0259$$

$$FC_R = \theta_{roof,C} \exp\left(-0.5 \cdot k \cdot \beta_{CR}^2/b\right) = 0.022 \exp\left(-0.5 \cdot 3.38 \cdot 0.20^2/1.12\right) = 0.0210$$

Since the FC of 0.0210 is less than the FD of 0.0259 the result is not satisfactory at 50% confidence level. Note that although the single-/double-stripe results match the IDA conclusion, the closeness of the stripes to collapse (see Figures 19.31a and 19.32b) means that the approximations inherent in the stripe methods can lead to a mediocre result. In such cases, it is better to resort to the MAF format (either with IDA or a stripe approach, Section 13.4.2.1) that can better handle collapse.

## 19.8 EXAMPLE 2: MOMENT-RESISTANT FRAME DESIGNED USING METHOD B

### 19.8.1 Performance objectives

A single PO was used for the design of this frame: the MAF of interstory drift during dynamic response exceeding 2% of the story height was limited to $p_o = 2.11 \times 10^{-3}\,year^{-1}$ (equivalent to a 10% probability in 50 years of exceeding 2% interstory drift), at 68% confidence.

### 19.8.2 Use of nonlinear response analysis in this example

To begin, a first-mode nonlinear static pushover analysis is used to check the performance of the initial design realization. NLRHA consisting of a single stripe (at $S_a(T_1)$ corresponding to the 10/50 hazard level) is used to evaluate performance.

### 19.8.3 System ductility limit

Design Step 1. Estimated roof displacement at yield: We use the results of the first-mode nonlinear static pushover analysis of the moment frame in Example 1 as the starting point for this example. Therefore, we estimate $D_y = 4.8$ in., $\Gamma_1 = 1.36$, and $\alpha_1 = 0.63$. The yield displacement of the ESDOF system is estimated as

$$D_y^* = \frac{D_y}{\Gamma_1} = \frac{4.8 \text{ in.}}{1.36} = 3.53 \text{ in.} = 0.090 \text{ m.}$$

Design Step 2. From Table 12.8 we estimate $\alpha_{3,\text{COD}} = 1.45$. Then, the roof displacement must be limited to approximately

$$D_{u,\text{drift}} = \frac{0.02}{\alpha_{3,\text{COD}}} h_n = \left(\frac{0.02}{1.45}\right)(54 \text{ ft} \cdot 12 \text{ in./ft}) = 8.94 \text{ in.}$$

Therefore, the ductility demand should not exceed $\mu = D_u/D_y = 8.94/4.8 = 1.86$ at an MAF of $p_o = 2.11 \times 10^{-3} \text{year}^{-1}$.

### 19.8.4 Assumptions required to generate YFS based on ASCE-7 UHS

The YFS for this site was generated using as a basis the ASCE-7 design spectrum (2/3 of the 2/50-year UHS taken as the 10/50 spectrum), for Site Class D soils. UHS for other hazard levels were generated by adopting the hazard curve log–slope values determined in Section 19.4.

The normalized capacity curve adopted is shown in Figure 19.33.

### 19.8.5 Required base shear strength

Design Step 3. Considering the YFS developed for this example, for $D_y^* = 3.53$ in. at 68% confidence (Figure 19.34), for $\mu = 1.86$ at $p_o = 2.11 \times 10^{-3} \text{year}^{-1}$, the required $C_y^* = 0.436$.

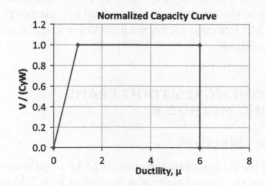

*Figure 19.33* Normalized capacity curve used for generating the YFS, Example 2.

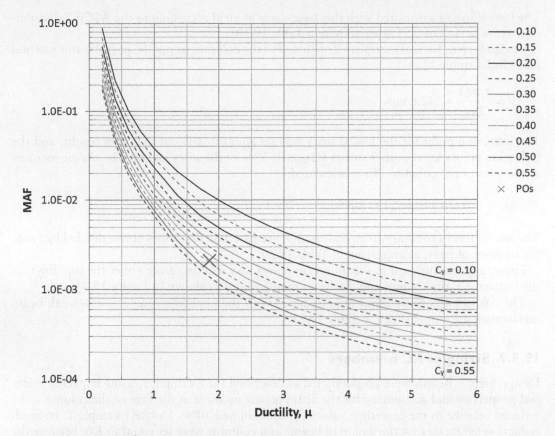

Figure 19.34   YFS for $D_y^* = 3.53$ in., at 68% confidence, Example 2.

The corresponding period of vibration is given by

$$T^* = 2\pi\sqrt{\frac{D_y^*}{C_y^* g}} = 0.91\,\text{s}$$

Design Step 4. Appling the relationship $C_y = \alpha_1 C_y^*$, the yield strength coefficient is estimated as $C_y = 0.63(0.436) = 0.275$. The design base shear at yield for a single frame is given by $V_y = C_y W/2 = 0.275(7,560\,\text{kips}) = 2,079\,\text{kips}$.

## 19.8.6 Design lateral forces and required member strengths

Design Step 5. The equivalent lateral force method of ASCE-7 is used in design (Equation 12.11). The effective height of the resultant of the ASCE-7 lateral forces, for a period of 0.91 s, is $0.771 h_n$. The height of the resultant of the first-mode forces, $h_1^*$, is estimated to be $0.72 h_n$ according to Table 12.1. Thus, the base shear to be used with the ASCE-7 lateral force distribution, to obtain the same resistance to overturning moment as is needed for response in the first mode, is estimated to be

$$V_y\,\frac{h_1^*}{h_{\text{eff,ASCE7}}} = 1,941\,\text{kips}$$

The lateral forces associated with this base shear at yield according to the ASCE 7 distribution (for a period of 0.91 s) are shown in Table 19.19.

Design Step 6. Lateral forces are distributed to the columns as per the portal frame method (used in Example 1).

$$V = \frac{1,941}{2n_b} = 242.6 \text{ kips}$$

The inflection point for the lowest story was set equal to 60% of the story height, and the inflection points for the other stories was set at 50% of the story height. The plastic moment at the base of the column, $M_c$, is calculated:

$$M_c = V(0.6)(15 \text{ ft}) = 2,183 \text{ kip-ft}$$

The lateral forces tributary to an end column are based on the values above divided by twice the number of bays, as appearing in Table 19.20.

Enforcing zero moment at the inflection points and working from either the top down or the bottom up, the column and the beam moments are as shown in Figure 19.35.

The column and beam moments associated with the implicit strong column–weak beam mechanism are summarized in Table 19.21.

## 19.8.7  Sizing of RC members

Design Step 7. Beams were proportioned as described for Example 1, using expected material properties and accounting for the beam plastic moment at the face of the column to be reduced relative to the centerline values (by an assumed 10%). Unlike Example 1, strength reduction ($\phi$) factors for the design of beams and columns were set equal to 1.0, because the design will be validated by nonlinear dynamic analysis.

The mathematical and preliminary design solutions for the beam sections appear in Tables 19.22 and 19.23.

Table 19.19  Lateral design forces at each floor level, Example 2

| Level | $F_i$ (kip) |
|---|---|
| 4 | 814 |
| 3 | 584 |
| 2 | 369 |
| 1 | 174 |
| Σ | 1,941 |

Table 19.20  Lateral forces tributary to an end column, Example 2

| Level | Lateral force tributary to end column (kips) |
|---|---|
| 4 | 101.6 |
| 3 | 73.0 |
| 2 | 46.3 |
| 1 | 21.8 |

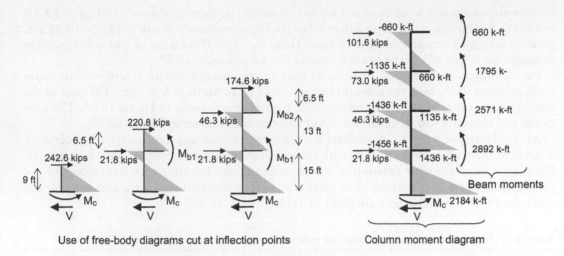

Use of free-body diagrams cut at inflection points        Column moment diagram

*Figure 19.35*   Moment diagram for an end column, Example 2.

*Table 19.21*   Equilibrium moments, Example 2

|  | Moment (kip-ft) |
| --- | --- |
| $M_c$ | 2,184 |
| $M_{b1}$ | 2,892 |
| $M_{b2}$ | 2,571 |
| $M_{b3}$ | 1,795 |
| $M_{b4}$ | 660 |

*Table 19.22*   Mathematical solution for beam dimensions and reinforcement, Example 2

| | Required strengths | | Mathematical solution | | | |
| --- | --- | --- | --- | --- | --- | --- |
| Level | $M_p^-$ (kip-in.) | $M_p^+$ (kip-in.) | $b_w$ (in.) | h (in.) | $A_{s,top}$ (in.²) | $A_{s,bot}$ (in.²) |
| 4 | 9,504 | 4,752 | 14.41 | 24.02 | 7.13 | 3.25 |
| 3 | 25,848 | 12,924 | 19.60 | 32.66 | 13.59 | 6.34 |
| 2 | 37,022 | 18,511 | 21.92 | 36.53 | 17.15 | 8.06 |
| 1 | 41,645 | 20,822 | 22.74 | 37.90 | 18.51 | 8.72 |

*Table 19.23*   Preliminary design (beam dimensions and reinforcement), Example 2

| Level | $b_w$ (in.) | h (in.) | Top bars | Bottom bars |
| --- | --- | --- | --- | --- |
| 4 | 14 | 24 | (7) No. 9 | (3) No. 9 |
| 3 | 20 | 32 | (14) No. 9 | (6) No. 9 |
| 2 | 22 | 36 | (18) No. 9 | (8) No. 9 |
| 1 | 22 | 38 | (18) No. 9 | (9) No. 9 |

Design Step 8. Critical column sections above the base were each designed for 6/5 of the moments at the beam centerline (i.e., without a reduction to the face of the beam). Beams frame into both sides of an intermediate column, with negative moments (equal to 4/3 of the beam moments shown in Figure 19.35) acting on one side and positive moments (equal to 2/3 of the moments in Figure 19.35) acting on the other side. Thus, critical sections of intermediate columns were designed for 6/5 of twice the moments shown in Figure 19.35. For the end columns, it is assumed that 4/3 of the beam moments shown in Figure 19.35 act, corresponding to negative beam moments; thus, the critical sections of end columns were designed for 6/5 of 4/3 of the beam moments shown in Figure 19.35.

Plastic hinges are intended to form as part of the desired strong column–weak beam mechanism at the base of the columns (Figure 19.15 from Method A design). The base of the intermediate columns was designed for twice the moments shown in Figure 19.35. The base of the end columns was designed for 4/3 of the moments shown in Figure 19.35.

As for Example 1, column sections were prismatic over any one story; reinforcement continued through each floor level, with splices located only at the midheight of each story. Column dimensions were determined for the intermediate columns with reinforcement not exceeding about 4%–5% of the gross section area. These dimensions were also used for the end columns. Column designs are given in Tables 19.24–19.27.

Table 19.24  Preliminary design of intermediate columns just above floor levels, Example 2

| Floor level | $M_c$ or $M_{pc}$ (k-in.) | D + 0.25L (k) | h (in.) | $A_s$ (in$^2$) | Bars | $A_{s,provided}$ (in.$^2$) | $P/A_g f_{ce}$ | $\rho_g$ % |
|---|---|---|---|---|---|---|---|---|
| 4 | NA | NA | NA | NA | NA | NA | NA | NA |
| 3 | 19,008 | 78.75 | 28 | 23.84 | (20) No 10 | 25.40 | 1.55% | 3.24% |
| 2 | 32,688 | 163.13 | 32 | 35.25 | (28) No. 10 | 35.56 | 2.45% | 3.47% |
| 1 | 41,357 | 247.50 | 36 | 37.50 | (32) No. 10 | 40.64 | 2.94% | 3.14% |
| Footing | 52,416 | 331.88 | 36 | 49.03 | (40) No. 10 | 50.80 | 3.94% | 3.92% |

Table 19.25  Preliminary design of intermediate columns just below floor levels, Example 2

| Floor level | $M_c$ (k-in.) | D + 0.25L (k) | h (in.) | $A_s$ (in.$^2$) | Bars | $A_{s,provided}$ (in.$^2$) | $P/A_g f_{ce}$ | $\rho_g$ % |
|---|---|---|---|---|---|---|---|---|
| 4 | 19,008 | 78.75 | 28 | 23.84 | (20) No. 10 | 25.40 | 1.55% | 3.24% |
| 3 | 32,688 | 163.13 | 32 | 35.25 | (28) No. 10 | 35.56 | 2.45% | 3.47% |
| 2 | 41,357 | 247.50 | 36 | 37.50 | (32) No. 10 | 40.64 | 2.94% | 3.14% |
| 1 | 41,933 | 331.88 | 36 | 37.12 | (32) No. 10 | 40.64 | 3.94% | 3.14% |
| Footing | NA | NA | NA | NA | NA | NA | NA | NA |

Table 19.26  Preliminary design of end columns just above floor levels, Example 2

| Floor level | $M_c$ or $M_{pc}$ (k-in.) | D + 0.25L (k) | h (in.) | $A_s$ (in$^2$) | Bars | $A_{s,provided}$ (in$^2$) | $P/A_g f_{ce}$ | $\rho_g$ % |
|---|---|---|---|---|---|---|---|---|
| 4 | NA | NA | NA | NA | NA | NA | NA | NA |
| 3 | 12,672 | 78.75 | 28 | 14.70 | (12) No. 10 | 15.24 | 1.55% | 1.94% |
| 2 | 21,792 | 163.13 | 32 | 21.53 | (20) No. 10 | 25.40 | 2.45% | 2.48% |
| 1 | 27,571 | 247.50 | 36 | 22.67 | (20) No. 10 | 25.40 | 2.94% | 1.96% |
| Footing | 34,944 | 331.88 | 36 | 29.43 | (24) No. 10 | 30.48 | 3.94% | 2.35% |

Table 19.27  Preliminary design of end columns just below floor levels, Example 2

| Floor level | $M_c$ (k-in.) | D + 0.25L (k) | h (in.) | $A_s$ (in.$^2$) | Bars | $A_{s,provided}$ (in.$^2$) | $P/A_g f_{ce}$ | $\rho_g$ % |
|---|---|---|---|---|---|---|---|---|
| 4 | 12,672 | 78.75 | 28 | 14.70 | (12) No. 10 | 15.24 | 1.55% | 1.94% |
| 3 | 21,792 | 163.13 | 32 | 21.53 | (20) No. 10 | 25.40 | 2.45% | 2.48% |
| 2 | 27,571 | 247.50 | 36 | 22.67 | (20) No. 10 | 25.40 | 2.94% | 1.96% |
| 1 | 27,955 | 331.88 | 36 | 21.95 | (20) No. 10 | 25.40 | 3.94% | 1.96% |
| Footing | NA | NA | NA | NA | NA | NA | NA | NA |

## 19.8.8  Preliminary evaluation of the initial design by nonlinear static (pushover) analysis

To quickly assess the basic characteristics of the preliminary design, a fiber element model was developed in SeismoStruct (Seismosoft, 2014). If the yield displacement and values of first-mode properties determined with this model suggest acceptable performance, a lumped plasticity model may be developed for use in detailed evaluation, in accordance with the design method. An eigenvalue analysis of the uncracked section model (assumes concrete material has tensile strength) determined $T_1 = 0.737\,\text{s}$, $\Gamma_1 = 1.35$, $\alpha_1 = 0.63$.

A first-mode nonlinear static (pushover) analysis was conducted as described in Example 1, with forces proportional to the vector given in Table 19.28, to obtain the capacity curve of Figure 19.36. The resultant of these first-mode forces is located at $0.777h_n$ above the base, which differs from the value of $0.72h_n$ assumed in design.

Just as for Example 1, during the pushover analysis, the moments in the columns above the base of the first floor did not exceed those associated with equilibrium for the intended beam-hinging mechanism, implying that the intended mechanism did occur in the first-mode nonlinear static analysis.

Table 19.28  First-mode force vector, Example 2

| Level | $F_{1st}$ (kip) |
|---|---|
| 4 | 889 |
| 3 | 627 |
| 2 | 388 |
| 1 | 175 |

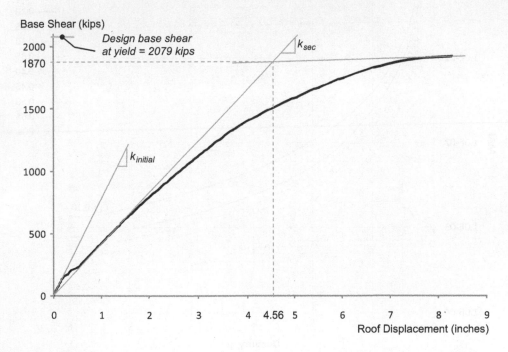

Figure 19.36   Capacity curve determined by first-mode pushover analysis of the first design realization of Example 2.

Based on the eigenvalue analysis, the period associated with the secant stiffness representative of cracked section behavior can be estimated as

$$T_{sec} = T_{initial}\sqrt{\frac{k_{initial}}{k_{sec}}} = 0.737 \text{ s}\sqrt{\frac{766 \text{ k/in.}}{411 \text{ k/in.}}} = 1.01 \text{ s}$$

which is a little high relative to $T^* = 0.91$ s. The base shear at yield, 1,870 kips, is a bit low compared with the design base associated with first-mode response of 2,079 kips. Much of this difference is attributed to the assumption of first-mode lateral forces having a resultant at $0.72h_n$. Had we used the value of $0.777h_n$ determined from the model, the base shear to use with the ASCE-7 force distribution would have been 2,079 kips $(0.777h_n/0.771h_n) = 2,095$ kips, which is 7.9% higher than the value of 1,941 kips used in design. Had we designed for this force level, yield strength in the first-mode pushover analysis of Figure 19.36 would have increased from 1,870 kips to approximately 2,018 kips, which is close to the first-mode target of 2,079 kips.

The results of the first-mode pushover analysis can be used to check the performance of this initial design realization. We estimate $D_y^* = D_y/\Gamma_1 = 4.56/1.35 = 3.38$ in. For $V_y = 1,870$ kips, $C_y = V_y/W = 1,870/7,560 = 0.247$. Therefore, $C_y^* = C_y/\alpha_1 = (0.247/0.63) = 0.393$. The YFS for this yield displacement is plotted in Figure 19.37, with new $C_y^*$ contours. Using the assumption that $\alpha_{3,COD} = 1.45$ and therefore the limit of 2% on interstory drift corresponds to a peak roof displacement of 8.94 in., the associated ductility limit is now $8.94/4.56 = 1.96$. According to the YFS, to limit $p_o$ to $2.11 \times 10^{-3}$ year$^{-1}$, the required $C_y^* = 0.420$, which

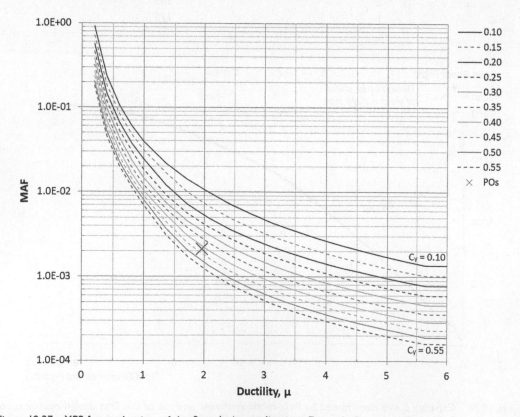

Figure 19.37    YFS for evaluation of the first design realization, Example 2.

is 7% higher than the current value of $C_y^* = 0.393$. (Alternatively, one may determine that the $p_o$ of exceeding a ductility of 1.96 is $2.36 \times 10^{-3}$year$^{-1}$ for $C_y^* = 0.393$.) A slightly higher base shear strength could be used to develop a second realization of the design, or the peak dynamic interstory drifts of the first realization could be established in a stripe analysis, noting that $\alpha_{3,COD}$ may differ from the assumed value of 1.45. We chose to evaluate the first design realization in detail in the following section.

## 19.8.9  Nonlinear modeling and acceptance criteria

Tables 19.29 and 19.30 summarize the parameters used for modeling the columns and beams, respectively. Tables 19.31 and 19.32 present the acceptance criteria for the columns and beams, respectively.

Table 19.29  Plastic hinge modeling parameters for columns, Example 2

| | End columns | | | | Intermediate columns | | | |
|---|---|---|---|---|---|---|---|---|
| Story | $M_y$ (kN·m) | a | b | c | $M_y$ (kN·m) | a | b | c |
| 4 | 1,703 | | | | 2,602 | | | |
| 3 | 3,154 | | | | 4,211 | | | |
| 2 | 3,753 | 0.030 | 0.044 | 0.20 | 5,554 | 0.030 | 0.044 | 0.20 |
| 1 | 3,861 (top) | | | | 5,658 (top) | | | |
| | 4,469 (bot) | | | | 6,847 (bot) | | | |

Table 19.30  Plastic hinge modeling parameters for beams, Example 2

| | Positive moment | | | | Negative moment | | | |
|---|---|---|---|---|---|---|---|---|
| Story | $M_y^+$ (kN·m) | a | b | c | $M_y^-$ (kN·m) | a | b | CP |
| 4 | 524 | 0.025 | 0.050 | | 1,228 | 0.021 | 0.033 | |
| 3 | 1,475 | 0.025 | 0.049 | 0.20 | 3,319 | 0.020 | 0.032 | 0.20 |
| 2 | 2,226 | 0.024 | 0.048 | | 4,827 | 0.020 | 0.031 | |
| 1 | 2,650 | 0.024 | 0.047 | | 5,159 | 0.020 | 0.032 | |

Table 19.31  Column plastic hinge rotation limits for IO, LS, and CP POs, Example 2

| | Positive rotation (rad) | | |
|---|---|---|---|
| Story | IO | LS | CP |
| $\theta_{p,max}$ (rad) | 0.005 | 0.034 | 0.044 |

Table 19.32  Beam plastic hinge rotation limits for IO, LS, and CP POs, Example 2

| | Positive rotation (rad) | | | Negative rotation (rad) | | |
|---|---|---|---|---|---|---|
| Story | IO | LS | CP | IO | LS | CP |
| 4 | 0.010 | 0.025 | 0.050 | 0.0059 | 0.0208 | 0.0334 |
| 3 | | | 0.049 | 0.0056 | 0.0202 | 0.0318 |
| 2 | 0.009 | 0.024 | 0.048 | 0.0055 | 0.0198 | 0.0308 |
| 1 | | | 0.047 | 0.0059 | 0.0199 | 0.0323 |

The static pushover capacity curve resulting from a first-mode-proportional lateral load pattern is presented in Figure 19.38, for both the distributed and lumped plasticity models. The fundamental period of the lumped plasticity model is $T_1 = 0.97\,s$; this exceeds the initial (uncracked) period of the fiber model of $T_1 = 0.79\,s$. As discussed in Section 19.7.7, matching the initial stiffness is not our target; matching the effective one is. Yield displacement is calculated according to Figure 19.39. Assuming a bilinear approximation of the capacity curve having the same effective stiffness as the lumped plasticity model yields $D_y = 3.97\,in$.

## 19.8.10 Performance evaluation of the initial design by nonlinear dynamic analysis

Although a constraint on interstory drift was the only PO used in design, the acceptability of peak dynamic interstory drifts and plastic hinge rotations was assessed at the 10/50 level by using the three methods introduced in Section 13.4.2, i.e., the MAF approach via IDA,

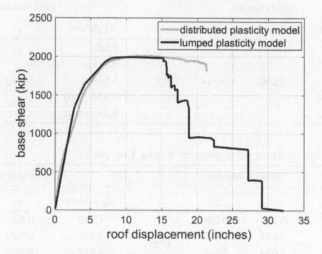

*Figure 19.38*   Static pushover curve of the distributed (gray color) and the lumped plasticity (black color) model, Example 2.

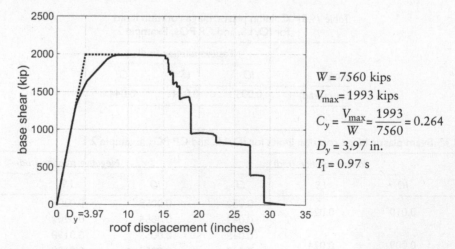

$W = 7560$ kips

$V_{max} = 1993$ kips

$C_y = \dfrac{V_{max}}{W} = \dfrac{1993}{7560} = 0.264$

$D_y = 3.97$ in.

$T_1 = 0.97$ s

*Figure 19.39*   Calculation of $D_y$, Example 2.

and the DCFD approach based on single- and double-stripe analysis. The results are summarized in Tables 19.33 and 19.34 for each PO. For IDA, the first two values of each PO correspond to estimated and allowable MAF, and for the stripe analyses they correspond to FD and FC, respectively. For the stripe analyses, plastic hinge rotations are expressed as DCRs.

Following the prescription of Design Method B, the results of the single-stripe analysis are shown in detail. Note that despite the simplification of the seismic hazard curve adopted for design, assessment will progress by using the actual hazard data, scaled to match the $10/50 \cdot S_a(T_1)$ value of the design spectrum (see Figure 19.40) to provide a fair comparison of the different methodologies.

Table 19.33 Verification of the maximum interstory drift and beam plastic hinge rotations (in terms of DCR) for Example 2

| | IDR[a] | | | Beam plastic hinge rotation | | |
|---|---|---|---|---|---|---|
| | IDA (year⁻¹) | Single (m/m) | Double (m/m) | IDA (year⁻¹) | Single (rad/rad) | Double (rad/rad) |
| Demand | 0.0016 | 0.0163 | 0.0169 | 0.0016 | 0.8179 | 0.8590 |
| Capacity | 0.0021 | 0.0190 | 0.0185 | 0.0021 | 0.9497 | 0.9297 |
| Check | ✓ | ✓ | ✓ | ✓ | ✓ | ✓ |
| Confidence | | 68% | | | 75% | |

[a] IDR = maximum interstory drift ratio over the height of the building.

Table 19.34 Verification of column plastic hinge rotations for Example 2

| | Column plastic hinge rotation | | |
|---|---|---|---|
| | IDA (year⁻¹) | Single (rad/rad) | Double (rad/rad) |
| Demand | 0.0008 | 0.2450 | 0.2079 |
| Capacity | 0.0021 | 0.9524 | 0.9750 |
| Check | ✓ | ✓ | ✓ |
| Confidence | | 75% | |

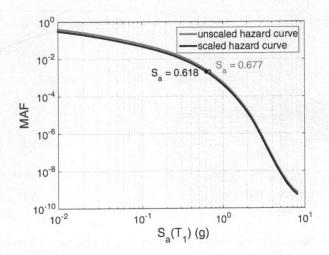

Figure 19.40 Seismic site hazard curve properly scaled to 10/50years for $S_a(T_1)$, Example 2.

For the stripe analysis, FD was calculated and compared to the FC. The interstory drift demand values were calculated for 17 records for $S_a^{design} = 0.62\,g$, as shown in Figure 19.41a. The points of the hazard curve that define the slope $k$ are presented in Figure 19.41b. FD and FC at 68% confidence are estimated as:

$$FD_{RPo} = \theta_{max,50} \exp\left(0.5 \cdot k \cdot \beta_{\theta max|Sa}^2/b\right) = 0.014 \exp\left(0.5 \cdot 2.65 \cdot 0.23^2/1\right) = 0.0148$$

$$FC_R = \theta_{max,C} \exp\left(-0.5 \cdot k \cdot \beta_{CR}^2/b\right) = 0.02 \exp\left(-0.5 \cdot 2.65 \cdot 0.20^2/1\right) = 0.0190$$

For a confidence level of 68%, the lognormal standard variate is $K_x = 0.468$ and the evaluation inequality becomes:

$$FC_R > FD_{RPo} \exp(K_x \beta_{TU}) = 0.0148 \exp(0.468 \cdot 0.20) = 0.0163$$

Since the FC of 0.0190 is greater than the FD of 0.0163, the result is satisfactory at the 68% confidence level.

The maximum DCR for beam plastic hinge rotations as calculated for 17 records for $S_a^{design} = 0.62\,g$, appears in Figure 19.42a and the corresponding hazard fit to define the slope $k$ is presented in Figure 19.42b. FD and FC are estimated as:

$$FD_{RPo} = DCR_{max,50} \exp\left(0.5 \cdot k \cdot \beta_{\theta max|Sa}^2/b\right) = 0.637 \exp\left(0.5 \cdot 2.58 \cdot 0.30^2/1.0\right) = 0.7147$$

$$FC_R = DCR_{max,C} \exp\left(-0.5 \cdot k \cdot \beta_{CR}^2/b\right) = 1.0 \exp\left(-0.5 \cdot 2.58 \cdot 0.20^2/1.0\right) = 0.9497$$

For a confidence level of 75%, the lognormal standard variate is $K_x = 0.674$ and the evaluation inequality becomes:

$$FC_R > FD_{RPo} \exp\left(K_x \cdot \beta_{TU}\right) = 0.7147 \exp(0.674 \cdot 0.20) = 0.8179$$

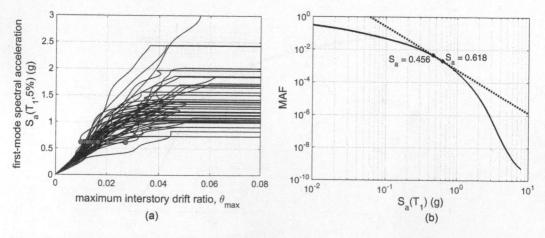

Figure 19.41    Single-stripe analysis method for estimating $F_{DRPo}$ and $F_{CR}$ for a 2% maximum interstory drift ratio, Example 2: (a) maximum interstory drifts calculated for $S_a^{design}$ and (b) the local fit of the seismic hazard curve.

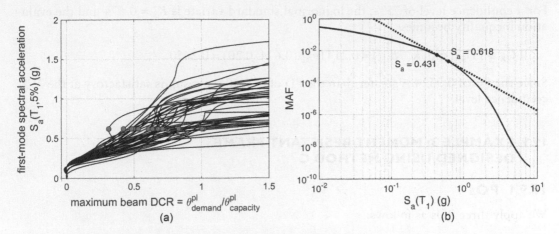

*Figure 19.42*    Single-stripe analysis method for estimating $F_{DRPo}$ and $F_{CR}$ for a DCR = 1 of beam plastic hinge rotations, Example 2: (a) maximum beam DCR calculated for $S_a^{design}$ and (b) the local fit of the seismic hazard curve.

Since the FC of 0.9497 is greater than the FD of 0.8179 the result is satisfactory at the 75% confidence level.

The maximum DCR for column plastic hinge rotations was calculated for 17 records for $S_a^{design} = 0.62\,g$, as shown in Figure 19.43a. The points of the hazard curve needed to estimate the slope $k$ are presented in Figure 19.43b. FD and FC are estimated as:

$$FD_{RPo} = DCR_{max,50} \exp\left(0.5 \cdot k \cdot \beta_{\theta max|Sa}^{2}/b\right) = 0.152\exp\left(0.5 \cdot 2.44 \cdot 0.53^2/1.0\right) = 0.2141$$

$$FC_R = DCR_{max,C} \exp\left(-0.5 \cdot k \cdot \beta_{CR}^{2}/b\right) = 1.0\exp\left(-0.5 \cdot 2.44 \cdot 0.20^2/1.0\right) = 0.9524$$

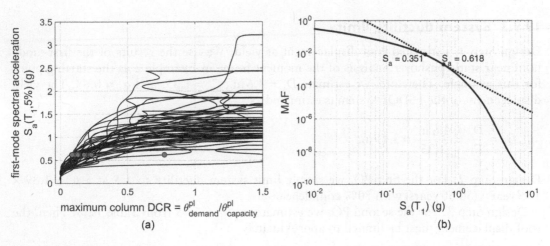

*Figure 19.43*    Single-stripe analysis method for estimating $F_{DRPo}$ and $F_{CR}$ for a DCR = 1 of column plastic hinge rotations, Example 2: (a) maximum column DCR calculated for $S_a^{design}$ and (b) the local fit of the seismic hazard curve.

For a confidence level of 75%, the lognormal standard variate is $K_x = 0.674$ and the evaluation inequality becomes:

$$FC_R > FD_{RPo} \exp(K_x \cdot \beta_{TU}) = 0.2141 \exp(0.674 \cdot 0.20) = 0.2450$$

Since the FC of 0.9524 is greater than the FD of 0.2450 the result is satisfactory at the 75% confidence level.

## 19.9 EXAMPLE 3: MOMENT-RESISTANT FRAME DESIGNED USING METHOD C

### 19.9.1 POs

We apply three POs as follows:

1. Limit the MAF of exceeding a system ductility demand of 1.5 to $p_o = 1.39 \times 10^{-2} \text{year}^{-1}$ (50/50 years).
2. Limit the MAF of exceeding an interstory drift ratio of 2% during dynamic response to $p_o = 2.11 \times 10^{-3} \text{year}^{-1}$ (10/50 years).
3. Limit the MAF of collapse to $p_o = 2.01 \times 10^{-4} \text{year}^{-1}$ (1/50 years). (Note: collapse of the ESDOF system is identified at the point that the constant $C_y$ curves transition to straight, horizontal lines on the YFS.)

We use a 70% confidence level for the system ductility demand and interstory drift POs, and a 90% level of confidence for the collapse PO.

### 19.9.2 Use of nonlinear response analysis in this example

A first-mode nonlinear static pushover analysis is used to do an initial check on the performance of the initial design realization. NLRHAs consisting of two stripes (at $S_a(T_1)$ and $1.10 \cdot S_a(T_1)$ are used to evaluate the acceptability of system ductility demand at the 50/50 hazard level and interstory drift at the 10/50 hazard level; IDA is used to evaluate collapse.

### 19.9.3 System ductility limits

Design Step 1. Estimated roof displacement at yield: We use the results of the first-mode nonlinear static pushover analysis of the moment frame in Example 2 as the starting point for this example. Therefore, we estimate $D_y = 4.6$ in., $\Gamma_1 = 1.35$, and $\alpha_1 = 0.63$. The yield displacement of the ESDOF system is estimated as

$$D_y^* = \frac{D_y}{\Gamma_1} = \frac{4.6 \text{ in.}}{1.35} = 3.41 \text{ in.}$$

Design Step 2. For the first PO, we simply limit system ductility to 1.5 at a $p_o = 1.39 \times 10^{-2} \text{year}^{-1}$ (50/50 years) with 70% confidence.

Design Step 3. For the second PO, we estimate $\alpha_{3,COD} = 1.45$ from Table 12.8. Then, the roof displacement must be limited to approximately

$$D_{u,\text{drift}} = \frac{0.02}{\alpha_{3,COD}} h_n = \left(\frac{0.02}{1.45}\right)(54 \text{ ft} \cdot 12 \text{ in./ft}) = 8.94 \text{ in.}$$

Therefore, the system ductility is limited to 8.94·in./4.6·in. = 1.94 at $p_o = 2.11 \times 10^{-3}$year$^{-1}$ (10/50 years) with 70% confidence.

Design Step 4. For the third PO, we determine the ductility limit at which point the constant $C_y$ curves "flatline" on the YFS, and determine the strength required for $p_o = 2.01 \times 10^{-4}$ (1/50 years) at 90% confidence.

### 19.9.4 Assumptions required to generate YFS

YFS are generated as discussed in Section 13.5.2 and based on the assumptions of Section 19.3. The assumed normalized capacity curve is given in Figure 19.33.

### 19.9.5 Required yield strength coefficient, $C_y^*$

Design Step 5. The YFS for $D_y^* = 3.41$ in. is plotted in Figures 19.44 and 19.45 for 70% and 90% confidence, respectively.

The associated performance points at the 70% and 90% confidence levels are shown in Table 19.35.

With reference to Figures 19.44 and 19.45, the collapse limit is indicated by "flatlining" of the $C_y$ contours on the YFS diagram. With reference to the normalized capacity curve (Figure 19.33), the flatlining would be expected at precisely a ductility of 6, but due to the use of discrete intervals in the generation of the YFS and the shallow slope of the curves, the YFS seem to indicate collapse at a ductility of approximately 5.6. This difference is of no consequence here, as the required ESDOF yield strength coefficient, $C_y^*$, remains the same at 0.408, at a 70% confidence level. Since we very much wish to avoid collapse, we chose a

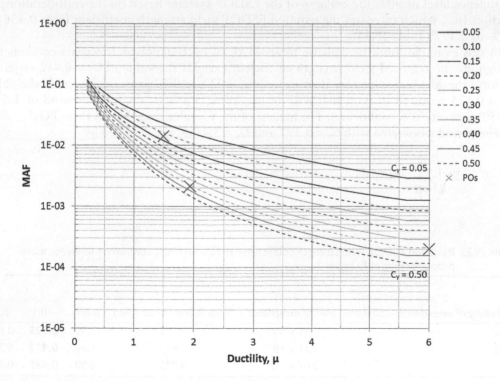

*Figure 19.44* YFS at 70% confidence for initial proportioning of Example 3.

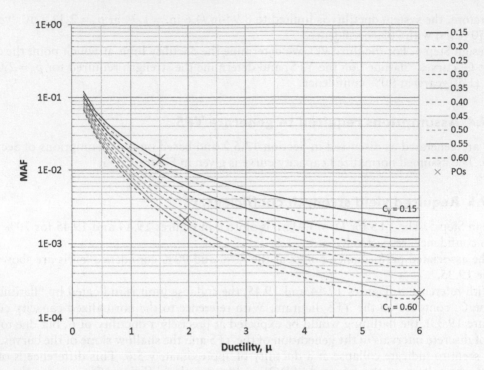

*Figure 19.45* YFS at 90% confidence for initial proportioning of Example 3.

confidence level of 90% (on collapse of the ESDOF system) based on the considerations in Section 18.7, which increases the required ESDOF yield strength coefficient from 0.408 by 40% to 0.577 (Table 19.35).

In this example, limiting collapse to an MAF of exceedance of 1/50 years controls the design, as the required yield strength coefficient exceeds the value of $C_y^* = 0.412$ required to limit the MAF of interstory drift exceeding 0.02 to 10/50 years, and exceeds the value $C_y^* = 0.125$ required to limit the MAF of exceeding a system ductility demand of 1.5 to 50/50 years (at 70% confidence). The highest of the $C_y^*$ values satisfies all three POs.

The period associated with $C_y^* = 0.577$ and $D_y^* = 3.41$ in. is calculated as

$$T^* = 2\pi\sqrt{\frac{D_y^*}{C_y^* g}} = 0.79 \text{ s}$$

*Table 19.35* Required ESDOF yield strength coefficients at 70% and 90% confidence levels for initial proportioning of Example 3

| Probability of exceedance in 50 years | MAF of exceedance | Mean Return Period (year) | μ-limit | $C_y^*$ 70% | 90% |
|---|---|---|---|---|---|
| 50% | $1.39 \times 10^{-2}$ | 72 | 1.50 | **0.125** | 0.158 |
| 10% | $2.11 \times 10^{-3}$ | 475 | 1.94 | **0.412** | 0.511 |
| 1% | $2.01 \times 10^{-4}$ | 4,975 | 6.00 | 0.408 | **0.577** |

Values associated with the specified confidence levels are indicated in bold.

### 19.9.6 Required base shear strength

Design Step 6. Applying the relationship $C_y = \alpha_1 C_y^*$, the base shear coefficient at yield is estimated as $C_y = 0.63(0.577) = 0.364$.

The design base shear at yield for a single frame is given by $V_y = C_y W = 0.364(7,560) = 2,752$ kips.

### 19.9.7 Design lateral forces and required member strengths

Design Step 7. In this example, we illustrate the use of the beta distribution of lateral forces for design. The effective height of the resultant lateral forces is given by $\Sigma(F_i h_i)/\Sigma(F_i) = 0.792 h_n$. The height of the resultant of the first-mode forces, $h_1^*$, according to Table 12.1, is estimated to be $0.72 h_n$ (=54·0.72 ft = 38.88 ft). Thus, the design base shear is modified to (2,752 kips)·(0.72/0.792) = 2,502 kips. This modified base shear is distributed over the height, as illustrated in Table 19.36 below.

Design Step 8. As in Example 2, the portal method is applied to distribute the story shears to the columns. Inflection points were assumed at 60% of the first-story height and 50% of the story heights for the second, third, and fourth stories. Then,

$$V = \frac{2,502}{2n_b} = 312.8 \text{ kips}$$

and the plastic moment at the base of an end column, $M_c$, is calculated as:

$$M_c = V(0.6)(15 \text{ ft}) = 2,815 \text{ kip-ft}$$

The lateral forces tributary to an end column are based on the values above divided by twice the number of bays, as shown in Table 19.37.

Enforcing zero moment at the inflection points and working from either the top down or the bottom up, the associated column and beam moments are as shown in Figure 19.46.

The column and beam moments associated with the implicit strong column–weak beam mechanism are given in Table 19.38.

Table 19.36 Beta distribution of lateral design forces at each floor level, Example 3

| $i$ | $w_i$ (kips) | $h_i$ (ft) | $w_i h_i$ (k-ft) | $\beta_i = V_i/V_y$ | $F_i$ (normalized) | $F_i h_i$ (normalized) | $F_i$ (kips) |
|---|---|---|---|---|---|---|---|
| 4 | 1,890 | 54.0 | 102,060 | 0.477 | 0.477 | 25.78 | 1,194 |
| 3 | 1,890 | 41.0 | 77,490 | 0.745 | 0.268 | 10.98 | 670 |
| 2 | 1,890 | 28.0 | 52,920 | 0.913 | 0.168 | 4.71 | 421 |
| 1 | 1,890 | 15.0 | 28,350 | 1.000 | 0.087 | 1.30 | 217 |
| $\Sigma$ | 7,560 | | 260,820 | | 1.000 | 42.76 | 2,502 |

Table 19.37 Lateral forces tributary to an end column

| Level | Lateral force tributary to end column (kips) |
|---|---|
| 4 | 149.3 |
| 3 | 83.8 |
| 2 | 52.6 |
| 1 | 27.1 |

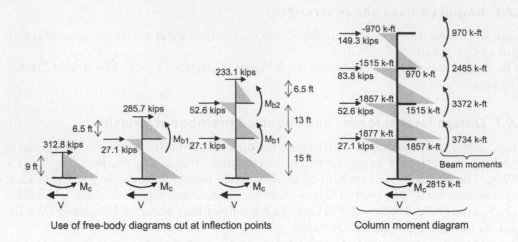

Figure 19.46 Moment diagrams for an end column, Example 3.

Table 19.38 Equilibrium moments, Example 3

|  | Moment (kip-ft) |
| --- | --- |
| $M_c$ | 2,815 |
| $M_{b1}$ | 3,734 |
| $M_{b2}$ | 3,372 |
| $M_{b3}$ | 2,485 |
| $M_{b4}$ | 970 |

## 19.9.8 Sizing of RC members

Design Step 9. Beams and columns were proportioned as described in Example 2. The mathematical and preliminary design solutions for the beam sections are given in Tables 19.39 and 19.40. Column designs are given in Tables 19.41–19.44.

Table 19.39 Mathematical solution for beam dimensions and reinforcement, Example 3

| | Required strengths | | Mathematical solution | | | |
| --- | --- | --- | --- | --- | --- | --- |
| Level | $M_p^-$ (kip-in.) | $M_p^+$ (kip-in.) | $b_w$ (in.) | h (in.) | $A_{s,top}$ (in.²) | $A_{s,bot}$ (in.²) |
| 4 | 13,968 | 6,984 | 16.21 | 27.01 | 9.14 | 4.20 |
| 3 | 35,784 | 17,892 | 21.69 | 36.15 | 16.79 | 7.87 |
| 2 | 48,557 | 24,278 | 23.86 | 39.77 | 20.45 | 9.66 |
| 1 | 53,770 | 26,885 | 24.64 | 41.07 | 21.86 | 10.33 |

Table 19.40 Preliminary design (beam dimensions and reinforcement), Example 3

| Level | $b_w$ (in.) | h (in.) | Top bars | Bottom bars |
| --- | --- | --- | --- | --- |
| 4 | 16 | 28 | (10) No. 9 | (5) No. 9 |
| 3 | 22 | 36 | (16) No. 9 | (8) No. 9 |
| 2 | 26 | 42 | (20) No. 9 | (10) No. 9 |
| 1 | 26 | 42 | (22) No. 9 | (10) No. 9 |

*Table 19.41* Preliminary design of intermediate columns just above floor levels, Example 3

| Floor level | $M_c$ or $M_{pc}$ (k-in.) | D + 0.25L (k) | h (in.) | $A_s$ (in.$^2$) | Bars | $A_{s,provided}$ (in.$^2$) | $P/A_{g}f_{ce}$ | $\rho_g$ % |
|---|---|---|---|---|---|---|---|---|
| 4 | NA | NA | NA | NA | NA | NA | NA | NA |
| 3 | 27,936 | 78.75 | 30 | 33.51 | (28) No. 10 | 35.56 | 1.35% | 3.95% |
| 2 | 43,632 | 163.13 | 34 | 45.01 | (36) No. 10 | 45.72 | 2.17% | 3.96% |
| 1 | 53,482 | 247.50 | 38 | 46.87 | (36) No. 10 | 45.72 | 2.64% | 3.17% |
| Footing | 67,560 | 331.88 | 38 | 61.01 | (48) No. 10 | 60.96 | 3.54% | 4.22% |

*Table 19.42* Preliminary design of intermediate columns just below floor levels, Example 3

| Floor level | $M_c$ (k-in.) | D + 0.25L (k) | h (in.) | $A_s$ (in.$^2$) | Bars | $A_{s,provided}$ (in.$^2$) | $P/A_{g}f_{ce}$ | $\rho_g$ % |
|---|---|---|---|---|---|---|---|---|
| 4 | 27,936 | 78.75 | 30 | 33.51 | (28) No. 10 | 35.56 | 1.35% | 3.95% |
| 3 | 43,632 | 163.13 | 34 | 45.01 | (36) No. 10 | 45.72 | 2.17% | 3.96% |
| 2 | 53,482 | 247.50 | 38 | 46.87 | (36) No. 10 | 45.72 | 2.64% | 3.17% |
| 1 | 54,058 | 331.88 | 38 | 46.48 | (36) No. 10 | 45.72 | 3.54% | 3.17% |
| Footing | NA | NA | NA | NA | NA | NA | NA | NA |

*Table 19.43* Preliminary design of end columns just above floor levels, Example 3

| Floor level | $M_c$ or $M_{pc}$ (k·in.) | D + 0.25L (k) | h (in.) | $A_s$ (in.$^2$) | Bars | $A_{s,provided}$ (in$^2$) | $P/A_{g}f_{ce}$ | $\rho_g$ % |
|---|---|---|---|---|---|---|---|---|
| 4 | NA | NA | NA | NA | NA | NA | NA | NA |
| 3 | 18,624 | 78.75 | 30 | 20.83 | (20) No. 10 | 25.40 | 1.35% | 2.82% |
| 2 | 29,088 | 163.13 | 34 | 27.75 | (24) No. 10 | 30.48 | 2.17% | 2.64% |
| 1 | 35,654 | 247.50 | 38 | 28.67 | (24) No. 10 | 30.48 | 2.64% | 2.11% |
| Footing | 33,780 | 331.88 | 38 | 25.70 | (20) No. 10 | 25.40 | 3.54% | 1.76% |

*Table 19.44* Preliminary design of end columns just below floor levels, Example 3

| Floor level | $M_c$ (k-in.) | D + 0.25L (k) | h (in.) | $A_s$ (in.$^2$) | Bars | $A_{s,provided}$ (in.$^2$) | $P/A_{g}f_{ce}$ | $\rho_g$ % |
|---|---|---|---|---|---|---|---|---|
| 4 | 18,624 | 78.75 | 30 | 20.83 | (20) No. 10 | 25.40 | 1.35% | 2.82% |
| 3 | 29,088 | 163.13 | 34 | 27.75 | (24) No.10 | 30.48 | 2.17% | 2.64% |
| 2 | 35,654 | 247.50 | 38 | 28.67 | (24) No. 10 | 30.48 | 2.64% | 2.11% |
| 1 | 36,038 | 331.88 | 38 | 27.94 | (24) No. 10 | 30.48 | 3.54% | 2.11% |
| Footing | NA | NA | NA | NA | NA | NA | NA | NA |

## 19.9.9 Preliminary evaluation of the initial design by nonlinear static (pushover) analysis

An initial check of the first design realization was made using a fiber element model in SeismoStruct (Seismosoft, 2014). An eigenvalue analysis of the uncracked section model identified first-mode properties to be $T_1 = 0.604$ s, $\Gamma_1 = 1.35$, and $\alpha_1 = 0.65$.

A capacity curve was determined by nonlinear static (pushover) analysis using forces proportional to the vector given in Table 19.45.

The resultant of these first-mode forces is located at $0.768h_n$ above the base, which differs from the value of $0.72h_n$ assumed in design.

Just as in Examples 1 and 2, during the pushover analysis, the moments in the columns above the base of the first floor did not exceed those associated with equilibrium associated

Table 19.45 First-mode force vector, Example 3

| Level | $F_{1st}$ (kip) |
|---|---|
| 4 | 1,141 |
| 3 | 824 |
| 2 | 533 |
| 1 | 254 |

with the intended beam-hinging mechanism, implying that the intended mechanism did occur in the first-mode nonlinear static analysis.

Based on the eigenvalue analysis, the period associated with the secant stiffness representative of cracked section behavior can be estimated (with reference to Figure 19.47) as

$$T_{sec} = T_{initial}\sqrt{\frac{k_{initial}}{k_{sec}}} = 0.604 \text{ s}\sqrt{\frac{1,132 \text{ k/in.}}{660 \text{ k/in.}}} = 0.79 \text{ s}$$

which matches the ESDOF period, $T^* = 0.79$ s. The base shear at yield, 2,550 kips, is a bit low compared with the design base shear associated with first-mode response, 2,752 kips. Much of this difference is attributed to the assumption of first-mode lateral forces having a resultant at $0.72h_n$. Had we used the value of $0.768h_n$ determined from the model, the base shear to use with the beta force distribution would have been 2,752 kips $(0.768h_n/0.792h_n) = 2,669$ kips, which is 6.7% higher than the value of 2,502 kips used in design. Had we designed for this force level, yield strength in the first-mode pushover analysis of Figure 19.47 would have increased from 2,550 kips to approximately 2,720 kips, which is very close to

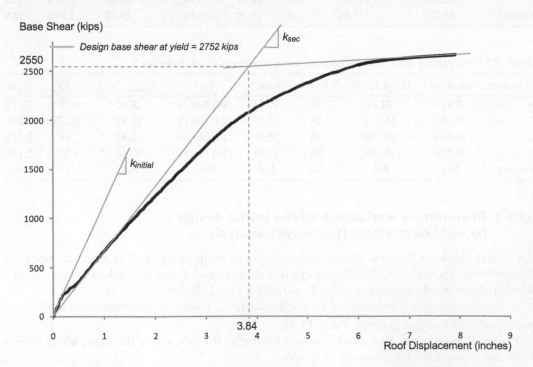

Figure 19.47 Capacity curve obtained in a first-mode pushover analysis of the initial design realization, Example 3.

the first-mode target of 2,752 kips. This increase of strength would have resulted in a small decrease in period (to about 0.76 s).

The results of the first-mode pushover analysis are used to evaluate the performance of this initial design realization. We estimate $D_y^* = D_y/\Gamma_1 = 3.84/1.35 = 2.84$ in. For $V_y = 2,550$ kips, $C_y = V_y/W = 2,550/7,560 = 0.337$. Therefore, $C_y^* = C_y/\alpha_1 = (0.337/0.65) = 0.518$.

For the interstory drift PO, the assumption that $\alpha_{3,COD} = 1.45$ and the limit on interstory drift of 2% leads to a peak roof displacement limit of 8.94 in.; the associated ductility limit is now 8.94/3.84 = 2.33. The YFS for this yield displacement at 70% confidence is plotted in Figure 19.48.

For the collapse PO, as discussed in Section 19.9.5, flatlining is seen to occur at a ductility of approximately 5.7 for the YFS plotted in Figure 19.49 at 90% confidence level. For a $p_o = 2.01 \times 10^{-4}$ year$^{-1}$, the required $C_y^*$ is 0.628 (Table 19.46), which is 8% above 0.577. (Alternatively, for a $p_o = 2.50 \times 10^{-4}$ year$^{-1}$ the required $C_y^*$ is 0.577.) To assure compliance with the stated POs, one may either redesign the frame to be stronger, or see if acceptable performance is obtained in a more precise assessment, e.g., using IDA on the full MDOF (rather than an ESDOF) to determine a better estimate of the MAF of collapse of the MDOF system. As is well known from FEMA P-695 (FEMA, 2009a), the latter approach allows us to introduce the effect of spectral shape, reducing the inherent conservatism in the assessment of collapse performance, and potentially making viable this more economic initial design. Therefore, this is the path that we shall take in the following.

### 19.9.10 Nonlinear modeling and acceptance criteria

Tables 19.47 and 19.48 summarize the modeling parameters for column and beam plastic hinges, respectively. Tables 19.49 and 19.50 present the acceptance criteria for column and beam plastic hinge rotations.

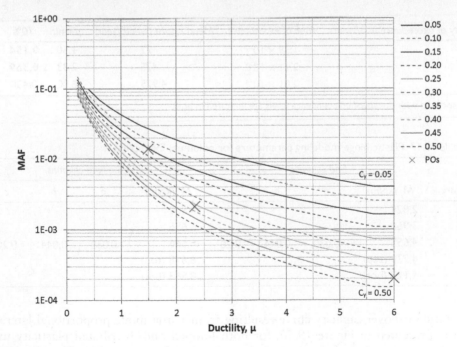

*Figure 19.48* YFS for $D_y^* = 2.84$ in at 70% confidence, for evaluation of the first design realization of Example 3.

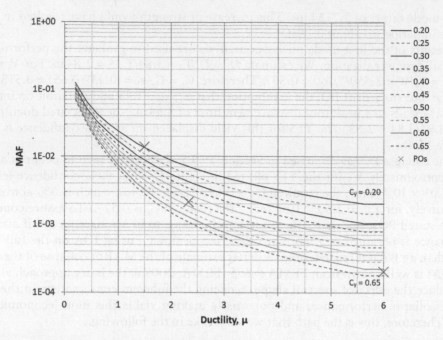

*Figure 19.49* YFS for $D_y^* = 2.84$ in at 90% confidence, for evaluation of the first design realization of Example 3.

*Table 19.46* Required ESDOF yield strength coefficients at 70% and 90% confidence levels for based on first design realization, Example 3

| Probability of exceedance in 50 years | MAF of exceedance | Mean return period (year) | $\mu$-limit | $C_y^*$ 70% | 90% |
|---|---|---|---|---|---|
| 50% | $1.39 \times 10^{-2}$ | 72 | 1.50 | **0.154** | 0.194 |
| 10% | $2.11 \times 10^{-3}$ | 475 | 2.33 | **0.369** | 0.468 |
| 1% | $2.01 \times 10^{-4}$ | 4,975 | 6.00 | 0.450 | **0.628** |

Values associated with the specified confidence levels are indicated in bold.

*Table 19.47* Plastic hinge modeling parameters for columns, Example 3

| | End columns | | | | Intermediate columns | | | |
|---|---|---|---|---|---|---|---|---|
| Story | $M_y$ (kNm) | a | b | c | $M_y$ (kNm) | a | b | c |
| 4 | 2,829 | | | | 3,809 | | | |
| 3 | 3,968 | | | | 5,655 | | | |
| 2 | 4,655 | 0.030 | 0.044 | 0.20 | 6,566 | 0.030 | 0.044 | 0.20 |
| 1 | 4,770 (top) | | | | 6,678 (top) | | | |
| | 4,128 (bot) | | | | 8,563 (bot) | | | |

The static pushover capacity curve resulting from a first-mode-proportional lateral load pattern is presented in Figure 19.50, for both lumped and distributed plasticity models. The fundamental period of the lumped plasticity model is $T_1 = 0.79$ s. This is larger than the initial period of the fiber model ($T_1 = 0.65$), and is appropriate for reasons explained in

Table 19.48 Plastic hinge modeling parameters for beams, Example 3

| Story | Positive moment | | | | Negative moment | | | |
|---|---|---|---|---|---|---|---|---|
| | $M_y^+$ (kN·m) | a | b | c | $M_y^-$ (kN·m) | a | b | CP |
| 4 | 1,065 | 0.025 | 0.050 | | 2,077 | 0.021 | 0.035 | |
| 3 | 2,226 | 0.024 | 0.049 | 0.20 | 4,348 | 0.020 | 0.031 | 0.20 |
| 2 | 3,274 | 0.025 | 0.049 | | 6,412 | 0.021 | 0.034 | |
| 1 | 3,274 | 0.024 | 0.048 | | 6,979 | 0.020 | 0.033 | |

Table 19.49 Column plastic hinge rotation limits, for IO, LS, and CP POs, Example 3

| Story | Positive rotation (rad) | | |
|---|---|---|---|
| | IO | LS | CP |
| $\theta_{p,max}$ (rad) | 0.005 | 0.034 | 0.044 |

Table 19.50 Beam plastic hinge rotation limits for IO, LS, and CP POs, Example 3

| Story | Positive rotation (rad) | | | Negative rotation (rad) | | |
|---|---|---|---|---|---|---|
| | IO | LS | CP | IO | LS | CP |
| 4 | 0.0100 | 0.0250 | 0.0500 | 0.0062 | 0.0212 | 0.0347 |
| 3 | 0.0092 | 0.0242 | 0.0485 | 0.0056 | 0.0199 | 0.0311 |
| 2 | 0.0096 | 0.0246 | 0.0492 | 0.0061 | 0.0208 | 0.0341 |
| 1 | 0.0088 | 0.0238 | 0.0477 | 0.0059 | 0.0201 | 0.0326 |

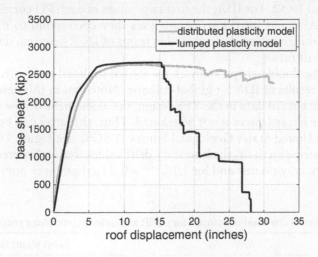

Figure 19.50    Static pushover curves of the distributed (gray color) and the lumped plasticity (black color) models of Example 3.

Section 19.7.7. Yield displacement is calculated according to Figure 19.51 assuming a bilinear approximation of the capacity curve having the same effective stiffness as the lumped plasticity model, resulting in $D_y = 3.5$ in.

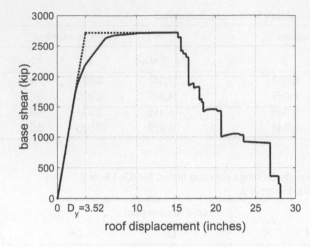

$W = 7560$ kips

$V_{max} = 2718$ kips

$C_y = \dfrac{V_{max}}{W} = \dfrac{2718}{7560} = 0.360$

$D_y = 3.50$ in.

$T_1 = 0.79$ s

*Figure 19.51*   Calculation of $D_y$ for Example 3.

## 19.9.11 Performance evaluation of the initial design by nonlinear dynamic analysis

Four performance criteria were checked: (i) peak dynamic interstory drifts at an MAF of 10/50 with confidence of 70%, plastic hinge rotations of beams and columns at an MAF of 10/50 at 75% confidence, a system ductility limit of 1.5 at an MAF of 50/50 at 70% confidence, and a global collapse LS at an MAF of 1/50 at 90% confidence.

All three NLRHA methods introduced in Chapter 13 were employed, i.e., IDA, and single- and double-stripe analyses. Results from double-stripe analysis are shown in detail, along with the use of IDA for evaluation of global collapse. Results for each PO are summarized in Tables 19.51 and 19.52. For IDA, the first two values of each PO correspond to estimated and allowable MAFs, and for the stripe analyses they correspond to FD and FC, respectively, with plastic hinge rotations expressed in terms of DCRs. System ductility is expressed in terms of roof drift ratio.

The results of the double-stripe analysis are shown in detail to illustrate Design Method C, along with the results of IDA for global collapse. Note that in this case, since we directly use the site-specific hazard data in the YFS design, the assessment will be fair by default, and thus, scaling of the hazard curve is not warranted. Thus, the latter can be employed directly as estimated from United States Geological Survey (USGS) (see Figure 19.52).

For the double-stripe analysis, the interstory drift values were calculated for 17 records for $S_a^{design} = 0.76$ g (dark gray points) and for $1.1 S_a^{design} = 0.83$ g (light gray points) (Figure 19.53a).

*Table 19.51*  Verification of the maximum interstory drift and beam plastic hinge rotations for Example 3

|  | IDR[a] | | | Beam plastic hinge rotation | | |
|---|---|---|---|---|---|---|
|  | IDA (year$^{-1}$) | Single (m/m) | Double (m/m) | IDA (year$^{-1}$) | Single (rad/rad) | Double (rad/rad) |
| Demand | 0.0008 | 0.0124 | 0.0124 | 0.0009 | 0.4849 | 0.4834 |
| Capacity | 0.0021 | 0.0189 | 0.0189 | 0.0021 | 0.9463 | 0.9496 |
| Check | ✓ | ✓ | ✓ | ✓ | ✓ | ✓ |
| Confidence |  | 70% |  |  | 75% |  |

[a] IDR = maximum interstory drift ratio over the height of the building.

*Table 19.52* Verification of column plastic hinge rotations, system ductility, and global collapse for Example 3

| | Column plastic hinge rotation | | | System ductility | | | Collapse |
|---|---|---|---|---|---|---|---|
| | IDA (year⁻¹) | Single (rad/rad) | Double (rad/rad) | IDA (year⁻¹) | Single (m/m) | Double (m/m) | IDA (year⁻¹) |
| Demand | 0.0004 | 0.2335 | 0.2220 | 0.0049 | 0.0051 | 0.0051 | 0.000571 |
| Capacity | 0.0021 | 0.9471 | 0.9738 | 0.0139 | 0.0079 | 0.0079 | 0.000201 |
| Check | ✓ | ✓ | ✓ | ✓ | ✓ | ✓ | ✗[a] |
| Confidence | | 75% | | | 70% | | 90% |

[a] Initial estimate. Refer to later discussion on improvements at the end of this section.

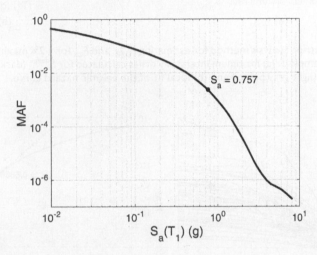

*Figure 19.52* Seismic site hazard curve as determined from USGS data for $S_a(T_1)$, Example 3.

The points of the hazard curve needed to estimate the slope $k$ are presented in Figure 19.53b. FD and FC at 70% confidence are estimated as:

$$FD_{RPo} = \theta_{max,50} \exp\left(0.5 \cdot k \cdot \beta_{\theta max|Sa}^2/b\right) = 0.011 \exp\left(0.5 \cdot 2.75 \cdot 0.17^2/1.01\right) = 0.0111$$

$$FC_R = \theta_{max,C} \exp\left(-0.5 \cdot k \cdot \beta_{CR}^2/b\right) = 0.02 \exp\left(-0.5 \cdot 2.75 \cdot 0.20^2/1.01\right) = 0.0189$$

For a confidence level of 70%, the lognormal standard variate is $K_x = 0.524$ and the evaluation inequality becomes:

$$FC_R > FD_{RPo} \exp(K_x \beta_{TU}) = 0.0111 \exp(0.524 \cdot 0.20) = 0.0124$$

Since the FC of 0.0189 is greater than the FD of 0.0124 the result is satisfactory at the 70% confidence level.

DCR values for beam plastic hinge rotations were also calculated for 17 records for $S_a^{design} = 0.76$g (dark gray points) and for $1.1 S_a^{design} = 0.83$g (light gray points) (Figure 19.54a). The points of the hazard curve needed to estimate the slope $k$ are presented in Figure 19.54b. FD and FC are estimated as:

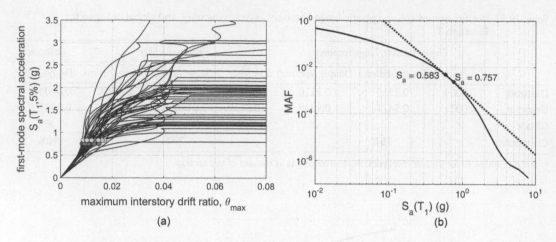

*Figure 19.53* Double-stripe analysis method for estimating $F_{DRPo}$ and $F_{CR}$ for a 2% maximum interstory drift ratio, Example 3: (a) maximum interstory drifts calculated for $S_a^{design}$ (dark gray points) and for $1.1S_a^{design}$ (light gray points) and (b) local fit of the seismic hazard curve.

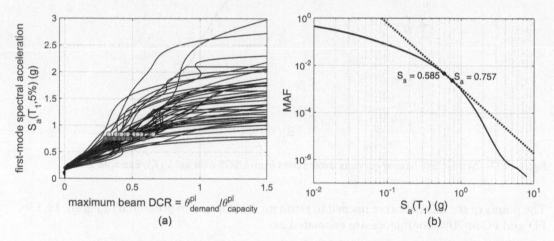

*Figure 19.54* Double-stripe analysis method for estimating $F_{DRPo}$ and $F_{CR}$ for a DCR = 1 of beam plastic hinge rotations, Example 3: (a) maximum beam DCR calculated for $S_a^{design}$ (dark gray points) and for $1.1S_a^{design}$ (light gray points), and (b) local fit of the seismic hazard curve.

$$FD_{RPo} = DCR_{max,50} \exp\left(0.5 \cdot k \cdot \beta_{\theta max|Sa}^{2}/b\right) = 0.404 \exp\left(0.5 \cdot 2.75 \cdot 0.18^2/1.06\right) = 0.4224$$

$$FC_R = DCR_{max,C} \exp\left(-0.5 \cdot k \cdot \beta_{CR}^{2}/b\right) = 1.0 \exp\left(-0.5 \cdot 2.75 \cdot 0.20^2/1.06\right) = 0.9496$$

For a confidence level of 75%, the lognormal standard variate is $K_x = 0.674$ and the evaluation inequality becomes:

$$FC_R > FD_{RPo} \exp\left(K_x \cdot \beta_{TU}\right) = 0.4224 \exp\left(0.674 \cdot 0.20\right) = 0.4834$$

As the FC of 0.9496 is greater than the FD of 0.4834 the result is satisfactory at the 75% confidence level.

Similarly, the DCR values for the column plastic hinge rotations were also calculated at $S_a^{design} = 0.76\,g$ (dark gray points) and $1.1S_a^{design} = 0.83\,g$ (light gray points) (Figure 19.55a). The points of the hazard curve needed to estimate the slope $k$ are presented in Figure 19.55b. FD and FC are estimated as:

$$FD_{RPo} = DCR_{max,50} \exp\left(0.5 \cdot k \cdot \beta_{\theta max|Sa}{}^2/b\right) = 0.185 \exp\left(0.5 \cdot 2.88 \cdot 0.27^2/2.17\right) = 0.1940$$

$$FC_R = DCR_{max,C} \exp\left(-0.5 \cdot k \cdot \beta_{CR}{}^2/b\right) = 1.0 \exp\left(-0.5 \cdot 2.88 \cdot 0.20^2/2.17\right) = 0.9738$$

For a confidence level of 75%, the lognormal standard variate is $K_x = 0.674$ and the evaluation inequality becomes:

$$FC_R > FD_{RPo} \exp\left(K_x \cdot \beta_{TU}\right) = 0.1940 \exp(0.674 \cdot 0.20) = 0.2220$$

Since the FC of 0.9738 is greater than the FD of 0.2220 the result is satisfactory at the 75% confidence level.

The acceptability of ductility limit of 1.5 was assessed at the 50/50 level. This ductility limit is equivalent to a roof drift ratio capacity of $\theta_{roof} = 1.5 \cdot D_y/H_{tot} = 1.5 \cdot 3.5$ in./648 in. = 0.0081. Values of $\theta_{roof}$ were calculated for 17 records for $S_a^{50\%/50} = 0.36\,g$ (dark gray points) and for $1.1S_a^{50\%/50} = 0.40\,g$ (light gray points) (Figure 19.56a). The points of the hazard curve needed to estimate the slope $k$ are presented in Figure 19.56b. FD and FC are estimated as:

$$FD_{RPo} = \theta_{roof,50} \exp\left(0.5 \cdot k \cdot \beta_{\theta max|Sa}{}^2/b\right) = 0.005 \exp\left(0.5 \cdot 1.58 \cdot 0.06^2/1.05\right) = 0.0047$$

$$FC_R = \theta_{roof,C} \exp\left(-0.5 \cdot k \cdot \beta_{CR}{}^2/b\right) = 0.0081 \exp\left(-0.5 \cdot 1.58 \cdot 0.20^2/1.05\right) = 0.0079$$

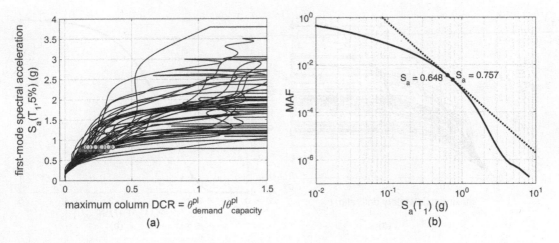

*Figure 19.55* Double-stripe analysis method for estimating $F_{DRPo}$ and $F_{CR}$ for a DCR = 1 for column plastic hinge rotations, Example 3: (a) maximum column DCR calculated for $S_a^{design}$ (dark gray points) and for $1.1S_a^{design}$ (light gray points) and (b) local fit of the seismic hazard curve.

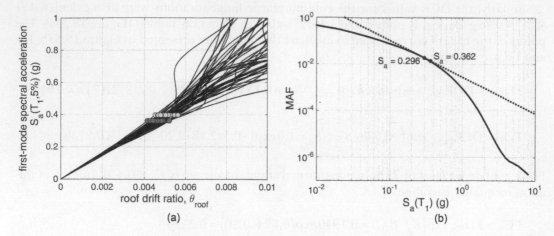

*Figure 19.56*   Double-stripe analysis method for estimating $F_{DRPo}$ and $F_{CR}$ for a ductility limit of 3.6, Example 3: (a) roof drift ratio calculated for $S_a^{50\%/50}$ (dark gray points) and for $1.1S_a^{50\%/50}$ (light gray points) and (b) local fit of the seismic hazard curve.

For a confidence level of 70%, the lognormal standard variate is $K_x = 0.524$ and the evaluation inequality becomes:

$$FC_R > FD_{RPo}\exp(K_x\beta_{TU}) = 0.0047\exp(0.524 \cdot 0.20) = 0.0051$$

Since the FC of 0.0079 is greater than the FD of 0.0051 the result is satisfactory at the 70% confidence level.

Global collapse is deemed to occur when numerical non-convergence appears or a large maximum interstory drift (taken as 8%) is exceeded, whichever occurs first. In this case, the first criterion governed, and the corresponding values of $S_a(T_1, 5\%)$ at the onset of collapse were employed to define the fragility curve (Figure 19.57). The median value of the lognormal distribution of these collapse values equals 1.92 g and the log standard deviation is 37%. By convolving with the hazard curve of Figure 19.52, the estimated MAF at 90% confidence

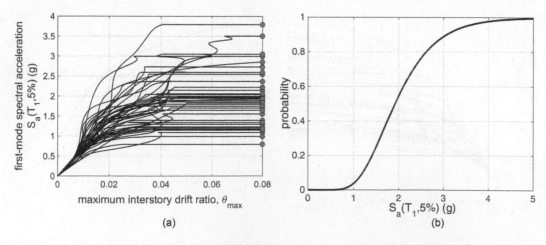

*Figure 19.57*   Analytical approach for estimating MAF of collapse for Example 3: (a) the $S_a(T_1, 5\%)$ values calculated for 8% maximum interstory drift ratio limit, $\theta_{max}$ and (b) corresponding fitted lognormal fragility curve.

equals 0.000571, which is about 2.5 times the required target of 1/50, or 0.000201, while only slightly greater than the 2/50 value of 0.000404. Note that there is an exponential relationship between the $C_y$ and the MAF; therefore, such large differences in MAF actually translate to much smaller difference in design strength. Obviously, this PO is not met. Here, the analyst has two choices, namely redesign or attempt an even more accurate (and elaborate) assessment. The reason for the latter is that IDA can be conservative close to global collapse (Luco and Bazzurro, 2007), as the simple amplitude scaling that it employs may not allow for capturing the appropriate spectral shape of high-intensity ground motions (Baker and Cornell, 2006). Specifically, FEMA P-695 (FEMA, 2009a) recommends employing a spectral shape factor (SSF), which happens to have been determined specifically for the suite of 44 ground motions used herein for IDA, to adjust the collapse capacity upwards. Although this remains a crude approximation and better methods do exist for achieving unbiased assessment of global collapse (e.g., see Lin et al., 2013; Kohrangi et al., 2017), the SSF can help us to refine the estimate of collapse capacity. For the case at hand, for a system ductility of 6, period $T_1 \approx 0.8$ s and assuming Seismic Design Category $D_{max}$, Table 7-1b of FEMA P695 suggests SSF = 1.35. The adjusted global collapse fragility has a median value of $1.35 \cdot 1.92$ g $= 2.60$ g and retains the same log standard deviation of 38%. This results in an improved MAF estimate of 0.00019, which now meets the PO of 0.00020 year$^{-1}$. It is noted that no such easy shortcut exists in current literature for LSs other than global collapse; therefore, if higher accuracy for other POs is sought, the aforementioned advanced methods involving record selection would need to be considered.

## 19.10 EXAMPLE 4: COUPLED WALL DESIGNED USING METHOD A

### 19.10.1 Coupled wall example plan and elevation

The wall cross section and coupling beam dimensions are shown in Figure 19.58, which illustrates the 12-story coupled wall. The coupled wall is part of the perimeter of a 12-story RC frame structure. The height of the first story is 4.5 m while the overlying stories are 3.4 m high, resulting in a total height of 41.9 m. The coupled wall consists of two rectangular section walls having a plan length of 4.5 m and thickness of 0.4 m. B500 steel reinforcement ($f_{yk} = 500$ MPa) and C-30 concrete ($f_{ck} = 30$ MPa) are used. The material safety factors used in the design are $\gamma_s = 1.0$ and $\gamma_c = 1.0$.

### 19.10.2 POs

Method A seeks performance comparable to that obtained in current codes. For this example, performance requirements comparable to those in EC-8 are used as follows:

1. Table 12.7 indicates the system ductility limit at the 10/50 level is equal to $3.3/\gamma_I = 3.3$ for coupled walls of high ductility (DCH).
2. Table 12.10 indicates the interstory drift limit is $0.010 \cdot h/\nu$, assuming that nonstructural elements do not interfere with structural deformation, or are not present.

### 19.10.3 Use of nonlinear response analysis in this example

A first-mode nonlinear static pushover analysis is used to evaluate the performance of the initial design realization. Although not required by Design Method A, nonlinear response history analysis will be employed for assessment.

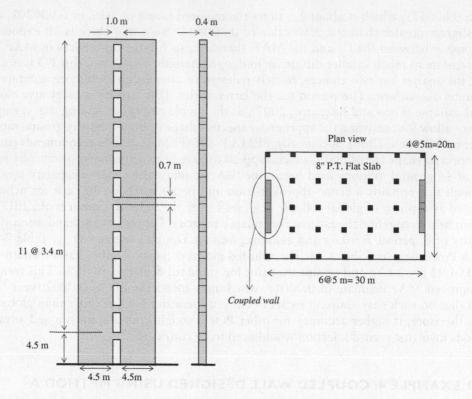

Figure 19.58   Elevation and plan views of 12-story coupled wall building, Example 4.

## 19.10.4 Required base shear strength

Design Step 1. The roof displacement at yield is estimated using Equation 9.11,

$$D_y = 0.68 \frac{\varepsilon_y}{D_{cw}} \frac{H^2}{3} = 0.68 \frac{500/200,000}{9.75} \frac{41.9^2}{3} = 0.102 \text{ m}$$

Design Step 2. To limit the interstory drift to $0.010 \cdot h/\nu$ (where $\nu = 0.5$ for buildings of ordinary importance), or equivalently, to enforce an interstory drift ratio of 2%, Equation 12.17 indicates the peak roof displacement should not exceed:

$$D_{u,\text{drift}} = C_1 \Gamma_1 S_d = C_1 \left( \frac{d_r}{h} \right) \left( \frac{1}{\lambda \gamma_1} \right) \left( \frac{\alpha_1}{\alpha_{3,\text{stat}}} \right) h_n$$

$$= 1.0 \left( \frac{0.010}{0.5} \right) \left( \frac{1}{0.85 \cdot 1.0} \right) \left( \frac{0.79}{1.1} \right) 41.9 \text{ m} = 0.71 \text{ m}$$

at the 10/50 level, where $\lambda = 0.85$ for buildings of three or more stories, $\gamma_1 = 1.0$ for buildings of ordinary importance; from Tables 12.1 and 12.8, we estimate $\alpha_1 = 0.79$ and $\alpha_{3,\text{stat}} = 1.1$.

Design Step 3. To limit the ductility demand to 3.3, the roof displacement must be limited to

$$D_u = 3.3(0.102 \text{ m}) = 0.337 \text{ m}.$$

Since we wish to exceed neither the drift nor the ductility limits, the more restrictive roof displacement limit of 0.337 m applies. The associated system ductility limit is $D_u/D_y = 0.337/0.102 = 3.3$, at the 10/50 level.

Design Step 4. Noting that the first-mode participation factor, $\Gamma_1$, should be approximately 1.45 for a coupled wall building of this height (Table 12.1), the associated yield displacement of an "equivalent" SDOF system is

$$D_y^* = \frac{D_y}{\Gamma_1} = \frac{0.102 \text{ m}}{1.45} = 0.070 \text{ m}$$

Design Step 5. The elastic design spectrum is calculated based on Eurocode 8. Figure 19.5 indicates the elastic horizontal spectral acceleration shape identified as Type 1.

Design Step 6. Yield point spectra (YPS) were generated considering the elastic response spectrum reduced by assumed values of the behavior factor $q$, representing different ductilities (EC-8, §3.2.2.5). The design spectrum is plotted with ordinate $S_d$ and abscissa yield displacement $D_y^*$, determined parametrically as a function of $T$. The spectral design acceleration $S_d(T)$ is given by EC-8(3.2.2.5), while $D_y^*(T)$ is given by:

$$D_y^*(T) = \frac{F_y}{k} = \frac{S_d(T)m}{\omega^2 m} = S_d(T)\left(\frac{T}{2\pi}\right)^2$$

Figure 19.59 shows the EC-8 design spectra in a YPS representation, for $q = 1, 2, 3.3,$ and 8.0, where $q$ is taken as the system ductility.

According to Figure 19.59 the required design spectral acceleration is $S_d = 1.17 \text{ m/s}^2$, which corresponds to $C_y^* = 1.17/9.81 = 0.12$. The associated period of vibration, applicable to both the SDOF system and the first mode of the MDOF system is

$$T = 2\pi\sqrt{\frac{D_y^*}{S_d}} = 1.54 \text{ s}$$

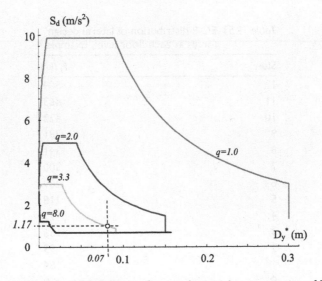

Figure 19.59  YPS representation of EC-8 Design Spectra for initial proportioning of Example 4.

Design Step 7. Assuming $\alpha_1 = 0.79$, the base shear coefficient at yield is given by $C_y = \alpha_1 C_y^* = 0.79(0.12) = 0.095$. The tributary mass per story is 234,000 kg; the total reactive weight is (234,000 kg)(12)(9.81 m/s²)(1 kN/1,000 N) = 27,546 kN. Therefore, the required base shear strength at yield (in a first-mode pushover analysis) is estimated as $C_y W = 0.095(27,546$ kN) = 2,615 kN.

## 19.10.5 Design lateral forces and required member strengths

Design Step 8. The horizontal seismic forces can then be calculated according to EC-8 §4.3.3.2.3(3), resulting in the values given in Table 19.53.

The overturning moment, $M_{OTM}$, at the base of the coupled wall, due to the horizontal seismic forces indicated in Table 19.53 is

$$M_{OTM} = \sum_i F_i h_i = 75,625 \text{ kN} \cdot \text{m}$$

In order to calculate the reinforcement in the members, the value of $\beta_{CB}$ is selected, typically between 0.25 and 0.75 (Priestley et al., 2007). In this example, we chose $\beta_{CB}$ to be equal to 0.45.

$$\beta_{CB} = \frac{M_{CB,b}}{M_{OTM}} = 0.45 \Rightarrow M_{CB,b} = 0.45 M_{OTM} = 34,030 \text{ kN} \cdot \text{m}$$

where $M_{CB,b}$ is the total moment of the coupling beams at the base. Assuming that the shear carried by all coupling beams is identical ($V_i$), with the coupling beams having dimensions $L_w = 4.5$ m and $L_{CB} = 1.0$ m, then $V_i$ is equal to 1,131 kN:

$$M_{CB,b} = \sum_{i=1}^{12} V_i \left( L_w/2 + L_{cb}/2 \right) \rightarrow V_i = \frac{0.45 \cdot 75,625}{12\left(4.5/2 + 1/2\right)} = 1,131 \text{ kN}$$

Table 19.53 EC-8 distribution of lateral design forces at each floor level, Example 4

| Story | $F_i$ (kN) |
|---|---|
| 12 | 504 |
| 11 | 463 |
| 10 | 422 |
| 9 | 381 |
| 8 | 341 |
| 7 | 300 |
| 6 | 259 |
| 5 | 218 |
| 4 | 177 |
| 3 | 136 |
| 2 | 95 |
| 1 | 54 |
| 0 | 0 |

At the same time, assuming that the ultimate behavior of the wall is described by the mechanism shown in Figure 19.60, the required flexural strength of each of the two walls at the base, $M_{CW,b}$, is

$$M_{CW,b} = \frac{M_{OTM}}{2} - M_{CB,b} = 3,738 \text{ kN} \cdot \text{m}$$

The moment at the base, $M_{CW,b}$, acts with an axial force in tension of 10,560 kN.

## 19.10.6 Sizing of RC members

Design Step 9. The cross section of Figure 19.61 is designed to satisfy EC-2 detailing requirements, and it contains 12Ø32 bars within each boundary.

Each coupling beam is designed to resist a shear $V_i = 1,131$ kN, along with a flexural moment of 565.5 kN·m at the face of the wall. The resulting reinforcement is indicated in Figure 19.62.

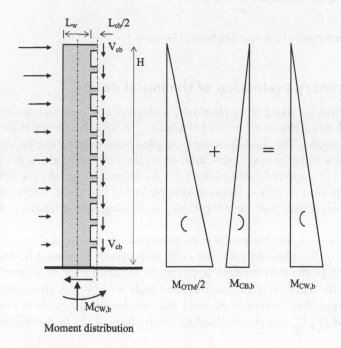

*Figure 19.60* Coupled wall mechanism analysis, Example 4.

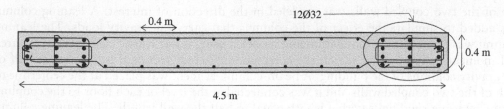

*Figure 19.61* Longitudinal reinforcement of a wall, Example 4.

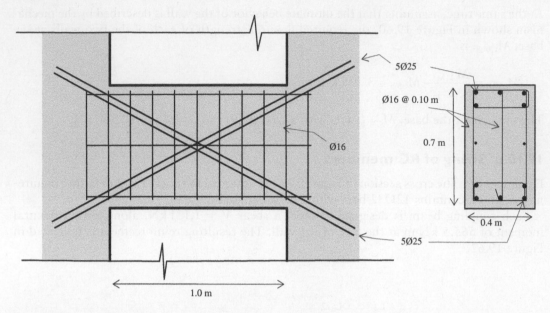

5Ø25

Ø16 @ 0.10 m

0.7 m

Ø16

0.4 m

5Ø25

1.0 m

Figure 19.62   Reinforcement of the coupling beams, Example 4.

## 19.10.7 Preliminary evaluation of the initial design

The walls were modeled using fiber elements, with steel and concrete materials represented at their expected strengths of $f_{ye} = 575\,\text{MPa}$ and $f_{ce} = 38\,\text{MPa}$ rather than at the nominal characteristic strengths. The models of each coupled wall were subjected to nonlinear static (pushover) analysis using lateral forces applied to the coupled wall in proportional to the first-mode forces. The applied force to story $i$ is $F_i$, defined by EC-8 (§ 4.3.3.2.3) [12].

The resulting period is 1.01 s. Because this period is less than 1.19 s, we are confident the spectral displacement will be acceptable, indicating the interstory drifts should be acceptable.

The solution of the eigenvalue problem by Seismostruct results in $\Gamma_1 = 1.47$ and $\alpha_1 = 0.62$, which compares to the values of 1.45 and 0.79, respectively, assumed in design. The reduction in $\alpha_1$, as well as the overstrength apparent in Figure 19.63, indicates that the system has ample strength. Thus, system ductility demands will be less than assumed in design. A reduction in base shear strength at yield and reinforcement content could be obtained using estimates of $D_y$, $\Gamma_1$, and $\alpha_1$ obtained in the analysis of the first realization.

## 19.10.8 Nonlinear modeling and acceptance criteria

A two-dimensional model of the structure was prepared, as shown in Figure 19.64. Only one of the two coupled walls was modeled in the direction of interest. A leaning column was added to simulate the effect of the columns that carry the gravity loads. The leaning column was pinned at the foundation and modeled using linear elastic elements having area and moment of inertia that match the corresponding cross-sectional properties of half of the gravity columns of the building. A beam-column element was placed at the centerline of each of the two coupled walls and it was connected at the level of each floor to the coupling beams by using rigid links with a length equal to half the wall length. The leaning column was constrained to the leftmost nodes of the coupled wall at each floor.

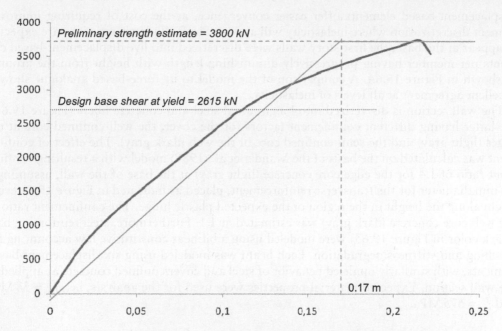

Figure 19.63  Capacity curve resulting from first-mode pushover analysis of the first design realization, Example 4.

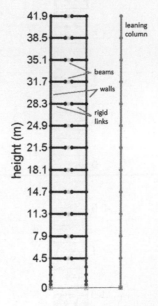

Figure 19.64  Two-dimensional structural model used for the analysis showing walls, beams, rigid links and the leaning column.

Each of the two coupled walls was modeled by using a single displacement-based distributed plasticity element per each story, with fiber section representation monitored at five integration points along the member's length. Force-based elements are usually preferable due to their improved capability at capturing plastic hinges along their length, yet at the same time they complicate the response history analysis and can introduce convergence issues.

Displacement-based elements offer easier convergence, at the cost of requiring improved element discretization where inelasticity will appear. Since all plastic rotations are expected to appear at the base, the first-story walls were discretized into five displacement-based elements per member having progressively diminishing length with height from the ground, as shown in Figure 19.64. A comparison of the model to its force-based analogue showed excellent agreement at all levels of inelasticity.

The wall section is discretized into longitudinal steel and concrete fibers (Figure 19.65), the latter having different confinement factors for the cover, the well-confined core at the edges (light gray) and the semi-confined core of the web (dark gray). The effect of confinement was calculated on the basis of the Mander et al. (1988) model with a resulting confinement ratio of 1.4 for the edge core concrete (light gray) at the base of the wall, assuming a 10-mm diameter for the transverse reinforcement, placed as indicated in Figure 19.61 every 10 cm along the height in the region of the expected plastic hinge. The confinement ratio of the web core concrete (dark gray) was estimated at 1.1. Furthermore, steel reinforcing bars (black color in Figure 19.65) were modeled using a bilinear constitutive law accounting for pinching and stiffness degradation. Each beam was modeled using six displacement-based elements, with similarly modeled behavior of steel and cover/confined concrete as applied in the wall section. Expected material properties were used for the analysis, i.e., $f_{ce} = 38$ MPa and $f_{ye} = 575$ MPa.

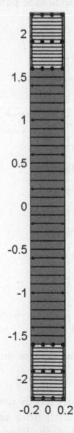

*Figure 19.65* Typical section fiber discretization used to model each wall of Example 4, showing reinforcing bars (black), semi-confined (dark gray) and confined (light gray) concrete fibers, (dimensions in m).

A number of different assumptions were made for the post-capping behavior of each material. In all cases, attempting to implement a realistic and severely degrading behavior of the steel or concrete fibers beyond their maximum strength was found to generate convergence problems. Thus, generous assumptions were made regarding the post-capping behavior to ease convergence. The resulting static pushover capacity curve for a first-mode-proportional lateral load pattern appears in Figure 19.66. The effect of the relaxed material deterioration appears in the residual plateau of nearly 40% the maximum strength that appears beyond a roof displacement of 0.6 m, or approximately a maximum interstory drift of 3%. To make sure the structural performance is not artificially boosted by this assumption, global collapse has been set at 3% drift, essentially assuming a vertical drop to zero strength, rather than an extended plateau beyond this deformation.

The initial (uncracked) period of the model is 0.88 s. A more useful effective period of the structure is calculated using the secant stiffness at 60% of the maximum base shear $V_{max}$ (Figure 19.67).

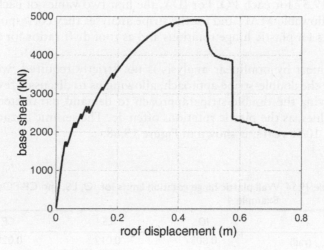

Figure 19.66  Static pushover curve, Example 4.

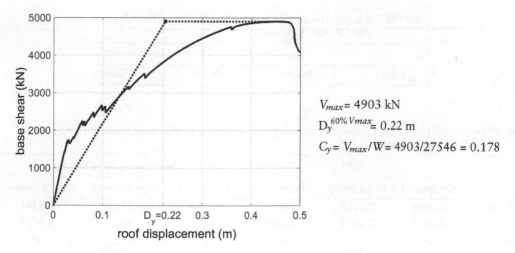

$V_{max} = 4903$ kN

$D_y^{60\% Vmax} = 0.22$ m

$C_y = V_{max}/W = 4903/27546 = 0.178$

Figure 19.67  Determination of nominal $D_y$ by fitting the elastic segment at the 60% $V_{max}$ point of the pushover, Example 4.

$$T_{sec} = T_1 \sqrt{\frac{k_{init}}{k_{eff}}} = 1.51 \text{ s}$$

matching fairly well with the 1.54 s estimated in design. Thus, in the following, the first-mode period will be taken as $T_1 = T_{sec}$.

In Tables 19.54 and 19.55 the acceptance criteria for wall and coupling beam plastic rotations, respectively, are presented.

## 19.10.9 Performance evaluation of the initial design by nonlinear dynamic analysis

The acceptability of interstory drift ratio of 2%, ductility limit of 3.3, and plastic hinge rotations was assessed versus a 10/50-year MAF at the 50%, 50%, and 60% confidence levels, respectively, using IDA, single- and double-stripe analysis. The results are summarized in Tables 19.56 and 19.57 for each PO. For IDA, the first two values of each PO correspond to estimated and allowable MAF, and for the stripe analyses they correspond to FD and FC (expressed as DCRs for plastic hinge rotations and as roof drift ratios for system ductility), respectively.

Although assessment by nonlinear analysis is not strictly required, we shall show the results in detail for the double-stripe approach, allowing us to discuss a few issues that may appear when applying the double-stripe approach to demand parameters that may have many near-zero values, as the plastic rotations often do. The seismic hazard curve properly scaled to $S_a(T_{eff})$ at 10/50 years, is shown in Figure 19.68.

Table 19.54 Wall plastic hinge rotation limits for IO, LS, and CP POs, Example 4

| PO | IO | LS | CP |
|---|---|---|---|
| $\theta_{p,max}$ (rad) | 0.005 | 0.017 | 0.020 |

Table 19.55 Coupling beam plastic hinge rotation limits for IO, LS, and CP POs, Example 4

| PO | IO | LS | CP |
|---|---|---|---|
| $\theta_{p,max}$ (rad) | 0.006 | 0.030 | 0.050 |

Table 19.56 Verification of the maximum interstory drift and beam plastic hinge rotations (in terms of DCR) for Example 4

| | IDR[a] | | | Beam plastic hinge rotation | | |
|---|---|---|---|---|---|---|
| | IDA (year⁻¹) | Single (m/m) | Double (m/m) | IDA (year⁻¹) | Single (rad/rad) | Double (rad/rad) |
| Demand | 0.0006 | 0.0118 | 0.0119 | 0.0006 | 0.7511 | 0.5341 |
| Capacity | 0.0021 | 0.0191 | 0.0189 | 0.0021 | 0.9614 | 0.9784 |
| Check | ✓ | ✓ | ✓ | ✓ | ✓ | ✓ |
| Confidence | | 50% | | | 60% | |

[a] IDR = maximum interstory drift ratio over the height of the building.

*Table 19.57* Verification of wall plastic hinge rotations (in terms of DCR) and system ductility (in terms of roof drift) for Example 4

|  | Wall plastic hinge rotation | | | System ductility | | |
|---|---|---|---|---|---|---|
|  | IDA (year⁻¹) | Single (rad/rad) | Double (rad/rad) | IDA (year⁻¹) | Single (m/m) | Double (m/m) |
| Demand | 0.0007 | 0.5255 | 0.5104 | 0.0005 | 0.0097 | 0.0097 |
| Capacity | 0.0021 | 0.9546 | 0.9762 | 0.0021 | 0.0165 | 0.0164 |
| Check | ✓ | ✓ | ✓ | ✓ | ✓ | ✓ |
| Confidence |  | 60% |  |  | 50% |  |

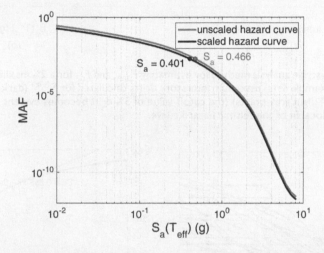

*Figure 19.68* Seismic site hazard curve properly scaled to 10/50 years for $S_a(T_{eff})$, Example 4.

Interstory drift values were calculated for 17 records for $S_a^{design} = 0.40$ g (dark gray points) and for $1.1 S_a^{design} = 0.44$ g (light gray points) (Figure 19.69a). The points of the hazard curve needed to estimate the slope $k$ are presented in Figure 19.69b. FD and FC at 50% confidence are estimated as:

$$FD_{RPo} = \theta_{max,50} \exp\left(0.5 \cdot k \cdot \beta_{\theta max|Sa}^2 / b\right) = 0.012 \exp\left(0.5 \cdot 2.34 \cdot 0.08^2 / 0.80\right) = 0.0119$$

$$FC_R = \theta_{max,C} \exp\left(-0.5 \cdot k \cdot \beta_{CR}^2 / b\right) = 0.02 \exp\left(-0.5 \cdot 2.34 \cdot 0.20^2 / 0.80\right) = 0.0189$$

Since the FC of 0.0189 is greater than the FD of 0.0119, the result is satisfactory at the 50% confidence level.

The coupling beam plastic rotation DCR values were also calculated at $S_a^{design} = 0.40$ g (dark gray points) and for $1.1 S_a^{design} = 0.44$ g (light gray points)—see Figure 19.70a.

It is noted here that in the two stripes, there are cases (i.e., records) where the coupling beams yield and others where they do not; thus, we often encounter zero (or near-zero) plastic rotations mixed together with nonzero ones. This does not abide with the lognormal assumption for demand that is required for utilizing the FC and FD approach for assessment. Actually, this may easily cause a gross overestimation of demand dispersion, often making it appear as larger than 100% when typical values of dispersion are less than 60%. This can cause the check to fail, even though it should actually pass quite easily. This can

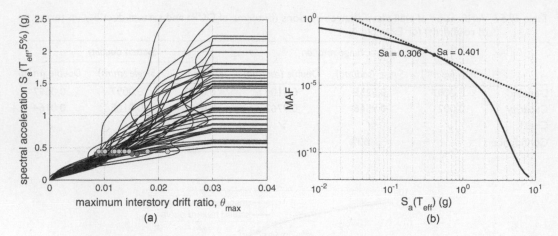

*Figure 19.69* Double-stripe analysis method for estimating $F_{DRPo}$ and $F_{CR}$ for a 2% maximum interstory drift ratio, Example 4: (a) maximum interstory drifts calculated for $S_a^{design}$ (dark gray points) and for $1.1S_a^{design}$ (light gray points) (the cutoff value of 3% drift becomes evident by the flatlines) and (b) the local fit of the seismic hazard curve.

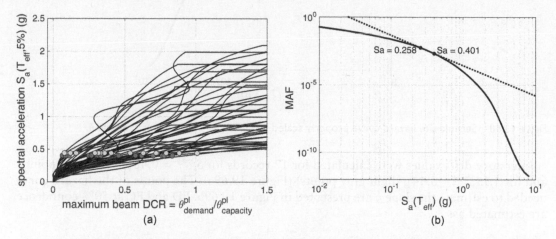

*Figure 19.70* Double-stripe analysis method for estimating $F_{DRPo}$ and $F_{CR}$ for beam plastic hinge rotation DCR = 1, Example 4: (a) maximum beam DCR calculated for $S_a^{design}$ (dark gray points) and for $1.1S_a^{design}$ (light gray points) and (b) the local fit of the seismic hazard curve.

be resolved by simply discarding such low values to perform a slightly more conservative assessment that is actually a lot more accurate under a lognormal assumption. In our case, all records with DCR < 0.03 were excluded from the determination of the dispersion, which still resulted in a high, yet manageable estimate of $\beta_{DCRmax} = 0.88$. A more aggressive cutoff could also have been implemented to achieve a more reliable (lower) dispersion estimate, but as we shall see in the following, this is not necessary for the case examined.

The points of the hazard curve needed to estimate the slope $k$ are presented in Figure 19.70b. FD and FC are estimated as:

$$FD_{RPo} = DCR_{max,50} \exp\left(0.5 \cdot k \cdot \beta_{DCRmax|Sa}{}^2/b\right) = 0.333 \exp\left(0.5 \cdot 2.23 \cdot 0.88^2/2.05\right) = 0.5077$$

$$FC_R = DCR_{max,C} \exp\left(-0.5 \cdot k \cdot \beta_{CR}{}^2/b\right) = 1.0 \exp\left(-0.5 \cdot 2.23 \cdot 0.20^2/2.05\right) = 0.9784$$

For a confidence level of 60%, the lognormal standard variate is $K_x = 0.253$ and the evaluation inequality becomes:

$$FC_R > FD_{RPo} \exp(K_x \cdot \beta_{TU}) = 0.5077 \exp(0.253 \cdot 0.20) = 0.5341$$

Since the FC of 0.9784 is greater than the FD of 0.5341, the result is satisfactory at the 60% confidence level.

Column DCR values were calculated for 17 records for $S_a^{design} = 0.40\,g$ (dark gray points) and for $1.1 S_a^{design} = 0.44\,g$ (light gray points)—see Figure 19.71a. The points of the hazard curve needed to estimate the slope $k$ are presented in Figure 19.71b. FD and FC are estimated as:

$$FD_{RPo} = DCR_{max,50} \exp\left(0.5 \cdot k \cdot \beta_{DCRmax|Sa}{}^2/b\right) = 0.470 \exp\left(0.5 \cdot 2.34 \cdot 0.23^2/1.94\right) = 0.4852$$

$$FC_R = DCR_{max,C} \exp\left(-0.5 \cdot k \cdot \beta_{CR}{}^2/b\right) = 1.0 \exp\left(-0.5 \cdot 2.34 \cdot 0.20^2/1.94\right) = 0.9762$$

For a confidence level of 60%, the lognormal standard variate is $K_x = 0.253$ and the evaluation inequality becomes:

$$FC_R > FD_{RPo}\exp(K_x \cdot \beta_{TU}) = 0.4852 \exp(0.253 \cdot 0.20) = 0.5104$$

Since the FC of 0.9762 is greater than the FD of 0.5104 the result is satisfactory at 60% confidence level.

To verify the limit on system ductility, $\theta_{roof}$ values were calculated for 17 records for $S_a^{design} = 0.40\,g$ (dark gray points) and for $1.1 S_a^{design} = 0.44\,g$ (light gray points)—see Figure 19.72a. The points of the hazard curve that define the slope $k$ are presented in Figure 19.72b. FD and FC at 50% confidence ($K_x = 0$) are estimated as:

$$FD_{RPo} = \theta_{roof,50} \exp\left(0.5 \cdot k \cdot \beta_{\theta roof|Sa}{}^2/b\right) = 0.009 \exp\left(0.5 \cdot 2.32 \cdot 0.20^2/0.89\right) = 0.0097$$

$$FC_R = \theta_{roof,C} \exp\left(-0.5 \cdot k \cdot \beta_{CR}{}^2/b\right) = 0.0225 \exp\left(-0.5 \cdot 2.32 \cdot 0.20^2/0.89\right) = 0.0164$$

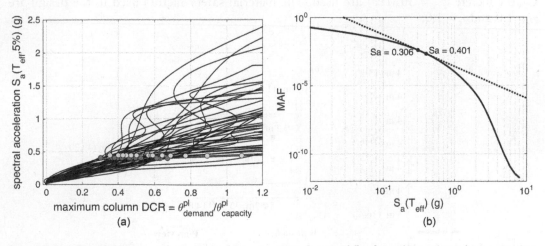

*Figure 19.71* Double-stripe analysis method for estimating $F_{DRPo}$ and $F_{CR}$ for column plastic hinge rotation DCR = 1: (a) maximum column DCR calculated for $S_a^{design}$ (dark gray points) and for $1.1 S_a^{design}$ (light gray points) and (b) the local fit of the seismic hazard curve.

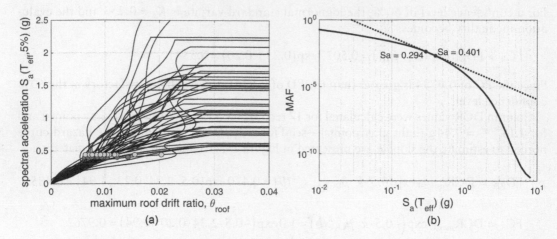

Figure 19.72 Double-stripe analysis method for estimating $F_{DRPo}$ and $F_{CR}$ for a ductility limit of 3.3, Example 4: (a) roof drift ratio calculated for $S_a^{design}$ (dark gray points) and for $1.1S_a^{design}$ (light gray points) and (b) the local fit of the seismic hazard curve.

Since the FC of 0.0164 is greater than the FD of 0.0097 the result is satisfactory at the 50% confidence level.

## 19.11 EXAMPLE 5: CANTILEVER SHEAR WALL DESIGNED USING METHOD B

### 19.11.1 Cantilever wall example plan and elevation

The wall is part of the perimeter of a seven-story RC structure (Figure 19.73). The height of the first story is 5 m while the overlying stories are 4 m high, resulting in a total height of 29 m. The building has eight walls of equal dimensions. The rectangular section walls have a plan length of 5.3 m and thickness of 0.3 m. B500 steel reinforcement ($f_{yk} = 500\,\text{MPa}$) and C-30 concrete ($f_{ck} = 30\,\text{MPa}$) are used. The material safety factors used in the design are $\gamma_s = 1.0$ and $\gamma_c = 1.0$.

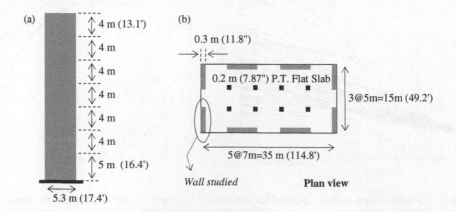

Figure 19.73 Seven-story building: (a) elevation and (b) plan.

## 19.11.2  POs

The seismic design POs are as follows:

1. the MAF of exceeding an interstory drift ratio of 0.02 is limited to 10/50 years (for $p_o = 2.11 \times 10^{-3}\text{year}^{-1}$), and
2. the MAF of plastic hinge rotation demand exceeding plastic hinge rotation capacity at the base of the wall is limited to 10/50 years ($p_o = 2.11 \times 10^{-3}\text{year}^{-1}$).

We assign a confidence level of 70% to the interstory drift limit and a confidence level of 85% to the plastic hinge rotation at the base of the wall. A single-stripe analysis at the 10/50 hazard level is used to evaluate the acceptability of the design.

## 19.11.3  Use of nonlinear response analysis in this example

A first-mode nonlinear static pushover analysis is used to evaluate the performance of the initial design realization. NLRHAs consisting of a single stripe (at $S_a(T_1)$ corresponding to the 10/50 hazard level) are used to evaluate the acceptability of plastic hinge rotations and peak interstory drifts.

## 19.11.4  System ductility limit

Design Step 1. Estimated roof displacement at yield:
According to expression 9.10, an estimate of the yield displacement of the roof is given by

$$\frac{D_y}{H} = \frac{H}{700 \cdot l_w} \rightarrow D_y = 0.23 \text{ m}$$

Design Step 2. Considering first the interstory drift limit, from Table 12.8 we estimate $\alpha_{3,\text{COD}} = 1.55$. Therefore, the roof displacement should not exceed approximately

$$D_{u,\text{drift}} = \frac{0.02}{1.55} 29 \text{ m} = 0.374 \text{ m}$$

at an MAF of exceedance of $p_o = 2.11 \times 10^{-3}\text{year}^{-1}$, with 70% confidence.
The associated ductility limit is $\mu = 0.374/0.23 = 1.63$ (to be achieved at a confidence level of 70%).
Design Step 3. To satisfy the plastic hinge rotation PO, we estimate a roof drift limit based on a plastic hinge rotation that is reduced from the expected plastic hinge rotation capacity to account for uncertainty. Table 17.4 provides a plastic hinge rotation capacity of 0.017 radians for the Life Safety performance level. We will estimate the roof displacement corresponding to a rotation of 0.01 radians. Allowing for a plastic hinge length equal to $l_w/2 = 5.3$ m/2 = 2.65 m, we estimate the roof displacement associated with a plastic hinge rotation of 0.01 radians as

$$D_{u,\text{ductility}} = D_y + \theta_p \left( h_n - \frac{l_p}{2} \right) = 0.24 \text{ m} + 0.01 \left( 29 \text{ m} - \frac{2.65 \text{ m}}{2} \right) = 0.517 \text{ m}$$

The associated system ductility limit is given as $\mu = 0.517/0.23 = 2.25$ (to be achieved at a confidence level of 85%).

### 19.11.5 Assumptions required to generate YFS based on EC-8 UHS

The YFS for this site was generated using the EC-8 spectrum described in Section 19.4. To generate the YFS, we assumed the normalized capacity curve of Figure 19.74.

### 19.11.6 Required base shear strength

Design Step 4: From Table 12.1, $\Gamma_1$ is estimated as 1.5. Therefore, $D_y^* = 0.23\,\mathrm{m}/1.5 = 0.15\,\mathrm{m}$. YFS for a yield displacement of $D_y^* = 0.15\,\mathrm{m}$ are plotted in Figures 19.75 and 19.76 for confidence levels of 70% and 85%, respectively. The results are summarized in Table 19.58. The highest required $C_y^* = 0.278$ (which also corresponds to the shorter of the two periods) is for the interstory drift limit, which therefore controls the design. The associated period is $T^* = 2\pi(D_y^*/C_y^* g)^{0.5} = 1.47\,\mathrm{s}$.

Design Step 5: The design base shear coefficient is $\alpha_1 C_y^* = 0.65(0.278) = 0.181$ (where $\alpha_1$ is estimated to be equal to 0.65 from Table 12.1). The tributary weight of the building in the direction where the wall is acting is 8,744 kN. Thus, the design base shear at yield is given by $V_y = C_y W = 0.181(8,744) = 1,583\,\mathrm{kN}$.

### 19.11.7 Design lateral forces and required member strengths

Design Step 6: The base shear is distributed over the height of the structure following the inverted triangular pattern of EC8 (CEN, 2005). Because the resultant is at a height of 20.76 m, while the resultant of the first-mode forces is estimated to be at a height of $0.77h_n$ (=22.33 m), the base shear used with the inverted triangular distribution is reduced to $1,583(20.76/22.33) = 1,471\,\mathrm{kN}$.

### 19.11.8 Sizing of RC members

Design Step 7: The longitudinal reinforcement at the base of the wall is calculated for an axial force $N = 1,172\,\mathrm{kN}$ (based on $D + 0.25L$) and a flexural moment $M = 1,471\cdot20.76 = 30,546\,\mathrm{kN\cdot m}$ (Figure 19.77). The shear reinforcement is calculated for 1,471 kN at the base.

The wall was proportioned using expected material properties: $f_{cm} = 38\,\mathrm{MPa}$ and $f_{ym} = 575\,\mathrm{MPa}$. Material safety factors were assumed equal to 1.0, since performance will be validated by nonlinear dynamic analysis.

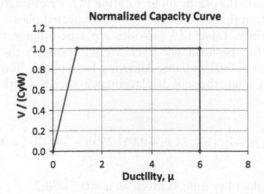

*Figure 19.74* Normalized capacity curve used for generating the YFS, Example 5.

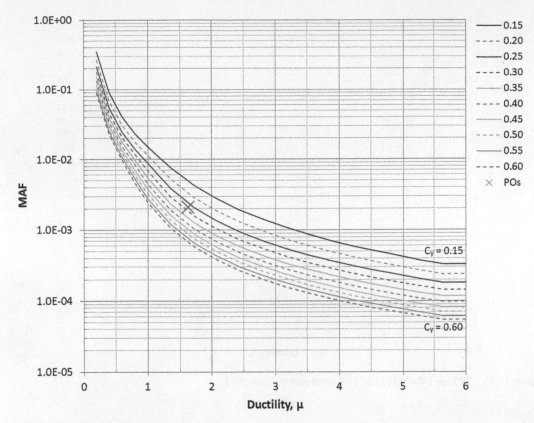

*Figure 19.75*   YFS for $D_y^* = 0.15$ m at 70% confidence, Example 5.

## 19.11.9  Preliminary evaluation of the initial design by nonlinear static (pushover) analysis

The preliminary design was modeled in SeismoStruct (Seismosoft, 2014) as described in Example 1. An eigenvalue analysis of the uncracked section model (which assumes concrete has tensile strength) determined $T_1 = 0.58$ s, $\Gamma_1 = 1.43$ and $\alpha_1 = 0.61$. This compares reasonably well with the assumed values of $\Gamma_1 = 1.5$ and $\alpha_1 = 0.65$.

A first-mode nonlinear static analysis was conducted, with forces proportional to the product of the mass and first-mode amplitude at each level, to obtain the capacity curve of Figure 19.78.

Based on the eigenvalue analysis, the period associated with the secant stiffness representative of cracked section behavior can be estimated as

$$T_{sec} = T_{initial}\sqrt{\frac{k_{initial}}{k_{sec}}} = 0.89 \text{ s}$$

which is a quite low compared with $T^* = 1.47$ s. The apparent strength in a first-mode pushover analysis, 1,910 kips, is significantly higher than the design base shear strength at yield, 1,471 kips. The yield displacement of approximately 0.186 m is lower than the estimated yield displacement of 0.23 m. Since the lower period will result in a smaller spectral

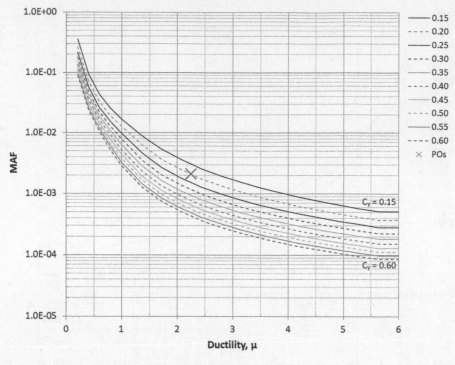

*Figure 19.76*　YFS for $D_y^* = 0.15$ m at 85% confidence, Example 5.

*Table 19.58* Performance criteria to be achieved at 10/50

| Performance criteria | Ductility limit | Confidence level | $C_y^*$ |
|---|---|---|---|
| Drift | 1.63 | 70% | 0.278 |
| Plastic hinge rotation | 2.25 | 85% | 0.198 |

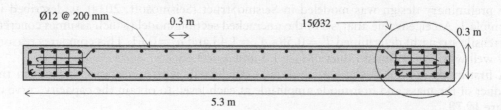

*Figure 19.77*　Reinforcement at the base of the cantilever shear wall, Example 5.

displacement, it is likely that the drift-controlled design will be adequate. The values of $D_y^*$, $\Gamma_1$, and $\alpha_1$ obtained in the analysis of the first realization could be used to refine the preliminary design.

## 19.11.10 Nonlinear modeling and acceptance criteria

A two-dimensional model of the structure was prepared, as shown in Figure 19.79. Only one of the four cantilever walls in the direction of interest was modeled. A leaning column

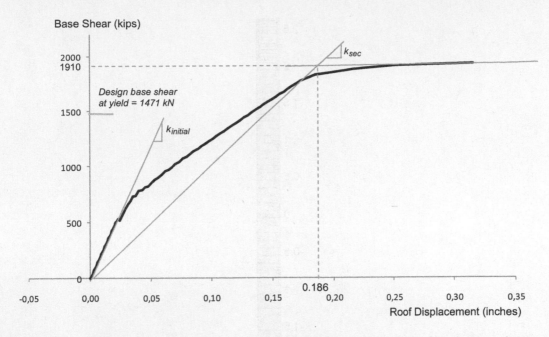

Figure 19.78   Capacity curve determined by first-mode pushover analysis of the first design realization, Example 5.

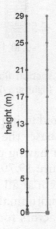

Figure 19.79   Two-dimensional structural model used for the analysis showing the well-discretized shear wall (left) and the leaning column (right), Example 5.

was added to simulate the effect of the columns that carry the gravity loads. The leaning column was pinned at the foundation and modeled using linear elastic elements having area and moment of inertia that match the corresponding cross-sectional properties of one-quarter of the gravity columns of the building plus one cantilever shear wall oriented along the other direction bending around its weak axis. A beam-column element was placed at the centerline of the wall. The leaning column was constrained to the nodes of the coupled wall at each floor.

The wall was modeled using a single displacement-based distributed plasticity element for each story, with fiber sections monitored at five integration points along the element length. Following a similar strategy as in the coupled-wall example (Section 19.10.8)

*Figure 19.80* Cantilever wall modeling: typical section fiber discretization showing rebars (black), semi-confined (dark gray) and confined (light gray) concrete fibers, Example 5.

displacement-based elements were preferred to improve convergence. Again, since all plastic rotations are expected to appear at the base, the first-story wall was discretized into five displacement-based elements per member having progressively diminishing length with the height from the ground, as shown in Figure 19.79. A comparison of the model to its force-based analogue showed excellent agreement at all levels of inelasticity.

The wall section is discretized into longitudinal steel and concrete fibers (Figure 19.80), the latter having different confinement factors for the cover, the well-confined core at the edges (light gray) and the semi-confined core of the web (dark gray). The effect of confinement was calculated on the basis of the Mander et al. (1988) model with a resulting confinement ratio of 1.5 for the edge core concrete (light gray) at the base of the wall, assuming 10-mm diameter transverse reinforcement, placed as indicated in Figure 19.77 every 10 cm along the height in the region of the expected plastic hinge. The confinement ratio of the web core concrete (dark gray) was estimated at 1.13. Furthermore, steel reinforcing bars (black color in Figure 19.80) were modeled using a bilinear constitutive law accounting for pinching and stiffness degradation. The expected material properties were used for the analysis, i.e., $f_{ce}$ = 38 MPa and $f_{ye}$ = 575 MPa.

Similar to the coupled wall example, a number of different assumptions were tested for the post-capping behavior of each material. To ease convergence, generous assumptions were made, resulting in the static pushover capacity curve of Figure 19.81 (for a first-mode-proportional lateral load pattern). To counter the effect of the relaxed material deterioration

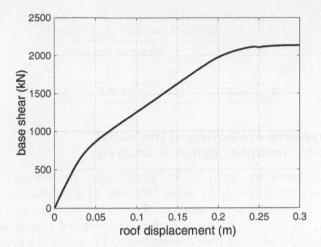

*Figure 19.81*    Static pushover curve, Example 5.

beyond a roof displacement of 0.3 m, global collapse was enforced at 3.5% drift, essentially assuming a vertical drop to zero strength, rather than an extended plateau beyond this deformation.

The initial (uncracked) period of the model is 0.83 s. A more useful effective period of the structure is calculated using the secant stiffness at 60% of the maximum base shear $V_{max}$ (Figure 19.82)

$$T_{sec} = T_1 \sqrt{\frac{k_{init}}{k_{eff}}} = 1.12 \text{ s}$$

This is stiffer than the value of 1.47 s assumed in design. The latter better matches the secant period at the point of maximum strength (100% $V_{max}$), equal to 1.46 s by reapplying the above calculation. Still, the stiffer period is a better representation of the MDOF system— thus in the following, the first-mode period will be taken as $T_1 = T_{sec} = 1.12$ s.

In Table 19.59, the acceptance criteria for wall plastic rotation are presented.

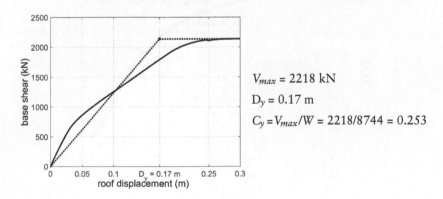

$V_{max}$ = 2218 kN

$D_y$ = 0.17 m

$C_y = V_{max}/W = 2218/8744 = 0.253$

*Figure 19.82*    Determination of nominal $D_y$ by fitting the elastic segment at the 60% $V_{max}$ point of the push-over, Example 5.

Table 19.59  Wall plastic hinge rotation limits for IO, LS, and CP POs, Example 5

| | Positive rotation (rad) | | |
|---|---|---|---|
| | IO | LS | CP |
| $\theta_{p,max}$ (rad) | 0.005 | 0.017 | 0.020 |

## 19.11.11 Performance evaluation of the initial design by nonlinear dynamic analysis

The acceptability of interstory drift ratio and plastic hinge rotation at the base of the wall was assessed versus a 10/50-year MAF at the 70% and 85% confidence levels, respectively, using IDA, single- and double-stripe analysis. Results are summarized in Table 19.60 for each PO. For IDA, the first two values of each PO correspond to estimated and allowable MAF, and for the stripe analyses they correspond to FD and FC (expressed as DCRs for plastic hinge rotations and interstory drift ratios, respectively).

According to Method B, a single-stripe analysis is used for the assessment. The seismic hazard curve scaled to $S_a(T_{eff})$ at 10/50 years, is shown in Figure 19.83.

Interstory drift values were calculated for 17 records for $S_a^{design} = 0.54$ g, as shown in Figure 19.84a. The points of the hazard curve needed to estimate the slope $k$ are presented in Figure 19.84b. FD and FC are estimated as:

Table 19.60  Verification of the maximum interstory drift and wall plastic hinge rotations (in terms of DCR) for Example 5

| | IDR[a] | | | Wall plastic hinge rotation | | |
|---|---|---|---|---|---|---|
| | IDA (year$^{-1}$) | Single (m/m) | Double (m/m) | IDA (year$^{-1}$) | Single (rad/rad) | Double (rad/rad) |
| Demand | 0.0008 | 0.0147 | 0.0147 | 0.0008 | 0.5116 | 0.4995 |
| Capacity | 0.0021 | 0.0190 | 0.0190 | 0.0021 | 0.9520 | 0.9627 |
| Check | ✓ | ✓ | ✓ | ✓ | ✓ | ✓ |
| Confidence | | 70% | | | 85% | |

[a]  IDR = maximum interstory drift ratio over the height of the building.

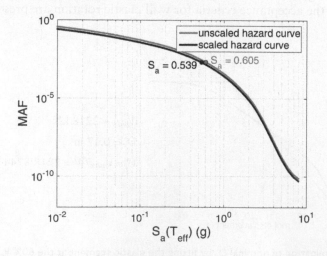

Figure 19.83   Seismic site hazard curve properly scaled to 10/50 years for $S_a(T_{eff})$, Example 5.

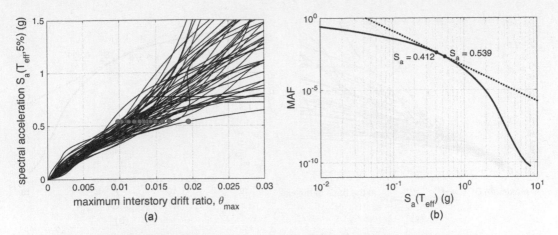

*Figure 19.84* Single-stripe analysis method for estimating $F_{DRPo}$ and $F_{CR}$ for a 2% maximum interstory drift ratio, Example 5: (a) maximum interstory drifts calculated for $S_a^{design}$ and (b) the local fit of the seismic hazard curve.

$$FD_{RPo} = \theta_{max,50} \exp\left(0.5 \cdot k \cdot \beta_{\theta max|Sa}{}^2/b\right) = 0.0128 \exp\left(0.5 \cdot 2.44 \cdot 0.18^2/1\right) = 0.0133$$

$$FC_{R} = \theta_{max,C} \exp\left(-0.5 \cdot k \cdot \beta_{CR}{}^2/b\right) = 0.02 \exp\left(-0.5 \cdot 2.44 \cdot 0.20^2/1\right) = 0.0190$$

For a confidence level of 70%, the lognormal standard variate is $K_x = 0.524$ and the evaluation inequality becomes:

$$FC_{R} > FD_{RPo} \exp\left(K_x \cdot \beta_{TU}\right) = 0.0133 \exp(0.524 \cdot 0.20) = 0.0147$$

Since the FC of 0.0190 is greater than the FD of 0.0147 the result is satisfactory at the 70% confidence level.

DCR values for the base of the wall were calculated for 17 records for $S_a^{design} = 0.53\,g$, as shown in Figure 19.85a. The points of the hazard curve that define the slope $k$ are presented in Figure 19.85b. FD and FC are estimated as:

$$FD_{RPo} = DCR_{max,50} \exp\left(0.5 \cdot k \cdot \beta_{\theta max|Sa}{}^2/b\right) = 0.3742 \exp\left(0.5 \cdot 2.46 \cdot 0.30^2/1.0\right) = 0.4159$$

$$FC_{R} = DCR_{max,C} \exp\left(-0.5 \cdot k \cdot \beta_{CR}{}^2/b\right) = 1.0 \exp\left(-0.5 \cdot 2.46 \cdot 0.20^2/1.0\right) = 0.9520$$

For a confidence level of 85%, the lognormal standard variate is $K_x = 1.036$ and the evaluation inequality becomes:

$$FC_{R} > FD_{RPo} \exp\left(K_x \cdot \beta_{TU}\right) = 0.4159 \exp(1.036 \cdot 0.20) = 0.5116$$

Since the FC of 0.9520 is higher than the FD of 0.5116 the result is satisfactory at the 85% confidence level.

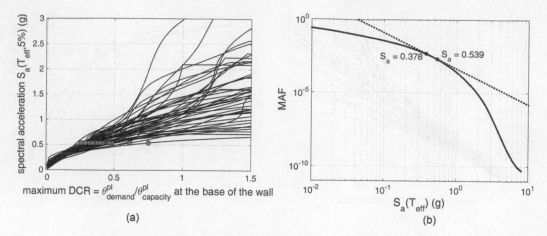

(a)

(b)

*Figure 19.85* Single-stripe analysis method for estimating $F_{DRPo}$ and $F_{CR}$ for a DCR = 1 of the plastic hinge rotation at the base of the wall, Example 5: (a) maximum DCR at the base of the wall calculated for $S_a^{design}$, and (b) the local fit of the seismic hazard curve.

## 19.12 EXAMPLE 6: UNBONDED POST-TENSIONED WALL DESIGNED USING METHOD C

### 19.12.1 Floor plan and elevation

An unbonded post-tensioned (UPT) wall provides lateral resistance for a four-story RC structure. The height of the first story is 5 m while the overlying stories are 4 m high, resulting in a total height of 17 m. The building has two post-tensioned walls in each direction, as shown in Figure 19.86. The walls are designed according to Eurocode 2 (CEN, 2004) and Eurocode 8 (CEN, 2005). Each wall section has a plan length of 6.0 m and thickness of 0.5 m. B500 steel reinforcement ($f_{yk}$ = 500 MPa) and C30 concrete ($f_{ck}$ = 30 MPa) are used. The material safety factors used in the design are $\gamma_s$ = 1.0 and $\gamma_c$ = 1.0.

A preliminary wall cross section was selected, as shown in Figure 19.87. The transverse hoops in the boundary zones at the ends of the wall section are 12-mm diameter, at a vertical spacing of 100 mm. According to the Mander et al. (1988) model, they provide confinement allowing the confined core concrete to develop a compressive strength ($f_{cc}$) of 47.85 MPa (based on characteristic values) at strain of $\varepsilon_{cc}$ = 0.0079 and with an ultimate strain of $\varepsilon_{cu}$ = 0.044. The wall contains five bundled tendons. Each tendon consists of some

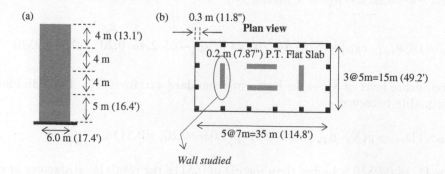

*Figure 19.86* Four-story RC building (a) elevation of the UPT RC wall and (b) plan of the building.

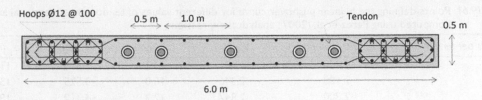

Hoops Ø12 @ 100    0.5 m  1.0 m    Tendon    0.5 m

6.0 m

*Figure 19.87* Initial configuration of the UPT wall.

number of individually greased and sheathed seven-wire strands; the precise quantity to be determined based on the seismic POs and site seismicity.

## 19.12.2 POs

To illustrate the design for multiple POs, we select three POs (somewhat arbitrarily).

1. Gap opening at the base of the wall equal to 2 cm, having an MAF of exceedance of 50/50 years ($p = 1.39 \times 10^{-2} \text{year}^{-1}$) with 50% confidence. This is equivalent to a roof displacement of 20 mm·17 m/6 m = 57 mm. This gap represents the maximum separation that will appear for this LS at one edge of the wall as it rotates around its distal edge. It has been selected as indicative of minor damage due to opening along a joint that has been designed for this purpose.
2. An interstory drift limit of 0.7% of the story height, having MAF of exceedance 30/50 years ($p = 0.71 \times 10^{-2} \text{year}^{-1}$) with 50% confidence.
3. The limit of response associated with crushing of the confined core (CCC behavior in Figure 15.35). We consider this simply as a point of incipient collapse, although the true sequence of events heralding collapse may be much more complicated. We assign an MAF of 1/50 years ($p = 2.01 \times 10^{-4} \text{year}^{-1}$) to this LS, with 90% confidence.

## 19.12.3 Use of nonlinear response analysis in this example

A first-mode nonlinear static pushover analysis is used to evaluate the performance of the initial design realization. All methods described in Chapter 13 are employed to assess the performance of the structure: single- and double-stripe analysis and IDA are used to evaluate the acceptability of the gap opening at the base of the wall at the 50/50 hazard level and of interstory drift ratio at the 30/50 hazard level; while only IDA is used to evaluate crushing of confined concrete at the 1/50 hazard level.

## 19.12.4 Effect of quantity of seven-wire strands on wall behavior

The walls may be modeled as described in Chapter 15. The quantity of seven-wire strand to provide for each tendon of Figure 19.87 will be determined to satisfy the seismic POs. Following the approach of Perez et al. (2007), $V_{\text{ell}}$, $\Delta_{\text{ell}}$, $V_{\text{llp}}$, and $\Delta_{\text{y}}$ are computed for different values of the tendon cross-sectional area, shown in Table 19.61 for 10, 15, and 20 strands per tendon. The corresponding pushover capacity curves are shown in Figure 19.88.

In all cases, the PT steel is prestressed to $0.55f_{\text{pu}}$, where $f_{\text{pu}}$ is the nominal ultimate strength of the prestressing steel, which is equal to 1,860 MPa. It can be seen that although the strength is controlled by the PT steel, the stiffness is determined primarily by the geometry of the wall and elastic modulus of concrete, and thus the stiffness of the curve in the linear range varies little. This fact suggests that the seismic design of this type of lateral system

*Table 19.61* Points defining the bilinear pushover curve for different values of tendon cross-sectional area computed using Perez et al. (2007) analytical expressions

| Strands per tendon | Area of tendon (mm²) | $V_{ell}$ (kN) | $\Delta_{ell}$ (mm) | $V_{llp}$ (kN) | $\Delta_y$ (mm) |
|---|---|---|---|---|---|
| 10 | 1,400 | 1,681 | 7.2 | 2,615 | 11.2 |
| 15 | 2,100 | 2,346 | 10.0 | 3,525 | 15.1 |
| 20 | 2,800 | 2,847 | 12.2 | 4,319 | 18.5 |

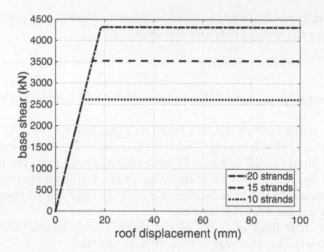

*Figure 19.88* Influence of area of post-tensioning reinforcement on the stiffness and strength of an UPT wall (using a simplified bilinear curve to characterize the response).

will converge rapidly if based on an assumed period, for a wall of a given section. Of course, the cross section may be revised if the structure is too flexible or overly stiff. Using a fiber model, with the commercial software SeismoStruct (Seismosoft, 2014), a period of 0.31 s was obtained. The Perez et al. (2007) model provided a similar estimate of the period 0.28 s. The latter was employed in the YFS design.

## 19.12.5 Design approach

The YFS for a constant period is used for the preliminary design of the structure. The normalized capacity curve needed for the YFS can be easily estimated having as a basis the pushover capacity curve computed using the Perez et al. (2007) analytical expressions. Assuming initially that 13 strands per tendon are needed, the computed pushover capacity curve using characteristic material properties is shown in Figure 19.89 where $\Delta_{ell}$ = 8.9 mm, $V_{ell}$ = 2,080 kN, $\Delta_y$ = 13.6 mm, $V_{llp}$ = 3,176 kN, $\Delta_{llp}$ = 362 mm, and $\Delta_{ccc}$ = 727 mm. Following the spirit of the suggestions by Perez et al. (2007), a reduction factor of $\psi$ = 0.9 was used to account for the differences between test results and the estimate of Mander et al. model for the confined concrete crushing (CCC) strain, reducing the estimated value of $\varepsilon_{cu}$ to 0.04.

The first two POs, namely $PO_1$ and $PO_2$, refer to low deformation values for which the post-tensioned wall is expected to behave as a nonlinear elastic oscillator as it rigidly rotates around its edges without appreciable material damage. For higher deformations, such as the one that $PO_3$ refers to, the toes of wall are expected to become damaged, resulting in a flag-shaped behavior (Chapter 4), exhibiting hysteresis and energy dissipation through material

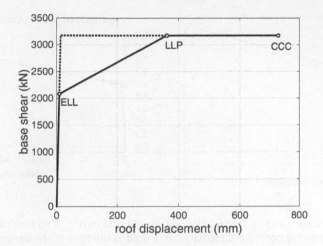

*Figure 19.89* Static pushover capacity curve estimated using Perez et al. analytical expressions and char-acteristicmaterialpropertiesfor13strandspertendon.Thiscurveisusedtogeneratethenormalized capacity curve needed for the YFS.

nonlinearity (see also the results of nonlinear analysis in Section 19.12.8). The SPO2IDA tool originally integrated in the YFS software (Vamvatsikos and Aschheim, 2016) does not account for the increase in displacement response expected for nonlinear elastic oscillators relative to those having fuller hysteretic loops. In our case, two different versions of the YFS-T tool, which employs a stable period assumption, will be used: for $PO_1$ and $PO_2$ we shall employ YFS-TNE[2] that incorporates the $R–\mu–T$ relationships of Bakalis et al. (2019) to capture the early nonlinear elastic response. For $PO_3$ where hysteresis is important, the conventional SPO2IDA-based YFS-T[3] software will be employed.

## 19.12.6  YFS based on an assumed normalized capacity curve

The normalized capacity curve used in the YFS for the first two POs is computed using the trilinear pushover curve of Figure 19.89 up to the LLP point, which corresponds to yielding of the PT steel. The ductility of the normalized curve ranges from 0 to $\mu_{LLP} = \Delta_{llp}/\Delta_{ell} = 362$ mm/8.9 mm = 40.75. Note that both the ductility and the strength of that curve are com-puted relative to the effective linear limit (ELL) point, thus the $C_y^*$ value estimated using the YFS-TNE will refer to $V_{ell}$ as well. The pushover capacity curve up to LLP point and the corresponding normalized capacity curve are shown in Figure 19.90a and b, respectively.

With regard to $PO_1$, the roof displacement of 57 mm is equivalent to a ductility limit of 57 mm/8.9 mm = 6.37 with respect to $\Delta_{ell}$. $PO_2$ is based on an interstory drift limit of 0.7% which is equivalent to a roof displacement limit of 0.007(17 m) = 0.119 m, assuming that $\theta_{max} = \theta_{roof}$, consistently with a rigid body rotation of the wall. This corresponds to a ductility of $\mu = 119$ mm/8.9 mm = 13.4 vis-à-vis $\Delta_{ell}$. The YFS-TNE for $T = 0.28$ s appear in Figure 19.91 for 50% confidence.

For the third PO, the normalized capacity curve required for YFS is computed using the bilinear approximation of the pushover curve of Figure 19.89 up to the CCC point. The strength and the ductility of the normalized capacity curve refer to the effective yield of the structure, with an ultimate ductility value of $\mu_{CCC} = \Delta_{ccc}/\Delta_y = 727$ mm/13.6 mm = 53.6

---

[2] http://users.ntua.gr/divamva/software/YFS-TNEapp.xls.
[3] http://users.ntua.gr/divamva/software/YFS-Tapp.xls.

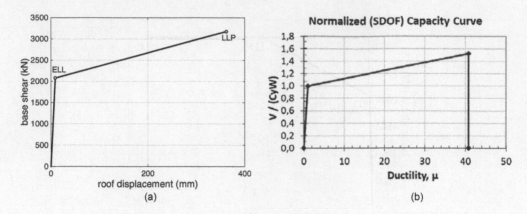

*Figure 19.90* (a) Pushover curve up to the LLP point for the case having 13 strands per tendon and (b) the corresponding normalized capacity curve used at the start of the design process for $PO_1$ and $PO_2$, for Example 6.

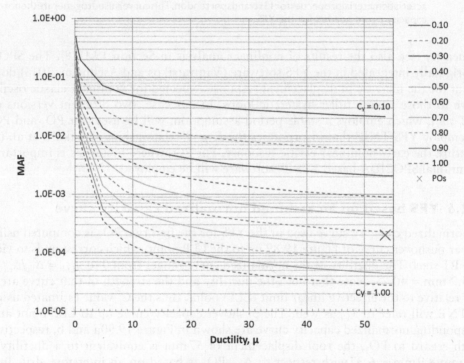

*Figure 19.91* YFS-TNE for $T = 0.28\,s$ at 50% confidence for the capacity curve of Figure 19.90, used to estimate $C_y$ for the first two POs, for Example 6.

at concrete crushing. The bilinear approximation of the pushover capacity curve is shown in Figure 19.92. The YFS-T generated for $T = 0.28\,s$ at 90% confidence appear in Figure 19.93.

## 19.12.7 Design strength

The $C_y^*$ values for the first two POs are estimated relative to $V_{ELL}$, thus the maximum $C_y^*$ value at 50% confidence, $\max\{0.381; 0.383\} = 0.383$, is equivalent to $C_y^* = 0.383\cdot V_{llp}/V_{ell} = 0.383\cdot3,176\,\text{kN}/2,080\,\text{kN} = 0.584$, vis-à-vis the effective yield point of the

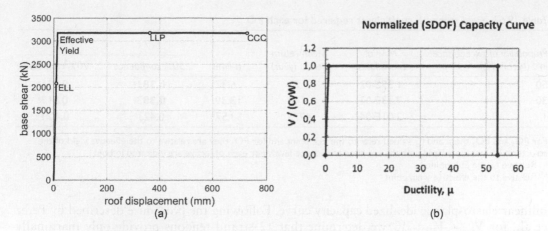

*Figure 19.92* (a) Bilinear approximation of the pushover curve up to the CCC point for the case of 13 strands per tendon and (b) corresponding normalized capacity curve used at in the YFS for PO₃ of Example 6.

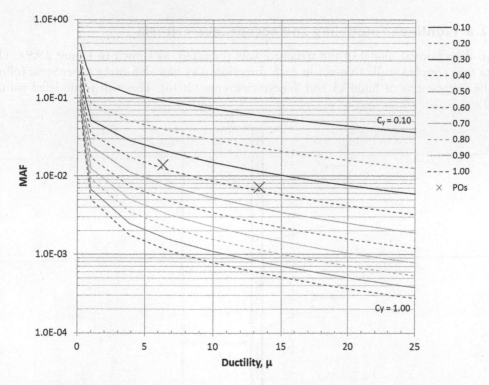

*Figure 19.93* YFS-T for $T = 0.28$ s at 90% confidence for the capacity curve of Figure 19.92, used to estimate $C_y^*$ for the third PO, for Example 6.

structure. The third PO resulted in $C_y^* = 0.621$, which is higher than 0.584 and thus controls the design (Table 19.62). Using Table 12.1 we estimate $\alpha_1 = 0.68$; thus, $C_y = 0.68(0.621) = 0.422$. With reference to Figure 19.92, the associated strength is given by $V_y = C_y W = 0.422(7,083\,\text{kN}) = 2,991\,\text{kN}$, which corresponds to the strength of the

*Table 19.62* Yield strength coefficients required for each PO

| Probability of exceedance in 50 years (%) | MAF of exceedance | Mean return period (year) | μ-limit | $C_y^*$ 50% confidence | 90% confidence |
|---|---|---|---|---|---|
| 50 | 1.39E-02 | 72 | 6.37[a] | **0.381**[a] | 0.491[a] |
| 30 | 7.13E-03 | 140 | 13.39[a] | **0.383**[a] | 0.519[a] |
| 1 | 2.01E-04 | 4975 | 53.57[b] | 0.423[b] | **0.621**[b] |

For $PO_1$ and $PO_2$ the $\mu$ and $C_y^*$ valued refer to the ELL point and for $PO_3$ they are relative to the effective yield of the structure. (Values associated with the specified confidence levels per each objective are indicated in bold.)
[a]  Relative to the ELL point.
[b]  Relative to the effective yield point.

bilinear elasto-plastic idealized capacity curve. Following the procedure described by Perez et al., for $V_{ccc} \approx V_{llp} = V_y$, we determine that 12-strand tendons provide only marginally higher strength than required ($V_{y,12strands} = 2,994\,kN$). Thus, 13-strand tendons are selected ($V_{y,13strands} = 3,176\,kN$). The pushover capacity curve estimated using the analytical expressions is the one presented in Figure 19.89 and is the one that was initially used to generate the YFS.

## 19.12.8 Nonlinear modeling and acceptance criteria

A two-dimensional model of the structure was prepared, as shown in Figure 19.94. One of the two coupled walls that act in each direction was modeled using OpenSees, following the suggestions of Buddika and Wijeyewickrema (2016). The wall is modeled using a

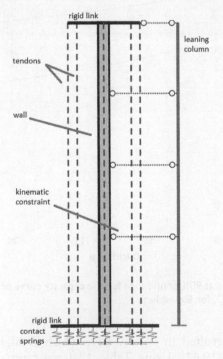

*Figure 19.94*  Two-dimensional structural model used for the analysis showing the wall (gray-black rectangle), the tendons (dashed), the leaning column, the contact springs, the rigid links, and the horizontal kinematic constraints (dotted).

single displacement-based distributed plasticity element for each story, with fiber section representation monitored at five integration points along the member length. Those elements were not further discretized or modeled using force-based elements, as the nonlinearity is concentrated at the rocking base. Rocking behavior at the base is modeled using no tension, concrete-like springs to connect the entire length of the wall to the foundation. Rigid beams were employed at the top and the bottom to define the outline of the wall. The tendons were modeled using tension-only corotational truss elements that were connected to the rigid top and to the foundation. A leaning column was added to simulate the effect of the columns that carry the gravity loads. The leaning column was pinned at the foundation and modeled using linear elastic elements having area and moment of inertia that match the corresponding cross-sectional properties of one half of the gravity columns of the building plus one post-tensioned wall oriented along the other direction bending around its weak axis, as only one half of the building was modeled in the direction of interest, due to symmetry. Rayleigh damping of 2% was assigned to the first and the second mode. Note that this value is lower than the typical value of 5% used for RC structures, but is considered realistic as cracking is directly incorporated in the concrete material models both at the base and in the wall itself, giving rise to hysteretic damping as soon as damage appears.

The wall section is discretized into longitudinal steel and concrete fibers (Figure 19.95), the latter having different confinement factors for the cover, the well-confined core at the edges (light gray) and the semi-confined core of the web (dark gray). The effect of confinement was calculated on the basis of the Mander et al. (1988) model with a resulting confinement ratio of 1.55 for the edge core concrete (light gray) based on the expected material properties and the detailing of Figure 19.87. The confinement ratio of the web

*Figure 19.95* Typical section fiber discretization used to model the wall of Example 6, showing rebars (black), semi-confined (dark gray) and confined (light gray) concrete fibers.

core concrete (dark gray) was estimated at 1.1. Furthermore, steel reinforcing bars (black color in Figure 19.95) were modeled using a bilinear constitutive law accounting for pinching and stiffness degradation. The expected material properties were used for the analysis, i.e., $f_{ce} = 38$ MPa and $f_{ye} = 575$ MPa.

The base contact springs were arranged as a single strip along the width of the wall, and further discretization was not required in the two-dimensional model. A force–displacement relationship was assigned to each spring that mimics the hysteresis of the corresponding concrete fibers in compression. To define the force, the concrete stress is multiplied by the relevant tributary area while the corresponding deformation is derived by the concrete strain multiplied by an effective height of twice the confined thickness of the wall, following the recommendations of Perez et al. (2007). The cross-sectional area of each tendon is equal to (140 mm²/strand)·(13 strands/tendon) = 1,820 mm²/tendon. A bilinear constitutive law was employed for the prestressing steel, with initial strain iteratively calculated to attain the initial pretension force of the tendons. The expected strand properties were used, thus $f_{pe} = f_{pk} + 66$ MPa = 1,926 MPa and $E_p = 195$ GPa, according to JCSS (2001).

The static pushover capacity curve resulting from a first-mode-proportional lateral load pattern is presented in Figure 19.96, along with the trilinear curve used for the design of the building. The two pushover curves may seem to match perfectly for this building, yet only under somewhat different initial assumptions: expected material properties were used for the former and characteristic material properties for the latter. The first two modes vibrate at $T_1 = 0.286$ s (fairly close to the initial estimate of 0.28 s) and 0.043 s. Figure 19.97 shows the loading-unloading pushover curve up to roof displacements of 57 mm, 119 mm, and 812 mm that correspond to $PO_1$, $PO_2$, and $PO_3$, respectively. Note that crushing of confined concrete at $PO_3$ corresponds to an interstory drift that is consistent with the loss of strength in the pushover analysis. It can be seen that for the first two POs, the system behaves as non-linear elastic and for the third one it develops hysteresis and becomes flag shaped, validating our use of two different hysteretic laws to develop the YFS-T.

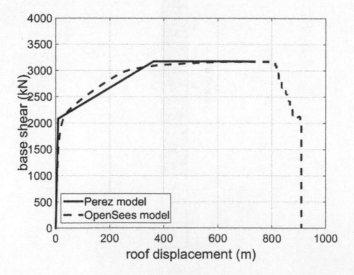

*Figure 19.96* Static pushover capacity curves for Example 6 in terms of roof displacement as created by an OpenSees model (using expected material properties) and by the Perez et al. (2007) analytical expressions (for characteristic material properties) for 13 strands per tendon.

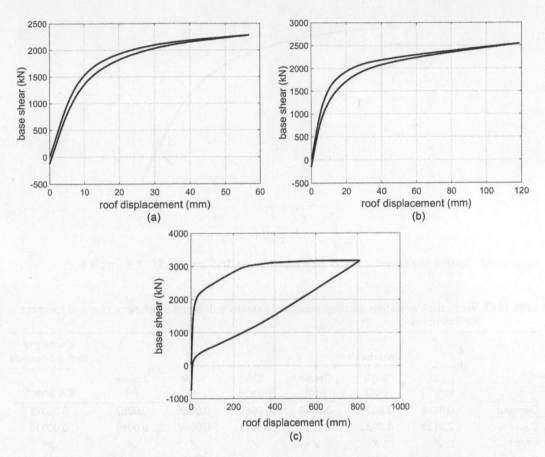

Figure 19.97   Loading–unloading pushover curves: (a) up to roof displacement of 57 mm for PO₁, (b) up to roof displacement of 119 mm for PO₂ and (c) up to 812 mm for PO₃, for Example 6.

## 19.12.9  Performance evaluation of the initial design by nonlinear response history analysis

All three NLRHA methods introduced in Chapter 13 were employed, i.e., IDA, and single- and double-stripe analyses. Results from IDA analysis are shown in detail. Note that in this case, since we directly use the site-specific hazard data in the YFS design, the assessment will be fair by default, and thus, scaling of the hazard curve is not warranted. Thus, the latter is employed directly as estimated from United States Geological Survey (USGS) (see Figure 19.98). Results for each PO are summarized in Table 19.63. For IDA, the first two values of each PO correspond to estimated and allowable MAFs, and for the stripe analyses they correspond to FD and FC, respectively. The system ductility is expressed in terms of roof drift ratio and crushing of confined concrete in terms of maximum interstory drift ratio.

The acceptability of a 2-cm gap opening at the base of the wall was assessed at the 50/50 level. This limit is equivalent to a roof displacement of 20 mm·17 m/6 m = 57 mm, or a roof drift limit of $\theta_{roof}$ = 57 mm/17,000 mm = 0.0033. IDA curves in terms of the first-mode spectral acceleration $S_a(T_1, 5\%)$ and roof drift ratio, $\theta_{roof}$ are presented in Figure 19.99a. The MAF of exceeding $\theta_{roof}$ = 0.0033 was estimated using the lognormal fragility curve that was fitted to the $S_a(T_1, 5\%)$ values that correspond to the roof drift limit (dark

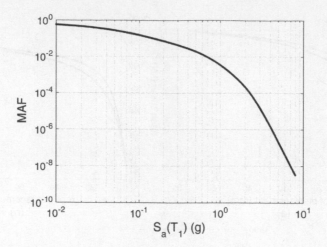

*Figure 19.98*   Seismic site hazard curve as determined from USGS data for $S_a(T_1)$, Example 6.

*Table 19.63* Verification of system ductility, maximum interstory drift and crushing of confined concrete for Example 6

|  | System ductility | | | IDR | | | Crushing of confined concrete |
|---|---|---|---|---|---|---|---|
|  | IDA (year⁻¹) | Single (rad) | Double (rad) | IDA (year⁻¹) | Single (−) | Double (−) | IDA (year⁻¹) |
| Demand | 0.0128 | 0.0033 | 0.0028 | 0.0062 | 0.0065 | 0.0052 | 0.00025 |
| Capacity | 0.0139 | 0.0032 | 0.0033 | 0.0071 | 0.0067 | 0.0069 | 0.00020 |
| Check | ✓ | ✗ | ✓ | ✓ | ✓ | ✓ | ✗ᵃ |
| Confidence |  | 50% |  |  | 50% |  | 90% |

ᵃ Initial estimate, revised later to account for spectral shape.

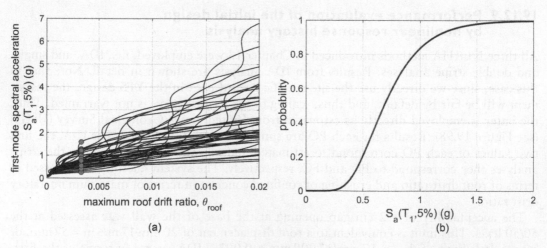

(a)

(b)

*Figure 19.99* Analytical approach for estimating MAF of exceeding maximum gap opening at the base of the wall equal to 2 cm, which is equivalent to roof drift ratio of $\theta_{roof} = 0.0033$, Example 6: (a) the $S_a(T_1, 5\%)$ values calculated for $\theta_{max} = 0.33\%$ and (b) corresponding fitted lognormal fragility curve.

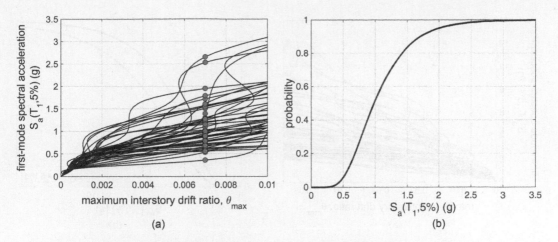

*Figure 19.100*   Analytical approach for estimating MAF of exceeding maximum interstory drift ratio of $\theta_{max} = 0.7\%$, Example 6: (a) the $S_a(T_1, 5\%)$ values calculated for $\theta_{max} = 0.7\%$ and (b) corresponding fitted lognormal fragility curve.

gray points in Figure 19.99a), and is presented in Figure 19.99b. The median value of the lognormal distribution equals 0.68 g and the log standard deviation is 38%. Following the approach of Section 13.4.2.1, the estimated MAF of exceeding $\theta_{roof} = 0.0033$ at 50% confidence is 0.0128, which is lower than the limit of 0.0139 (50/50 years). This means that the PO was met.

The MAF of exceeding the maximum interstory drift ratio of $\theta_{max} = 0.7\%$ was estimated using a lognormal fragility curve that was fitted to the $S_a(T_1, 5\%)$ values that correspond to this limit, according to the IDA results (Figure 19.100a). The median value of the lognormal distribution equals 0.98 g and the log standard deviation is 43% (Figure 19.100b). The estimated MAF of exceeding $\theta_{max} = 0.7\%$ at 50% confidence equals 0.0062, which is lower than the corresponding code value of 0.0071 (30% probability of exceeding PO in 50 years); thus, the PO is met. Note that according to Table 19.63, the double-stripe approach achieves the same conclusions as IDA, while the single-stripe approach marginally indicates an unsatisfactory design for $PO_1$. Of course, this is the simplest of the three approaches and it is particularly disadvantaged for this short-period system where the equal displacement rule does not apply. Given how close all the checks are for the other two methods, receiving such a marginally erroneous result still indicates relatively good predictive ability.

Crushing of confined concrete is deemed to occur when numerical non-convergence appears, or a maximum interstory drift of 0.054 is exceeded, whichever occurs first. The first criterion is *numerically* indicative of global dynamic instability (Sections 4.8.6 and 5.3.4), closely corresponding to the catastrophic loss of strength appearing at the end of the plateau in the pushover curve of Figure 19.96. The second criterion corresponds to an interstory drift that is *physically* consistent with the loss of strength in the pushover analysis, used to guard against any excessive deformations due to some low residual strength that may appear. The values of $S_a(T_1, 5\%)$ that correspond to $\theta_{max} = 0.054$ are employed to define the fragility curve for $PO_3$ (dark gray points in Figure 19.101). Their median value equals 4.36 g and the log standard deviation is 52%. By convolving with the hazard curve of Figure 19.98, the estimated MAF at 90% confidence equals 0.00025, which is higher than the required target of 1/50, or 0.000201, thus the PO is not met. In general, this MAF estimate for global collapse is conservative, as the simple amplitude scaling that IDA employs and the use of $S_a(T_1, 5\%)$ as the IM do not allow for capturing the appropriate spectral shape of high-intensity ground

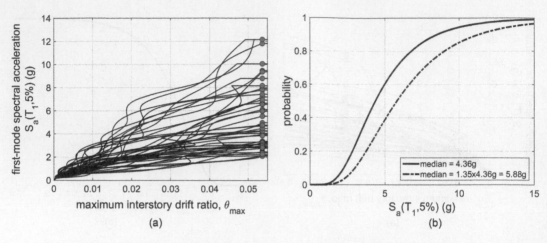

*Figure 19.101* Analytical approach for estimating MAF of crushing of confined concrete which is equivalent to a maximum interstory drift ratio of $\theta_{max}$ = 5.4%: (a) the $S_a(T_1, 5\%)$ values calculated for $\theta_{max}$ = 5.4%, and (b) corresponding fitted lognormal fragility curve.

motions. Following the same strategy as in Example 3 (Section 19.9.11), for a global ductility greater than 6, period $T_1$ < 1.0 s and assuming Seismic Design Category $D_{max}$, Table 7-1b of FEMA P695 suggests an adjustment factor of SSF = 1.35. Now, the adjusted global collapse fragility has a median value of 1.35·4.36 g = 5.88 g with the same standard deviation of 52%. This results in an improved MAF estimate of 0.00005 that meets the PO of 0.00020.

## REFERENCES

ACI (2014). *Building Code Requirements for Structural Concrete and Commentary*, ACI 318-14, American Concrete Institute, Farmington Hills, MI.

ASCE (2010). *Minimum Design Loads for Buildings and Other Structures*, ASCE/SEI 7-10, American Society of Civil Engineers, Reston, VA.

Bakalis, K., Kazantzi, A. K., Vamvatsikos, D., and Fragiadakis, M. (2019). Seismic performance evaluation of liquid storage tanks using nonlinear static procedures, *ASME Journal of Pressure Vessel Technology*, doi:10.1115/1.4039634.

Baker, J. W., and Cornell, C. A. (2006). Spectral shape, epsilon and record selection, *Earthquake Engineering and Structural Dynamics*, 35(9):1077–1095.

Buddika, H. S., and Wijeyewickrema, A. C. (2016). Seismic performance evaluation of posttensioned hybrid precast wall-frame buildings and comparison with shear wall-frame buildings, *Journal of Structural Engineering*, 142(6):04016021.

CEN (2004). *Eurocode 2: Design of Concrete Structures—Part 1-1: General Rules and Rules for Buildings*, European Committee for Standardization, Brussels.

CEN (2005). *Eurocode 8: Design of Structures for Earthquake Resistance—Part 1: General Rules, Seismic Actions and Rules for Buildings*, European Committee for Standardization, Brussels.

FEMA (2009a). Quantification of building seismic performance factors, FEMA P-695, *prepared by the Applied Technology Council for the Federal Emergency Management Agency*, Washington, DC.

FEMA (2009b). FEMA P695 Far field ground motion set, http://users.ntua.gr/divamva/resourcesRC-book/FEMA-P695-FFset.zip.

Haselton, C. B. (2006). Assessing seismic collapse safety of modern reinforced concrete moment frame buildings, *Ph.D. Dissertation*, Stanford, CA.

JCSS (2001). Probabilistic model code, part III: Resistance models, Joint Committee on Structural Safety. www.jcss.byg.dtu.dk/Publications/Probabilistic_Model_Code.

Kennedy, R. C., and Short, S. A. (1994). *Basis for Seismic Provisions of DOE-STD-1020 (No. UCRL-CR--111478; BNL--52418)*, Lawrence Livermore National Lab., Livermore, CA, Brookhaven National Lab., Upton, NY.

Kohrangi, M., Bazzurro, P., Vamvatsikos, D., and Spillatura, A. (2017). Conditional spectrum based ground motion record selection using average spectral acceleration, *Earthquake Engineering and Structural Dynamics*, 46(10):1667–1685.

Lin, T., Haselton, C. B., and Baker, J. W. (2013). Conditional spectrum-based ground motion selection. Part I: Hazard consistency for risk-based assessments, *Earthquake Engineering and Structural Dynamics*, 42(12):1847–1865.

Luco, N., and Bazzurro, P. (2007). Does amplitude scaling of ground motion records result in biased nonlinear structural drift responses? *Earthquake Engineering and Structural Dynamics*, 36(13):1813–1836.

Mander, J. B., Priestley, M. J. N., and Park, R. (1988). Theoretical stress–strain model for confined concrete, *Journal of Structural Engineering*, 114(8):1804–1826.

McKenna, F., Fenves, G., Jeremic, B., and Scott, M. (2000). Open system for earthquake engineering simulation, http://opensees.berkeley.edu.

Perez, F. J., Sause, R., and Pessiki, S. (2007). Analytical and experimental lateral load behavior of unbonded posttensioned precast concrete walls, *Journal of Structural Engineering*, 133(11):1531–1540.

Priestley, M. J. N., Calvi, G. M., and Kowalsky, M. J. (2007). *Displacement-Based Seismic Design of Structures*, IUSS Press, Pavia.

Seismosoft (2014). SeismoStruct v7.0 – A computer program for static and dynamic nonlinear analysis of framed structures, www.seismosoft.com.

Vamvatsikos, D., and Aschheim, M. A. (2016). Performance-based seismic design via yield frequency spectra, *Earthquake Engineering and Structural Dynamics*, 45(11):1759–1778.

# Appendix 1

## Design charts for rectangular and barbell section walls

## A1.1 INTRODUCTION

Design charts (Figures A1.1–A1.13) were developed to provide accurate estimates of nominal flexural strength and yield curvature, for materials having actual strengths equal to the nominal values identified. The basis used to derive the charts is described in Section A1.2, while the application of the charts for use with expected material properties is described in Section A1.3.

## A1.2 ASSUMPTIONS USED IN DEVELOPING DESIGN CHARTS

Concrete and steel strengths were taken equal to their nominal values ($f_c' = 4$, 5, and 6 ksi, and $f_y = 60$ ksi), respectively. Concrete was assumed to have zero tensile capacity; reinforcing steel was modeled as elastic-plastic. Section analyses follow ACI 318 requirements; constitutive relationships, strain compatibility, and equilibrium were satisfied, assuming plane sections remain plane. Nominal flexural strength, $M_n$, was determined per ACI 318 at an extreme fiber strain of 0.003.

In most of the cases considered, the curvature corresponding to first yield, $\phi_y'$ was taken equal to the curvature at the instant in the plane sections analysis that the extreme tension reinforcement reached a strain of $\varepsilon_y$ ($= f_y/E_s$). At this curvature the extreme concrete fiber stress was less than $f_c'$. In some cases with relatively high levels of axial load and/or high reinforcement ratios, the steel would remain elastic while the concrete reached its strength. In these cases, the "yield" curvature was defined as that corresponding to the extreme concrete fiber reaching a stress of $f_c'$. Thus, the curvature at first yield was defined by the first event to occur: the extreme tensile reinforcement reaching $f_y$ or the extreme concrete fiber reaching $f_c'$. The corresponding moment was termed the yield moment $M_y'$.

The effective yield curvature, $\phi_y$, for the cross section was determined by extrapolating the first yield value to the point where the moment reaches the nominal strength level:

$$\phi_y' = \frac{M_n}{M_y'}\phi_y' \tag{A1.1}$$

where $M_n$ = moment resistance corresponding to a concrete strain of 0.003 at the extreme compression fiber and $M_y'$ = moment resistance when longitudinal boundary reinforcement strain reaches $\varepsilon_y$ (or where the extreme concrete fiber reaches $f_c'$).

The effective yield curvatures plotted below are consistent with the observations of Paulay (2002), who suggested effective yield curvatures of $1.8\varepsilon_y/l_w$ and $2.0\varepsilon_y/l_w$ for rectangular section walls with and without longitudinal boundary (end) reinforcement, respectively. These two values correspond to $0.0037/l_w$ and $0.0041/l_w$, respectively, for Grade 60 reinforcement. Paulay also suggested effective yield curvatures for other cross sections: $2.0\varepsilon_y/l_w$ for I- and C-shaped sections, and $1.4\varepsilon_y/l_w$ for T-shaped sections in which the flange is in compression and $1.8\varepsilon_y/l_w$ for T-shaped sections in which the flange in tension.

The parametric analyses are based on lumping the longitudinal boundary reinforcement $A_s = \rho(2d't_w)$ at a distance $d'$ from the edge of a rectangular section, and $A_s = \rho(t_f t_w)$ at the centroid of the boundary element. The longitudinal web reinforcement was assumed to be uniformly distributed as a thin sheet. The web area, $A_w$ is given equivalently by $t_f \cdot t_w$.

## A1.3 USE OF DESIGN CHARTS TO ESTIMATE EXPECTED STRENGTH AND YIELD CURVATURE

The use of expected material strengths ($f'_{ce}$ and $f_{ye}$) rather than nominal material strengths ($f'_c$ and $f_y$) affects the effective yield curvature and expected flexural strength. Suggestions for using the design charts, which are based on nominal material strengths, are provided below.

Expected effective yield curvature is estimated as the product of the nominal value plotted below and the ratio $f_{ye}/f_y$.

The charts for nominal flexural strength may be used by maintaining the same value of $\rho f_y$, for a constant value of axial load, $P$, for a chart developed based on the correct concrete strength. Cross-sectional dimensions are held constant.

To estimate the strength of a known cross section, enter a chart developed for a compressive strength $f'_c$ approximately equal to $f'_{ce}$ but using an amplified value of $\rho$, equal to $\rho_{actual}(f_{ye}/f_y)$. The axial load acting on the section should be maintained, so the axial load index is defined as $P_{actual}/f'_c A_w$, where $f'_c$ is the value of $f'_c$ used to develop the chart. The expected flexural strength is estimated as the flexural strength index value times $f'_c t_w l_w{}^2$, where $f'_c$ is the value of $f'_c$ used to develop the chart.

To determine the reinforcement to provide a section with adequate expected strength, enter a chart developed for a compressive strength $f'_c$ approximately equal to $f'_{ce}$. The axial load index should be calculated as $P_{actual}/f'_c A_w$, where $f'_c$ is the value of $f'_c$ used to develop the chart. The flexural strength index is determined as the required expected strength divided by $f'_c t_w l_w{}^2$, where $f'_c$ is the value of $f'_c$ used to develop the chart. The required reinforcement ratio is given by $\rho(f_y/f_{ye})$, where $\rho$ is the value of $\rho$ read from the chart.

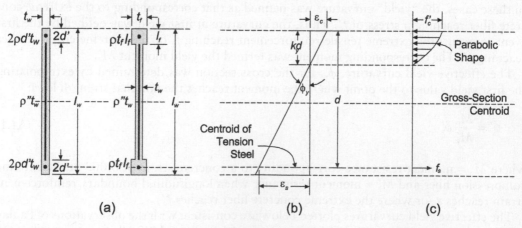

*Figure A1.1* (a) Definition of rectangular and barbell parametric cross sections, (b) assumed strain profile, and (c) assumed stress distributions.

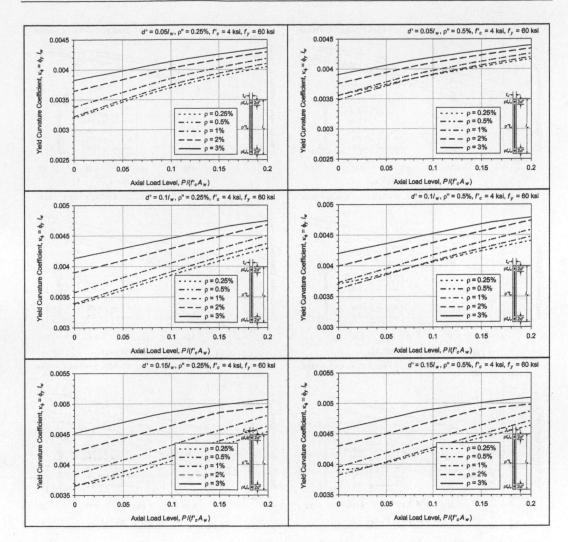

*Figure A1.2* Nominal effective yield curvature in normalized form, for rectangular sections having $f_c' = 4$ ksi and $f_y = 60$ ksi; individual plots are for $\rho'' = 0.25\%$ (left column) and 0.50% (right column); $d'/l_w = 0.05$ (top row), 0.10 (middle row), and 0.15 (bottom row).

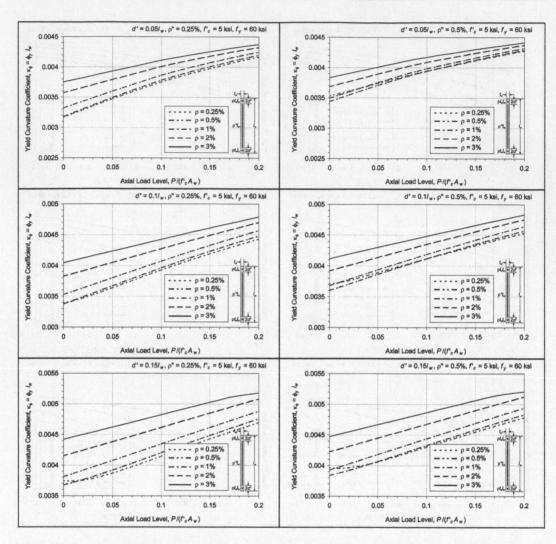

*Figure A1.3* Nominal effective yield curvature in normalized form, for rectangular sections having $f'_c = 5$ ksi and $f_y = 60$ ksi; individual plots are for $\rho'' = 0.25\%$ (left column) and 0.50% (right column); $d'/l_w = 0.05$ (top row), 0.10 (middle row), and 0.15 (bottom row).

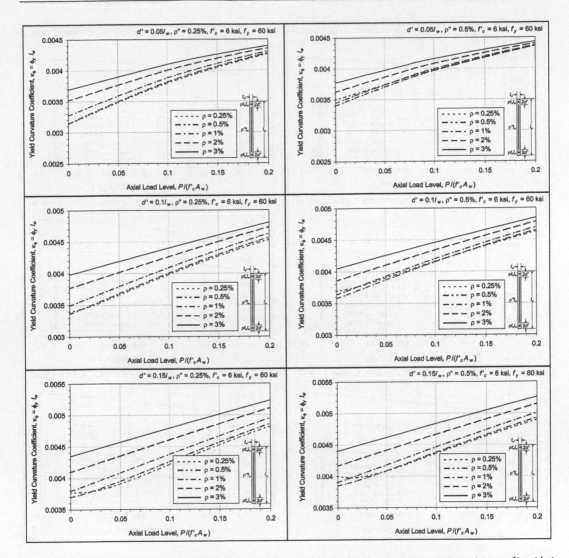

*Figure A1.4* Nominal effective yield curvature in normalized form, for rectangular sections having $f'_c = 6$ ksi and $f_y = 60$ ksi; individual plots are for $\rho'' = 0.25\%$ (left column) and $0.50\%$ (right column); $d'/l_w = 0.05$ (top row), $0.10$ (middle row) and $0.15$ (bottom row).

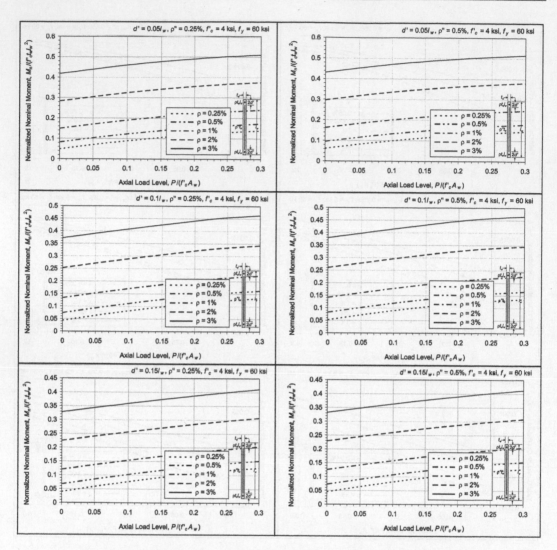

*Figure A1.5* Nominal flexural strength calculated according to ACI 318, in normalized form, for rectangular sections having $f_c' = 4$ ksi and $f_y = 60$ ksi; individual plots are for $\rho'' = 0.25\%$ (left column) and 0.50% (right column); $d'/l_w = 0.05$ (top row), 0.10 (middle row), and 0.15 (bottom row).

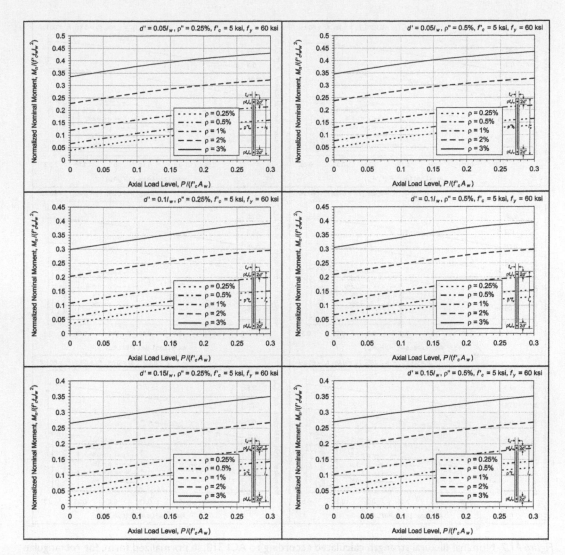

*Figure AI.6* Nominal flexural strength calculated according to ACI 318, in normalized form, for rectangular sections having $f_c' = 5$ ksi and $f_y = 60$ ksi; individual plots are for $\rho'' = 0.25\%$ (left column) and 0.50% (right column); $d'/l_w = 0.05$ (top row), 0.10 (middle row), and 0.15 (bottom row).

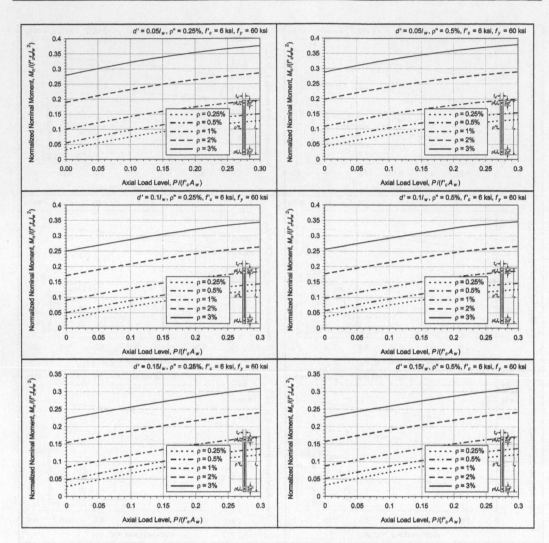

*Figure AI.7* Nominal flexural strength calculated according to ACI 318, in normalized form, for rectangular sections having $f_c' = 6$ ksi and $f_y = 60$ ksi; individual plots are for $\rho'' = 0.25\%$ (left column) and 0.50% (right column); $d'/l_w = 0.05$ (top row), 0.10 (middle row), and 0.15 (bottom row).

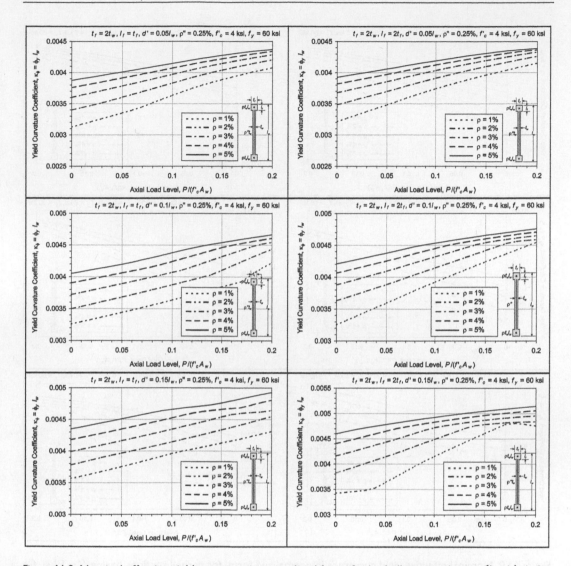

*Figure A1.8* Nominal effective yield curvature in normalized form, for barbell sections having $f'_c = 4$ ksi, $f_y = 60$ ksi, $t_f = 2\ t_w$, and $\rho'' = 0.25\%$; individual plots are for $l_f/t_f = 1$ (left column) and 2 (right column); $d'/l_w = 0.05$ (top row), 0.10 (middle row), and 0.15 (bottom row).

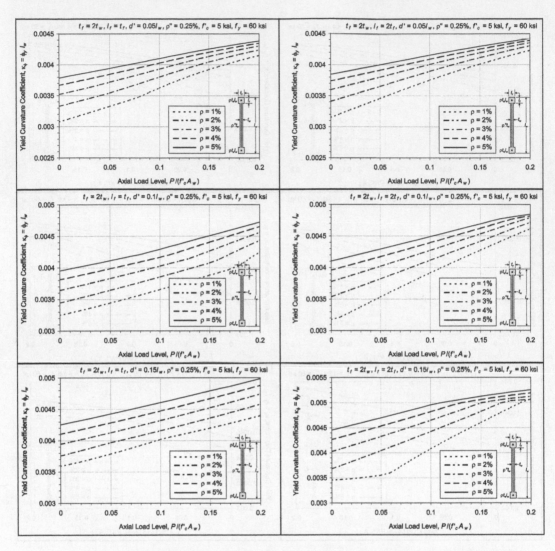

*Figure A1.9* Nominal effective yield curvature in normalized form, for barbell sections having $f'_c = 5$ ksi, $f_y = 60$ ksi, $t_f = 2 t_w$, and $\rho'' = 0.25\%$; individual plots are for $l_f/t_f = 1$ (left column) and 2 (right column); $d'/l_w = 0.05$ (top row), 0.10 (middle row), and 0.15 (bottom row).

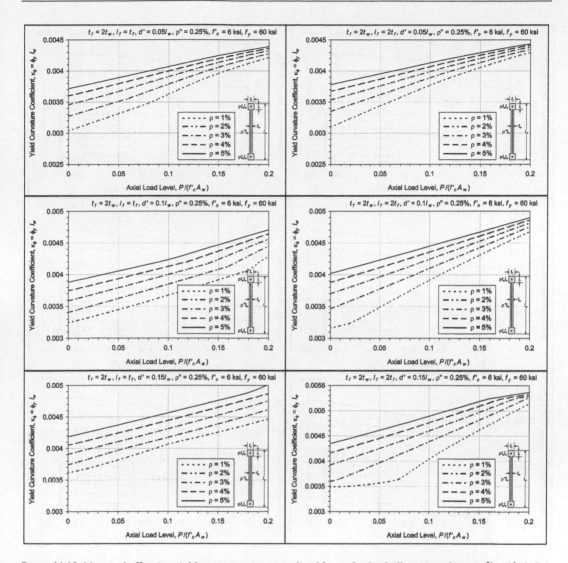

*Figure A1.10* Nominal effective yield curvature in normalized form, for barbell sections having $f'_c = 6$ ksi, $f_y = 60$ ksi, $t_f = 2\ t_w$, and $\rho'' = 0.25\%$; individual plots are for $l_f/t_f = 1$ (left column) and 2 (right column); $d'/l_w = 0.05$ (top row), 0.10 (middle row), and 0.15 (bottom row).

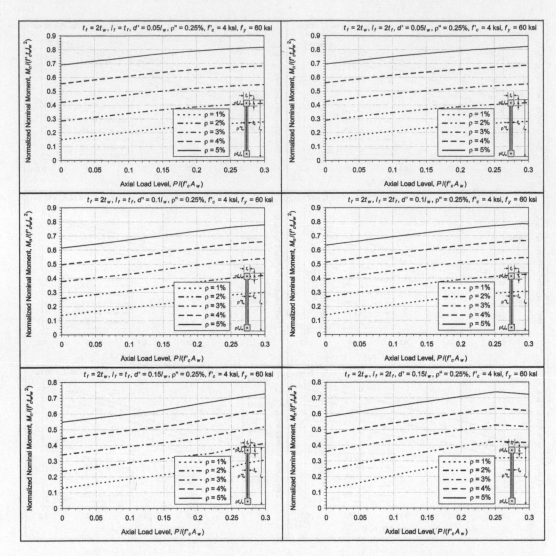

*Figure A1.11* Nominal flexural strength calculated according to ACI 318, in normalized form, for barbell sections having $f'_c = 4$ ksi, $f_y = 60$ ksi, $t_f = 2\ t_w$, and $\rho'' = 0.25\%$; individual plots are for $l_f/t_f = 1$ (left column) and 2 (right column); $d'/l_w = 0.05$ (top row), 0.10 (middle row), and 0.15 (bottom row).

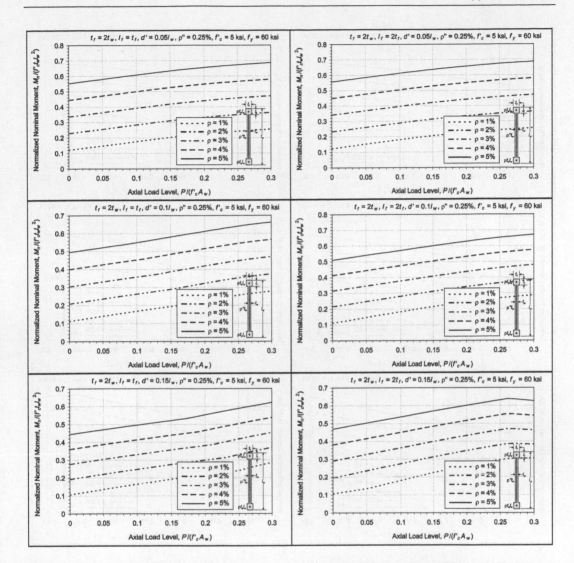

*Figure A1.12* Nominal flexural strength calculated according to ACI 318, in normalized form, for barbell sections having $f'_c = 5$ ksi, $f_y = 60$ ksi, $t_f = 2\ t_w$, and $\rho'' = 0.25\%$; individual plots are for $l_f/t_f = 1$ (left column) and 2 (right column); $d'/l_w = 0.05$ (top row), 0.10 (middle row), and 0.15 (bottom row).

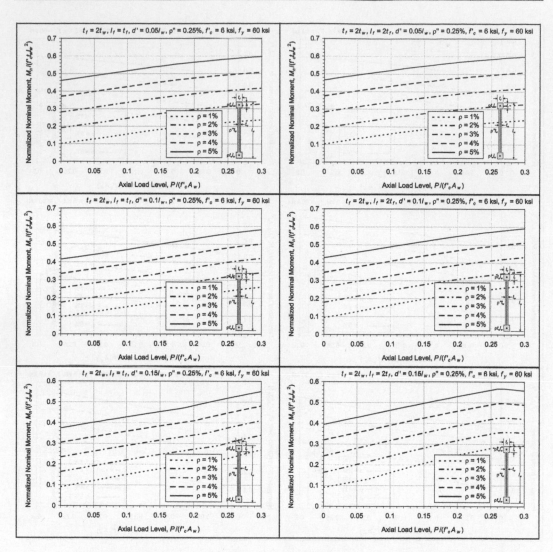

*Figure A1.13* Nominal flexural strength calculated according to ACI 318, in normalized form, for barbell sections having $f_c' = 6$ ksi, $f_y = 60$ ksi, $t_f = 2\ t_w$, and $\rho'' = 0.25\%$; individual plots are for $l_f/t_f = 1$ (left column) and 2 (right column); $d'/l_w = 0.05$ (top row), 0.10 (middle row), and 0.15 (bottom row).

# REFERENCES

Paulay, T. (2002). An estimation of displacement limits for ductile systems, *Earthquake Engineering and Structural Dynamics*, 31(3):583–599.

# Appendix 2
## Plan torsion

## A2.1 PURPOSE AND OBJECTIVES

This appendix provides background to understand the mechanics of plan torsion, beginning with one-story structures and extending to considerations for multistory buildings. An understanding of conditions promoting and preventing significant plan torsion is essential for providing torsionally resistant buildings.

## A2.2 INTRODUCTION

Plan torsion involves the rotational response of building floors in their plane over the height of the building. The typical cause is due to an eccentricity between the center of mass, where the horizontal inertial force resultant is located, and the center of resistance to lateral force, at one or more floor levels. Figure A2.1 illustrates a single-story structure in which the center of mass ($c_m$) is at the center of the floor plate (or roof) and, owing to the high strength and stiffness of the wall along Line A, the center of resistance ($c_r$) is offset from $c_m$ by a significant distance. When subjected to excitation in the $y$ direction, the floor plate undergoes displacements in the $y$ direction ($\Delta_{cr}$) as well as rotations ($\theta_{cr}$) about $c_r$ in order to mobilize resistance to the induced lateral force.

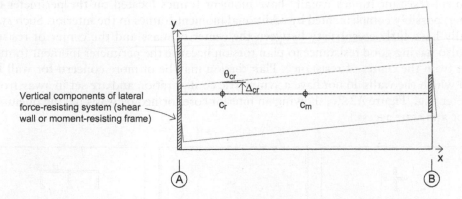

*Figure A2.1* Eccentricity between resultant seismic force (at center of mass, $c_m$) and center of resistance, $c_r$ results in both displacement and rotation when mobilizing resistance to the induced lateral forces. (Vertical components required for resisting excitation in the $x$-direction are not shown for clarity, although they may contribute to torsional resistance depending on their configuration.)

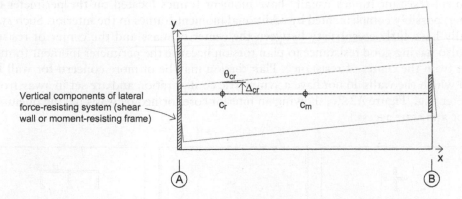
Vertical components of lateral force-resisting system (shear wall or moment-resisting frame)

In special cases, torsional response has been associated with the three-dimensional interaction of coupled translational and torsional motions, complicated by beating or modal interference associated with systems having closely spaced periods of vibration (Mahin et al., 1989).

Although we customarily consider only the translational components of ground motion, rotational components of motion about each of the three Cartesian coordinates are also present. Thus, some torsional excitation is present, although this source of excitation usually is neglected in structural design.

Torsional demands are important in torsionally irregular buildings because (i) the plan rotation induces interstory drift at locations distant from the $c_r$, (ii) this drift induces additional lateral force into vertical components (e.g., columns and walls) of the lateral force-resisting system, and (iii) the plan rotation induces twist and torsion into each of these vertical components.

## A2.3 CONFIGURATION OF VERTICAL ELEMENTS OF THE LATERAL FORCE-RESISTING SYSTEM

Vertical components of the lateral force-resisting system are subjected to translations in both $x$ and $y$ directions as well as rotations about a vertical axis (e.g., Figure A2.1). These elements (e.g., structural walls or moment-resistant frames) usually have in-plane stiffness much higher than their out-of-plane and torsional stiffnesses. Thus, they are usually considered to resist load only in their in-plane direction.

The vertical elements of the lateral force-resisting system may be poorly configured as shown in Figure A2.2a. This configuration may provide sufficient lateral strength in the $x$ and $y$ directions, but provides little resistance to rotation in plan. By relocating the vertical elements to the perimeter of the floor plate (Figure A2.2b), resistance to torsional rotation is provided, since any rotation in plan will mobilize the lateral stiffness of vertical elements. In addition, by providing multiple vertical elements, resistance to plan torsion is maintained even if some vertical elements should soften or fail, providing redundancy. Buildings that have multiple vertical elements of the lateral force-resisting system distributed around the building perimeter have inherently superior resistance to plan torsion compared with having these elements closer to the $c_m$ or having fewer perimeter elements.

Moment-resistant frames usually have moment frames located on the perimeter of the building, possibly complemented by additional moment frames in the interior. Such systems generally have little eccentricity between the center of mass and the center of resistance, while also having good resistance to plan torsion because the perimeter moment frames are distant from the center of resistance. Plan torsion may be of more concern for wall buildings in which the walls do not have a symmetric configuration and are set in away from the perimeter (e.g., Figure A2.2c, showing an interior core formed by walls having openings for elevator and stair access.)

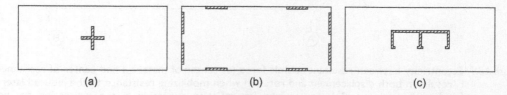

(a)   (b)   (c)

*Figure A2.2* Illustrative configurations of vertical components of lateral force-resisting system: (a) cruciform pattern lacks resistance to plan torsion; (b) perimeter lateral force-resisting systems maximize resistance to plan torsion; and (c) use of flanged walls in elevator or stair core walls.

## A2.4 CENTER OF RESISTANCE

The center of resistance is the location of the resultant of lateral forces developed as the floor plate undergoes horizontal translation. The $x$-coordinate of the center of resistance, $x_{cr}$, can be determined by imposing a translation in the $y$ direction without allowing rotation, determining the forces developed in the vertical components of the lateral force-resisting system, and determining the location of the resultant of these forces. Similarly, a translation imposed in the $x$-direction can be used to determine $y_{cr}$.

For linear elastic response, the center of resistance is equivalent to the *center of stiffness*, also termed the *center of rigidity*, and often denoted by the term $c_k$. However, as nonlinearity develops in the load–deformation response of the vertical elements of the lateral force-resisting system, the center of resistance may change. In the case of regular, symmetric plans, softening of the load–deformation response is synchronized with translation and the center of resistance is stable. For the special case that the component inelastic response is idealized as perfectly plastic, the center of resistance when all vertical elements of the lateral force-resisting system are yielding is termed the *center of strength*, often denoted $c_v$. The determination of $c_k$ and $c_v$ is illustrated for a single story in Box A2.1.

### BOX A2.1   DETERMINATION OF CENTER OF STIFFNESS, CENTER OF STRENGTH, AND TORSIONAL MOMENT OF INERTIA

Figure A2.3a illustrates a one-story building having a nonsymmetric disposition of walls in the y direction. Wall stiffnesses and strengths are as indicated for strong-axis bending. Columns, as well as walls acting in weak-axis bending, are assumed to have a negligible contribution.

As shown in Figure A2.3(b), the center of stiffness may be determined by imposing a unit displacement of the floor (or roof) plate in one direction and determining the forces acting on the floor plate by the vertical components of the lateral force-resisting system. The center of stiffness is the location of the resultant of these forces, or equivalently, the location of an external force required to keep the floor plate in static equilibrium under pure translation. The x and y coordinates of the center of stiffness are found by independent analyses of the floor plate displaced in each orthogonal direction.

As shown in Figure A2.3(c), a force applied at the center of mass (e.g., associated with relative accelerations) can be resolved into an equivalent force ($V$) applied at $c_r$ and torque ($T$), where $T = Ve = V(x_{cm} - x_{cr})$. The displacement resulting from $V$ is given by $\Delta_y = V/k$ where $k = k_A + k_B$. The forces in each wall due to the displacement are given by $V_i = k_i \cdot \Delta_y = k_i \cdot V/k$.

As shown in Figure A2.3(d), imposing a unit rotation about $c_k$ allows determination of the displacements $\Delta_i$ in the walls and resulting forces imposed on the floor plate; the associated torque is required for equilibrium. The rotation resulting from the torque is given by $\theta = T/J$, where $J$ = the polar moment of inertia = $\Sigma(k_{yi}(x_i)^2) + \Sigma(k_{xi}(y_i)^2)$ and $x_i$ and $y_i$ are distances measured perpendicular from the line of action of each wall to the center of stiffness. Thus, the forces in each wall due to the rotation are given by $V_i = k_i \cdot \Delta_i = k_i \cdot (\theta)(x_i \text{ or } y_i) = k_i (T/J)(x_i \text{ or } y_i)$.

This analysis is applicable to linear elastic response; the forces in each wall due to displacement and rotation can be superposed.

As shown in Figure A2.3(e), the center of strength is determined by a similar process, except that the walls are assumed to be ductile enough that a displacement sufficient to mobilize their strengths can be imposed. The center of strength is the location of the resultant of the forces acting on the floor plate, or equivalently, the location of the resultant required to provide for static equilibrium.

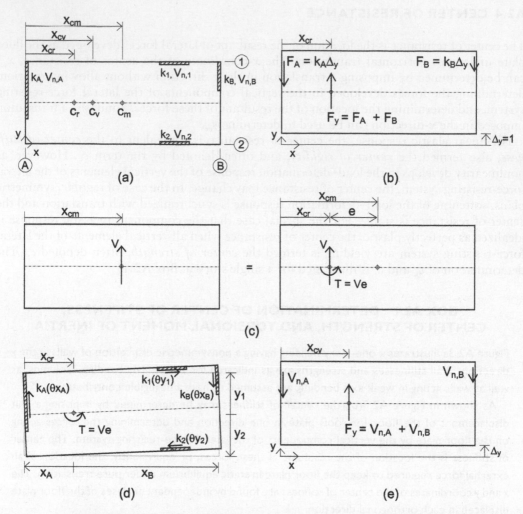

*Figure A2.3* (a) Illustration of centers of stiffness, strength, and mass; (b) free-body diagram showing forces acting on the floor plate; the center of stiffness is the location of resultant when subjected independently to $\Delta_x = 1$ and $\Delta_y = 1$; (c) equivalent force system; (d) rotation allows forces due to $V$ and $T$ to be evaluated in each vertical component of the lateral force-resisting system; (e) free-body diagram showing forces and torque acting on the floor plate; and (f) free-body diagram showing forces acting on the floor plate; the center of strength is the location of the resultant when all vertical components have developed their strength.

## A2.5 COMPLICATIONS WITH NONSYMMETRIC CONFIGURATIONS, NONSYMMETRIC CROSS SECTIONS, AND MULTIPLE STORIES

Single-story buildings in which the vertical components of the lateral load-resisting system have symmetric response (for positive and negative displacements) can illustrate concepts such as the center of stiffness and the eccentricity between the center of mass and center of stiffness. While these concepts have intuitive value for all systems, nonsymmetric configurations, nonsymmetric cross sections, and multistory construction introduce some complications.

1. In the case of initially symmetric systems, if yield strengths differ among the vertical elements of the lateral force-resisting system, torsional response can result (e.g., Bruneau and Mahin, 1992).
2. Where walls are not located in a symmetric configuration, softening and yielding of the walls may occur in a nonsynchronized fashion under pure translation. This will cause the center of resistance to change location.
3. In the case of nonsymmetric walls (e.g., flanged walls), the strength and stiffness depend on the direction of loading (e.g., flange in tension or in compression). Thus, if nonsymmetric walls are arranged in a way that lacks overall symmetry, centers of stiffness ($c_k$) and strength ($c_v$) will differ depending on the direction of loading.
4. For a wall undergoing lateral response, the presence of gravity load may cause closure of cracks and an increase in lateral stiffness at small lateral displacements. Where walls of different lengths (in plan) are present and/or axial loads vary, such changes in stiffness may occur at different lateral displacements, inducing short duration shifts in the center of resistance.
5. Determination of the center of resistance for multistory construction is not straightforward. For any vertical element of the lateral force-resisting system, forces applied at any floor level induce displacements at floors above and below this level. It is not clear what the vertical profile of horizontal translations of the floors should be when determining the center of resistance of each floor level. Goel and Chopra (1993) observe that there is no generally accepted definition of the center of resistance and that locations of the center of resistance depend on the height-wise distribution of lateral forces.
6. The yield displacements of individual walls having different lengths in plan will differ. As the building undergoes lateral translation, the longest walls yield first (e.g., by forming a plastic hinge at the base) while shorter walls may still be elastic. If the walls were uncoupled (freestanding), each wall would display a different displacement profile, with the longest walls hinging at their base. Thus, as noted by Beyer et al. (2014), patterns of incremental deformations among the walls will differ. Because the walls are joined together by the floor slabs, significant forces may develop in the floor diaphragm to enforce compatibility of wall deformations; these floor diaphragm forces must be in equilibrium with the shears in the walls. Clearly, this complicates the determination of wall shears. In the case of systems lacking symmetrically disposed walls, the altered wall shears can be expected to affect the relative stiffnesses of the walls and hence the location of the center of resistance at each floor.

In principle, these complications could be addressed with high-fidelity structural models. We suggest it is preferable to avoid significant torsional problems by configuring building systems to have little eccentricity between centers of mass and resistance, and with vertical elements of the lateral force-resisting system located far from the center of resistance, to provide redundant systems with inherent resistance to torsional rotation.

## A2.6 CODE TREATMENTS

Current code treatments of plan torsion come from an era of limited understanding about inelastic dynamic response. Significant attention is given to demands computed from linear elastic response, consideration of an "accidental" eccentricity, and amplification of these demands for structures classified as torsionally irregular.

In ASCE-7, the accidental eccentricity is defined as an offset of the center of mass (or location where lateral force is applied) of 5% of the horizontal plan dimension. Since the lateral offset can be in either (+ or −) direction, there are many possible permutations to consider in the analysis of multistory buildings. Engineers typically consider just two cases; one with positive eccentricities at each floor and one with negative eccentricities at each floor.

ASCE-7 defines both a "torsional irregularity" and an "extreme torsional irregularity." A torsional irregularity is present where the maximum story drift at one end of the structure is more than 1.2 times the average of the story drifts at the two ends of the structure, while an "extreme torsional irregularity" is present where the maximum story drift is 1.4 times the average of the story drifts at the two ends of the structure. Drifts are computed including accidental torsion but without amplification for torsional irregularity. Where either torsional irregularity is present, the influence of the accidental eccentricity is amplified, when using the equivalent lateral force method; no amplification is imposed for linear dynamic analysis. (Note that structures with extreme torsional irregularities are not permitted in the highest seismic regions.)

As noted by Paulay (1996), code approaches for plan torsion are largely irrelevant. Fortunately, more recent studies of the inelastic dynamic response of multistory buildings are available for design guidance.

## A2.7 DESIGN CONSIDERATIONS

To minimize torsional response, designers should minimize demand, by minimizing eccentricities between the center of mass and the centers of stiffness and strength, and should maximize resistance, by positioning the vertical elements of the lateral force-resisting system far from the center of mass.

More specifically, Priestley et al. (2007) focus on the location of the center of strength. Based on the results of Beyer (2007) and Castillo et al. (2004), torsional response is kept to a minimum by simply distributing the design base shear to the vertical elements of the lateral force-resisting system in such a way that the center of strength coincides with the center of mass. For buildings with identical floor plans over the height, the lateral forces used for design at any level would have a resultant that coincides with the center of mass of the floor. The intended mechanism relies on developing a plastic flexural hinge at the base of each wall.

If it is not possible to locate the center of strength at the center of mass, the design system ductility used for the determination of the design base shear may be modified. In applying the design procedure described by Priestley et al. (2007), 90% of the design system ductility would be used for determination of the design base shear for buildings in which drift limits control the design, and 110% of the design system ductility would be used where system ductility limits control the design. As noted in Chapter 18, three-dimensional nonlinear dynamic analyses should be used to confirm the acceptability of the resulting design.

## REFERENCES

Aschheim, M., and Maurer, E. (2007). Dependency of COD on ground motion intensity and stiffness distribution, *Structural Engineering and Mechanics*, 27(4):425–438.

Beyer, K. (2007). Seismic Design of Torsionally Eccentric Buildings with RC U-Shaped Walls. *PhD thesis*, Rose School, IUSS, Pavia.

Beyer, K., Simonini, S., Constantin, R., and Rutenberg, A. (2014). Seismic shear distribution among interconnected cantilever walls of different lengths, *Earthquake Engineering and Structural Dynamics*, 43(10):1423–1441.

Bruneau, M., and Mahin, S. A. (1992). Normalizing inelastic seismic response of structures having eccentricities in plan, *Journal of Structural Engineering*, American Society of Civil Engineers, 116(12):3358–3379.

Castillo, R. (2004). Seismic Design of Asymmetric Ductile Systems. *PhD Thesis*, University of Canterbury, Christchurch, New Zealand.

Goel, R. K., and Chopra, A. K. (1993). Seismic code analysis of buildings without locating centers of rigidity, *Journal of Structural Engineering*, American Society of Civil Engineers, 119(10):3039–3055, Oct.

Mahin, S. A., Boroschek, R., and Zeris, C. (1989). Engineering interpretation of the responses of three instrumented buildings in San Jose, Seminar on Seismological and Engineering Implications of Recent Strong Motion Data, SMIP89, Strong Motion Instrumentation Program, Sacramento, CA, 1989, pp. 12–1 to 12–11.

Paulay, T. (1996). Seismic design for torsional response of ductile buildings, *Bulletin of the New Zealand Society for Earthquake Engineering*, 29(3):178–198.

Priestley, M. J. N., Calvi, G. M., and Kowalsky, M. J. (2007). *Direct Displacement Based Design of Structures*, IUSS Press, Pavia, 721 pages.

# Appendix 3
## Validation of column flexural stiffness model

## A3.1 INTRODUCTION

Although the effective flexural stiffness is often estimated as a fixed percentage of the gross stiffness, fundamental mechanics indicate that the flexural stiffness must depend on both longitudinal reinforcement and axial load. This dependency has to be modeled to properly represent the influence of strength on stiffness and yield displacement. This Appendix provides background for the modeling approach recommended in Section 17.8.2.1.

## A3.2 VALIDATION OF COLUMN MODELING

In this section, yield displacements are estimated for a series of cantilever columns and compared with estimates made using models developed by Haselton (2006) and Biskinis and Fardis (2008). These models were developed based on the review of a very large number of experimental test results. However, the models differ in that the Haselton model makes no accounting of the effect of reinforcement content on stiffness, while the Biskinis and Fardis model represents the effect of reinforcement content and axial load through their effect on the yield curvature calculated from first principles.

In the approach developed here and recommended in Section 17.8.2.1, as well as the approach recommended by Biskinis and Fardis (2008), deformations consisting of flexure, shear, and anchorage slip ($\Delta_{flex}$, $\Delta_{sh}$ and $\Delta_{sl}$ in Figure A3.1) can be modeled for a cantilever column. The simpler approach recommended by Haselton (2006) represents these sources of deformation using a reduced value of $EI$, for use in a model that considers only flexural deformation. Haselton's approach was adopted by PEER/ATC 72-1 (2010). ASCE SEI-41 (2013) adopted a similar approach, in which a reduced value of $EI$ is used to account for flexural deformations and anchorage slip, while shear deformation is modeled explicitly, using a shear stiffness equal to $0.4E_cA_g$.

In order to compare the predictions of the models, yield displacements were estimated for a 20 in. square cantilever column having different heights and longitudinal reinforcing, over a range of axial load given by $0.0 < P/A_gf_c' < 0.7$. Heights are represented by the cantilever height (or shear span, $L_s$) normalized by the overall section depth, for $L_s/H = 2, 4,$ and $6$. Total longitudinal reinforcement normalized by the gross section area is represented for $\rho_g = 0.92\%, 3.0\%,$ and $4.7\%$. Results are provided in the left panels of Figure A3.2.

As shown in the left panels of Figure A3.2, trends identified in the Haselton model are also manifested by the Biskinis and Fardis model. Initially, $\phi_y$ was taken equal to $2.1\epsilon_y/d$ as

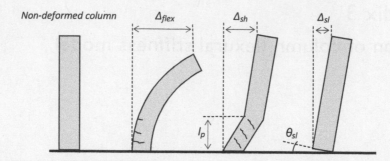

*Figure A3.1* Deformations (shown as separate components) in a cantilever column, exaggerated for clarity.

recommended by Biskinis and Fardis (2008), However, this led to substantially larger yield displacement estimates than could be justified based on Haselton, and these were independent of axial load ratio. As an improvement, the plots on the left side of Figure A3.2 use the estimate of yield curvature given by Equation A3.1.

$$\phi_y = \left(1.8 - 1.3\frac{P}{A_g f_c'} + 9\left(\rho_g - 0.025\right)\right)\frac{\varepsilon_y}{d} \tag{A3.1}$$

One may interpret the left panels of Figure A3.2 as indicating that the Haselton model (perhaps inadvertently) exaggerates the effect of longitudinal reinforcement content on yield displacement—because the $EI$ value does not vary with reinforcement content, stronger sections necessarily have a larger yield displacement.[1] The Biskinis and Fardis model, as modified herein, generally follows the trend with $P/A_g f_c'$ determined by Haselton for the intermediate value of $\rho_g$, for aspect ratios $L_s/H$ equal to 4 and 6. Differences emerge for $L_s/H = 2$.

The plots of the right side of Figure A3.2 show that the model recommended here and in Section 17.8.2.1 tends to follow the trends with axial load ratio present in the Haselton and modified Biskinis and Fardis models. Unlike the Haselton model, effects of longitudinal reinforcement on yield displacement are represented.

The recommended model is expected to have prediction uncertainty comparable to or better than the Haselton model (for which the logarithmic standard deviation $\sigma_{LN} = 0.28$ provides relatively good prediction uncertainty) while being simpler to apply than the Biskinis and Fardis model.

---

[1] The effect of using an $EI$ value that is invariant with longitudinal reinforcement ratio is illustrated in Figure 17.19.

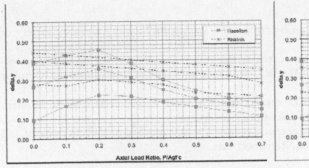

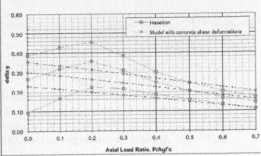

(a) L_s/H = 2

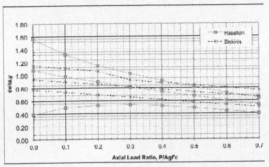

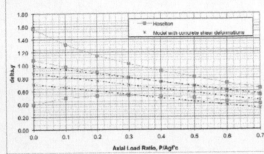

(b) L_s/H = 4

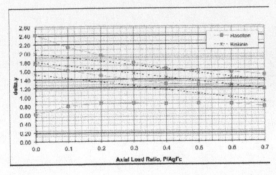

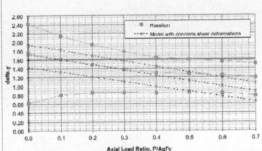

(c) L_s/H = 6

*Figure A3.2* Yield displacement estimates for cantilever columns obtained using different models, for columns of different shear span/depth ratios. Each plot contains curves for different longitudinal reinforcement ratios: $\rho_g = 0.92\%$ (lowest curve), 3.0% (middle curve), and 4.7% (upper curve). The left panels compare results based on the Biskinis and Fardis (2008) model, modified to use the yield curvature estimate of Equation A3.1, with results obtained using the Haselton (2006) effective stiffness relationship. Plots on the right side compare the recommended model (with concrete shear deformations) with estimates obtained using the Haselton effective stiffness relationship (2008).

## REFERENCES

ASCE SEI-41 (2013). *Seismic Evaluation and Retrofit of Existing Buildings*, Amercian Society of Civil Engineers, Reston, VA.

Biskinis and Fardis (2008). Cyclic deformation capacity, resistance and effective stiffness of RC members with or without retrofitting, *14th World Conference on Earthquake Engineering*, October 12–17, 2008, Beijing, China.

Haselton, C. B. (2006). Assessing Seismic Collapse Safety of Modern Reinforced Concrete Moment Frame Buildings. *Doctoral dissertation*, Stanford University.

PEER/ATC 72-1 (2010). Modeling and acceptance criteria for seismic design and analysis of tall buildings, Applied Technology Council, Redwood City.

# Index